Seeds

The Ecology of Regeneration in Plant Communities

2nd Edition

———————————————

Seeds

The Ecology of Regeneration in Plant Communities

2nd Edition

_____

Edited by

Michael Fenner

School of Biological Sciences
University of Southampton, UK

CABI *Publishing*

CABI *Publishing* is a division of CAB *International*

CABI Publishing
CAB International
Wallingford
Oxon OX10 8DE
UK

Tel: +44 (0)1491 832111
Fax: +44 (0)1491 833508
Email: cabi@cabi.org
Web site: http://www.cabi.org

CABI Publishing
10 E 40th Street
Suite 3203
New York, NY 10016
USA

Tel: +1 212 481 7018
Fax: +1 212 686 7993
Email: cabi-nao@cabi.org

A catalogue record for this book is available from the British Library, London, UK.

Library of Congress Cataloging-in-Publication Data
Seeds : the ecology of regeneration in plant communities / edited by M. Fenner.-- 2nd ed.
    p. cm.
  Includes bibliographical references.
  ISBN 0-85199-432-6 (alk. paper)
    1. Seeds--Ecology. 2. Plants--Reproduction.   I. Fenner, Michael, 1949-
QK661 .S428 2000
575.6′8--dc21

                                   00-041388

ISBN 0 85199 432 6 ✓

Typeset by Columns Design Ltd, Reading.
Printed and bound in the UK by Biddles Ltd, Guildford and King's Lynn.

Contents

Contributors

David D. Ackerly, Department of Biological Sciences, Stanford University, Stanford, CA 94305, USA

Fakhri A. Bazzaz, Department of Organismic and Evolutionary Biology, Harvard University, Cambridge, MA 02138, USA

James M. Bullock, NERC Centre for Ecology and Hydrology Dorset, Winfrith Technology Centre, Dorchester, Dorset DT2 8DH, UK

Michael J. Crawley, Department of Biology, Imperial College of Science, Technology and Medicine, Silwood Park, Ascot, Berkshire SL5 7PY, UK

Richard H. Ellis, Department of Agriculture, University of Reading, Earley Gate, PO Box 236, Reading, Berkshire RG6 6AT, UK

Michael Fenner, School of Biological Sciences, University of Southampton, Southampton SO16 7PX, UK

C.J. Fotheringham, Organismic Biology, Ecology and Evolution, University of California, Los Angeles, CA 90095, USA

J. Philip Grime, Unit of Comparative Plant Ecology, Department of Animal and Plant Sciences, University of Sheffield, Sheffield S10 2TN, UK

Yitzchak Gutterman, The Jacob Blaustein Institute for Desert Research and Department of Life Sciences, Ben-Gurion University of the Negev, Sede Boker Campus 84993, Israel

Henk W.M. Hilhorst, Laboratory of Plant Physiology, Wageningen Agricultural University, Arboretumlaan 4, 6703 BD Wageningen, The Netherlands

Susan H. Hillier, Unit of Comparative Plant Ecology, Department of Animal and Plant Sciences, University of Sheffield, Sheffield S10 2TN, UK

Pedro Jordano, Estación Biológica de Doñana, CSIC, Apdo. 1056, E-41080 Sevilla, Spain

Cees M. Karssen, Laboratory of Plant Physiology, Wageningen Agricultural University, Arboretumlaan 4, 6703 BD Wageningen, The Netherlands

Jon E. Keeley, US Geological Survey Biological Resources Division, Western Ecological Research Center, Sequoia-Kings Canyon Field Station, Three Rivers, CA 93271, USA

Kaoru Kitajima, Botany Department, University of Florida, Gainesville, Florida, USA

Michelle R. Leishman, Department of Biological Sciences, Macquarie University, Sydney, New South Wales 2109, Australia

Angela T. Moles, Department of Biological Sciences, Macquarie University, Sydney, New South Wales 2109, Australia

Alistair J. Murdoch, Department of Agriculture, University of Reading, Earley Gate, PO Box 236, Reading, Berkshire RG6 6AT, UK

Thijs L. Pons, Department of Plant Biology, Utrecht University, PO Box 800.84, 3508 TB Utrecht, The Netherlands

Robin J. Probert, Seed Conservation Department, Royal Botanic Gardens, Kew, Wakehurst Place, Ardingly, West Sussex RH17 6TN, UK

Edward G. Reekie, Department of Biology, Acadia University, Patterson Hall, University Avenue, Wolfville, Nova Scotia, Canada B0P 1X0

Edmund W. Stiles, Department of Biological Sciences, Rutgers University, PO Box 1059, Piscataway, NJ 08855–1059, USA

Ken Thompson, Unit of Comparative Plant Ecology, Department of Animal and Plant Sciences, University of Sheffield, Sheffield S10 2TN, UK

Anna Traveset, Institut Mediterrani d'Estudis Avançats (CSIC-UIB), c/ Miguel Marqués 21, 07190-Esporles, Balearic Islands, Spain

Mark Westoby, Department of Biological Sciences, Macquarie University, Sydney, New South Wales 2109, Australia

Mary F. Willson, 5230 Terrace Place, Juneau, AK 99801, USA

Ian J. Wright, Department of Biological Sciences, Macquarie University, Sydney, New South Wales 2109, Australia

Preface

Michael Fenner

School of Biological Sciences, University of Southampton, Southampton, UK

The aim of the second edition of this book remains the same as the first, namely, to provide an overview of current thinking in plant reproductive ecology as it relates to seeds. The contributors were asked to revise their chapters by making whatever changes (additions, omissions, modifications) were required to bring them up to date. The 8 years that have elapsed since the first edition have seen advances on all fronts in this field, and this has necessitated considerable reworking of the text. Some major areas that were poorly represented in the first edition are now covered by the addition of four new chapters – on seed size, seedling establishment, the role of gaps, and regeneration from seed after fire. The volume now has a total of 27 authors, all of them eminent biologists with specialist knowledge and expertise on the topic of their chapter.

Research into plant reproduction from seed continues to provide a wide range of opportunities for investigating fundamental aspects of ecology. The study of allocation patterns in plants has revealed some important insights into the true costs of reproduction. The mysteries of dormancy and germination will provide the physiological ecologist with ample scope for experiments for the foreseeable future. There is still a rich seam of potential material for population ecologists to uncover in

order to determine the causes of mortality in seeds and seedlings. We still do not know the fate of the majority of seeds in the soil or the cause of death of most seedlings. Regeneration studies also touch on numerous aspects of community ecology. Animal–plant interactions are especially well represented in the process of plant reproduction, because of the roles that animals play as pollinators, seed predators, fruit eaters, seed dispersers and agents of physical disturbance. These interactions often impose fascinating trade-offs as a response to selective pressures on both the plant and animal. What is for convenience referred to as 'seed ecology' thus covers a very broad field, in which general ecological principles can be studied.

Although a great many excellent field experiments have been carried out in recent years, we are still much better informed about the laboratory responses of seeds than about their behaviour under natural conditions. It could be argued that this is not unreasonable or even undesirable. A knowledge of responses to highly controlled conditions enables us to tease out the effects of each component of the natural environment, in which a range of factors are acting simultaneously. Chapters 9 to 12 deal with seed responses to different environmental factors. These show that many recent laboratory experiments incorporate interactions between factors, and these are proving to be very useful in enabling us to interpret field behaviour with more confidence. Laboratory experiments are not a substitute for fieldwork, but are an important complement to it.

The new chapters in this edition focus on topics which, although mentioned in the first edition, are of sufficient importance to merit separate coverage in their own right. For example, the significance of seed size is a theme that crops up in many contexts throughout this book. Leishman *et al.* in Chapter 2 provide a neat summary of this multifaceted topic, exploring the many evolutionary compromises that have resulted from (often opposing) selection pressures favouring successful dispersal and establishment. The use of game-theory models to answer such questions as 'why is there such a wide range of seed sizes in the same habitat?' is a development that has great potential for producing new insights into some of the more intractable problems in this area. The new chapter on seedling establishment (Kitajima and Fenner, Chapter 14) deals with the trade-offs required for growth and survival of very young plants in the field, concentrating on causes of mortality in the early stages. Successful regeneration clearly includes surmounting these early hurdles. The special problems associated with establishment from seed in fire-prone habitats is also the subject of a new chapter (Keeley and Fotheringham, Chapter 13), which considers some interesting cases of species that appear to be 'fire-dependent' for regeneration. The new concluding chapter (Bullock, Chapter 16) reviews the burgeoning literature on the role of disturbance in creating opportunities for regeneration from seed. Differences in the abilities of species to colonize gaps of different sizes and shapes in closed vegetation is well established, but the long-term effects of these differences on species composition and community diversity are often less clear. Bullock provides a synthesis of current thinking on this complex aspect of regeneration.

My aim as editor has been to allow the authors complete freedom to treat their topics in their own way. Their only instruction was to produce a wide-ranging review of current ideas in the allocated subject. My policy has been to interfere as little as possible with what the various authors have wanted to say, even if I did not agree with them. So, apart from making a few suggestions and correcting minor errors, I requested very few changes in the content of the manuscripts submitted. This has led, no doubt, to occasional instances of overlap in coverage of topics between the chapters. None of the topics is entirely self-contained. Seed size, for example, has implications for allocation, dispersal, predation, longevity and seedling establishment. Seed dispersal by animals touches on their roles as seed predators and frugivores. However, I regard any occasional instances of overlap as providing useful links between chapters, giving alternative perspectives on the same facts. I have tried to ensure that appropriate cross-references are included in the text where relevant.

Another consequence of exercising a high level of editorial restraint is that some inconsistency may occur between the authors on the question of what constitutes dormancy. Indeed, the first edition of this book was criticized on this account by Vleeshouwers *et al.* (1995) in their excellent paper on redefining seed dormancy (*Journal of Ecology* 83, 1031–1037). In correspondence with various authors, I have realized that the issues are more complex than I first thought. I have found it impossible to obtain an agreed answer to the question of how to distinguish conditions required for dormancy-breaking from those required for germination-induction. I therefore leave it to the reader to judge between the subtle differences of approach to dormancy between the authors of Chapters 8 to 12. It seemed to me that no useful purpose would be served by insisting on a uniform viewpoint on this matter throughout the book. The differences found here are an honest reflection of the differences of opinion currently held by various researchers in the field. Happily, there seems to be no disagreement on the ecological significance of dormancy as a mechanism to prevent inopportune germination.

One of the great pleasures of 'seed ecology' is that so much useful work (especially fieldwork) can still be carried out without the use of difficult techniques or sophisticated equipment. Fundamental questions can often be answered by merely counting the number of seeds dispersed, eaten or germinated, or by measuring the weight of seeds produced, the distance dispersed, the size of gaps, the time taken to germinate, or the relative growth rate of the seedlings. Crawley's recent experiments to determine whether the regeneration of various species was seed-limited exemplify the use of a simple technique to answer a fundamental question. Extra seeds were simply added to the community to see if this resulted in increased recruitment. There is particular scope for well-designed experiments to investigate such questions as the proportion of seeds which germinate too deep to emerge, the effect of mast seeding on recruitment, the influence of gaps made by different agents on the differential regeneration of species in a community, the role of nearest neighbours on seedling establishment, and possible trade-offs between growth rate and defence in seedlings.

This book is essentially a review of current data, concepts and theories about the regeneration of plants. So much information has accumulated on this subject over the last two decades that there is a real need for some sort of overview to guide us through the maze of data and allow us to focus on the key questions. This is what the contributors to this volume have attempted to do. I hope the book will prove to be useful and interesting to a wide range of readers, regardless of their level of study. I especially hope it will give pleasure to the non-specialist, who may be surprised by the complexities of the subject and enjoy the ingenuity of the experiments designed to unravel them. For the researcher, I hope the book will inspire some new interpretations of the current data, stimulate some new hypotheses and suggest fresh lines of research.

Michael Fenner
Southampton
September 2000

Chapter 1
Reproductive Allocation in Plants

Fakhri A. Bazzaz,[1] David D. Ackerly[2] and Edward G. Reekie[3]

[1]*Department of Organismic and Evolutionary Biology, Harvard University, Cambridge, Massachusetts, USA;* [2]*Department of Biological Sciences, Stanford University, Stanford, California, USA;* [3]*Department of Biology, Acadia University, Wolfville, Nova Scotia, Canada*

Introduction

The completion of a plant's life cycle and the regeneration and establishment of plant populations depend on the process of reproduction: the production of physiologically independent individuals. Yet there is a great deal of variation among higher plant species in the quantity and timing of reproduction; individuals may concentrate all their reproductive output in a single episode, after lifespans ranging from as short as 3 weeks to as long as several decades, or they may reproduce repeatedly, at regular or intermittent intervals. These variations in life history are influenced by both ecological and evolutionary factors, and a great deal of research on plant biology in the last 30 years has focused on physiological and demographic aspects of reproductive strategies (Harper, 1967; Silvertown and Lovett Doust, 1993).

The theory of allocation, borrowed from microeconomics, first introduced to biology by MacArthur (cited in Cody, 1966) and extended to the study of plants by Harper (Harper, 1967; Harper and Ogden, 1970), has provided the principal conceptual framework for linking individual physiology and life-history theory. The study of allocation in biology assumes that organisms have a limited supply of some critical resource (e.g. energy, time, biomass or nutrients), which they must generally divide between several competing functions, broadly defined as growth, maintenance and reproduction. These functions are further assumed to be mutually exclusive, such that allocation to one function necessarily leads to a decrease in the simultaneous allocation to other functions, and as a result an optimal pattern of allocation will exist that maximizes some output parameter. In evolutionary ecology, this output is a measure of fitness, an individual's contribution to future generations, which depends on total reproductive output and the timing and frequency of reproduction (Willson, 1983). In the study of plants, this approach has been widely utilized, because the various functions of the plant can be approximately assigned to discrete structures: carbon gain is primarily the function of leaves, nutrient uptake of roots and reproduction of inflorescences, seeds and ancillary structures (but see Bazzaz, 1993). This rough equivalence opens up numerous avenues of empirical research. Unlike the animal ecologist's study of the allocation of an organism's time, the plant ecologist's assessment of allocation to various structures can be made by harvesting plants at single points in the life cycle, and studies can be conducted in both the field and the laboratory. Proportional allocation to different structures is thus used as a measure of investment in corresponding functions, following the scheme shown in Fig. 1.1a.

The two critical assumptions of allocational theory are that the resource in question is in fixed supply and that allocation among competing functions is mutually exclusive, thus generating trade-offs between functions. In its application to the study of reproductive strategies, these two assumptions are not always observed, for two principal reasons. First, several processes, such as the photosynthesis of reproductive parts, can lead to an increase in total resource supply associated with reproduction. And, secondly, plant structures can contribute to more than one function: stems can simultaneously provide support for both leaves and fruits. If the plant reproduces vegetatively, roots, stems and leaves may all contribute to reproduction as well as to their vegetative functions. As a result, measures of allocation to structures do not always reflect the plant's investment in function; Fig. 1.1b shows a more complete picture of the relationship between the two. The concept of reproductive effort (RE), which is often equated with reproductive allocation (RA), should refer specifically to the individual's net investment of resources in reproduction, which is diverted from vegetative activity (Tuomi *et al.*, 1983; Bazzaz and Reekie, 1985). In this chapter, we discuss several factors that decouple RA and RE, and argue that these two must be conceptually distinguished in future research. We address this difference within ecological and evolutionary contexts, in relation to plant size, seed production and life-history evolution.

As mentioned above, the process of reproduction can be defined as the production of physiologically independent individuals. In plants, this can be accomplished either by the production of seed, or through vegetative production of genetically identical offspring. The relative importance of these two mechanisms varies widely in different plant communities (see Grubb, 1977). Establishment from seed dominates in early successional communities, following major disturbances, and also in most forests, in which regenera-tion occurs intermittently throughout the vegetational mosaic. In various low-stature communities, such as savannahs, grasslands and arctic/alpine systems, regeneration is primarily by clonal growth. The study of allocation is considerably more difficult when clonal reproduction is considered. Clonal reproduction occurs by the production of vegetative tissues that contribute to their own growth, so it is very difficult to assess the allocation of the parent plant to the growth of ramets (Ogden, 1974). Numerous studies demonstrate that the daughter ramets are dependent on the parent for a varying length of time (see review by Pitelka and Ashmun, 1985); because of this physiological dependence, it can be difficult to distinguish between growth and reproduction. In this chapter, we shall primarily consider allocation to seed production, but with the reminder that much of what is said may apply to clonal growth as well.

Assessment of resource allocation to plant structures

Before embarking on a more detailed discussion of the relationship between structure and function and the decoupling of RA and RE, it is necessary to clarify the means of assessing allocation to structures themselves. In most studies, RA is assessed as the proportion of standing biomass or other resources found in reproductive structures at the time of harvest. This measure may be an inaccurate reflection of the allocation of total production, for several reasons. First, plants may be harvested before the end of the season, before final reproductive output has been expressed. This problem is particularly acute in plants with indeterminate reproductive development, which continues until frost or drought prevents further growth. When such plants are grown experimentally, the results may be strongly influenced by the time at which the experiment was terminated (e.g. Geber, 1990; Clauss and Aarssen, 1994). Secondly, for perennial, iteroparous plants there is a carry-over of standing bio-

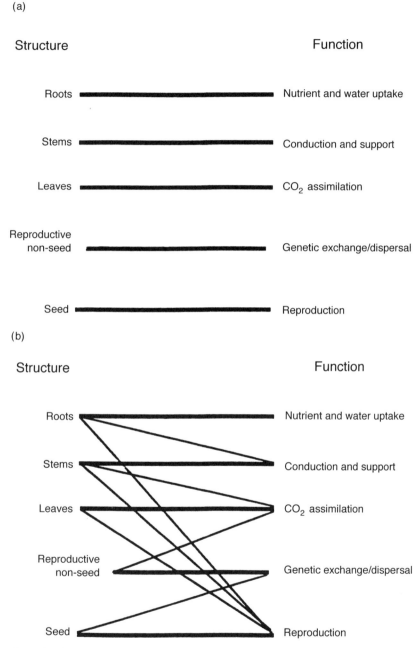

Fig. 1.1. Relationship between structure and function in plants. The functions represent biological processes that most plants must accomplish to complete their life cycle. The relative importance of different functions will vary with life history and environment. (a) In this highly simplified view, each structure corresponds to a single function; this is the basis for the application of resource allocation theory to the study of plant function. (b) A more complex view considers the contribution of structures to various functions. Roots, stems and leaves can all contribute to reproduction through various forms of clonal growth; roots contribute to support, stems and reproductive structures can be photosynthetic, and seed characteristics contribute to dispersal as one component of reproduction. The relative importance of each link in this diagram will vary in different situations, leading to varying degrees of decoupling between structure and function.

mass from year to year in woody tissue, such that standing RA underestimates RA based on annual production (Willson, 1983). Finally, any loss of a resource during growth will lead to discrepancies between the standing pattern of allocation and the total acquisition and allocation. Loss from vegetative organs will lead to overestimates of RA, while loss from reproductive organs will result in underestimates of RA. This loss may occur by many processes, and the amount of loss relative to the standing pool will differ for different resources. Mineral resources probably have the lowest loss rate relative to the amount present in the plant under most circumstances; losses of minerals, as well as total biomass and associated energy content, will be due primarily to damage, herbivory, abscission of senesced parts and dispersal of reproductive structures. These losses are fairly easily assessed in experimental situations by repeated collection of abscised parts (e.g. Harper and Ogden, 1970).

In addition to these structural losses, carbon and energy are also dissipated by respiratory activity associated with tissue construction and maintenance. Respiratory costs will only alter patterns of RA if vegetative and reproductive tissue have markedly different respiratory activity. In a study of *Agropyron repens*, Reekie and Bazzaz (1987a) measured whole-plant respiration associated with reproduction and found that rates of respiration for vegetative and reproductive structures were apparently fairly similar, such that measures of RA that included respiration were very similar to those based solely on biomass. Laporte and Delph (1996), however, report that in the dioecious plant, *Silene latifolia*, male plants have a respiratory rate that is twice as high as that of female plants. In this case, ignoring the respiratory costs associated with reproduction may seriously underestimate resource allocation to male function (see also Goldman and Willson, 1986; Marshall *et al.*, 1993). Of all the resources, water has the highest loss rate, through transpiration. Measuring the plant's allocation of water requires elaborate measurements of transpiration, and

has rarely been considered on a whole-plant basis. In general, for those resources with a large flux through the plant relative to the amount present at one point in time, measures of standing allocation will correlate poorly with total allocation of the resource pool.

Three different measures of resource allocation to reproductive structures must therefore be carefully distinguished: (i) standing RA is the proportion of a resource contained in reproductive structures; (ii) short-term RA is a measure of proportional allocation of a resource over some short interval, relative to plant lifetime, such as a growing season for a tree, or a period of days or weeks for an annual; and (iii) lifetime RA is the proportion of total available resource invested in reproductive structures over the entire lifespan of an individual. This last measure is of most general interest when comparing individuals with different life-history strategies or from different environments; many studies report lifetime RA for annuals and other semelparous taxa, but virtually no data exist for iteroparous perennials. Owing to the factors discussed above, standing RA will reflect short-term and lifetime allocation only under a limited set of conditions. For comparative purposes, measures of RA should be based on the net production of the plant, including structural losses. The inclusion of respiration is more difficult and it has been considered in so few studies that no general conclusion is at present possible regarding its contribution to RA.

The currency of allocation

In addition to these factors, it is important to bear in mind that the patterns of resource allocation will differ for different resources. In one population of *Verbascum thapsus*, for example, RA was 40% based on biomass, but varied from less than 5% to almost 60% when calculated for various mineral elements (Abrahamson and Caswell, 1982; see also Fenner, 1985a, b; Reekie and Bazzaz, 1987b; Benner and Bazzaz, 1988). Which resource is the appropriate currency of allocation? The most difficult aspect of this problem arises from the assumption that there is a crucial resource that limits both vegetative and reproductive functions, such that its allocation will determine individual performance. It has long been recognized that different resources limit growth in different environments, or even at different times during the plant's lifetime. In addition, different functions of a single plant might be limited by different resources. This problem has been cast in a new light in recent years by the observation that plant growth is not always limited by a single resource, but tends to be limited by several resources simultaneously (Chapin *et al.,* 1987). The plasticity of plant development allows individuals to alter the investment in the uptake of different resources – by altering root:shoot ratios, for example. It is often observed that allocation to roots increases in response to low nutrient and water availability, and allocation to leaves increases in low light. Theoretically, growth will be maximized when investment of a gram of biomass, or other resource, in the acquisition of any single resource from the environment leads to an equivalent increase in growth (Bloom *et al.,* 1985). The acquisition of different resources is thus scaled to their relative availability or to their need, such that the concentrations of resources in the plant are essentially uncorrelated with their availability in the environment (Abrahamson and Caswell, 1982).

If growth and reproduction are not limited by a single, crucial resource, then, which resource is the appropriate currency of allocation? We suggest that the answer may be that any of the various limiting resources is appropriate. It is true that absolute allocation to reproduction may differ for various currencies. For example, the proportion of whole-plant nitrogen that is allocated to reproductive structures is often higher than the proportion of biomass, i.e. carbon (Williams and Bell, 1981; Abrahamson and Caswell, 1982; Reekie and Bazzaz, 1987b; Garbutt *et al.,* 1990; Benech Arnold *et al.,* 1992). However, pro-

viding these resources are in fact limiting growth, relative rankings among species, populations, genotypes and environments should be similar, regardless of the currency used (Reekie, 1999). To use the above example, even though the proportion of nitrogen allocated to reproduction may be higher than that of carbon, when comparing the reproductive allocation of two plants (e.g. two genotypes from contrasting populations), the plant with the higher biomass allocation to reproduction should also have the higher nitrogen allocation to reproduction. If both resources are in fact limiting growth, their allocation will be subject to selection and, if selection favours high RA, this should be reflected in both currencies. The limited data available appear to support this suggestion. Reekie and Bazzaz (1987b) grew three genotypes of *Agropyron repens* in seven different nutrient environments (i.e. various levels of nitrogen and phosphorus availability) and calculated RA in terms of biomass, nitrogen and phosphorus. Correlations between RA calculated in terms of the three currencies across the 21 genotype $\times$ environment combinations were uniformly high. Hemborg and Karlsson (1998) compared reproductive allocation of 13 plant populations representing eight species, using biomass, nitrogen and phosphorus as the currency of allocation. Correlations among RA calculated in terms of the three different currencies across the various populations showed that the allocation of one resource closely reflected the allocation of the other resources. Ashman (1994) compared the resource allocation of hermaphrodite and female plants of *Sidalcea oregana*, using biomass, nitrogen, phosphorus and potassium as currencies, and found that conclusions regarding the relative allocation patterns of the plants were unaffected by the currency chosen.

A second aspect of the currency question that has been raised in recent years is the hypothesis that in some species growth and reproduction may be limited primarily by the availability of meristems, rather than abiotic resources. In a study of population growth in the water hyacinth,

Eichhornia crassipes, Watson (1984) compared two populations growing in enclosed tanks, one of which flowered while the other remained vegetative. In this species, the rate of meristem production is fairly constant and every axillary meristem that is produced is elaborated into either a vegetative module, a new ramet, or an inflorescence. As a result, in the population that flowered, the rate of population growth, in terms of cumulative ramet number, declined during flowering and the saturation population size in the tank was lower than in the vegetative population. A more detailed study of potential meristem limitation in *Polygonum arenastrum*, an indeterminate annual, was presented by Geber (1990). In *P. arenastrum*, like *E. crassipes*, every axillary meristem develops into either a branch or a terminal inflorescence. Geber compared growth and reproduction, based on commitment of meristems, in 26 full-sib families from a natural population grown in a common garden experiment. Analysis of genetic and phenotypic correlations demonstrated that early commitment of meristems to reproduction was negatively correlated with later reproductive output and growth. In addition, a strong positive genetic correlation was observed between the number of vegetative meristems and the number of reproductive meristems that were produced late in development, which was interpreted as evidence for meristem limitation – see Geber (1990) for discussion of this interpretation. Both of the above studies examined species whose meristems develop into either vegetative or reproductive shoots. In many species, however, meristems can have a third fate: they can remain dormant for prolonged periods of time and, indeed, may never give rise to an actively growing shoot. Lehtila *et al.* (1994) examined the question of meristem limitation in one such species, *Betula pubescens*. In this species, meristems may develop into vegetative shoots, generative long shoots with male catkins or short shoots with female catkins or remain dormant. In spite of the commitment of the shoot apices to catkin production, male and female shoots had approximately the same bud production rate as vegetative shoots, because new axillary buds developed from dormant meristems to compensate for the lost shoot apices. This implies that this species is not limited by meristem availability.

Studies that focus on meristem allocation make an important contribution to research on reproductive strategies by focusing attention on the developmental context of growth and reproduction. However, it must also be kept in mind that, even in species whose meristems always develop into either vegetative or reproductive shoots, developmental programmes evolve in the context of the reproductive strategies. Are rosette plants semelparous because they develop only a single apical meristem or do they only develop one meristem because they are semelparous? Many of these plants, whose growth is apparently constrained by their single apical meristem, also grow vegetatively from rhizomes; in one study of *Agave deserti*, 95% of new plants were derived from rhizomatous production of vegetative ramets (Nobel, 1977). As is the case for different abiotic resources, the availability of meristems in most plants is partially scaled to the availability of resources. The rate of module initiation by apical meristems usually increases with increasing nutrient, light and, occasionally, carbon dioxide availability and decreases with increasing density (Bazzaz and Harper, 1977; Harper and Sellek, 1987; Ackerly *et al.*, 1992). Meristem activity, carbon gain and nutrient uptake are all strongly temperature-dependent as well. If meristem availability scales to resource availability, the complete elaboration of developed meristems does not necessarily indicate that meristem numbers limit growth or reproduction. Testing the hypothesis of meristem limitation requires further research in two directions: simultaneous comparisons of meristem and resource allocation, and responses of meristem allocation to gradients of resource availability.

Reproductive allocation (RA) versus reproductive effort (RE)

RA is defined as the proportion of the total resource supply devoted to reproductive structures. RE, on the other hand, is used to refer to the investment of a resource in reproduction, which results in its diversion from vegetative activity (Tuomi *et al.*, 1983; Reekie and Bazzaz, 1987c; Fig. 1.2). As we have mentioned, several factors can decouple these two parameters in an individual

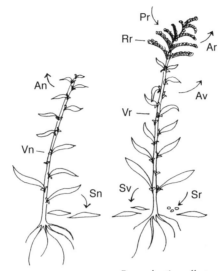

Reproductive allocation

1) $\dfrac{Rr}{Tr}$

2) $\dfrac{Rr + Rv}{Tr}$

3) $\dfrac{Rr + Rv + Sr + Ar}{Tr + Sv + Av}$

Components of resource pool
Non-reproductive plant
Vn = Vegetative size
Sn = Structural losses
An = Atmospheric losses
     (respiration/transpiration)

Reproductive effort

Direct

4) $\dfrac{(Rr + Rv + Sr + Ar) - Pr}{(Tr + Sv + Av) - Pr}$

Indirect

5) $\dfrac{Vr - Vn}{Vn}$

6) $\dfrac{(Vr + Sv + Av) - (Vn + Sn + An)}{(Vn + Sn + An)}$

Reproductive plant
Tr = Total standing pool
Vr = Vegetative pool
Rr = Reproductive pool
Rv = Vegetative biomass attributable to
     reproduction
Sv = Structural losses from vegetative organs
Sr = Structural losses from reproductive organs
Av = Atmospheric losses from vegetative organs
Ar = Atmospheric losses from reproductive organs
Pr = Enhancement of total resource supply due to
     reproduction

Fig. 1.2. Definitions of reproductive allocation (RA) and reproductive effort (RE). The resource pool of an individual plant is divided into various components; these components can then be used to calculate various measures of RA and RE. The relative importance of different components will differ in different studies, so several alternatives are provided for both RA and RE, depending on circumstances. Equations 1, 2, 4, 5 and 6 are equivalent to the terms defined as RE2, RE3, RE5, RE6 and RE7, respectively, by Reekie and Bazzaz (1987a, c). Equations 1 and 5 are equivalent to RE/E and RE_s as defined by Tuomi *et al.* (1983). Artwork by K. Norweg.

plant. Three factors are considered here: (i) reproduction can lead to an increase in resource supply, either through direct uptake by reproductive structures or by enhancement of uptake rates by vegetative structures; (ii) vegetative growth may be greater in reproductive plants, such that some of the vegetative biomass should be attributed to the function of reproduction; and (iii) resources may be moved between vegetative and reproductive structures and, as a result, allocation of a resource to one function does not prevent its subsequent allocation to another.

Reproductive enhancement of the resource supply

Many flowers and fruits are known to be photosynthetic, and any carbon that is supplied *in situ* reduces the investment in reproduction that is required from the vegetative structures of the plant. In a study of *Ambrosia trifida*, Bazzaz and Carlson (1979) reported that the contribution of *in situ* photosynthesis to the carbohydrate demands of male and female inflorescences was 41% and 57%, respectively. A comparative study of 15 temperate tree species found that reproductive photosynthesis contributed 2.3–64.5% of carbohydrate needs for production of female flowers and seeds (Bazzaz *et al.*, 1979). Jurik (1985) included respiratory costs associated with the construction and maintenance of various reproductive organs in *Fragaria virginiana*, and found that *in situ* photosynthesis contributed as much as 54.8% of carbohydrate to flowers, but less than 10% to flower stalks, flower-buds and fruits. For all organs combined, photosynthesis contributed 3.6–8.9% of carbon costs. The achenes of *Ranunculus adoneus* are green and maintain a positive net assimilation rate throughout fruit maturation (Galen *et al.*, 1993). Shading the infructescences reduced the weight of the mature achenes by 16–18%.

Reproduction may also contribute indirectly to the plant's total resource pool, due to enhancement of leaf photosynthetic rates. Changes in leaf-level physiology, due to the enhancement in sink strength during reproduction, have been observed in fruit-trees and crop species (Neales and Incoll, 1968; DeJong, 1986). In the study of *Agropyron repens* by Reekie and Bazzaz (1987a), changes in leaf photosynthesis ranged from −33 to 64%, depending on genotype and nutrient treatment. Leaves of pollinated female plants in the dioecious species *Silene latifolia* have light-saturated photosynthetic rates 30% higher than those of unpollinated females 28 days after flowering (Laporte and Delph, 1996). Interestingly, male plants of *S. latifolia* have higher photosynthetic rates than pollinated females, even though they have a lower reproductive allocation and, apparently, a lower sink strength (Gehring and Monson, 1994; Laporte and Delph, 1996). Similar sexual differences have also been observed for another dioecious species, *Phoradendron juniperinum* (Marshall *et al.*, 1993). These apparent differences in the effect of sink strength on photosynthetic rate may be related to sexual differences in nutrient and water use (Gehring, 1993; Gehring and Monson, 1994).

In addition to direct photosynthesis by reproductive structures and sink-induced enhancement of leaf photosynthesis, reproduction can enhance carbon gain through changes in canopy structure and allocation patterns. In many plants, reproduction is associated with stem elongation (i.e. bolting). Elongation of the stem facilitates both the receipt of pollen and the subsequent dispersal of seeds. This stem elongation can also have beneficial effects on carbon gain, in that it may reduce self-shading and may improve the capacity of the plant to compete for light in a closed canopy. In *Oenothera biennis*, bolting has no effect on growth in an open canopy, but increases growth in a closed canopy (Reekie *et al.*, 1997). Reproduction has also been shown to enhance carbon uptake through increases in leaf area ratio, either through increases in allocation to leaves (Reekie and Bazzaz, 1992) or through changes in leaf morphology (i.e. specific leaf area) (Reekie and Reekie, 1991).

Although the effect of reproduction on carbon uptake has been studied to a greater extent, there is evidence that reproduction can also enhance the uptake of mineral resources. By artificially manipulating levels of reproduction by removal of inflorescences and comparing the total resource pool in vegetative versus reproductive plants, it is possible to examine the impact of reproduction on the uptake of a variety of resources. Using this technique, Thoren *et al.* (1996) found that reproductive plants of three different species of carnivorous plants in the genus *Pinguicula* accumulated more biomass, nitrogen and phosphorus than equivalent vegetative plants. The mechanisms by which reproduction may enhance nutrient uptake are not entirely clear, but Karlsson *et al.* (1994) found that, in *Pinguicula vulgaris*, reproductive individuals captured almost twice as many prey as vegetative individuals. This could account for as much as 58% of the increased nitrogen uptake of reproductive plants (Thoren *et al.*, 1996). However, there was no evidence of increased rates of prey uptake with reproduction in the other two species, in spite of the fact that reproduction did enhance both nitrogen and phosphorus uptake. Although reproduction may enhance the uptake of mineral nutrients, positive effects of reproduction on carbon uptake may be more widespread. Hemborg and Karlsson (1998), using the same technique as Thoren *et al.* (1996), examined the effect of reproduction on biomass, nitrogen and phosphorus uptake in eight subarctic species. They found that reproductive plants always accumulated more biomass (mostly carbon) than vegetative plants, but that this was not the case for nitrogen and only in a few cases did reproductive plants accumulate more phosphorus.

As a result of these direct and indirect influences of reproduction on the resource budget, the cost of reproductive organs will differ from that reflected in measures of resource distribution.

Reproductive support structures

Reproduction can also lead to increases in support structures beyond those necessary to support leaves. This reproductive support biomass may be readily apparent, such as the stalks of rosette plants, which bolt when they flower, or the upper stems of many herbaceous plants, which do not bear leaves. Thompson and Stewart (1981) have suggested that any structures which are not found on vegetative plants should be considered part of reproductive biomass. Bazzaz and Reekie (1985) have also noted that some existing vegetative organs will increase in size in reproductive plants; they applied allometric relationships between leaf biomass and various support structures in vegetative plants in order to determine the fraction of support biomass that is attributable to reproduction. In contrast to the enhanced resource uptake due to reproduction, which decreases the carbon cost to the vegetative plant, the consideration of reproductive support biomass may significantly increase measures of RA.

In order to determine vegetative and reproductive biomass experimentally, Antonovics (1980) suggested the study of plants in which flowering can be induced by small variation in photoperiod, allowing the comparison of flowering and non-flowering plants under essentially identical resource conditions. Reekie and Bazzaz (1987a, c) followed this approach in detailed analysis of the carbon budget of *A. repens*, considering the influence on RE of reproductive photosynthesis, respiration and changes in support biomass due to reproduction. Under various nutrient and light conditions, inflorescence biomass in this species ranged from 0 to 10% of total biomass. Inclusion of reproductive support biomass increased RA to 20–45%. Consideration of respiration had little effect, as discussed above. The inclusion of reproductive photosynthesis, however, decreased RE to below 20% and, in one of the six genotypes, RE was negative in several treatments, suggesting that there was a net supply of photosynthate to the vegetative plant as a result of reproduction.

A final problem is introduced by the reallocation and translocation of resources within the plant. Several models and empirical studies have considered the consequences of carbon storage for patterns of RA (Schaffer *et al.*, 1982; Chiariello and Roughgarden, 1984). These models allow the plant to store biomass for later deployment to reproduction, but they still assume that any given unit of resource may be allocated to only one function. In contrast, water and many mineral resources may be recycled within the plant. In the semelparous, desert species *Agave deserti*, inflorescence development requires 18 kg of water, all of which can be supplied by reallocating water from the leaves (Nobel, 1977). In this case, the allocation of a resource at one point in time does not reflect its functional utilization within the plant. Ashman (1994) examined the importance of resorption of nitrogen and phosphorus from reproductive structures in calculating reproductive allocation in *Sidalcea oregana*. Two measures of RA were calculated, one based upon the average nutrient content of the reproductive structures ('initial' *sensu* Chapin, 1989) and one based upon nutrient content after any resorption from senescing calyces, corollas, etc. had taken place ('final' *sensu* Chapin, 1989). The two estimates differed substantially. Reproductive allocation was reduced by as much as 70% by nutrient resorption from senescing reproductive structures. Furthermore, measures of RA based upon resorption were more highly correlated with the effect of reproduction upon subsequent performance than measures that failed to take resorption into consideration. Although this study provides an indication of the possible magnitude of the error involved in ignoring nutrient reallocation, ideally, reproductive allocation should also take into consideration the amount of time a nutrient ion is allocated to a particular function. Empirical determination of the extent of reallocation would require a complete resource budget for a plant throughout its ontogeny. We are aware of no theoretical

treatment that provides a framework for including reallocation among structures in studies of RA patterns (and consequently for measures of RE), and suggest that it may provide a fruitful avenue for future research.

There are many factors that must be included in evaluation of RA and RE. There is no single definition of these terms that can both incorporate all of these factors and be usable for research projects of different scale and motivation. We do suggest, however, that each of the terms be used to refer to an exclusive set of concepts. RA refers to the proportion of a resource devoted to reproduction relative to the total resource pool. Measurement is based on the quantity of a resource in reproductive structures as a fraction of the total resource pool, plus consideration of reproductive support tissues and losses of the resource through senescence, respiration, etc. whenever possible. The direct measurement of RE, on the other hand, requires calculation of the change in the resource supply due to reproduction. This quantity is removed both from the investment in reproduction and from the total quantity available (Bazzaz and Reekie, 1985). Indirect measurements of RE can be calculated if the total size of non-reproductive and reproductive plants in the same resource environment are known. RE is then defined as the proportionate change in vegetative biomass resulting from reproduction (Tuomi *et al.*, 1983; Reekie and Bazzaz, 1987c). In calculating indirect measures of RE, only standing resource pools need be considered, excluding tissue and respiratory losses, in order to focus on the resources available for future growth and reproduction. These definitions are formalized in Fig. 1.2. This distinction between RA and RE is critical if the concepts are to retain their usefulness both in ecological studies, relating allocation to fecundity, and in the study of life-history evolution, which is based on potential trade-offs between reproduction and future survival and fecundity.

RA and reproductive output (RO)

If two plants of equal size differ in their allocation to reproduction, the individual with the higher allocation would be expected to have higher reproductive success in terms of contribution to future regeneration of the population. However, RA is only one component of fecundity and, in order to link allocation and RO, it must be considered in relation to plant size. RA is a measure of the proportion of resources devoted to reproduction, while RO is a measure of the total quantity of reproduction. The absolute quantity of reproductive biomass will be the product of total plant biomass and RA. In addition, the quantity of seed biomass will depend on the seed fraction relative to other structures; the number of seeds will equal the total seed mass divided by the mean individual seed mass, and the number of successful offspring will depend on germination percentage and seedling survivorship. The latter components of successful reproduction, germination and establishment are treated in later chapters of this book; we shall focus on recent studies of the relative contributions of plant size and RA to total fecundity. Variation in these parameters can be considered at three levels: phenotypic differences among individuals resulting from environmental influences; genotypic and ecotypic differences among individuals of a population or populations of a species; and variation among species across habitat types and life-forms.

Environmental influences

Owing to the indeterminate nature of development in most higher plants, the reproductive individuals in a population can vary enormously in size (Harper, 1977). Clearly, at this broad scale, variation in RA cannot compensate for such large size differences. A plant that weighs 10 g and devotes 20% of its biomass to seeds will necessarily have a higher reproductive output than another individual which weighs 1 g, regardless of the latter's pattern of allocation. Variation in both size and RA can arise from fine-scale environmental heterogeneity within the range of a population, and from competitive interactions, which lead to inequalities in the distribution of resources and result in population size hierarchies (Weiner and Solbrig, 1984).

In annual species, the majority of the studies seem to indicate that variation in RO is correlated more closely with variation in size than with RA. For example, in experimental monocultures of *Ambrosia artemisiifolia*, final biomass of reproductive individuals varied from 1.5 to 12 g, but RA did not vary significantly with size; as a result, most variation in reproductive output was due to variation in plant weight (Ackerly and Jasienski, 1990). The influences of resource availability on size versus allocation are also demonstrated in various experimental studies under controlled conditions. In their pioneering study of RA, Harper and Ogden (1970) grew *Senecio vulgaris* in three pot sizes. In the medium pots, total production was reduced by 85% relative to plants in the large pots, while seed allocation was reduced by only 21%; as a result, most of the reduction in RO was due to the decrease in size. In the smallest pots, however, most individuals failed to set seed, so reproduction failed completely. In a study of several *Polygonum persicaria* clones grown along a light gradient, Sultan and Bazzaz (1993a) observed changes of more than two orders of magnitude in both total and fruit biomass, while RA only changed from about 10% to 25%. Cheplick (1989) investigated allocation to above- and below-ground seed production in the annual grass *Amphicarpum purshii* grown at two nutrient levels. Variation in fecundity was mostly due to variation in plant size, but allocation to below-ground seed production was fairly constant at the two nutrient levels, while allocation to above-ground seeds increased. Thus, in most cases, variation in RA is less significant than variation in biomass in annuals, such that RO in many populations is strongly correlated with plant size (Harper, 1977; Solbrig and Solbrig, 1984). This conclusion is supported by studies that utilize path analysis to determine the contribution of different traits to reproductive success in natural populations of annuals (Farris and Lechowicz, 1990; Mitchell-Olds and Bergelson, 1990; Schwaegerle and Levin, 1990). These studies have all demonstrated the importance of early size differences and subsequent variation in growth rate to variation in final size and fecundity. Similar conclusions emerge from studies on the role of seed size and emergence time in determining competitive hierarchies of size and fecundity (Stanton, 1985; Waller, 1985).

The relative importance of size versus RA in determining RO in perennials is less clear. Annual plants have only one opportunity to reproduce; therefore, it is not surprising that variation in RA tends to be relatively low. All annuals should have a uniformly high RA at maturity and it is to be expected that most variation in RO will be associated with size. Variation in RA in perennials, on the other hand, should be greater, given that individuals may have opportunities to reproduce in future years. There is, indeed, a great deal of evidence that RA in perennials varies with a variety of environmental factors (e.g. Wankhar and Tripathi, 1990; Dale and Causton, 1992; Williams, 1994; but see Benech Arnold et al., 1992). This variation can have a significant impact on total RO. For example, in a study of the perennial grass Agropyron repens, Reekie and Bazzaz (1987c) found that RO measured as the weight of the infructescences increased over fivefold as light and nutrient levels increased from minimal to maximal levels. Increases in plant size could only account for 50% of the increase in RO; the remaining 50% was accounted for by an increase in RA. The role of variation in RA in explaining lifetime RO, however, has not been studied to the same extent as it has in annuals, because of the difficulties in following individuals over an indeterminate lifespan (see Alvarez-Buylla and Martínez-Ramos (1992) for a detailed study of lifetime RO in an early successional tropical tree).

Given that environmentally induced variation in RA may be important in determining RO in perennial, if not annual, species, there are relatively few studies that have attempted to explain this variation. Instead, variation in RA has been interpreted, for the most part, as an indirect consequence of the effect of the environment on plant size. RA of perennials, calculated on the basis of annual production, is frequently observed to vary with plant size (Piñero et al., 1982; Samson and Werk, 1986; Bazzaz et al., 1987; Oyama and Dirzo, 1988; Weiner, 1988; see Klinkhamer et al. (1990) for a discussion of statistical problems associated with the analysis of size-dependent allocations). Samson and Werk (1986) suggest that this correlation results from underlying allometric constraints. For example, if flowers are borne in the axils of leaves, increasing the number of leaves will also increase the number of flowers. They go on to demonstrate that, if this dependence results in a linear relationship between RO and plant size, RA will increase, decrease or remain constant, depending upon whether the x-intercept is positive, negative or goes through the origin (see Fig. 1.3). They therefore argue that size-dependent changes in RA are simply a reflection of morphological constraints and should be ignored when trying to determine whether or not there has been direct selection for phenotypic plasticity in RA. Instead, it is suggested that it is necessary to correct for differences in plant size among environments by comparing the linear relationship between RO and plant size among environments. Only if the slope or intercept of this relationship changes among environments can it be concluded that there are differences in allocation patterns that may be the result of selection. This approach has been widely adopted in recent studies of RA in plants (e.g. Hartnett, 1991; Aarssen and Taylor, 1992; deRidder and Dhondt, 1992; Korpelainen, 1992; Schmid and Weiner, 1993; Clauss and Aarssen, 1994; Pickering, 1994). These studies demonstrate that much of the variation in RA among environments can indeed be correlated with differences in size. Although there are often shifts in the

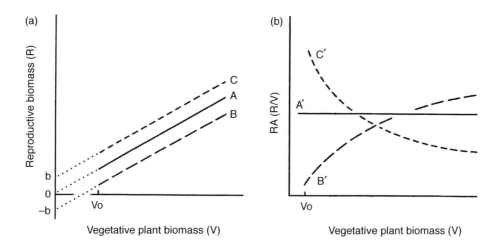

Fig. 1.3. The implications of a linear relationship between reproductive biomass and plant size (a) on the relationship between reproductive allocation and plant size (b). Depending on whether the *y*-intercept of the relationship between reproductive biomass and plant size is zero, negative or positive, RA will not vary (A), increase (B) or decrease (C) with plant size, respectively. (From Samson and Werk, 1986.)

relationship between RO and size among environments, these effects on RA are often less important than the size-correlated effects. For example, in a study of natural populations of several *Centaurium* species, Schat *et al.* (1989) found that RA increased more rapidly with size in infertile sites, but the level of allocation was higher in fertile sites because of the larger overall plant size.

The above approach has a great deal of merit, in that it explicitly recognizes that changes in RA may not be the result of direct selection but simply a consequence of morphological or allometric constraints. Unfortunately, it also has a major disadvantage, in that the 'constraints' are described only, in somewhat vague terms, as any size-correlated change in RA. The idea that size-related changes in allocation patterns are a direct result of constraints is originally derived from the zoological literature. In particular, it stems from studies of skeletal structure, where there is a fixed relationship among the sizes of various bones, which can be described by a linear relationship between the logarithm of the size of one bone and the logarithm of another

(i.e. an allometric relationship). This linear relationship appears to hold true across environments and developmental stages and even among species. The fact that the relationship is constant implies that it is 'constrained' in some fashion. The constraint in the case of skeletal structure is quite apparent. Take the bones of a skull, for example; if you increase the size of one of the bones and do not change the size of the other bones in the skull in proportion, it is unlikely that the various bones will still be able to fit together (i.e. the skull will fall apart). Allometric relationships in plants, however, are much more variable. They vary among environments, developmental stages and species (Hartnett, 1991; deRidder and Dhondt, 1992; Korpelainen, 1992; Clauss and Aarssen, 1994; Welham and Setter, 1998). Given that plants have a modular structure and indeterminate growth, as compared with the very determinate growth patterns of most animals, this difference should not be surprising. To use an earlier example, RA in a particular plant species may be 'constrained' by the number of leaves in inflorescences if the flowers are produced in the axils of leaves.

But, given that in most plants a large proportion of axillary buds remain dormant and never develop into inflorescences or branches, the individual still has a great deal of flexibility in RA, in that it can control what proportion of the population of buds actually develops into inflorescences (Lehtila *et al.*, 1994). The fact that allometric relationships in plants can vary implies that RA is not closely constrained by the size of the plant and that it is free (i.e. variation is present at a given size) to be acted upon by selection. We cannot automatically assume that size-dependent variation in RA is a function of allometric constraints on the basis of a principle imported from the zoological literature that is not applicable to plants. Rather, we must critically examine the question of whether there are any size-related constraints based upon an understanding of the biology of the particular species in question.

One such explanation is offered by Weiner (1988). He argues that many plants must reach a minimum size before reproduction is possible. This minimum size is necessary because the plant must have enough resources to construct the reproductive support structures (e.g. the flower stalk) before the first seed can be produced. Further, as plant size is increased beyond the minimum size, there should be a more or less linear increase in RO, as plant size reflects the accumulated resources available for reproduction. As a result, there should be a linear relationship between RO and plant size, with a positive x-intercept that represents the minimum size required for reproduction (i.e. maximum plant size when RO is still zero). Given this relationship, RA will initially increase with size and eventually approach an asymptote (Fig. 1.3). With this model, variation in RA with size is a simple consequence of the necessity of reaching a minimum size before reproduction is possible. One problem with this model is that, although the relationship between RO and plant size often has a positive intercept, there are also many cases in which the intercept is negative (Samson and Werk, 1986; Shipley and Dion, 1992). This implies that the mini-

mum size required for reproduction is less than zero. Since this is clearly not possible, the simple interpretation of the x-intercept as the minimum size required for reproduction may not be correct. Klinkhamer *et al.* (1992) offer a possible solution to this dilemma by suggesting that the relationship between plant size and RO is, in many cases, curvilinear. This could result if there were economies of scale such that fewer resources were required to produce a seed as more seeds are produced. Fitting a linear relationship to such data would result in an underestimate of the x-intercept. They provide an alternative model that incorporates both a minimum size required for reproduction and the potential for a curvilinear increase in RO with plant size. This is an attractive model, in that it explains size-dependent variation in RA on the basis of constraints that are directly related to plant biology. To date, however, there has been little attempt to test the model rigorously to see if it truly describes the relationship between RO and plant size for a variety of species (e.g. see Schmid and Weiner, 1993).

Given that it is unclear to what extent size-dependent variation in RA is a function of constraints, it would be unwise to simply dismiss the possibility that this variation may in fact be adaptive. There are probably a number of reasons why it may be adaptive for RA to vary with plant size. For example, Hara *et al.* (1988) have presented a model of optimal allocation strategies which predicts that the relationship between biomass and RA in herbaceous plants will depend on habitat. In species of open, disturbed environments, they predict a constant allocation to RA among individuals of a population, as observed in several examples above, but, for plants of closed habitats, such as forest understorey, they predict a decrease in RA with increasing biomass. The difference between the two types of plants emerges in part from the assumptions about the relationship between individual plant size and the length of the growing season. For plants of closed environments, growing-season length is assumed to be constant, regardless of plant size. For plants of open, disturbed environments, however, small size is assumed to be associated with a shorter growing season, as smaller plants are often shaded early in growth by larger individuals in the population. The predicted patterns in relation to habitat are supported by data for several species in Japanese plant communities.

Genotypic variation

Relatively few studies have focused on genotypic variation in size, RA and fecundity within natural populations, despite its central role in evolutionary dynamics. Furthermore, most of these studies have focused on annual species, where one would expect relatively little variation in RA. Geber (1990) studied 26 families of full siblings from a population of *Polygonum arenastrum* grown in a common garden to minimize environmental influences. She observed a twofold variation in total size, measured as either biomass or meristem number, and found strong genetic and environmental correlations between size and total reproduction. Sultan and Bazzaz (1993a, b, c) grew clones of 20 genotypes from two populations of *P. persicaria* on three resource gradients, namely, nutrients, light and moisture. Along all three gradients, fecundity was positively correlated with size. Genetic variation for RA was found in response to all three factors, but genetic variation for total size and total fruit biomass was present only in the moisture experiment. Clauss and Aarssen (1994) grew three genotypes of *Arabidopsis thaliana* on light, nutrient and pot-volume gradients. Fecundity increased with plant size, but analysis of covariance revealed that the relationship between size and fecundity (i.e. RA) varied both among genotypes and among environments. Sugiyama and Bazzaz (1997) examined the relative importance of size and RA in determining fecundity in *Abutilon theophrasti* by growing eight families of full siblings across a nutrient gradient. They found that variation in RA was more important in determining seed output

at low nutrient levels, while variation in plant size was more important at high nutrient levels. Also relevant here are studies of yield determination in crop cultivars where the harvested portion is a seed or fruit. These studies examine how increased yields (i.e. RO) have been achieved through crop selection. Both modern and historic cultivars are grown in a common environment and growth (i.e. size), harvest index (i.e. RA) and yield (i.e. RO) are compared. These studies show that the substantial increases in yield that have been achieved by crop selection are to a large extent the result of increases in RA rather than an increase in size (e.g. Wells *et al.*, 1991).

Collectively, the above studies indicate that genetic differences in both RA and size can be important in determining reproductive output, but their relative importance varies. Based upon the limited data, it is difficult to make generalizations as to when size versus RA will be more important. However, it seems logical that size should be relatively more important whenever there is asymmetric competition that results in marked size hierarchies, such as at high nutrient levels and plant densities, where competition for light is the predominant factor influencing plant size (e.g. at high soil fertility in the Sugiyama and Bazzaz (1997) study). On the other hand, RA should be relatively more important when plant densities are low (e.g. cultivated plants grown at spaced intervals) or when competition is symmetric, as is the case when nutrients are limiting (e.g. at low soil fertility in the Sugiyama and Bazzaz (1997) study).

Variation among populations

Studies on a variety of plant species demonstrate differences in RO among populations that are attributable to a combination of genetic and environmental factors. In some cases, populations with greater biomass also have higher allocation to reproduction, and both contribute to greater reproductive biomass (e.g. *Solidago speciosa*: Abrahamson and Gadgil, 1973).

In contrast, in five populations of *Polygonum cascadense* growing along a moisture gradient, RA decreased while vegetative biomass increased in successively wetter sites; the increase in plant size was greater than shifts in allocation, so that RO as such increased with size. When two of these populations were grown under common conditions, the size differences persisted, but allocation to reproduction was similar (Hickman, 1975). On the other hand, Reekie (1991) found that differences in RA among populations of *Agropyron repens* persisted in a common garden experiment, populations from disturbed sites having a lower RA than populations from less disturbed sites. DeRidder and Dhondt (1992) compared the relationship between fecundity and plant size among populations of *Drosera intermedia* growing in disturbed versus undisturbed sites and found that the slope of this relationship was steeper in more disturbed sites. In *Plantago lanceolata*, Primack and Antonovics (1982) observed a range of RA among eight field populations, from 17 to 65%, which was attributable primarily to environmental influences. No data are provided on relative plant size, however, in order to evaluate the contribution of RA to fecundity. Hartnett *et al.* (1987) compared two populations of *Ambrosia trifida* from fields of different successional age. When grown in a common garden, the plants from the older field produced 78% more seed biomass than those from an annually disturbed field, but this difference was primarily due to a 55% increase in total size and secondarily to a smaller increase in RA. Scheiner (1989) compared populations of the grass *Danthonia spicata* from five sites that differed in the time since the last burning. Plants in the most recently burnt site had the highest fecundity, due to the larger size, a higher percentage of flowering stalks and more spikelets per stalk. All of these factors were reduced in populations from intermediate age, but in the older site, which was burned 80 years previously, a slight increase in the percentage of flowering stalks resulted in an increase in fecundity. Hartnett (1991) conducted a similar study with the tall-grass prairie forb *Ratibida columnifera*, but found that plants in the most recently burned sites had the lowest fecundity. Increases in fecundity with time from last burning were largely the result of increases in plant size. However, there were shifts in the relationship between fecundity and plant size among sites, such that RA increased with time from last burning. Ostertag and Menges (1994) present a series of models that predict contrasting patterns of RA with time since last fire, depending upon fire frequency and its predictability.

In general, variation in RO among populations appears to be primarily due to differences in total size and, to a lesser extent, to changes in allocational patterns. At the same time, however, it is also clear that there are often marked differences among populations in RA. Some of these differences are correlated with differences in size, but size-independent effects are also common. Furthermore, these differences in RA can persist when plants from different populations are grown in a common environment. This suggests that selection for differences in RA have taken place, even though it seems to have a relatively insignificant effect on RO when populations are compared in the field. This raises an interesting question. Why would there be selection for differences in RA if it did not have a significant effect on fitness? This apparent contradiction can be resolved if you consider that fitness should be measured relative to competitors within a given environment. Although there may be large differences in size among populations growing in different environments, within a given population the environment is likely to be less variable and the variation in size correspondingly less. As a result, within a given population, variation in RA may be more significant in determining RO than variation in size.

Variation among species and communities

Many studies have demonstrated that RA is higher in annuals than in perennials, and higher in herbaceous plants of open habitats than in those of closed habitats (e.g. Gadgil and Solbrig, 1972; Pitelka, 1977; Abrahamson, 1979; see reviews by Evenson, 1983; Hancock and Pritts, 1987). Allocation of community-level production to sexual reproduction also decreases with successional age; in a recent study of old-field succession in the American Midwest, RA of community biomass declined from about 8% in recently abandoned fields to below 1% in fields abandoned for 20 or more years (Gleeson and Tilman, 1990). In a study of bog vegetation in England, Forrest (1971) found that total allocation of production to sexual reproduction was only 0.5%. RA of production in forest trees is approximately 5% in both temperate and tropical forests, and can be as low as 2% (Whittaker and Marks, 1975; Kira, 1978).

Silvertown and Dodd (1996) point out that comparisons of RA (and other traits) among species should control for phylogeny. Otherwise, it is impossible to know whether the trait correlations that are observed (e.g. differences in RA between annuals versus perennials) result from convergent evolution or simply from common descent. They reanalysed data from the literature and used phylogenetically independent contrasts to examine whether annual versus perennial and early versus late successional species differ in RA. Their analysis supports the conclusions of the earlier studies, in that they found annuals and species of early succession have greater RA than perennials and species of later succession.

In general, comparisons of total fecundity of different species or life-forms are difficult to interpret, because of variation in life-history strategies. A general discussion of plant reproductive capacity is presented by Salisbury (1942).

Allocation of reproductive biomass to seeds

The proportion of reproductive biomass allocated to seeds varies widely. Lloyd (1988) reviews data for several species; in 15 outcrossing taxa, seed and fruit alloca-
tion ranges from 34 to 83% of reproductive biomass, while, in females of the dioecious species *Silene alba* and in *Impatiens* spp., which have cleistogamous flowers, these values are greater than 90%. In a comparative study of numerous species with various breeding systems, Cruden and Lyon (1985) found that most outcrossing species exhibit greater investment in male than in female reproductive structures in the flowers, but following fertilization the majority of reproductive biomass is devoted to seeds and fruits. These data suggest that the primary factor influencing the evolution of reproductive allocation is pollinator limitation, in the case of male function, and the necessity of provisioning and dispersing the offspring, in the case of female function (Antos and Allen, 1994). Because seed filling is considerably more expensive than flower production in most species, the seed weight fraction of reproductive biomass commonly decreases when resources are limiting. Failure to set seed may result from limitation by pollinators or resources, and there is little consensus regarding their relative importance in natural systems (Lee, 1988; Calvo and Horvitz, 1990; Lawrence, 1993; Ramsey, 1997). Many plants vary the relative investment in male versus female reproductive structures as a function of plant size; in most species, relative investment in female function increases with size (DeJong and Klinkhamer, 1989), though the opposite pattern is observed for several wind-pollinated species: e.g. *Ambrosia trifida* (Abul-Fatih *et al.*, 1979); *Xanthium strumarium* (Solomon, 1989); *A. artemisiifolia* (McKone and Tonkyn, 1986; Ackerly and Jasienski, 1990). In most cases, total RO increases with size, even when relative allocation to male or female shifts. However, in the case of genotypes of *Plantago major* grown in a common environment, it was found that RO did not increase with size, because larger genotypes allocated a larger proportion of their resources to reproductive support structures; i.e. the length of the flowering culm increased with size, decreasing the resources available for seed production (Reekie, 1998a). The larger

genotypes were isolated from habitats with a relatively tall canopy; therefore, their longer flowering culms may increase fitness by ensuring pollination (it is a wind-pollinated species) and effective seed dispersal.

Considerable attention has also been given to the trade-off between seed size and number for any quantity of seed biomass (Harper *et al.*, 1970). Seed size is usually small in species from more disturbed habitats, which exhibit high levels of dispersal in time and space. In a comparison of nine temperate tree species, Grime and Jeffrey (1965) demonstrated that seedling survivorship in the shade was positively correlated with seed size. Foster (1986) reviews the value of large seed size for trees of moist tropical forests. Leishman and Westoby (1994) suggest that large seed size may be an advantage in the shade for two reasons. First, large seeds, with their greater initial reserves, allow seedlings to survive longer, which may permit seedlings to survive until a gap in the canopy is created. Secondly, large-seeded species exhibit a greater etiolation response than small-seeded species, and this height difference will give large seeds an advantage where there is a steep light gradient, such as when seeds germinate below litter. Grubb and Metcalfe (1996) compared seed size of shade-tolerant and intolerant species at both the intergeneric level and the intrageneric level in the Australian tropical lowland rainforest flora. They found that large seeds were associated with shade tolerance at the intergeneric level but not at the intrageneric level. They interpreted these results to mean that seed size is fairly low in the hierarchy of characteristics enabling plants to become established in shade. Seed size also increases along a gradient of decreasing moisture in plant communities in California, which is interpreted in light of the requirements for rapid early growth in dry environments in order to reach water-supplies (Baker, 1972). Armstrong and Westoby (1993) have demonstrated that large-seeded species tolerate defoliation better than small-seeded species in the initial stages of growth. In

species of composites with larger seeds, however, a greater fraction of the seed is invested in seed-coat, and seedlings from larger seeds have lower initial relative growth rates (Fenner, 1983). Shipley and Peters (1990) have compiled data on 204 species and found a significant negative correlation between seed size and seedling relative growth rate. Gross (1984), in a study of seedling establishment of six monocarpic perennials, found that this correlation is environment-dependent, and was reversed when the seeds germinated under a cover of *Poa annua*. In contrast, in a field experiment of colonization of various size gaps in a grass canopy, seed size differences among four annual species did not correlate with performance in different gap sizes (McConnaughay and Bazzaz, 1987).

Seed size is frequently considered the least plastic component of fecundity, in comparison with plant size and seed allocation (Harper *et al.*, 1970). This apparent constancy results in part from the tendency to determine the mean weight of large numbers of seeds, rather than the distribution of individual seed weights (Fenner, 1985c). Recent studies have shown that there is frequently considerable variation in seed size, even within individual plants; much of this variation can be attributed to positional effects during development, exemplified by the sizes of successive seeds in pods of many legumes (Wulff, 1986). Thompson (1984) reported eightfold variation in size within individuals of *Lomatium grayi* (*Umbelliferae*) and 15-fold variation within a natural population. Manasse (1990) reported a range in seed size from 0.1 to 66 g in the tropical perennial herb *Crinum erubescens* (*Amaryllidaceae*). Sultan and Bazzaz (1993b) observed decreases in seed size of *Polygonum persicaria* along a moisture gradient paralleling the community-level patterns found by Baker, but the consequences of this variation for germination and establishment are not known. In response to elevated carbon dioxide levels, seed weight of *Abutilon theophrasti* increases and seed number decreases (Garbutt and Bazzaz,

1984). Large seed size has been shown to result in competitive advantages in the course of intraspecific competition in populations of *Raphanus raphanistrum* (Stanton, 1985); in contrast to the patterns observed between species, variation in seed size within a population of *R. raphanistrum* did not influence initial growth rate (Choe *et al.*, 1988). When relative timing of emergence was controlled experimentally, seed size also influenced the outcome of interspecific competition between two annual species grown in controlled conditions (Bazzaz *et al.*, 1989).

RE and life-history evolution

As we stated above, the concept of RE is not synonymous with reproductive resource allocation. In order to understand the evolution of reproductive strategies, RE must be considered not only in physiological terms, but in relation to demographic components of fitness. Willson (1983) defines the three primary components of reproductive performance as clutch size, timing of reproduction and frequency of reproduction. RE can be defined physiologically either in terms of the supply of resources invested in reproduction which is derived from the vegetative plant, or by examining changes in vegetative biomass resulting from reproduction (Tuomi *et al.*, 1983; Reekie and Bazzaz, 1987a, c). In order to relate these physiological measures of RE to a life-history measure, the relationship between investment in vegetative function and future reproduction must be established. Additionally, trade-offs between current and future reproduction (residual reproductive value (Williams, 1966)) will only contribute to evolution of life history if they have a genetic basis, but there are still few studies that demonstrate negative genetic correlations between life-history parameters or directly assess responses to selection (Reznick, 1985).

In demographic studies, reductions in survival or future fecundity resulting from reproductive activity are frequently termed the cost of reproduction. In evolutionary terms, however, the essential parameter is total contribution to future generations, not survival of individual organisms. Consequently, we suggest that the negative correlations observed between present reproduction and residual reproductive value be identified as trade-offs. The consequences of different reproductive schedules must be considered in terms of total reproductive success, and a cost of reproduction is incurred if investment in reproduction at a particular time leads to a decrease in lifetime reproduction relative to other members of the population. Geber's (1990) study of *Polygonum arenastrum* illustrates this distinction: she observed negative genetic correlations between early and late reproduction, but no genetic correlation between early and total reproduction. Thus, there is a trade-off between early and late reproduction, but in this experiment there was no extra cost for early reproduction.

Patterns of RA and RE are most commonly studied in annuals; for annual plants that live in seasonal environments and produce only one generation each year, the only component of reproductive success is total fecundity. Because all seeds must remain dormant until the following year, there is no advantage to early reproduction due to exponential increase of precocious offspring, as there is in populations with overlapping generations (Cole, 1954). Timing will, however, be important if the length of the growing season is unpredictable and reproduction is indeterminate (King and Roughgarden, 1982b; Geber, 1990). In determinate annuals, a physiological measure of RE can be obtained from a careful study of the individual's resource budget. However, the optimal value of RE will not depend on the interaction of demographic parameters, but on the developmental problem of maximizing the product of plant size and RA. A number of models exist that predict the optimal pattern of allocation for annual plants. By assuming that the structure and function of vegetative and reproductive organs are strictly distinguished, the models can derive constraint functions for

growth and fecundity, and then apply optimization techniques (Fox, 1992). In environments with a fixed growing season, the
optimal allocation strategy involves a single switch from purely vegetative to strictly
reproductive growth (Cohen, 1971). If the
loss of vegetative and reproductive biomass
during the growing season is considered,
multiple switches between vegetative and
reproductive activity may be optimal (King
and Roughgarden, 1982a). In contrast, if
the length of the growing season varies
unpredictably, a gradual transition from
vegetative to reproductive allocation is predicted (King and Roughgarden, 1982b;
Kozlowski, 1992) and, if seasonality influences growth and reproduction differentially, plants should initially store resources
and subsequently allocate them to reproduction (Schaffer et al., 1982; Chiariello
and Roughgarden, 1984). Several of these
models have also explored the factors that
may control the time of the switch from
vegetative to reproductive growth. Cohen
(1976) demonstrated that early reproduction would be advantageous if relative
growth rate decreases with size, if the probability of mortality increases with time or if
the reproductive structures are photosynthetic. High rates of vegetative tissue loss
will also favour an early switch to reproduction (Kozlowski, 1992). On the other
hand, late reproduction will be favoured if
a large size confers particular advantages,
such as improving the ability of the plant
to compete for light (Schaffer, 1977) or
improving access to pollinators (Cohen,
1976). Late reproduction will also be
favoured if flowering uses reserves stored
during the vegetative phase (Chiariello and
Roughgarden, 1984; Kozlowski and
Wiegert, 1986).

Trade-offs between current reproduction and residual reproductive value have
been demonstrated for several taxa. For
example, Sarukhan, Piñero and their co-
workers have conducted detailed studies of
resource budgets and demographic parameters in the tropical palm Astrocaryum mexicanum (Martínez-Ramos et al., 1988).
Using matrix methods, they have estimated
lifetime reproductive schedules and

demonstrated a strong negative correlation
between fecundity at various ages and
residual reproductive value (Piñero et al.,
1982; see also Oyama and Dirzo, 1988).
Primack and Hall (1990) manipulated level
of reproduction in Cypripedium acaule by
controlling pollination and examined the
impact of reproduction on subsequent survival, growth and reproduction over a 4-
year period. Reproduction decreased
growth and reduced the probability of
flowering in subsequent years. In trees,
negative correlations are frequently
observed between reproductive activity
and vegetative growth within a season
(Kozlowski, 1971). Newell (1991) found
that branches of Aesculus californica that
fruited in one season had greatly reduced
leaf-area development the following season, but it was not possible to examine
interactions at the whole-plant level.
Cipollini and Whigham (1994) experimentally manipulated reproduction in the
woody shrub Lindera benzoin. Fruit thinning enhanced fruit production the following year.

Although there are many studies that
demonstrate negative correlations between
current reproduction and residual reproductive value, there are also many studies
that fail to find any relationship. For example, Horvitz and Schemske (1988)
directly manipulated reproductive output
by removing flower-buds in a population of
the perennial, tropical herb Calathea ovandensis. In contrast to the studies above,
reproduction did not result in any significant changes in survival, growth or reproduction in the following season. Similar
studies have been conducted with a wide
variety of taxa, including Pinguicula alpina
(Karlsson et al., 1990), Viscaria vulgaris
(Jennersten, 1991), Alaria nana (Pfister,
1992) and Blandfordia grandiflora
(Ramsey, 1997), with identical results. This
apparent lack of any trade-off between current reproduction and residual reproductive value is puzzling. True, there are
known physiological mechanisms that may
reduce the impact of reproduction on
growth (see literature discussed above),
but, if current reproduction really has no

effect on residual reproductive value, then why has there been no selection to increase RA to the point where there is a trade-off? Some other factor must be constraining or limiting RA. A partial answer to this question may be pollen limitation (Calvo and Horvitz, 1990). Reproduction in some of these species is known to be limited by pollen availability (e.g. see Jennersten, 1991; Ramsey, 1997), but this cannot explain the lack of trade-off in all of these species. It is also possible that RA is constrained by extreme events, rather than 'normal' or 'average' conditions. Several studies have shown that the trade-off between current reproduction and residual reproductive value varies with environment (Zimmerman, 1991; Syrjanen and Lehtila, 1993; Agren and Willson, 1994; Ramadan *et al.*, 1994; Saikkonen *et al.*, 1998). For example, Primack *et al.* (1994) found that, in *Cypripedium acaule*, plants had to be placed under severe physiological stress by extensive defoliation before experimentally increasing level of reproduction by hand pollination had any negative effects on growth or flowering the following year. If RA is limited by these extreme events, it is not surprising that it may be difficult to detect trade-offs under many circumstances. It will also be difficult to detect trade-offs when reproduction is largely dependent upon stored reserves. Several studies have shown that stored reserves can provide a buffer that will mask effects of reproduction on subsequent growth and reproduction in the short term (Westley, 1993; Cunningham, 1997; Geber *et al.*, 1997).

Direct tests of the principles of life-history theory require that a genetic basis exists for the trade-offs between current and residual reproduction found within individuals of a population. Few studies of plants are available that consider genetic correlations among life-history traits in iteroparous species. Those studies that are available suggest that, as in the case of phenotypic correlations, the evidence for a negative correlation between current and residual reproduction is variable. Law (1979) studied variation in fecundity in two seasons among genetic families from two populations of *Poa annua*, in which most individuals actually reproduce over 2 years. Using a population model to evaluate the contribution of each reproductive episode, he found a strong negative genetic correlation between reproductive value in successive years. Genetic trade-offs between current reproduction and future reproduction have also been found in *Gladiolus* (Rameau and Gouyon, 1991) and *Pseudotsuga menziesii* (El-Kassaby and Barclay, 1992). However, there are also studies that have failed to find any evidence of a genetic trade-off. Galen (1993) examined the effect of early reproduction on subsequent growth and survival in the alpine perennial *Polemonium viscosum*. Measurements were made on 34 maternal half-sib families over a period of 4 years. Approximately half (62%) of the families did not exhibit early flowering, but there was little evidence of a genetic trade-off between early reproduction and subsequent survival. Jackson and Dewald (1994) examined the effect of a mutation that increased reproductive allocation in the perennial grass *Tripsacum dactyloides*. Although the mutation substantially increased RA (plants produce only pistillate flowers), there were no effects on vegetative growth. As with phenotypic trade-offs, there is evidence that genetic trade-offs vary with environment. Reekie (1998b) grew 15 maternal half-sib families of *Plantago major* in mown versus unmown grassland sites. Variation in RA among families had no effect on growth the following year in the mown sites, but there was a negative correlation between RA and growth in the unmown sites.

It must be remembered that failure to detect a significant effect in an experiment does not necessarily mean that there is no effect; it only means that the effect, if it exists, is too small to be detected, given the level of variance within the study. Therefore, the failure of a number of studies to detect significant trade-offs between current reproduction and residual reproductive value does not mean that there is no trade-off. However, the wide variation, both among and within studies, in whether a trade-off is detected does indicate that the trade-off is highly variable and depends upon environment. Variation in the magnitude of the trade-off is likely to select for different patterns of RA. For example, in the case of *P. major*, the lower genetic trade-off in mown sites as compared with unmown sites (Reekie, 1998b) may explain why genotypes isolated from mown sites have a higher RA than genotypes from unmown sites (Warwick and Briggs, 1980). Variation in the phenotypic trade-off between current and residual reproduction, on the other hand, will select for phenotypic plasticity in RA. Individuals should increase RA in those environments where the phenotypic trade-off is minimal and reduce RA in those environments where the trade-off function is steeper. To date, there has been little attempt to test this prediction in the field. However, several studies do present evidence that suggests the timing of reproduction in some species coincides with environmental conditions that minimize the trade-off (Brewer, 1995; Reekie *et al.*, 1997; Shitaka and Hirose, 1998).

Evolutionary models incorporate the negative correlations between life-history components as physiological constraints underlying life-history evolution (e.g. Gadgil and Bossert, 1970; Schaffer, 1974). It must be remembered, however, that the genetic variation observed in natural populations may arise from the selective neutrality of genotypes resulting from negatively correlated life-history components (Falconer, 1952; Lande, 1982) or from the interaction of phenotypic plasticity and environmental heterogeneity (Bazzaz and Sultan, 1987; Sultan, 1987). The variation revealed by quantitative genetic analysis can only predict evolutionary trajectories in the immediate future. The most interesting evolutionary change may result not from evolution within these physiological constraints, but from the evolution of the constraints themselves.

Summary

The theory of resource allocation has been widely used to examine the physiological basis for variation in RO and the evolution of life-history strategies in plants. Application of this theory depends on the assumption that the resource supply is fixed and that allocation of resources to various plant structures is equivalent to allocation to corresponding functions. Neither of these assumptions is valid for many (or possibly any) plants, and various physiological processes (such as photosynthesis of reproductive parts and internal reallocation of resources) decouple structure and function. As a result, measures of RA are not adequate for assessing RE, and these two concepts must be distinguished. In an ecological context, RO can be considered as the product of plant size and RA. The relative importance of these two terms in determining RO appears to vary with species and environmental conditions; plant size is more important in annual than perennial species, particularly when there is asymmetric competition for resources, resulting in marked size hierarchies. In perennial species, RA exhibits a great deal of phenotypic plasticity as well as genetic differentiation among populations. Studies that examine the evolution of different patterns of RA have focused on genetic differentiation among populations and have largely ignored the importance of phenotypic plasticity. In the study of life-history evolution, particularly for iteroparous species, RE needs to be related to demographic parameters, such as changes in future survival and RO. Studies where this has been done suggest that both the phenotypic and genetic trade-off between current reproduction and future performance is highly variable. Models of life-history evolution have usually considered trade-offs between fitness components as consequences of physiological constraints, but little is known about the evolution of the constraints themselves and how these constraints may be modified by the environment. To explain both phenotypic plasticity in RA and genetic differentiation in RA among populations, it will be necessary to address this gap in our knowledge.

References

Aarssen, L.W. and Taylor, D.R. (1992) Fecundity allocation in herbaceous plants. *Oikos* 65, 225–232.

Abrahamson, W.G. (1979) Patterns of resource allocation in wildflower populations of fields and woods. *American Journal of Botany* 66, 71–79.

Abrahamson, W.G. and Caswell, H. (1982) On the comparative allocation of biomass, energy, and nutrients in plants. *Ecology* 63, 982–991.

Abrahamson, W.G. and Gadgil, M. (1973) Growth form and reproductive effort in Goldenrods (*Solidago*, Compositae). *American Naturalist* 107, 651–661.

Abul-Fatih, H.A., Bazzaz, F.A. and Hunt, R. (1979) The biology of *Ambrosia trifida* L. III. Growth and biomass allocation. *New Phytologist* 83, 829–838.

Ackerly, D.D. and Jasienski, M. (1990) Size-dependent variation of gender in high density stands of the monoecious annual, *Ambrosia artemisiifolia* (Asteraceae). *Oecologia* 82, 474–477.

Ackerly, D.D., Coleman, J.S., Morse, S.R. and Bazzaz, F.A. (1992) CO_2 and temperature effects on leaf area production in two annual plant species. *Ecology* 73, 1260–1269.

Agren, J. and Willson, M.F. (1994) Cost of seed production in the perennial herbs *Geranium maculatum* and *G. sylvaticum*: an experimental field study. *Oikos* 70, 35–42.

Alvarez-Buylla, E.R. and Martínez-Ramos, M. (1992) Demography and allometry of *Cecropia obtusifolia*, a neotropical pioneer tree – an evaluation of the climax–pioneer paradigm for tropical rainforests. *Journal of Ecology* 80, 275–290.

Antonovics, J. (1980) Concepts of resource allocation and partitioning in plants. In: Staddon, J.E.R. (ed.) *Limits to Action: the Allocation of Individual Behavior*. Academic Press, New York, pp. 1–25.

Antos, J.A. and Allen, G.A. (1994) Biomass allocation among reproductive structures in the dioecious shrub *Oemleria cerasiformis*: a functional interpretation. *Journal of Ecology* 82, 21–29.

Armstrong, D.P. and Westoby, M. (1993) Seedlings from large seeds tolerate defoliation better: a test using phylogenetically independent contrasts. *Ecology* 74, 1092–1100.

Ashman, T.L. (1994) Reproductive allocation in hermaphrodite and female plants of *Sidalcea oregana* ssp. *spicata* (Malvaceae) using four currencies. *American Journal of Botany* 81, 21–29.

Baker, H.G. (1972) Seed weight in relation to environmental conditions in California. *Ecology* 53, 997–1010.

Bazzaz, F.A. (1993). Use of plant growth analysis in global change studies: modules, individuals and populations. In: Schulze, E.D. and Mooney, H.A. (eds) *Design and Execution of Experiments on CO_2 Enrichment*. Commission of the European Communities, Luxembourg, pp. 53–71.

Bazzaz, F.A. and Carlson, R.W. (1979) Photosynthetic contribution of flowers and seeds to reproductive effort of an annual colonizer. *New Phytologist* 82, 223–232.

Bazzaz, F.A. and Harper, J.L. (1977) Demographic analysis of the growth of *Linum usitatissimum*. *New Phytologist* 78, 193–208.

Bazzaz, F.A. and Reekie, E.G. (1985) The meaning and measurement of reproductive effort in plants. In: White, J. (ed.) *Studies on Plant Demography: a Festschrift for John L. Harper*. Academic Press, London, pp. 373–387.

Bazzaz, F.A. and Sultan, S. (1987) Ecological variation and the maintenance of plant diversity. In: Urbanska, K. (ed.) *Differentiation in Higher Plants*. Academic Press, London, pp. 69–93.

Bazzaz, F.A., Carlson, R.W. and Harper, J.L. (1979) Contribution to reproductive effort by photosynthesis of flowers and fruits. *Nature* 279, 554–555.

Bazzaz, F.A., Chiariello, N.R., Coley, P.D. and Pitelka, L.F. (1987) Allocation resources to reproduction and defense. *Bioscience* 37, 58–67.

Bazzaz, F.A., Garbutt, K., Reekie, E.G. and Williams, W.E. (1989) Using growth analysis to interpret competition between a C_3 and a C_4 annual under ambient and elevated CO_2. *Oecologia* 79, 223–235.

Benech Arnold, R.L., Fenner, M. and Edwards, P.J. (1992) Mineral allocation to reproduction in *Sorgum bicolor* and *Sorgum halpense* in relation to parental nutrient supply. *Oecologia* 92, 138–144.

Benner, B.L. and Bazzaz, F.A. (1988) Carbon and mineral element accumulation and allocation in two annual plant species in response to timing of nutrient addition. *Journal of Ecology* 76, 19–40.

Bloom, A.J., Chapin, F.S., III and Mooney, H.A. (1985) Resource limitation in plants – an economic analogy. *Annual Review of Ecology and Systematics* 16, 363–392.

Brewer, J.S. (1995) The relationship between soil fertility and fire-stimulated floral induction in two populations of grass-leaved golden aster, *Pityopsis graminifolia*. *Oikos* 74, 45–54.

Calvo, R.N. and Horvitz, C.C. (1990) Pollinator limitation, cost of reproduction, and fitness in plants: a transition matrix demographic approach. *American Naturalist* 136, 499–516.

Chapin, F.S., III (1989) The cost of tundra plant structures: evaluation of concepts and currencies. *American Naturalist* 133, 1–19.

Chapin, F.S., III, Bloom, A.J., Field, C.B. and Waring, R.H. (1987) Plant responses to multiple environmental factors. *Bioscience* 37, 49–57.

Cheplick, G.P. (1989) Nutrient availability, dimorphic seed production, and reproductive allocation in the annual grass *Amphicarpum purshii*. *Canadian Journal of Botany* 67, 2514–2521.

Chiariello, N. and Roughgarden, J. (1984) Storage allocation in seasonal races of an annual plant: optimal versus actual allocation. *Ecology* 65, 1290–1301.

Choe, H.S., Chu, C., Koch, G., Gorham, J. and Mooney, H.A. (1988) Seed weight and seed resources in relation to plant growth rate. *Oecologia* 76, 158–159.

Cipollini, M.L. and Whigham, D.F. (1994) Sexual dimorphism and cost of reproduction in the dioecious shrub *Lindera benzoin* (Lauraceae). *American Journal of Botany* 81, 65–75.

Clauss, M.J. and Aarssen, L.W. (1994) Patterns of reproductive effort in *Arabidopsis thaliana*: confounding effects of size and developmental stage. *Ecoscience* 1, 153–159.

Cody, M.L. (1966) A general theory of clutch size. *Evolution* 20, 174–184.

Cohen, D. (1971) Maximizing final yield when growth is limited by time or by limiting resources. *Journal of Theoretical Biology* 33, 299–307.

Cohen, D. (1976) The optimal timing of reproduction. *American Naturalist* 110, 801–807.

Cole, L.C. (1954) The population consequences of life history phenomena. *Quarterly Review of Biology* 29, 103–137.

Cruden, R.W. and Lyon, D.L. (1985) Patterns of biomass allocation to male and female functions in plants with different mating systems. *Oecologia* 66, 299–306.

Cunningham, S.A. (1997) The effect of light environment, leaf area, and stored carbohydrates on inflorescence production by a rainforest understory palm. *Oecologia* 111, 36–44.

Dale, M.P. and Causton, D.R. (1992) The ecophysiology of *Veronica chamaedrys*, *Veronica montana* and *Veronica officinalis*. III. Effects of shading on the phenology of biomass allocations: a field experiment. *Journal of Ecology* 80, 505–515.

DeJong, T.J. (1986) Effects of reproductive and vegetative sink activity on leaf conductance and water potential in *Prunus persica* cultivar Fantasia. *Scientific Horticulture* 29, 131–138.

DeJong, T.J. and Klinkhamer, P.G.L. (1989) Size-dependency of sex-allocation in hermaphoditic, monocarpic plants. *Functional Ecology* 3, 201–206.

deRidder, F. and Dhondt, A.A. (1992) The reproductive behaviour of a clonal herbaceous plant, the long-leaved sundew *Drosera intermedia*, in different heathland habitats. *Ecography* 15, 144–153.

El-Kassaby, Y.A. and Barclay, H.J. (1992) Cost of reproduction in Douglas-fir. *Canadian Journal of Botany* 70, 1429–1432.

Evenson, W.E. (1983) Experimental studies of reproductive energy allocation in plants. In: Jones, C.E. and Little, R.J. (eds) *Handbook of Experimental Pollination Biology*. Van Nostrand Reinhold, New York, pp. 249–274.

Falconer, D.S. (1952) The problem of environment and selection. *American Naturalist* 86, 293–298.

Farris, M.A. and Lechowicz, M.J. (1990) Functional interactions among traits that determine reproductive success in a native annual plant. *Ecology* 71, 548–557.

Fenner, M. (1983) Relationships between seed weight, ash content and seedling growth in twenty-four species of Compositae. *New Phytologist* 95, 697–706.

Fenner, M. (1985a) The allocation of minerals to seeds in *Senecio vulgaris* plants subjected to nutrient shortage. *Journal of Ecology* 74, 385–392.

Fenner, M. (1985b) A bioassay to determine the limiting minerals for seeds from nutrient-deprived *Senecio vulgaris* plants. *Journal of Ecology* 74, 497–506.

Fenner, M. (1985c) *Seed Ecology*. Chapman and Hall, London.

Forrest, G.I. (1971) Structure and production of North Pennine blanket bog vegetation. *Journal of Ecology* 59, 453–479.

Foster, S.A. (1986) On the adaptive value of large seeds for tropical moist forest trees, a review and synthesis. *Botanical Review* 53, 260–299.

Fox, G.A. (1992) Annual plant life histories and the paradigm of resource allocation. *Evolutionary Ecology* 6, 482–499.

Gadgil, M. and Bossert, W.H. (1970) Life historical consequences of natural selection. *American Naturalist* 104, 1–24.

Gadgil, M. and Solbrig, O.T. (1972) The concept of r- and K-selection: evidence from wild flowers and some theoretical considerations. *American Naturalist* 106, 14–31.

Galen, C. (1993) Cost of reproduction in *Polemonium viscosum*: phenotypic and genetic approaches. *Evolution* 47, 1073–1079.

Galen, C., Dawson, T.E. and Stanton, M.L. (1993) Carpels as leaves: meeting the carbon cost of reproduction in an alpine buttercup. *Oecologia* 95, 187–193.

Garbutt, K. and Bazzaz, F.A. (1984) The effects of elevated CO_2 on plants. III. Flower, fruit and seed production and abortion. *New Phytologist* 98, 433–446.

Garbutt, K., Williams, W.E. and Bazzaz, F.A. (1990) Analysis of differential response of five annuals to elevated CO_2 during growth. *Ecology* 71, 1185–1194.

Geber, M.A. (1990) The cost of meristem limitation in *Polygonum arenastrum*: negative genetic correlations between fecundity and growth. *Evolution* 44, 799–819.

Geber, M.A., De Kroon, H. and Watson, M.A. (1997) Organ preformation in mayapple as a mechanism for historical effects on demography. *Journal of Ecology* 85, 211–223.

Gehring, J.L. (1993) Temporal patterns in sexual dimorphisms in *Silene latifolia* (Caryophylaceae). *Bulletin of the Torrey Botanical Club* 120, 405–416.

Gehring, J.L. and Monson, R.K. (1994) Sexual differences in gas exchange and response to environmental stress in dioecious *Silene latifolia* (Caryophyllaceae). *American Journal of Botany* 81, 166–174.

Gleeson, S.K. and Tilman, D. (1990) Allocation and the transient dynamics of succession on poor soils. *Ecology* 71, 1144–1155.

Goldman, D.A. and Willson, M.F. (1986) Sex allocation in functionally hermaphroditic plants: a review and critique. *Botanical Review* 52, 157–194.

Grime, J.P. and Jeffrey, D.W. (1965) Seedling establishment in vertical gradients of sunlight. *Journal of Ecology* 53, 621–642.

Gross, K. (1984) Effects of seed size and growth form on seedling establishment of six monocarpic perennial plants. *Journal of Ecology* 72, 369–387.

Grubb, P.J. (1977) The maintenance of species richness in plant communities: the importance of the regeneration niche. *Biological Reviews* 52, 107–145.

Grubb, P.J. and Metcalfe, D.J. (1996) Adaptation and inertia in the Australian tropical lowland rainforest flora: contradictory trends in intergeneric and intrageneric comparisons of seed size in relation to light demand. *Functional Ecology* 10, 512–520.

Hancock, J.F. and Pritts, M.P. (1987) Does reproductive effort vary across different life forms and seral environments? A review of the literature. *Bulletin of the Torrey Botanical Club* 114, 53–59.

Hara, T., Kawano, S. and Nagai, Y. (1988) Optimal reproductive strategy of plants, with special reference to the modes of reproductive resource allocation. *Plant Species Biology* 3, 43–59.

Harper, J.L. (1967) A Darwinian approach to plant ecology. *Journal of Ecology* 55, 247–270.

Harper, J.L. (1977) *The Population Biology of Plants*. Academic Press, London.

Harper, J.L. and Ogden, J. (1970) The reproductive strategy of higher plants. I. The concept of strategy with special reference to *Senecio vulgaris* L. *Journal of Ecology* 58, 681–698.

Harper, J.L. and Sellek, C. (1987) The effects of severe mineral nutrient deficiencies on the demography of leaves. *Proceedings of the Royal Society of London, Series B* 232, 137–157.

Harper, J.L., Lovell, P. and Moore, K. (1970) The shapes and sizes of seeds. *Annual Review of Ecology and Systematics* 1, 327–356.

Hartnett, D.C. (1991) Effects of fire in tallgrass prairie on growth and reproduction of prairie coneflower (*Ratibida columnifera*: Asteraceae). *American Journal of Botany* 78, 429–435.

Hartnett, D., Hartnett, B. and Bazzaz, F.A. (1987) Persistence of *Ambrosia trifida* populations in old fields and responses to successional changes. *American Journal of Botany* 74, 1239–1248.

Hemborg, A.M. and Karlsson, P.S. (1998) Somatic costs of reproduction in eight subarctic plant species. *Oikos* 82, 149–157.

Hickman, J.C. (1975) Environmental unpredictability and plastic energy allocation strategies in the annual *Polygonum cascadense*. *Journal of Ecology* 63, 689–701.

Horvitz, C.C. and Schemske, D.W. (1988) Demographic cost of reproduction in a neotropical herb, an experimental approach. *Ecology* 69, 1741–1745.

Jackson, L.L. and Dewald, C.L. (1994) Predicting evolutionary consequences of greater reproductive effort in *Tripsacum dactyloides* a perennial grass. *Ecology* 75, 627–641.

Jennersten, O. (1991) Cost of reproduction in *Viscaria vulgaris* (Caryophyllaceae): a field experiment. *Oikos* 61, 197–204.

Jurik, T.W. (1985) Differential costs of sexual and vegetative reproduction in wild strawberry populations. *Oecologia* 66, 394–403.

Karlsson, P.S., Svensson, B.M., Carlsson, B.A. and Nordell, K.O. (1990) Resource investment in reproduction and its consequences in three *Pinguicula* species. *Oikos* 59, 393–398.

Karlsson, P.S., Thoren, L.M. and Haslin, H.M. (1994) Prey capture by three *Pinguicula* species in a subarctic environment. *Oecologia* 99, 188–193.

King, D. and Roughgarden, J. (1982a) Multiple switches between vegetative and reproductive growth in annual plants. *Theoretical Population Biology* 21, 194–204.

King, D. and Roughgarden, J. (1982b) Graded allocation between vegetative and reproductive growth for annual plants in growing season of random length. *Theoretical Population Biology* 22, 1–16.

Kira, T. (1978) Community architecture and organic matter dynamics in tropical lowland rainforests of Southeast Asia with special reference to Pasoh Forest, West Malaysia. In: Tomlinson, P.B. and

Zimmermann, M.H. (eds) *Tropical Trees as Living Systems*. Cambridge University Press, Cambridge, pp. 561–590.

Klinkhamer, P.G.L., de Jong, T.J. and Meelis, E. (1990) How to test for proportionality in the reproductive effort of plants. *American Naturalist* 135, 291–300.

Klinkhamer, P.G.L., Meelis, E., de Jong, T.J. and Weiner, J. (1992) On the analysis of size-dependent reproductive output in plants. *Functional Ecology* 6, 308–316.

Korpelainen, H. (1992) Patterns of resource allocation in male and female plants of *Rumex acetosa* and *Rumex acetosella*. *Oecologia* 89, 133–139.

Kozlowski, J. (1992) Optimal allocation of resources to growth and reproduction: implications for age and size at maturity. *Trends in Ecology and Evolution* 7, 15–19.

Kozlowski, J. and Wiegert, R.G. (1986) Optimal allocation of energy to growth and reproduction. *Theoretical Population Biology* 29, 16–37.

Kozlowski, T.T. (1971) *Growth and Development of Trees*, Vol. 2. Academic Press, New York.

Lande, R. (1982) A quantitative theory of life history evolution. *Ecology* 63, 607–615.

Laporte, M.M. and Delph, L.F. (1996) Sex-specific physiology and source–sink relations in the dioecious plant *Silene latifolia*. *Oecologia* 106, 63–72.

Law, R. (1979) The cost of reproduction in annual meadow grass. *American Naturalist* 113, 3–16.

Lawrence, W.S. (1993) Resource and pollen limitation: plant size-dependent reproductive patterns in *Physalis longifolia*. *American Naturalist* 141, 296–313.

Lee, T.D. (1988) Patterns of fruit and seed production. In: Lovett Doust, J. and Jovett Doust, L. (eds) *Plant Reproductive Ecology: Patterns and Strategies*. Oxford University Press, New York, pp. 179–202.

Lehtila, K., Tumori, J. and Sulkinoja, M. (1994) Bud demography of the mountain birch *Betula pubescens* ssp. *tortuosa* near the tree line. *Ecology* 75, 945–955.

Leishman, M.R. and Westoby, M. (1994) The role of large seed size in shaded conditions: experimental evidence. *Functional Ecology* 8, 205–214.

Lloyd, D.G. (1988) Benefits and costs of biparental and uniparental reproduction in plants. In: Michod, R.E. and Levin, B.R. (eds) *The Evolution of Sex*. Sinauer Associates, Sunderland, Massachusetts, pp. 233–252.

McConnaughay, K.D.M. and Bazzaz, F.A. (1987) The relationship between gap size and performance of several colonizing annuals. *Ecology* 68, 411–416.

McKone, M.J. and Tonkyn, D.W. (1986) Intrapopulation gender variation in common ragweed (Asteraceae, *Ambrosia artemisiifolia* L.), a monoecious, annual herb. *Oecologia* 70, 63–67.

Manasse, R.S. (1990) Seed size and habitat effects on the water-dispersed perennial, *Crinum erubescens* (Amaryllidaceae). *American Journal of Botany* 77, 1336–1342.

Marshall, J.D., Dawson, T.E. and Ehleringer, J.R. (1993). Gender-related differences in gas exchange are not related to host quality in xylem-tapping mistletoe, *Phoradendron juniperinum* (Viscaceae). *American Journal of Botany* 80, 641–645.

Martínez-Ramos, M., Sarukhan, J. and Piñero, D. (1988) The demography of tropical trees in the context of forest gap dynamics: the case of *Astrocaryum mexicanum* at Los Tuxtlas tropical rainforest. In: Davy, A.J., Hutchings, M.J. and Watkinson, A.R. (eds) *Plant Population Ecology*. Blackwell Scientific Publications, Oxford, pp. 315–342.

Mitchell-Olds, T. and Bergelson, J. (1990) Statistical genetics of an annual plant, *Impatiens capensis*. II. Natural selection. *Genetics* 124, 417–421.

Neales, T.F. and Incoll, L.O. (1968) The control of leaf photosynthesis rate by the level of assimilate concentration in the leaf: a review of the hypothesis. *Annals of Botany* 30, 349–363.

Newell, E.A. (1991) Direct and delayed costs of reproduction in *Aesculus californica*. *Journal of Ecology* 79, 365–378.

Nobel, P.S. (1977) Water relations of flowering of *Agave deserti*. *Botanical Gazette* 138, 1–6.

Ogden, J. (1974) The reproductive strategy of higher plants. II. The reproductive strategy of *Tussilago farfara* L. *Journal of Ecology* 62, 291–324.

Ostertag, R. and Menges, E.S. (1994) Patterns of reproductive effort with time since last fire in Florida scrub plants. *Journal of Vegetation Science* 5, 303–310.

Oyama, K. and Dirzo, R. (1988) Biomass allocation in the dioecious tropical palm *Chamaedorea tepejilote* and its life history consequences. *Plant Species Biology* 3, 27–33.

Pfister, C.A. (1992) Costs of reproduction in an intertidal kelp: patterns of allocation and life history consequences. *Ecology* 73, 1586–1596.

Pickering, C.M. (1994) Size-dependent reproduction in Australian alpine *Ranunculus*. *Australian Journal of Ecology* 19, 336–344.

Piñero, D., Sarukhan, J. and Alberdi, P. (1982) The costs of reproduction in a tropical palm, *Astrocaryum mexicanum*. *Journal of Ecology* 70, 473–481.

Pitelka, L.F. (1977) Energy allocation in annual and perennial lupines (*Lupinus*, Leguminosae). *Ecology* 58, 1055–1065.

Pitelka, L.F. and Ashmun, J. (1985) Physiology and integration of ramets in clonal plants. In: Jackson, J.B.C., Buss, L.W. and Cook, R.E. (eds) *Population Biology and Evolution of Clonal Organisms.* Yale University Press, New Haven, pp. 399–436.

Primack, R.B. and Antonovics, J. (1982) Experimental ecological genetics in *Plantago.* VII. Reproductive effort in populations of *P. lanceolata* L. *Evolution* 36, 742–752.

Primack, R.B. and Hall, P. (1990) Costs of reproduction in pink lady's slipper orchid: a four year experimental study. *American Naturalist* 136, 638–656.

Primack, R.B., Miao, S.L. and Becker, K.R. (1994) Costs of reproduction in the pink lady's slipper orchid (*Cypripedium acaule*): defoliation, increased fruit production and fire. *American Journal of Botany* 81, 1083–1090.

Ramadan, A.A., El-Keblawy, A., Shaltout, K.H. and Lovett-Doust, J. (1994) Sexual polymorphism, growth, and reproductive effort in Egyptian *Thymelaea hirsuta* (Thymelaeceae). *American Journal of Botany* 81, 847–857.

Rameau, C. and Gouyon, P.H. (1991) Resource allocation to growth, reproduction and survival in *Gladiolus*: the cost of male function. *Journal of Evolutionary Biology* 4, 291–308.

Ramsey, M. (1997) No evidence of demographic costs of seed production in the pollen-limited perennial herb *Blandfordia grandiflora* (Liliaceae). *International Journal of Plant Sciences* 158, 785–793.

Reekie, E.G. (1991) Cost of seed versus rhizome poduction in *Agropyron repens. Canadian Journal of Botany* 69, 2678–2683.

Reekie, E.G. (1998a) An explanation for size-dependent reproductive allocation in *Plantago major. Canadian Journal of Botany* 76, 43–50.

Reekie, E.G. (1998b) An experimental field study of the cost of reproduction in *Plantago major* L. *Ecoscience* 5, 200–206.

Reekie, E.G. (1999) Resource allocation, trade-offs, and reproductive effort in plants. In: Vuorisalo, T.O. and Mutikainen, P.K. (eds) *Life History Evolution in Plants.* Kluwer Academic Publishers, The Netherlands, pp. 173–194.

Reekie, E.G. and Bazzaz, F.A. (1987a) Reproductive effort in plants. 1. Carbon allocation to reproduction. *American Naturalist* 129, 876–896.

Reekie, E.G. and Bazzaz, F.A. (1987b) Reproductive effort in plants. 2. Does carbon reflect the allocation of other resources? *American Naturalist* 129, 897–906.

Reekie, E.G. and Bazzaz, F.A. (1987c) Reproductive effort in plants. 3. Effect of reproduction on vegetative activity. *American Naturalist* 129, 907–919.

Reekie, E.G. and Bazzaz, F.A. (1992) Cost of reproduction in genotypes of two congeneric plant species with contrasting life histories. *Oecologia* 90, 21–26.

Reekie, E.G. and Reekie, J.Y.C. (1991) An experimental investigation of the effect of reproduction on canopy structure, allocation and growth in *Oenothera biennis. Journal of Ecology* 79, 1061–1071.

Reekie, E., Parmiter, D., Zebian, K. and Reekie, J. (1997) Trade-offs between reproduction and growth influence time of reproduction in *Oenothera biennis. Canadian Journal of Botany* 75, 1897–1902.

Reznick, D. (1985) Costs of reproduction, an evaluation of the empirical evidence. *Oikos* 44, 257–267.

Saikkonen, K., Koivunen, S., Vuorisalo, T. and Mutikainen, P. (1998) Interactive effects of pollination and heavy metals on resource allocation in *Potentilla anserina* L. *Ecology* 79, 1620–1629.

Salisbury, E.J. (1942) *The Reproductive Capacity of Plants.* Bell, London.

Samson, D.A. and Werk, K.S. (1986) Size-dependent effects in the analysis of reproductive effort in plants. *American Naturalist* 124, 667–680.

Schaffer, W.M. (1974) Optimal reproductive effort in fluctuating environments. *American Naturalist* 108, 783–790.

Schaffer, W.M. (1977) Some observations on the evolution of reproductive rate and competitive ability in flowering plants. *Theoretical Population Biology* 11, 90–104.

Schaffer, W.M., Inouye, R.S. and Whittam, T.S. (1982) Energy allocation by an annual plant when the effects of seasonality on growth and reproduction are decoupled. *American Naturalist* 120, 787–815.

Schat, H., Ouborg, J. and DeWit, R. (1989) Life history and plant architecture: size-dependent reproductive allocation in annual and biennial *Centaurium* species. *Acta Botanica Neerlandica* 38, 183–201.

Scheiner, S.M. (1989) Variable selection along a successional gradient. *Evolution* 43, 548–562.

Schmid, B. and Weiner, J. (1993) Plastic relationships between reproductive and vegetative mass in *Solidago altissima. Evolution* 47, 61–74.

Schwaegerle, K.E. and Levin, D.A. (1990) Environmental effects on growth and fruit production in *Phlox drummondii. Journal of Ecology* 78, 15–26.

Shipley, B. and Dion, J. (1992) The allometry of seed production in herbaceous angiosperms. *American Naturalist* 139, 467–483.

Shipley, B. and Peters, R.H. (1990) The allometry of seed weight and seedling relative growth rate. *Functional Ecology* 4, 523–529.

Shitaka, Y. and Hirose, T. (1998) Effects of shift in flowering time on the reproductive output of *Xanthium canadense* in a seasonal environment. *Oecologia* 114, 361–367.

Silvertown, J. and Dodd, M. (1996) Comparing plants and connecting traits. *Philosophical Transactions of the Royal Society of London B – Biological Sciences* 351, 1233–1239.

Silvertown, J.W. and Lovett Doust, J. (1993) *Introduction to Plant Population Biology*. Blackwell Scientific Publications, Oxford.

Solbrig, O.T. and Solbrig, D.J. (1984) Size inequalities and fitness in plant populations. *Oxford Survey in Evolutionary Biology* 1, 141–159.

Solomon, B.P. (1989) Size-dependent sex ratios in the monoecious, wind-pollinated annual, *Xanthium strumarium*. *American Midland Naturalist* 121, 209–218.

Stanton, M. (1985) Seed size and emergence time within a stand of wild radish (*Raphanus raphanistrum* L.): the establishment of a fitness hierarchy. *Oecologia* 67, 524–531.

Sugiyama, S. and Bazzaz, F.A. (1997) Plasticity of seed output in response to soil nutrients and density in *Abutilon theophrasti*: implications for maintenance of genetic variation. *Oecologia* 112, 35–41.

Sultan, S. (1987) Evolutionary implications of phenotypic plasticity in plants. *Evolutionary Biology* 21, 127–178.

Sultan, S. and Bazzaz, F.A. (1993a) Phenotypic plasticity in *Polygonum persicaria*. I. Diversity and uniformity in genotypic norms of reaction to light intensity. *Evolution* 47, 1009–1031.

Sultan, S. and Bazzaz, F.A. (1993b) Phenotypic plasticity in *Polygonum persicaria*. II. Norms of reaction to soil moisture and the maintenance of genetic diversity. *Evolution* 47, 1032–1049.

Sultan, S. and Bazzaz, F.A. (1993c) Phenotypic plasticity in *Polygonum persicaria*. III. The evolution of ecological breadth for nutrient environment. *Evolution* 47, 1050–1071.

Syrjanen, K. and Lehtila, K. (1993) The cost of reproduction in *Primula veris*: differences between two adjacent populations. *Oikos* 67, 465–472.

Thompson, J.N. (1984) Variation among individual seed masses in *Lomatium grayi* (Umbelliferae) under controlled conditions: magnitude and partitioning of variance. *Ecology* 65, 626–631.

Thompson, K. and Stewart, A.J.A. (1981) The measurement and meaning of reproductive effort in plants. *American Naturalist* 117, 205–211.

Thoren, L.M., Karlsson, P.S. and Tuomi, J. (1996) Somatic cost of reproduction in three carnivorous *Pinguicula* species. *Oikos* 76, 427–434.

Tuomi, J., Hakala, T. and Haukioja, E. (1983) Alternative concepts of reproductive effort, costs of reproduction, and selection in life-history evolution. *American Zoologist* 23, 25–34.

Waller, D.M. (1985) The genesis of size hierarchies in seedling populations of *Impatiens capensis*. *New Phytologist* 100, 243–260.

Wankhar, B. and Tripathi, R.S. (1990) Growth and reproductive allocation of *Centella asiatica* raised from stem cuttings of different sizes in relation to light regimes, soil texture and soil moisture. *Acta Oecologica* 11, 683–692.

Warwick, S.I. and Briggs, D. (1980) The genecology of lawn weeds. V. The adaptive significance of different growth habits in lawn and roadside populations of *Plantago major* L. *New Phytologist* 85, 289–300.

Watson, M.A. (1984) Developmental constraints, effect of population growth and patterns of resource allocation in a clonal plant. *American Naturalist* 123, 411–426.

Weiner, J. (1988) The influence of competition on plant reproduction. In: Lovett-Doust, J. and Lovett-Doust, L. (eds) *Plant Reproductive Ecology*. Oxford University Press, Oxford, pp. 228–245.

Weiner, J. and Solbrig, O.T. (1984) The meaning and measurement of size hierarchies. *Oecologia* 61, 334–336.

Welham, C.V.J. and Setter, R.A. (1998) Comparison of size-dependent reproductive effort in two dandelion (*Taraxacum officinale*) populations. *Canadian Journal of Botany* 76, 166–173.

Wells, R., Anderson, W.F. and Wynne, J.C. (1991) Peanut yield as a result of fifty years of breeding. *Agronomy Journal* 83, 957–961.

Westley, L.C. (1993) The effect of inflorescence bud removal on tuber production in *Helianthus tuberosus* L. (Asteraceae). *Ecology* 74, 2136–2144.

Whittaker, R.H. and Marks, P.L. (1975) Methods of assessing terrestrial productivity. In: Lieth, H. and Whittaker, R.H. (eds) *Primary Productivity of the Biosphere*. Springer-Verlag, New York, pp. 55–118.

Williams, G.C. (1966) Natural selection, the costs of reproduction, and a refinement of Lack's principle. *American Naturalist* 100, 687–690.

Williams, M.J. (1994) Reproductive resource allocation in rhizoma peanut. *Crop Science* 34, 477–482.

Williams, R.B. and Bell, K.L. (1981) Nitrogen allocation in Mojave desert winter annuals. *Oecologia* 48, 145–150.

Willson, M.F. (1983) *Plant Reproductive Ecology.* Wiley, New York.

Wulff, R.D. (1986) Seed size variation in *Desmodium paniculatum.* I. Factors affecting seeds size. *Journal of Ecology* 74, 87–97.

Zimmerman, J.K. (1991) Ecological correlates of labile sex expression in the orchid *Catasetum viridiflavum. Ecology* 72, 597–608.

Chapter 2
The Evolutionary Ecology of Seed Size

Michelle R. Leishman, Ian J. Wright, Angela T. Moles and Mark Westoby
*Department of Biological Sciences, Macquarie University, Sydney,
New South Wales, Australia*

Introduction

Seed mass is a trait that occupies a pivotal position in the ecology of a species. It links the ecology of reproduction and seedling establishment with the ecology of vegetative growth, strategy sectors that are otherwise largely disconnected (Grime *et al.*, 1988; Shipley *et al.*, 1989; Leishman and Westoby, 1992).

There is a startling diversity of shapes and sizes of seeds among the plant species of the world. Seeds range from the dust seeds of the *Orchidaceae* and some saprophytic and parasitic species (around 10^{-6} g), across ten orders of magnitude to the double coconut *Lodoicea seychellarum* (10^4 g) (Harper *et al.*, 1970). Within species, seed size typically spans less than half an order of magnitude (about fourfold: Michaels *et al.*, 1988). Most within-species variation occurs within plant rather than among plants or populations (Michaels *et al.*, 1988; Obeso, 1993; Vaughton and Ramsey, 1998), indicating environmental effects during development rather than genetic differences between mothers. This chapter is concerned with the differences in seed size among species, and the consequences for vegetation dynamics and community composition.

During the last 10–15 years, there has been considerable progress in the ecology of seed mass. Unlike many other areas of comparative plant ecology, we have substantial published information from several different scales and research styles. As well as field experiments and demographic studies with a few species at a time, we have simple experiments with larger numbers of species (ten to 50), quantification of seed mass and its correlates in whole-vegetation types (hundreds of species) and tests of consistency across different continents. The wide-scale quantification began as early as Salisbury (1942) and Baker (1972), but has been much added to and consolidated over the past 10 years (e.g. Mazer, 1989, 1990; Leishman and Westoby, 1994a; Leishman *et al.*, 1995; Eriksson and Jakobsson, 1998). The work spanning large numbers of species is complementary to detailed experiments involving only a few species, giving a stronger sense of how widely the results from particular experiments can be generalized.

Much of the literature examines how natural selection on seed size might be influenced by various environmental factors. In this context, it is at first glance surprising that seed size varies within communities across a remarkable five to six orders of magnitude (Leishman *et al.*, 1995; Fig. 2.1). Further, there is strong overlap of seed-size distributions between quite different habitats. Within the temperate zone,

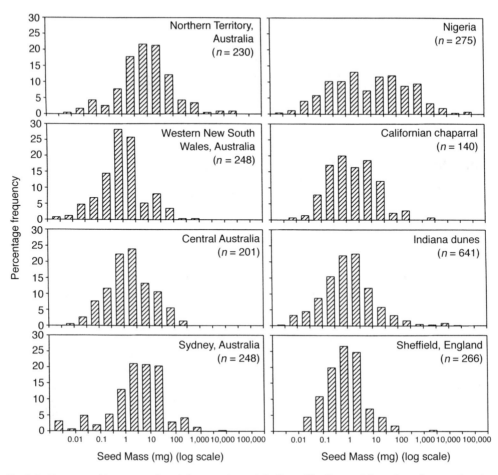

Fig. 2.1. Frequency histograms of seed dry mass from eight floras. The floras originate from four continents and include representatives from tropical and temperate biomes and a diversity of environmental conditions, vegetation types and phylogenetic histories. Data from western New South Wales, central Australia, Sydney (Leishman *et al.*, 1995); Northern Territory, Nigeria (Lord *et al.*, 1997); Californian chaparral (Keeley, 1991); Indiana dunes (Mazer, 1989); and Sheffield (Grime *et al.*, 1988). Seed masses are grouped into half-log classes.

differences between communities account for only about 4% of the variation in seed size between species (Leishman *et al.*, 1995). Differences between the tropics and the temperate zone are somewhat larger (Lord *et al.*, 1995), but variation within a habitat remains a very large component of overall between-species variation. Alternative mechanisms that might shape this wide within-habitat variation have not yet been fully formulated theoretically, much less exposed to strong experimental hypothesis tests.

Components and measurement of seed size

Seeds consist of an embryo plus endosperm (sometimes termed the seed reserve), plus a protective seed-coat or testa. Many seeds have distinctive dispersal appendages attached to the seed, such as plumes and hairs for wind dispersal, hooks and barbs for adhesion dispersal, elaiosomes for ant dispersal and arils or flesh for vertebrate dispersal. These dispersal appendages plus the seed are termed the diaspore. The mass of dispersal structure and the proportion of seed mass that is seed-coat can vary considerably between species.

There is no single measure of seed size that is ideal for all purposes. For discussing seedling establishment, seed reserve mass best reflects the resources available to the seedling. For discussing the size of the object that has to be moved by a given dispersal mechanism, seed mass including the seed-coat is most relevant. For discussing costs to the mother per seed produced, mass of the whole diaspore is better than seed mass, though still not a complete measure of all costs of reproduction. Westoby (1998) recommended dry seed mass, including seed-coat but excluding dispersal structures, as an ecological strategy axis, partly as a compromise among alternative measures, partly because it is easiest to measure and partly to maintain comparability with the majority of existing data. At the same time, increasing numbers of studies go to the trouble to dissect diaspores into components where it seems relevant (e.g. Westoby *et al.*, 1990; Jurado and Westoby, 1992; Leishman and Westoby, 1994b, c). Fortunately, in data sets spanning a wide range of seed mass, the alternative measures will be strongly correlated. Suppose, for example, that two species have dispersal structure mass 0 and 300% of seed mass: this can only reverse diaspore-mass ranking, relative to seed-mass ranking, for species whose seeds differ in mass by less than a factor of three. Thus, in data sets spanning orders of magnitude of seed mass, there is a strong positive relationship between log dry diaspore mass and log dry seed mass: e.g. in western New South Wales, $n = 243$, $r^2 = 0.71$; central Australia, $n = 199$, $r^2 = 0.83$; Sydney, $n = 286$, $r^2 = 0.97$. On the other hand, among sets of species spanning only (say) three- to fourfold in seed mass, diaspore-mass ranking could be substantially different from seed-mass ranking. Similarly, log seed mass and log seed reserve mass tend to be closely correlated, even though substantial variation exists in the proportion of mass due to the seed-

coat. For example, for the Sydney data set (Westoby *et al.*, 1990) $r^2 = 0.92$, $P < 0.0005$, while the percentage of dry seed mass due to the seed-coat varied between 1.2 and 96%, with a mean value of 43%. In a smaller data set of woody perennials from a range of habitats in New South Wales (Wright and Westoby, 1999), seed mass and reserve mass were again tightly correlated ($r^2 = 0.99$, $P < 0.0005$), while % coat varied from 7 to 57%, with a mean of 30%. In neither data set was there a relationship between % coat and seed mass.

To the extent that mineral nutrients as well as energy are decisive during seedling establishment, the mineral nutrient content of the seed would be just as informative as seed mass. The nutrient content is the product of the nutrient concentration and seed mass, but, since there is much greater cross-species variation in seed mass than in nutrient concentration (e.g. 6.7 versus 1.7 orders of magnitude in the *c.* 1500 species of Barclay and Earle, 1974), in large data sets seed mass and nutrient content tend to be correlated (even if mass and concentration themselves are not). Some authors have reported a negative association between seed mass and nutrient concentration (e.g. Fenner (1983) for 24 species of *Asteraceae*; Grubb and Burslem (1998) within species for the majority of 12 South-East Asian trees; Grubb *et al.* (1998) for 194 species from lowland tropical rainforest) while others have found no correlation (e.g. Kitajima (1996a) for 12 tropical woody species of *Bignoniaceae*, *Bombacaceae*, *Leguminosae*; Grubb and Burslem (1998) across the 12 South-East Asian tree species; Milberg *et al.* (1998) for 21 *Eucalyptus*, *Banksia* and *Hakea* species). Thus, no consistent relationship has emerged between seed mass and nutrient concentration, although evidence is beginning to emerge that variation in these attributes should be considered simultaneously with measures of allocation to seed defence structures, such as seed-coats (Grubb *et al.*, 1998).

The seed size of a species represents the amount of maternal investment in an individual offspring, or how much 'packed lunch' an embryo is provided with to start its journey in life. Seed size represents a fundamental trade-off, within the strategy of a species, between producing more small seeds versus fewer larger seeds from a given quantity of resource allocated to reproduction. The trade-off and its consequences were formalized in the model by Smith and Fretwell (1974). There is always selection pressure to produce more seeds, since more seeds represent more offspring (although there may be a lower limit to the seed size that permits a functional seedling to be produced (Raven, 1999)). On the other hand, larger, better-provisioned offspring have a greater chance of successful establishment, described by the Smith–Fretwell function in Fig. 2.2. The best outcome from the mother's point of view is to maximize the ratio of seedling establishment chance to provisions invested in each seed, and this occurs where the steepest possible line from the origin just touches the Smith–Fretwell function (Fig. 2.2). Thus a key prediction of

the model is that, if a mother plant is in a position to allocate more resources to seed output, it should produce more seeds of the same size. The physiological machinery of seed provisioning should have been selected to approximate this outcome, rather than increasing the size of a fixed number of seeds. In order to generate this prediction, the exact shape of the Smith–Fretwell function is not important. All that is required is for there to be some minimum size for a seed to have any chance of establishing and for there to be diminishing returns at some stage as seed mass increases further. The curvature of the Smith–Fretwell function ensures that, if resources are reallocated such that one seed has higher seed mass than the optimum while another has less, the gain in fitness in the augmented seed is smaller than the loss in fitness in the diminished seed. The moderate observed variation in seed mass within a species can be attributed either to the machinery of seed provisioning having limited capacity to deliver a

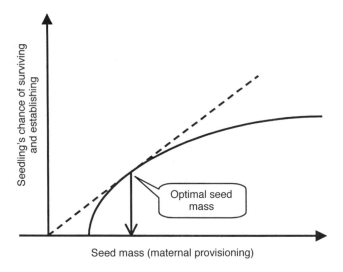

Fig. 2.2. Optimal allocation across seeds of a limited total maternal expenditure on seed provisioning, after Smith and Fretwell (1974). The curved Smith–Fretwell function describes how a seedling's prospects respond to maternal provisioning, the curvature reflecting diminishing returns beyond some point. The best allocation to each seed, from the point of view of genes in the mother, is where the steepest possible straight line from the origin just touches the Smith–Fretwell function. The mother should aim to produce as many seeds as possible of this size. At this point on the curve, if some resource were to be transferred from one seed to another, the fitness gain in the enhanced seed would be smaller than the fitness loss in the diminished seed; hence, maternal fitness would decrease.

completely standardized seed mass or to variability in the Smith–Fretwell function that seedlings are exposed to.

In the Smith–Fretwell treatment, all factors affecting a seedling's chance of establishing and growing to adulthood are amalgamated into the Smith–Fretwell function, including the effects of competition from established vegetation or from other seedlings. Subsequent game-theoretical models (Geritz, 1995; Rees and Westoby, 1997) address frequency-dependent effects among species directly. The question whether these models might be capable of accounting for the wide spread of seed mass observed between species within a habitat is taken up later.

Much of the variation in seed size among species is associated with taxonomy, such as family membership (Hodgson and Mackey, 1986; Mazer, 1989, 1990; Peat and Fitter, 1994; Lord *et al.*, 1995). Some authors believe variation correlated with phylogeny should not be regarded as interpretable in relation to ecology. They regard phylogenetic or correlated-divergence methods of data analysis as compulsory, superseding cross-species correlations rather than complementing them (Kelly and Purvis, 1993; Rees, 1993, 1996; Harvey *et al.*, 1995a, b; Kelly, 1995, 1997). In our view (Leishman *et al.*, 1995; Lord *et al.*, 1995; Westoby *et al.*, 1995a, b, c), phylogenetic and ecological accounts of seed size variation should not be considered mutually exclusive. An important mode of evolution is phylogenetic niche conservatism: a process whereby, because ancestors have a particular constellation of traits, their descendants tend to be most successful using similar ecological opportunities, and so natural selection tends to maintain the same traits among most, if not all, descendant lineages. Niche conservatism: is at least as likely a cause of similarity among related species as constraint – more likely for quantitative traits. It is a process that is phylogenetic and also invokes ecological functionality continuing into the present day. Thus it is simplistic to treat phylogenetic patterns of seed mass as somehow alternatives to ecological patterns.

Dormancy, seed banks and seed mass

In herbaceous vegetation of north-western Europe, persistence in the soil is associated with small and rounded seeds (Thompson, 1987; Thompson *et al.*, 1993; Eriksson, 1995; Bakker *et al.*, 1996; Bekker *et al.*, 1998). Evolutionary divergences in seed size are also correlated with evolutionary divergences in dormancy (Rees, 1993, 1996; Hodkinson *et al.*, 1998) and small seeds dominate the seed bank (Eriksson and Eriksson, 1997). However, it is clear that this pattern is not universal in all floras and vegetation types. Although a few studies in different floras have found similar patterns (e.g. Dalling *et al.* (1997) for species of tropical forest in Panama; Leck (1989) for wetland seed banks; Price and Joyner (1997) for seed banks of the Mojave desert flora of California), other authors have not found this pattern elsewhere. Leishman and Westoby (1994a, 1998) found that small seeds were not more likely to be dormant among species of western New South Wales or among a wide range of Australian species. Lunt (1995) found no relationship between seed size and longevity for six species of Australian grassland. Garner and Witkowski (1997) showed that seeds of three South African woody savannah species were both large and persistent in the soil. Finally, Moles *et al.* (2000) found that dormant species did not consistently have smaller seeds in a data set of 47 native New Zealand species. The relationship between seed size, shape and dormancy for three different floras is shown in Fig. 2.3.

There are two classes of prospective explanations as to why dormancy may be associated with small seed size. First, theory about bet-hedging against zero survivorship predicts that the higher the level of bet-hedging via one mechanism (for example, better dispersal to other sites), the weaker the selection for other mechanisms (for example, bet-hedging seed dormancy) (Venable and Brown, 1988; Philippi and Seger, 1989). Venable and Brown (1988) and Rees (1996) regarded large-seededness

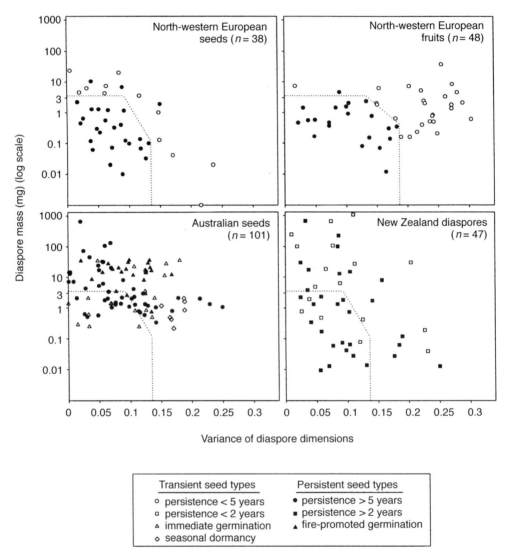

Fig. 2.3. Relationship between mass and variance of dimensions of seeds, fruits or diaspores from north-western Europe (Thompson *et al.*, 1993), Australia (Leishman and Westoby, 1998) and New Zealand (Moles *et al.*, 2000). Variance was calculated from length, width and breadth measurements (standardized such that length equals 1); hence, a perfect sphere has zero variance and a needle or thin disc has a maximum variance of 0.33. Note that, for the Australian data, 'seed' was defined as the diaspore minus any dispersal structures (and thus includes one-seeded fruits, such as achenes); for the New Zealand data, measurements were taken on the part of the diaspore that was most likely to be incorporated into the soil profile (dispersal structures were included if they were permanently attached); and, in the European data set, most 'fruits' were entire diaspores (exceptions are detailed in Thompson *et al.*, 1993).

as contributing to bet-hedging by permitting establishment under a wider range of seedling establishment conditions. Consequently, they predicted that smaller-seeded species would be more strongly selected for bet-hedging dormancy, as is actually observed in England.

Secondly, although seeds of all sizes are subject to predation from a variety of predators while on the soil surface (Abbott and Van Heurck, 1985; Meiners and Stiles, 1997; Reader, 1997), only small seeds (which are consumed by invertebrates) can escape predation by quickly becoming incorporated into the soil. In contrast, large seeds are not protected from predation by burial, as vertebrates can find buried seeds (Thompson, 1987). Work by van Tooren (1988) and Chambers *et al.* (1991) has shown that small seeds are both buried more easily and incorporated more quickly into the soil than large seeds. Thompson *et al.* (1994) suggested that small seeds are also more likely to be taken down through the soil profile by earthworms. Thus the proposed mechanism for the association between small seed size and persistence is that only small seeds are able to persist as they are immune from predation once they are beneath the soil surface. Clearly it would be an advantage for persistent seeds to enter the seed bank quickly and thus small seeds may be rounded to facilitate burial. If this is the correct explanation for persistent seeds tending to be small and rounded in European herbaceous vegetation, presumably mechanisms of burial and disturbance are different in Australia (Leishman and Westoby, 1998) and New Zealand (Moles *et al.*, 2000).

Community patterns of seed size variation

Much of the literature on seed size variation considers seed size differences among species in terms of different environmental conditions that seedlings face during establishment. Large seeds are generally considered to be adaptive under harsh establishment conditions (Willson, 1983; Westoby *et al.*, 1992, 1996), and there is considerable evidence for this (discussed later).

The strongest pattern of seed-mass variation in relation to environmental factors is the relationship between large seeds and shaded habitats. This pattern was recognized long ago by Salisbury (1942, 1975), who showed that seed size increased with increasing shadiness of the habitat for

British species. The same pattern has been reported for studies on other British species (Hodgson and Mackey, 1986; Hodkinson *et al.*, 1998), other European communities (Luftensteiner, 1979), tropical woody species in Malaysia (Ng, 1978), Peru (Foster and Janson, 1985; Hammond and Brown, 1995), Guyana and Panama (Hammond and Brown, 1995), tropical species in Singapore (Metcalfe and Grubb, 1995) and Australia (between-genera comparisons only: Grubb and Metcalfe, 1996), the Indiana dunes flora of the USA (Mazer, 1989, 1990), annual communities of California (Marañón and Bartolome, 1989) and angiosperm tree species of temperate North America (but not gymnosperms: Hewitt, 1998). The pattern is strongly associated with phylogeny, with particular genera, families or orders tending to contribute more species in shaded than in unshaded situations (Mazer, 1990; Grubb and Metcalfe, 1996; Hodkinson *et al.*, 1998). The pattern is not absolute; for example, some of the smallest-seeded species (< 1 mg) in rainforest sites are very shade-tolerant and are successful in establishing on steep litter-free slopes (Metcalfe and Grubb, 1995, 1997; Grubb and Metcalfe, 1996).

Evidence for an association between large seeds and dry habitats is quite limited. Several within-species studies have reported a tendency for larger seeds in drier habitats (Schimpf, 1977; Sorenson and Miles, 1978; Stromberg and Patten, 1990). In phylogenetically independent contrasts comparing related high- and low-rainfall species, Wright and Westoby (1999) found larger seed mass in three of the five high-rainfall species. The most widely cited study is that of Baker (1972), who compared seed weights of over 2500 Californian taxa with moisture availability of the habitat. He found a positive correlation between moisture stress and seed size among herbaceous species, but not among trees or shrubs. Westoby *et al.* (1992) have argued that Baker's data should be viewed cautiously, as the relationship between seed size and dryness of the habitat is due to very small seeds in flood-prone sites rather than large seeds in dry sites. Mazer (1989) found no evidence for a relationship between seed size and moisture for 648 species of the Indiana dunes. Similarly, Telenius and Torstensson (1991) found no seed size–moisture relationship among 48 species of the genus *Spergularia*, and Long and Jones (1996) found no relationship in 14 oak species. However, it is not clear that measurements of annual rainfall are a good indicator of moisture stress during establishment for seedlings. Many seedlings only germinate in particular seasons or after suitable rains, and thus establishment conditions for the few weeks after germination may be equivalent in areas of different annual rainfall.

Evidence for larger seeds in low-nutrient soils is also very limited. Patterns of large seed size associated with lower-fertility soils have been found among 12 species of *Chionochloa* in New Zealand (Lee and Fenner, 1989) and for two species pairs in the *Proteaceae* (Mustart and Cowling, 1992) (although, for these habitats, soil moisture and fertility effects could not be separated). In contrast, Grubb and Coomes (1997) found smaller mean seed size among 27 Amazonian forest species on poorer compared with richer soils, while Westoby *et al.* (1990), Hammond and Brown (1995) and Wright and Westoby (1999) found no relationship between seed size and soil types of varying fertilities. Thus the evidence for a relationship between seed size and soil nutrient availability remains equivocal.

In summary, there is a clear and consistent pattern of larger seeds being associated with shaded habitats. However, any association between large seeds and dry or low-nutrient soils appears much more marginal.

Experimental evidence for the role of large seed size during seedling establishment

Seedlings face a variety of hazards during establishment. Mortality rates are often very high (Harper, 1977); consequently,

natural selection may operate strongly during this early stage of a plant's life cycle. Many studies have shown that initial seedling size is positively related to seed size, both within species (Dolan, 1984; Wulff, 1986; Zhang and Maun, 1991; Moegenburg, 1996) and among species (Stebbins (1976) for 15 Mediterranean annuals; Jurado and Westoby (1992) for 32 central Australian species; Seiwa and Kikuzawa (1991, 1996) for Japanese tree species; Cornelissen (1999) for 58 semi-woody British species). Within particular establishment sites, larger seeds have better seedling survival, again both within species (Stanton, 1984; Morse and Schmitt, 1985; Winn, 1988; Tripathi and Khan, 1990; Wood and Morris, 1990) and among species (Marshall, 1986; Chambers, 1995; Greene and Johnson, 1998).

For natural selection to favour larger seeds under particular hazards, it is not sufficient that larger-seeded species have better seedling survival. The relative advantage of larger-seeded species has to be greater under the hazard than in its absence. In this section, we review evidence from manipulative experiments that the advantage of larger-seeded over smaller-seeded species is indeed greater in the presence than in the absence of particular hazards.

Competition

In most manipulative experiments on the role of seed size in competitive environments, adult plant cover is removed to reduce competition. These experiments have shown that small-seeded species are less successful below closed canopies than large-seeded species (Gross and Werner, 1982; Gross, 1984; McConnaughay and Bazzaz, 1987; Reader, 1993; Ryser, 1993) in a variety of (mostly herbaceous) environments. Similarly, Burke and Grime (1996) and Eriksson and Eriksson (1997) have shown that small-seeded species are more dependent on disturbance (and hence reduced competition) than large-seeded species. There are some exceptions: both

Fenner (1978) and Reader (1991) have reported no correlation between seed size and competitive ability among a range of species.

Only two experiments have been reported where seedlings from different seed sizes competed soon after germination and survivorship was traced. Black (1958) grew seedlings of *Trifolium subterraneum* with different initial seed sizes in swards and showed that large seeds were more successful. Leishman (2001) grew multi-species mixtures of three seed size classes and showed that there is a competitive hierarchy among seedlings based on seed size, such that large seeds consistently win over smaller seeds.

Shade

Early work by Grime and Jeffrey (1965) and Hutchinson (1967) showed that small-seeded species suffer higher seedling mortality in shaded conditions. Grime and Jeffrey (1965) grew seedlings in vertical tubes with varying light gradients and found that, among nine tree species, longevity in deep shade was greatest for large-seeded species. Hutchinson (1967) grew seedlings in the dark and also found a correlation between seed size and longevity. More recent experiments that have examined mortality of seedlings during early life and in dense shade have also found that seed size is positively related to longevity (e.g. Leishman and Westoby (1994b) for 23 Australian species grown in the glasshouse for 6 weeks under 1% photosynthetically active radiation (PAR); Saverimuttu and Westoby (1996) for 11 pairs of Australian species grown in the glasshouse under < 1% PAR; Walters and Reich (2000) for ten North American species grown in the glasshouse under 2% PAR). Seiwa and Kikuzawa (1996) grew seedlings in large gaps, in small gaps and in the forest understorey and found higher mortality of small-seeded species in the forest understorey after canopy closure. Many other experiments have measured growth of seedlings in non-lethal shade

and have shown consistently that growth of smaller-seeded species is relatively more depressed than that of larger-seeded species (e.g. Piper, 1986; Seiwa and Kikuzawa, 1991; Osunkoya *et al.*, 1993, 1994; Leishman and Westoby, 1994b).

Other experiments where longevity in shade has not been correlated with seed mass have extended over longer periods, and have applied shading levels permitting at least some photosynthesis. For example, Augspurger (1984) found for 18 neotropical tree species that survival measured over 1 year was not correlated with seed mass, but rather with seedling characteristics, such as density of wood and leaf tissue. Experiments by Saverimuttu and Westoby (1996) and Walters and Reich (2000) deliberately compared outcomes during the cotyledon stage with outcomes during the later stages of seedling growth, and showed that seed mass was influential during early but not during later stages. This makes sense in terms of mechanism (discussed below), and seems capable of accounting for the discrepancy between the Augspurger study and the others cited.

Low soil moisture

There is limited and equivocal experimental evidence about the advantage of large seeds for establishment under low soil moisture conditions. Within-species studies have shown that larger seeds had better (Wulff, 1986) or worse (Hendrix *et al.*, 1991) survival in drier conditions. Buckley (1982) studied four arid dune-crest species and showed that larger seeds resulted in lower post-emergence mortality in the field. Leishman and Westoby (1994c) found, in a field experiment using 18 species from semi-arid Australia, that seedlings from large seeds had higher percentage emergence and survival than small-seeded species, but there was no evidence of a relatively greater advantage of large seeds in less-watered treatments. However, the climatic conditions were particularly harsh during the field experiment, so that survival was low even in watered treatments. In a repeat experiment in the glasshouse, using 23 species, larger seeds did confer a relatively greater advantage under increasingly dry soil conditions, and longevity was positively related to seed size.

Nutrient deprivation

Within-species (Krannitz et al., 1991) and between-species studies (Jurado and Westoby, 1992) have shown that seedlings from large seeds survive longer in conditions of nutrient deprivation. Stock et al. (1990) found among five Proteaceae species that seedling survival in nutrient-deficient soils was not associated with seed size. However, the small range of seed sizes among these five species (10–30 mg) may have made a significant effect less likely.

Burial

Given that large seeds produce large seedlings, it is not surprising that seedlings from large seeds are able to emerge from greater soil depths. This has been shown experimentally in several studies (Maun and Lapierre, 1986; Gulmon, 1992; Jurado and Westoby, 1992; Jurik et al., 1994) for a variety of species and habitats. The relative ability of seedlings to emerge through leaf litter may also be an important determinant of species composition in some habitats (Sydes and Grime, 1981; Bergelson, 1991; Facelli and Pickett, 1991; Facelli and Facelli, 1993; Facelli, 1994). As for depth of burial, large robust seedlings would be expected to emerge more successfully through litter. Experiments by Gulmon (1992), Vazquez-Yanes and Orozco-Segovia (1992) and Seiwa and Kikuzawa (1996) have shown that seedling emergence through litter is also positively associated with seed size. Buckley (1982) argued that for desert sand-dunes, larger seeds permitted germination from deeper in the soil, where moisture conditions are more favourable.

Herbivory

A few experiments have suggested that seedlings from large seeds tolerate defoliation (simulating herbivory) better than small-seeded seedlings. Armstrong and Westoby (1993) showed that capacity to survive removal of 95% of cotyledons was positively associated with seed size within genera and families, but not across all species, for 40 Australian species. Bonfil (1998) removed the entire cotyledons from two species of Quercus and found that seed mass was positively correlated with survival and growth for both species. In a slightly different approach, Harms and Dalling (1997) removed the entire shoot at 1 cm above the soil surface, at first-leaf stage, for 13 neotropical woody species. They found that only the largest-seeded species (at least 5 g) were capable of resprouting, while smaller-seeded species, which failed to resprout, died after clipping.

Mechanisms for tolerating establishment hazards

Thus, in experiments, larger-seeded species perform better under a diversity of adverse establishment conditions, including competition, shade, low soil moisture and nutrients, burial and herbivory. Might a common mechanism underlie these results, or are different effects or correlates of large-seededness responsible under different conditions or at different developmental stages? Under shading, at least, the advantage of larger seeds is confined to cotyledon-stage seedlings and does not persist into later seedling life, so any common mechanism would need to account for that. Westoby et al. (1996) distinguished three mechanisms by which larger seed mass might translate into greater success in the face of various hazards:

1. Seedling size effect: larger seeds result in larger seedlings, enabling better access to light (through penetration of soil or litter layer or relative to competing vegetation) and/or a reliable water-supply (via a longer radicle).

2. Reserve effect (Westoby et al., 1996), also called cotyledon functional morphology hypothesis (Hladik and Miquel, 1990; Garwood, 1995; Kitajima, 1996a, b) or larger-seed–slower-deployment hypothesis (Kidson and Westoby, 2000): obviously larger-seeded species will have larger total resources in their seeds. But this will not sustain their seedlings longer under a carbon deficit unless more resources remain uncommitted at a given time after germination, not just absolutely, but relative to the functional size of the seedling.

3. Metabolic effect: lower relative growth rate (RGR) and perhaps lower respiration rate in larger-seeded species enables longer survival under adverse conditions.

Seedling size effects

Seedlings need to reach light, whether by penetrating the soil or a litter layer or by overtopping competing vegetation. At shading below the compensation point, any seedling, whether large- or small-seeded, will eventually die. In the field, there may be steep gradients of light and soil water within a few centimetres of the soil surface, and under these circumstances centimetres or even millimetres of extra shoot or root length could be important. Light gradients near the ground would be steepest in closed herbaceous vegetation. Larger seedlings with larger root systems (e.g. Evans and Etherington, 1991; Jurado and Westoby, 1992) may gain access to soil moisture at deeper levels.

Seed mass is the largest influence on a seedling's initial reach above and below the ground, but etiolation should also be considered. The ability to etiolate under low light (by increased extension of the hypocotyl relative to that occurring under high light) is roughly similar in both small- and large-seeded species. However, the apparent cost is greater in smaller-seeded species, as etiolation is achieved via a greater drop in hypocotyl tissue density and a proportionally greater decrease in root mass and length (Ganade and Westoby, 1999). Analogous to the etiolation response

in shoots, several studies have shown that some species have a root elongation response in low soil moisture conditions (Osonubi and Davies, 1981; Molyneux and Davies, 1983; Sydes and Grime, 1984; Evans and Etherington, 1991).

The seedling size effect cannot account for increased seedling survivorship under hazards where there is no gradient of resource (e.g. light or soil moisture) away from the soil surface. Although resource gradients may be common in the field, experiments with nutrient deficiency and defoliation have not provided such gradients, nor have most shading experiments, with the exception of those of Grime and Jeffrey (1965). Consequently, seedling size effects cannot provide a universal mechanism accounting for the better survival of seedlings from larger seeds under hazards.

Larger-seed–slower-deployment effect

A spectrum of cotyledon types exists from thin, leaf-like, high-specific leaf area (SLA), primarily photosynthetic cotyledons to thick, low-SLA, non-photosynthetic storage organs (Hladik and Miquel, 1990; Garwood, 1995; Kitajima, 1996a, b; Wright and Westoby, 1999). Thicker cotyledons are generally found in larger seeds. The extreme case of non-photosynthetic cotyledons is cryptocotylar cotyledons, which remain protected within the testa. A strong association has been found between large seed size and cryptocotyly (Ng, 1978; Wright *et al.*, 2000), although small-seeded cryptocotylar species are also known.

Storage tissue is not always in cotyledons. Sometimes it is in the endosperm, and occasionally in the hypocotyl. There are no publications that compare seedling survivorship under hazards specifically across a broad range of endospermic species. Consequently, the following discussion will be couched in terms of cotyledon reserve storage. It seems reasonable that the same mechanisms would apply to storage in other locations, but there is little evidence to discuss on this issue.

The reserve effect requires not only

that the cotyledons have a storage role but that, across species, storage mass increases at a greater rate than seedling mass. Thus, for large-seeded species, greater proportions of the seed reserve remain uncommitted at any given stage during deployment. Consequently, the resources available to support respiration under carbon deficit would tend to be greater, relative to the autotrophic functioning parts of the seedling, in larger seeds. This would remain true up until the stage when all stored resources had been deployed into the functioning structures of the seedling. Recent quantitative surveys confirm that seedlings from larger-seeded species do indeed tend to have greater reserves relative to functioning parts of the seedling (Ganade and Westoby, 1999; Kidson and Westoby, 2000), in both cross-species and evolutionary divergences (phylogenetic analyses).

If species with large stored seed reserves committed all those resources very quickly into leaves, roots or other fixed structures of the seedling, the resources would no longer be available to support the seedling in adversity. The phrase 'larger-seed–slower-deployment' serves as a reminder that it is not sufficient for large-seeded species to have proportionately greater stored reserves: these reserves must also be held back from commitment to fixed structures over a longer period. In large seeds, the period of reserve transfer may be very considerable (e.g. a full year for *Chlorocardium rodiei* (ter Steege *et al.*, 1994)). Transfer may (Kitajima, 1996b) or may not (ter Steege *et al.*, 1994) be accelerated if the seedling is growing in the light, but in either event the slow deployment of reserves represents an opportunity cost if the seedling is not shaded or subjected to some other hazard. That is, if the reserves had been transferred quickly, the initial leaf area and root length of the seedling would have been greater, and it would have had a continuing growth advantage in the light. The same would, of course, have been true if the resources had been built into potentially autotrophic parts of the embryo within the seed, instead of placed in storage tissues.

In dense shade below the compensation point, slow deployment from reserves confers greater longevity in large-seeded species, but the longevity is not indefinite, so how might this translate into improved fitness? Longer survival would increase the chance of a seedling surviving until the formation of a tree-fall gap. However, not all forests have dynamics driven by tree-fall gaps. Also, Thompson (1987) and Seiwa and Kikuzawa (1996) point out that many deciduous forest seedlings germinate in winter or early spring, when the overstorey is leafless.

Further, in experiments where dense shade is applied, a tenfold increase in seed mass confers a gain in longevity that is considerably less than tenfold (e.g. Saverimuttu and Westoby, 1996). In other words, considering the effects of shading alone, the total seedling-days produced in the understorey per gram of seed decreases with seed mass. This should mean that the average number of seedlings still alive when a gap opens above them, per gram of seed produced, should be lower for larger-seeded species. Most probably, shading experiments in shade houses or growth chambers overestimate the likely longevity of small-seeded species. In the field, seedlings 'waiting' in shaded understorey will also be exposed to herbivores, pathogens and physical damage, and the slowly deployed reserves in large seeds will be beneficial in surviving these hazards, as well as in supporting respiration below the compensation point (Metcalfe and Grubb, 1995, 1997; Grubb and Metcalfe, 1996).

Metabolic effect and correlates of slow metabolism

Across species, a negative correlation has commonly been found between the seed mass and the potential relative growth rate (RGR) of seedlings grown under favourable conditions (e.g. Grime and Hunt, 1975; Shipley and Peters, 1990; Jurado and Westoby, 1992; Marañón and Grubb, 1993; Rincón and Huante, 1993; Osunkoya *et al.*,

1994). However, there is no known mechanism through which larger seed mass might directly cause lower potential RGR. Rather, low potential RGR and large seed mass appear to be part of a trait syndrome also involving sturdy tissue construction (low-SLA and specific root length) (Reich *et al.*, 1998; Wright and Westoby, 1999; Fig. 2.4) and low rates of tissue turnover (Bongers and Popma, 1990; Seiwa and Kikuzawa, 1991). Conversely, smaller-seeded species generally have higher potential RGR under near-optimal conditions, which is due in part to the seedlings being constructed of thinner or lower-density tissue with high turnover rates. Under unfavourable growth conditions (e.g. under low light or nutrients), the realized RGR of large- and small-seeded species may be similar, and (larger-seeded) species with more robust leaf and root tissue may survive longer than smaller-seeded species (e.g. Gross, 1984; Seiwa and Kikuzawa, 1991; Leishman and Westoby, 1994b; Walters and Reich, 2000).

Summary about mechanisms

In experiments that expose seedlings to hazards, a striking feature is the variety of different hazards under which larger seed mass confers improved survival. To the extent that this is underpinned by a common mechanism, this must be the larger-seed–slower-deployment effect. At the same time, effects of initial seedling size are clearly important in certain situations.

It has become clear that there are two sets of mechanisms in the determination of seedling survival under adverse conditions. The two sets operate at different stages. Early after germination, for cotyledon-stage seedlings, survival is influenced mostly by stored reserves and by initial seedling size. The experiments of Grime and Jeffrey (1965), Hutchinson (1967), Leishman and Westoby (1994b) and Saverimuttu and Westoby (1996) for cotyledon-stage seedlings provide good examples of cases in which these mechanisms are likely to be operating under shaded condi-

tions. Later on, growth and survival are largely determined by seedling morphology and physiology (e.g. interactions of light and nutrient availability with leaf-area ratio, net assimilation rate, RGR, etc.), and seed size is no longer directly relevant. The studies by Augspurger (1984), Kitajima (1984) and Saverimuttu and Westoby (1996) for leaf-stage seedlings and by Grubb and Metcalfe (Metcalfe and Grubb, 1995, 1997; Grubb and Metcalfe, 1996) provide good evidence of the operation of this second stage in determining seedling survival. By recognizing the two sets of mechanisms and the two phases of seedling life, some apparent inconsistencies in experimental results are resolved.

Associations between seed size and other plant attributes

Dispersal mode

It has long been recognized that the dispersal mode employed by seeds is associated with seed size (Harper *et al.*, 1970; Primack, 1987). Early work by Foster and Janson (1985) showed that, among 203 tropical woody plants of Peruvian forest, species with mammal-dispersal syndrome had significantly larger seeds than species with bird-dispersal syndrome. These results have been confirmed in other tropical forest woody species by Hammond and Brown (1995). In arid and semi-arid habitats of Australia, Jurado *et al.* (1991) and Leishman and Westoby (1994a) found that animal-syndrome (including both ant- and vertebrate-syndrome) seeds were, on average, significantly larger than unassisted, wind- or adhesion-syndrome seeds. Leishman *et al.* (1995) found that the relationship between seed mass and dispersal mode was broadly similar in five very different temperate habitats (central Australia, western New South Wales and Sydney in Australia, Indiana dunes in the USA and Sheffield, UK). Dispersal mode was associated with 29% of variation in seed mass between species. While the pattern of association of seed size with

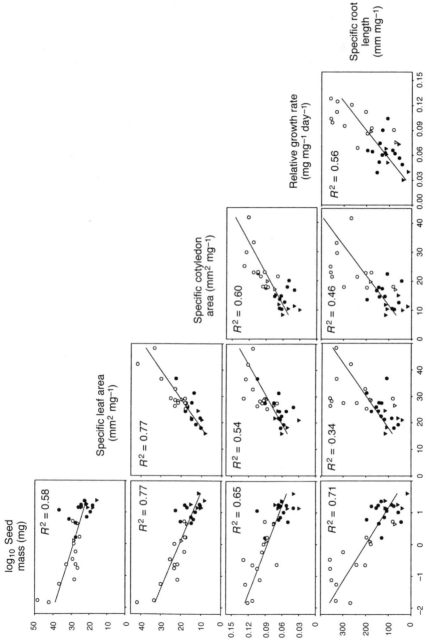

Fig. 2.4. Relationships between five seedling attributes for 33 woody species from New South Wales, Australia (data from Wright and Westoby, 1999). Low relative growth rate and large seed mass appear to be part of a trait syndrome also involving sturdy tissue construction (low area per unit dry mass of cotyledons and seedling leaves, and low root length per unit dry mass), although the advantages of such traits may operate at different stages of a seedling's life (see text). ●, *Fabales*; ○, *Myrtaceae*; ▼, *Proteaceae*; ▽, species from other clades. All relationships are significant at $P < 0.01$.

dispersal mode was different among the five floras (dispersal mode × flora interaction, $r^2 = 0.03$), this effect was about ten times smaller than the overall seed size/dispersal mode relationship ($r^2 = 0.29$). Note that the r^2 of 0.29 means that 71% of all seed-mass variation is within rather than between dispersal modes.

Hughes *et al.* (1994) have shown that the nature of this relationship is that seeds larger than about 100 mg tend to be adapted for dispersal by vertebrates, and seeds smaller than 0.1 mg tend to be unassisted, but between 0.1 and 100 mg many dispersal modes are feasible (Fig. 2.5). The fact that seeds < *c.* 0.1 mg tend to be unassisted presumably comes about both because the effectiveness of unassisted dispersal decreases above this size, and also because the relative cost of attaching an

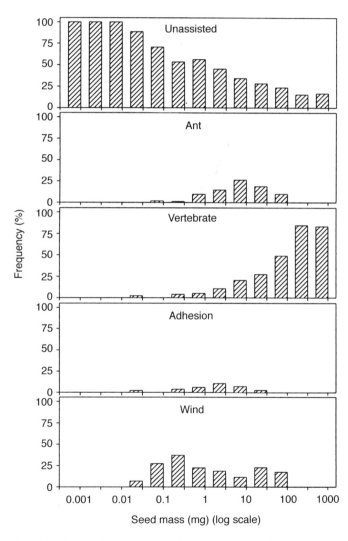

Fig. 2.5. Seed size distributions for unassisted and ant-, vertebrate-, adhesion- and wind-dispersed seeds from five floras: western New South Wales (Australia), central Australia, Sydney (Australia), Indiana dunes (USA) and Sheffield (England). Data compiled by Hughes *et al.* (1994). Note that all dispersal modes occur for seeds between 0.1 and 100 mg, while the smallest seeds tend to be unassisted and the largest seeds unassisted or adapted for dispersal by vertebrates.

effective dispersal structure is high below this size (Leishman and Westoby, 1994a).

Most information about distance travelled by seeds is anecdotal (Hughes *et al.*, 1994); consequently, an r^2 between seed mass and distance travelled cannot be given. It seems to be widely believed that smaller seeds travel further, because seeds falling freely during dispersal have a slower terminal velocity if small and consequently would be expected to travel further in a given wind. However, a number of factors counteract this. Larger-seeded species tend to have wings or hairs that slow the rate of fall if wind-dispersed. Among species with wings or hairs, typically the wing area increases with seed mass. Other larger-seeded species use animals or ballistic mechanisms for dispersal (Fig. 2.5). Even among unassisted species, those with larger seeds tend to be taller. That is, the height from which seeds are released tends to be greater, as does the time to reach the ground and the lateral distance travelled at a given wind speed. Taking these factors together, there is little evidence for any coherent relationship between seed mass and distance travelled across the full range of species and dispersal modes (Hughes *et al.*, 1994).

Growth form and plant height

A relationship between seed size and growth form has been well documented for a variety of floras from a range of habitats. Seed size generally increases from forbs and grasses through shrubs to trees and vines (Salisbury, 1942; Baker, 1972; Foster and Janson, 1985; Mazer, 1989; Jurado *et al.*, 1991; Leishman and Westoby, 1994a; Hammond and Brown, 1995; Leishman *et al.*, 1995; Metcalfe and Grubb, 1995; Fig. 2.6). Leishman *et al.* (1995) showed that this association between growth form and seed size was reasonably consistent across five temperate floras. The average relationship between seed size and growth form accounted for 20% of seed-size variation, while differences between floras (the growth form × flora interaction) accounted for only 2%.

Given that there is a correlation between growth form and plant height, it is not surprising that a positive correlation between seed size and plant height has also been reported consistently for different floras (Foster and Janson, 1985; Thompson and Rabinowitz, 1989; Leishman and Westoby, 1994a; Peat and Fitter, 1994; Leishman *et al.*, 1995; Rees, 1996). Interestingly, work by Leishman *et al.* (1995) on five quite different temperate floras showed that, although there is a large overlap in the amount of seed-size variation accounted for by plant height and growth form, both variables are able to account for a small but significant amount of variation after each other, in each of the floras.

What prospective explanations might there be for a relationship between seed size and growth form or plant height? Thompson and Rabinowitz (1989) invoked allometry, but, in the absence of a plausible mechanism, whether developmental or evolutionary, this merely restates the correlation. One plausible mechanism of natural selection might be that a greater height of release can compensate for a larger seed size in achieving a given dispersal distance. Under this argument, we would expect the plant height–seed size relationship to hold for species dispersed by wind or gravity, but not for animal-dispersed species. In fact, Leishman *et al.* (1995) found that the relationship between seed size and growth form/plant height was just as strong for animal-dispersed species as for wind-assisted and unassisted species. Similar results have also been found by Thompson and Rabinowitz (1989) for herbaceous species of *Asteraceae* and *Fabaceae*. In summary, the present situation is that no plausible mechanism is known that might account for the strong association between seed size and plant height or growth form.

Longevity of parent plant

In some studies, longer-lived plants tend to have larger seeds (Baker, 1972; Silvertown, 1981; Telenius and Torstensson, 1991).

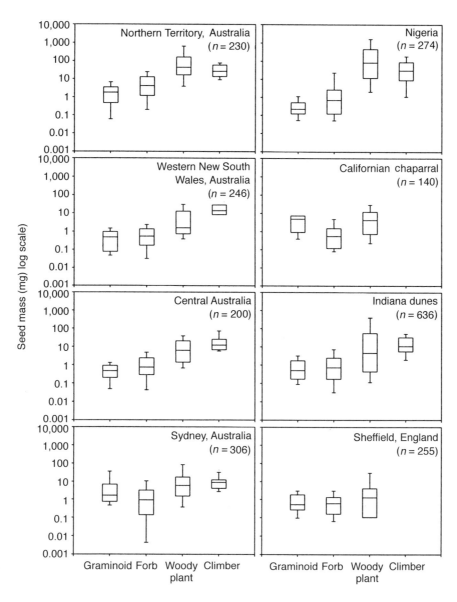

Fig. 2.6. Seed dry masses in four growth forms in eight floras: western New South Wales, central Australia, Sydney (Leishman *et al.*, 1995); Northern Territory, Nigeria (Lord *et al.*, 1997); Californian chaparral (Keeley, 1991); Indiana dunes (Mazer, 1989); and Sheffield (Grime *et al.*, 1988). The boxes span from the 25th percentile to the 75th percentile, with the line inside the box representing the sample median. Whiskers indicate the 10th and 90th percentiles. Outliers have not been shown.

Other studies have found no such relationship (Thompson, 1984; Mazer, 1989; Rydin and Borgegard, 1991). In studies where a number of plant attributes have been considered simultaneously, it has been shown that any relationship between seed size and plant longevity can be explained via secondary correlations of both seed size and plant longevity with plant height/growth form (Jurado *et al.*, 1991; Leishman and Westoby, 1994a; Leishman *et al.*, 1995).

Genome size

Larger-seeded species have been found to have a larger 2C DNA content, among mainly herbaceous and north-temperate species. Thompson (1990) found a positive association between seed size and 2C DNA values for 131 herbaceous species. Within-genus studies have also found consistent positive associations between seed size and 2C DNA content (e.g. Davies (1977) for 12 species of *Vicia*; Bennet (1987) and Peat and Fitter (1994) for a range of genera; Marañón and Grubb (1993) within 12 *Poaceae* but not within seven *Asteraceae*). In contrast, Lawrence (1985) found no such relationship within *Senecio*. Grassland species of the Sheffield flora that have large seeds and large nuclear DNA contents have longer cell cycles and tend to be more frost-resistant than seeds with smaller genomes, as cell division is less constrained by low temperatures (Thompson, 1990; Macgillivray and Grime, 1995).

Other miscellaneous seed-size associations

Several macroscale patterns of seed-size associations are of interest. Lord *et al.* (1997) compared data from five floras (two Australian temperate, one Australian tropical, two tropical from other continents). Seed size of tropical species was consistently larger than that of temperate species, independent of growth form and dispersal-mode differences between the floras. They suggested that the higher temperatures of tropical systems might result in higher metabolic costs of seedling growth and hence select for larger seed size.

Two large cross-species studies have found positive correlations between geographical range size and seed number (Peat and Fitter, 1994; Eriksson and Jakobsson, 1998); hence, given the seed size/number trade-off, larger-seeded species have smaller range sizes. However, both Oakwood *et al.* (1993) and Edwards and Westoby (1996) showed that, although larger seed size was associated with smaller range size, this correlation could be

understood as arising from secondary correlations via growth form and dispersal mode.

There is conflicting evidence on the association between seed size and mycorrhizal infection. Janos (1980) found that late successional species of tropical forests (which tend to have large seeds) were more likely to have mycorrhizas. Peat and Fitter (1994) found similar results for the British flora analysed at the family level. Westoby *et al.* (1992) found no evidence at species level for a correlation between seed size and mycorrhizal infection among the British grassland species of Grime *et al.* (1988). Allsopp and Stock (1992, 1995) found that, among South African species, mycorrhizal dependency increases with smaller seed size. Thus, the evidence for any association between seed size and tendency to mycorrhizal infection remains equivocal.

Summary about correlations with other plant attributes

Growth form, plant height and dispersal mode are the only attributes known to be correlated with substantial variation in seed size, independently of other attributes. There appears to be a positive correlation between seed size and genome size, but the potential for this to be understood as a secondary correlation via some third attribute has not yet been investigated, and the pattern itself has yet to be generalized across habitats other than cool-temperate grasslands. There is also an intriguing pattern of larger seeds in tropical floras, independent of growth form, height and dispersal mode, but as yet we have no understanding of the selection processes resulting in this pattern.

The broad spread of seed size among coexisting species

A striking and consistent pattern is the broad span of seed sizes within assemblages (typically five to six orders of magnitude), together with a strong overlap in seed sizes among assemblages (Fig. 2.1). Among five very different temperate communities from three continents, ranging from arid woodlands through coastal rainforest and sclerophyll woodlands to closed herbaceous communities, Leishman *et al.* (1995) showed that differences between the floras accounted for only 4% of seed-size variation. Much of the research literature on seed size has been directed towards understanding the hazards that seedlings face during establishment. However, the observation that such different habitats have very similar seed-size distributions points to the possibility that the prevalence of different physical conditions for establishment is not the main influence on seed size.

Grime *et al.* (1997) have shown for 43 British species that most of the trait variation for the regenerative stage is accounted for by the trade-off between seed size and number. Coexisting plant species are spread along a spectrum of different solutions, or strategies, along the seed size–number trade-off. The question is, are they spread along this spectrum simply because different solutions are equally competent and there is nothing to prevent them spreading? Or are there frequency-dependent processes that spread coexisting species out along the spectrum, with a high abundance of small-seeded species favouring an increase of larger-seeded species, and vice versa?

Frequency-dependent or game-theoretical models are indeed capable of predicting an evolutionarily stable strategy (ESS) consisting of a broad mix of seed-size strategies (Geritz, 1995; Rees and Westoby, 1997). In this, they contrast with the prediction of a single optimal seed-size strategy, in a given establishment environment, from the Smith–Fretwell model (Smith and Fretwell, 1974). The difference arises because the game-theoretical models express the idea that a seed's chance of producing an adult depends on what other seeds, of what sizes, are present in the competing mixture. Specifically, the models are driven by competition between

seedlings and by a colonization–competition trade-off. The adult that establishes at each patch or living site is assumed to come from the largest seed reaching the patch. Hence, a strategy mixture can be invaded by larger seeds because they will win in competition with small seeds. On the other hand, since smaller seeds are produced in larger numbers, they will reach some patches that are not reached by any larger seeds, and thus can persist in the mixture.

These game-theoretical models predict the broad spread of seed size that is actually observed among coexisting species, but how likely is it that they are a true description of the processes involved? One key mechanism, that large seeds win in competition with small seeds, appears generally true during early seedling competition (reviewed above). There is also consistent evidence that larger-seeded species produce fewer seeds per unit biomass than smaller-seeded species (Shipley and Dion, 1992; Greene and Johnson, 1994; Eriksson and Jakobsson, 1998; Leishman, 2001; M.L. Henery and M. Westoby, unpublished data). How widely might it be true that the species mixture (and hence the seed-size mixture) is decided mainly by competition between seedlings? This seems plausible in assemblages of annuals, in fire-prone and arid assemblages and in vegetation with gap dynamics. In these assemblages, most seedling establishment occurs at a common time (after fire, rain or gap creation), and growth to adulthood is arguably decided by competition among seedlings rather than with adults. But this list does not cover all possible vegetation types.

At ESS in the game-theoretical-strategy-mixture models, more establishment opportunities (vacant patches) must be reached by small-seeded species than by large-seeded species. There is little quantitative evidence available that small seeds dominate the seed rain. Spence (1990) found that small seeds dominated the seed rain of four New Zealand alpine communities, but Leishman (2001) found the opposite for a calcareous grassland community in the UK. Several studies have shown that small seeds dominate the seed bank (e.g. Leck, 1989; Eriksson and Eriksson, 1997; Price and Joyner 1997), which may also contribute to colonization opportunities. Clearly, this is an area where additional data are needed.

Also at ESS in the game-theoretical-strategy-mixture models, a negative correlation is expected between seed size and species abundance, measured as biomass or cover (M.R. Leishman and B.R. Murray, unpublished data). This is because the smaller-seeded species have to produce enough seeds to disperse successfully not just to patches not reached by one of their larger-seeded competitors, but to patches not reached by any of their larger-seeded competitors. This requires high abundance, as well as large numbers of seed produced from each unit of biomass or cover. The evidence is against this prediction, on the whole. Rabinowitz (1978) and Mitchley and Grubb (1986) found positive correlations between seed size and abundance for limited subsets of species within US tall-grass prairie and UK chalk grassland communities, respectively. However, other studies have found a negative correlation (Rees, 1995) or none (Eriksson and Jakobsson, 1998). M.R. Leishman and B.R. Murray (unpublished data) examined 12 different communities from four geographical regions, including both tropical and temperate, and found no evidence for consistent seed size/abundance patterns. In the four communities where significant seed size/abundance correlations were found, the relationships were positive.

In summary, we do not yet have a satisfactory biological interpretation for the consistently broad spread of seed mass within vegetation types. Existing game-theoretical models are driven by competition among seedlings (which in reality is unlikely to decide the species mixture in all vegetation types) and make some predictions that do not seem to be satisfied with any consistency. The logically possible alternatives are: (i) that, within each assemblage, there is a broad variety of establishment conditions (i.e. a variety of Smith–Fretwell functions), and each seed size does best in its own specific situation; (ii) that, within assemblages, many species occur as sink populations, supported by dispersal from source populations in other habitats; and (iii) that game-theoretical or frequency-dependent processes spread out the ESS mixture of some other species attribute, which in turn is correlated with seed size.

Conclusion

The last 10 years have seen considerable progress in understanding the evolutionary ecology of seed size. Substantial numbers of experiments have accumulated comparing ten to 50 species at a time. Correlative information across hundreds of species on several continents serves to place the experiments in context and to characterize the field distribution of seed mass and its relationships to other traits.

Larger seeds tend to produce seedlings with a greater proportion of their mass as stored reserve relative to autotrophically functioning structure of the seedling. These larger reserves are deployed progressively rather than immediately into structures. The consequence must be that, at any given stage during deployment, seedlings of larger-seeded species tend to have more reserves uncommitted and available to compensate for various hazards. Experiments have compared species across a range of seed sizes, both across species and as phylogenetically independent contrasts, and have applied competing vegetation, dense shade, drought, mineral nutrient deficiency, clipping and burial under litter and soil. Larger-seeded species have been advantaged under all these hazards. The larger initial seedling size is important in some situations, as well as the holding of uncommitted reserves. Later in seedling life, after reserves have been fully deployed into seedling structures, larger seed size no longer confers any direct advantage. Rather, shade tolerance is conferred by slow turnover of well-defended tissues, which may be loosely correlated with larger seed mass.

Seed mass is correlated with height, growth form and dispersal mode. These relationships have similar form in different temperate vegetation types from different continents. Although larger seed mass has been shown experimentally to confer improved tolerance against a wide variety of hazards, the only clear habitat pattern in the field is that species establishing under shade tend to have larger seeds. Seeds tend to be larger in the tropics, independent of growth form and dispersal mode. The reason for this remains unclear.

There is a markedly wide spread of seed mass among species within vegetation types. Recent game-theoretical models predict this, but the limited empirical evidence suggests that some key mechanisms in the models are not realistic in the field. In other words, the models make the right predictions for the wrong reasons. Understanding the broad spread of coexisting seed-mass strategies remains an outstanding challenge.

References

Abbott, I. and Van Heurck, P. (1985) Comparison of insects and vertebrates as removers of seed and fruit in a Western Australian forest. *Australian Journal of Ecology* 10, 165–168.

Allsopp, N. and Stock, W.D. (1992) Mycorrhizas, seed size and seedling establishment in a low nutrient environment. In: Read, D.J., Lewis, D.H., Fitter, A.H. and Alexander, I.J. (eds) *Mycorrhizas in Ecosystems*. CAB International, Wallingford, UK, pp. 59–64.

Allsopp, N. and Stock, W.D. (1995) Relationships between seed reserves, seedling growth and mycorrhizal responses in 14 related shrubs (Rosidae) from a low-nutrient environment. *Functional Ecology* 9, 248–254.

Armstrong, D.P. and Westoby, M. (1993) Seedlings from large seeds tolerate defoliation better: a test using phylogenetically independent contrasts. *Ecology* 74, 1092–1100.

Augspurger, C.K. (1984) Light requirements of neotropical tree seedlings: a comparative study of growth and survival. *Journal of Ecology* 72, 777–795.

Baker, H.G. (1972) Seed weight in relation to environmental conditions in California. *Ecology* 53, 997–1010.

Bakker, J.P., Poschlod, P., Strykstra, R.J., Bekker, R.M. and Thompson, K. (1996) Seed banks and seed dispersal: important topics in restoration ecology. *Acta Botanica Neerlandica* 45, 461–490.

Barclay, A.S. and Earle, F.R. (1974) Chemical analyses of seeds III. Oil and protein content of 1253 species. *Economic Botany* 28, 178–236.

Bekker, R.M., Bakker, J.P., Grandin, U., Kalamees, R., Milberg, P., Poschlod, P., Thompson, K. and Willems, J.H. (1998) Seed size, shape and vertical distribution in the soil: indicators of seed longevity. *Functional Ecology* 12, 834–842.

Bennet, M. (1987) Variation in genomic form in plants and its ecological implications. *New Phytologist* 106 (Suppl.), 177–200.

Bergelson, J. (1991) Competition between plants, before and after death. *Trends in Ecology and Evolution* 6, 378–379.

Black, J.N. (1958) Competition between plants of different initial seed sizes in swards of subterranean clover (*Trifolium subterraneum* L.) with particular reference to leaf area and the light microclimate. *Australian Journal of Agricultural Research* 9, 299–318.

Bonfil, C. (1998) The effects of seed size, cotyledon reserves, and herbivory on seedling survival and growth in *Quercus rugosa* and *Q. laurina* (Fagaceae). *American Journal of Botany* 85, 79–87.

Bongers, F. and Popma, J. (1990) Leaf dynamics of seedlings of rainforest species in relation to canopy gaps. *Oecologia* 82, 122–127.

Buckley, R.C. (1982) Seed size and seedling establishment in tropical arid dunecrest plants. *Biotropica* 14, 314–315.

Burke, M.J.W. and Grime, J.P. (1996) An experimental study of plant community invasibility. *Ecology* 77, 776–790.

Chambers, J.C. (1995) Relationship between seed fates and seedling establishment in alpine ecosystems. *Ecology* 76, 2124–2133.

Chambers, J.C., MacMahon, J.A. and Haefner, J.H. (1991) Seed entrapment in alpine ecosystems: effects of soil particle size and diaspore morphology. *Ecology* 72, 1668–1677.

Cornelissen, J.H.C. (1999) A triangular relationship between leaf size and seed size among woody species: allometry, ontogeny, ecology and taxonomy. *Oecologia* 118, 248–255.

Dalling, J.W., Swaine, M.D. and Garwood, N.C. (1997) Soil seed bank community dynamics in seasonally moist lowland tropical forest, Panama. *Journal of Tropical Ecology* 13, 659–680.

Davies, D.R. (1977) DNA contents and cell number in relation to seed size in the genus *Vicia*. *Heredity* 39, 153–163.

Dolan, R.W. (1984) The effect of seed size and maternal source on individual size in a population of *Ludwigia leptocarpa* (Onagraceae). *American Journal of Botany* 71, 1302–1307.

Edwards, W. and Westoby, M. (1996) Reserve mass and dispersal investment in relation to geographic range of plant species: phylogenetically independent contrasts. *Journal of Biogeography* 23, 329–338.

Eriksson, A. and Eriksson, O. (1997) Seedling recruitment in semi-natural pastures: the effects of disturbances, seed size, phenology and seed bank. *Nordic Journal of Botany* 17, 469–482.

Eriksson, O. (1995) Seedling recruitment in deciduous forest herbs – the effects of litter, soil chemistry and seed bank. *Flora* 190, 65–70.

Eriksson, O. and Jakobsson, A. (1998) Abundance, distribution and life histories of grassland plants: a comparative study of 81 species. *Journal of Ecology* 86, 922–933.

Evans, C.E. and Etherington, J.R. (1991) The effect of soil water potential on seedling growth of some British plants. *New Phytologist* 118, 571–579.

Facelli, J.M. (1994) Multiple indirect effects of plant litter affect the establishment of woody seedlings in old fields. *Ecology* 75, 1727–1735.

Facelli, J.M. and Facelli, E. (1993) Interactions after death: plant litter controls priority effects in a successional plant community. *Oecologia* 95, 277–282.

Facelli, J.M. and Pickett, S.T.A. (1991) Plant litter: light interception and effects on an old-field plant community. *Ecology* 72, 1024–1031.

Fenner, M. (1978) A comparison of the abilities of colonizers and closed-turf species to establish from seed in artificial swards. *Journal of Ecology* 66, 953–963.

Fenner, M. (1983) Relationships between seed weight, ash content and seedling growth in twenty-four species of Compositae. *New Phytologist* 95, 697–706.

Foster, S.A. and Janson, C.H. (1985) The relationship between seed size and establishment conditions in tropical woody plants. *Ecology* 66, 773–780.

Ganade, G. and Westoby, M. (1999) Seed mass and the evolution of early-seedling etiolation. *The American Naturalist* 154, 469–480.

Garner, R.D. and Witkowski, E.T.F. (1997) Variations in seed size and shape in relation to depth of burial in the soil and predispersal predation *in Acacia nilotica, A. tortilis* and *Dichrostachys cinerea. South African Journal of Botany* 63, 371–377.

Garwood, N.C. (1995) Functional morphology of tropical tree seedlings. In: Swaine, M.D. (ed.) *The Ecology of Tropical Forest Tree Seedlings*. Parthenon, New York, pp. 59–129.

Geritz, S.A.H. (1995) Evolutionarily stable seed polymorphism and small-scale spatial variation in seedling density. *The American Naturalist* 146, 685–707.

Greene, D.F. and Johnson, E.A. (1994) Estimating the mean annual seed production of trees. *Ecology* 75, 642–647.

Greene, D.F. and Johnson, E.A. (1998) Seed mass and early survivorship of tree species in upland clearings and shelterwoods. *Canadian Journal of Forest Research – Journal Canadien de la Recherche Forestière* 28, 1307–1316.

Grime, J.P. and Hunt, R. (1975) Relative growth-rate: its range and adaptive significance in a local flora. *Journal of Ecology* 63, 393–422.

Grime, J.P. and Jeffrey, D.W. (1965) Seedling establishment in vertical gradients of sunlight. *Journal of Ecology* 53, 621–642.

Grime, J.P., Hodgson, J.G. and Hunt, R. (1988) *Comparative Plant Ecology: a Functional Approach to Common British Species*. Unwin-Hyman, London.

Grime, J.P., Thompson, K., Hunt, R., Hodgson, J.G., Cornelissen, J.H.C., Rorison, I.H., Hendry, G.A.F., Ashenden, T.W., Askew, A.P., Band, S.R., Booth, R.E., Bossard, C.C., Campbell, B.D., Cooper, J.E.L., Davison, A.W., Gupta, P.L., Hall, W., Hand, D.W., Hannah, M.A., Hillier, S.H., Hodkinson, D.J., Jalili, A., Liu, Z., Mackey, J.M.L., Matthews, N., Mowforth, M.A., Neal, A.M., Reader, R.J., Reiling, K., Ross Fraser, W., Spencer, R.E., Sutton, F., Tasker, D.E., Thorpe, P.C. and Whitehouse, J. (1997) Integrated screening validates primary axes of specialisation in plants. *Oikos* 79, 259–281.

Gross, K.L. (1984) Effects of seed size and growth form on seedling establishment of six monocarpic perennial plants. *Journal of Ecology* 72, 369–387.

Gross, K.L. and Werner, P.A. (1982) Colonizing abilities of 'biennial' plant species in relation to ground cover: implications for their distributions in a successional sere. *Ecology* 63, 921–931.

Grubb, P.J. and Burslem, D.F.R.P. (1998) Mineral nutrient concentrations as a function of seed size within seed crops: implications for competition among seedlings and defence against herbivory. *Journal of Tropical Ecology* 14, 177–185.

Grubb, P.J. and Coomes, D.A. (1997) Seed mass and nutrient content in nutrient-starved tropical rainforest in Venezuela. *Seed Science Research* 7, 269–280.

Grubb, P.J. and Metcalfe, D.J. (1996) Adaptation and inertia in the Australian tropical lowland rainforest flora: contradictory trends in intergeneric and intrageneric comparisons of seed size in relation to light demand. *Functional Ecology* 10, 512–520.

Grubb, P.J., Metcalfe, D.J., Grubb, E.A.A. and Jones, G.D. (1998) Nitrogen-richness and protection of seeds in Australian tropical rainforest: a test of plant defence theory. *Oikos* 82, 467–482.

Gulmon, S.L. (1992) Patterns of seed germination in Californian serpentine grassland species. *Oecologia* 89, 27–31.

Hammond, D.S. and Brown, V.K. (1995) Seed size of woody plants in relation to disturbance, dispersal, soil type in wet neotropical forests. *Ecology* 76, 2544–2561.

Harms, K.E. and Dalling, J.W. (1997) Damage and herbivory tolerance through resprouting as an advantage of large seed size in tropical trees and lianas. *Journal of Tropical Ecology* 13, 617–621.

Harper, J.L. (1977) *Population Biology of Plants*. Academic Press, New York.

Harper, J.L., Lovell, P.H. and Moore, K.G. (1970) The shapes and sizes of seeds. *Annual Review of Ecology and Systematics* 1, 327–356.

Harvey, P., Read, A. and Nee, S. (1995a) Why ecologists need to be phylogenetically challenged. *Journal of Ecology* 83, 535–536.

Harvey, P., Read, A. and Nee, S. (1995b) Further remarks on the role of phylogeny in comparative ecology. *Journal of Ecology* 83, 735–736.

Hendrix, S.D., Nielsen, E., Nielsen, T. and Schutt, M. (1991) Are seedlings from small seeds always inferior to seedlings from large seeds? Effects of seed biomass on seedling growth in *Pastinaca sativa* L. *New Phytologist* 119, 299–305.

Hewitt, N. (1998) Seed size and shade-tolerance: a comparative analysis of North American temperate trees. *Oecologia* 114, 432–440.

Hladik, A. and Miquel, S. (1990) Seedling types and plant establishment in an African rainforest. In: Bawa, K.S. and Hadley, M. (eds) *Reproductive Ecology of Tropical Plants*. UNESCO/Parthenon, Paris/Carnforth, pp. 261–282.

Hodgson, J.G. and Mackey, J.M.L. (1986) The ecological specialization of dicotyledonous families within a local flora: some factors constraining optimization of seed size. *New Phytologist* 104, 497–515.

Hodkinson, D.J., Askew, A.P., Thompson, K., Hodgson, J.G., Bakker, J.P. and Bekker, R.M. (1998) Ecological correlates of seed size in the British flora. *Functional Ecology* 12, 762–766.

Hughes, L., Dunlop, M., French, K., Leishman, M.R., Rice, B., Rodgerson, L. and Westoby, M. (1994) Predicting dispersal spectra: a minimal set of hypotheses based on plant attributes. *Journal of Ecology* 82, 933–950.

Hutchinson, T.C. (1967) Comparative studies of the ability of species to withstand prolonged periods of darkness. *Journal of Ecology* 55, 291–299.

Janos, D. (1980) Mycorrhizae influence tropical succession. *Biotropica* 12 (Suppl.), 56–95.

Jurado, E. and Westoby, M. (1992) Seedling growth in relation to seed size among species of arid Australia. *Journal of Ecology* 80, 407–416.

Jurado, E., Westoby, M. and Nelson, D. (1991) Diaspore weight, dispersal, growth form and perenniality of Central Australian plants. *Journal of Ecology* 79, 811–830.

Jurik, T.W., Wang, S.C. and Vanderwalk, A.G. (1994) Effects of sediment load on seedling emergence from wetland seed banks. *Wetlands* 14, 159–165.

Keeley, J.E. (1991) Seed germination and life history syndromes in the California chaparral. *Botanical Review* 57, 81–116.

Kelly, C.K. (1995) Seed size in tropical trees – a comparative study of factors affecting seed size in Peruvian angiosperms. *Oecologia* 102, 377–388.

Kelly, C.K. (1997) Seed mass, habitat conditions and taxonomic relatedness – a re-analysis of Salisbury (1974). *New Phytologist* 135, 169–174.

Kelly, C.K. and Purvis, A. (1993) Seed size and establishment conditions in tropical trees: on the use of taxonomic relatedness in determining ecological patterns. *Oecologia* 94, 356–360.

Kidson, R. and Westoby, M. (2000) Seed mass and seedling dimensions in relation to seedling establishment. *Oecologia* 125(1), 11–17.

Kitajima, K. (1984) Relative importance of photosynthetic traits and allocation patterns as correlates of seedling shade tolerance of 13 tropical trees. *Oecologia* 98, 419–428.

Kitajima, K. (1996a) Cotyledon functional morphology, patterns of seed reserve utilization and regeneration niches of tropical tree seedlings. In: Swaine, M.D. (ed.) *The Ecology of Tropical Forest Tree Seedlings*. Parthenon, New York, pp. 193–210.

Kitajima, K. (1996b) Ecophysiology of tropical tree seedlings. In: Mulkey, S.S., Chazdon, R.L. and Smith, A.P. (eds) *Tropical Forest Plant Ecophysiology*. Chapman and Hall, New York, pp. 559–596.

Krannitz, P.G., Aarssen, L.W. and Dow, J.M. (1991) The effect of genetically based differences in seed size on seedling survival in *Arabidopsis thaliana* (Brassicaceae). *American Journal of Botany* 78, 446–450.

Lawrence, M. (1985) *Senecio* L. in Australia: nuclear DNA amounts. *Australian Journal of Botany* 33, 221–232.

Leck, M.A. (1989) Wetland seed banks. In: Leck, M.A., Parker, V.T. and Simpson, R.L. (eds) *Ecology of Soil Seed Banks*. Academic Press, San Diego, pp. 283–305.

Lee, W.G. and Fenner, M. (1989) Mineral nutrient allocation in seeds and shoots of twelve *Chionochloa* species in relation to soil fertility. *Journal of Ecology* 77, 704–716.

Leishman, M.R. (2001) Does the seed size/number trade-off model determine plant community structure? An assessment of the model mechanisms and their generality. *Oikos* (in press).

Leishman, M.R. and Westoby, M. (1992) Classifying plants into groups on the basis of associations of individual traits – evidence from Australian semi-arid woodlands. *Journal of Ecology* 80, 417–424.

Leishman, M.R. and Westoby, M. (1994a) Hypotheses on seed size: tests using the semiarid flora of western New South Wales, Australia. *The American Naturalist* 143, 890–906.

Leishman, M.R. and Westoby, M. (1994b) The role of large seed size in shaded conditions: experimental evidence. *Functional Ecology* 8, 205–214.

Leishman, M.R. and Westoby, M. (1994c) The role of large seeds in seedling establishment in dry soil conditions – experimental evidence from semi-arid species. *Journal of Ecology* 82, 249–258.

Leishman, M.R. and Westoby, M. (1998) Seed size and shape are not related to persistence in soil in Australia in the same way as in Britain. *Functional Ecology* 12, 480–485.

Leishman, M.R., Westoby, M. and Jurado, E. (1995) Correlates of seed size variation: a comparison among five temperate floras. *Journal of Ecology* 83, 517–530.

Long, T.J. and Jones, R.H. (1996) Seedling growth strategies and seed size effects in fourteen oak species native to different soil moisture habitats. *Trees – Structure and Function* 11, 1–8.

Lord, J., Westoby, M. and Leishman, M.R. (1995) Seed size and phylogeny in six temperate floras: constraints, niche-conservatism and adaptation. *The American Naturalist* 146, 349–364.

Lord, J., Egan, J., Clifford, H.T., Jurado, E., Leishman, M.R., Williams, D. and Westoby, M. (1997) Larger seeds in tropical floras: consistent patterns independent of growth form and dispersal mode. *Journal of Biogeography* 24, 205–211.

Luftensteiner, H.W. (1979) The eco-sociological value of dispersal spectra of two plant communities. *Vegetatio* 41, 61–67.

Lunt, I.D. (1995) Seed longevity of six native forbs in a closed *Themeda triandra* grassland. *Australian Journal of Botany* 43, 439–449.

McConnaughay, K.D.M. and Bazzaz, F.A. (1987) The relationship between gap size and performance of several colonizing annuals. *Ecology* 68, 411–416.

Macgillivray, C.W. and Grime, J.P. (1995) Genome size predicts frost resistance in British herbaceous plants: implications for rates of vegetation response to global warming. *Functional Ecology* 9, 320–325.

Marañón, T. and Bartolome, J.W. (1989) Seed and seedling populations in two contrasted communities: open grassland and oak (*Quercus agrifolia*) understorey in California. *Acta Oecologia* 10, 147–158.

Marañón, T. and Grubb, P.J. (1993) Physiological basis and ecological significance of the seed size and relative growth rate relationship in Mediterranean annuals. *Functional Ecology* 7, 591–599.

Marshall, D.L. (1986) Effect of seed size on seedling success in three species of *Sesbania* (Fabaceae). *American Journal of Botany* 73, 457–464.

Maun, M. and Lapierre, J. (1986) Effects of burial by sand on seed germination and seedling emergence of four dune species. *American Journal of Botany* 73, 450–455.

Mazer, S.J. (1989) Ecological, taxonomic and life history correlates of seed mass among Indiana Dune angiosperms. *Ecological Monographs* 59, 153–175.

Mazer, S.J. (1990) Seed mass of Indiana Dune genera and families: taxonomic and ecological correlates. *Evolutionary Ecology* 4, 326–357.

Meiners, S.J. and Stiles, E.W. (1997) Selective predation on the seeds of woody plants. *Journal of the Torrey Botanical Society* 124, 67–70.

Metcalfe, D.J. and Grubb, P.J. (1995) Seed mass and light requirements for regeneration in southeast Asian rain forest. *Canadian Journal of Botany – Revue Canadienne de Botanique* 73, 817–826.

Metcalfe, D.J. and Grubb, P.J. (1997) The responses to shade of seedlings of very small-seeded tree and shrub species from tropical rain forest in Singapore. *Functional Ecology* 11, 215–221.

Michaels, H.J., Benner, B., Hartgerink, A.P., Lee, T.D., Rice, S., Willson, M.F. and Bertin, R.I. (1988) Seed size variation: magnitude, distribution, and ecological correlates. *Evolutionary Ecology* 2, 157–166.

Milberg, P., Pérez-Fernández, M.A. and Lamont, B.B. (1998) Growth responses to added nutrients of seedlings from three woody genera depend on seed size. *Journal of Ecology* 86, 624–632.

Mitchley, J. and Grubb, P.J. (1986) Control of relative abundance of perennials in chalk grassland in southern England. *Journal of Ecology* 74, 1139–1166.

Moegenburg, S.M. (1996) *Sabal palmetto* seed size – causes of variation, choices of predators, and consequences for seedlings. *Oecologia* 106, 539–543.

Moles, A.T., Hodson, D.W. and Webb, C.J. (2000) Do seed size and shape predict persistence in soil in New Zealand? *Oikos* 89, 541–545.

Molyneux, D.E. and Davies, W.J. (1983) Rooting pattern and water relations of three pasture grasses growing in drying soil. *Oecologia* 58, 220–224.

Morse, D.H. and Schmitt, J. (1985) Propagule size, dispersal ability and seedling performance in *Asclepias syriaca*. *Oecologia* 67, 372–379.

Mustart, P.J. and Cowling, R.M. (1992) Seed size: phylogeny and adaptation in two closely related Proteaceae species-pairs. *Oecologia* 91, 292–295.

Ng, F.S.P. (1978) Strategies of establishment in Malayan forest trees. In: Tomlinson, P.B. and Zimmerman, M.H. (eds) *Tropical Trees as Living Systems*. Cambridge University Press, Cambridge, pp. 129–162.

Oakwood, M., Jurado, E., Leishman, M.R. and Westoby, M. (1993) Geographic ranges of plant species in relation to dispersal morphology, growth form and diaspore weight. *Journal of Biogeography* 20, 563–572.

Obeso, J.R. (1993) Seed mass variation in the perennial herb *Asphodelus albus*: sources of variation and position effect. *Oecologia* 93, 571–575.

Osonubi, O. and Davies, W.J. (1981) Root growth and water relations of oak and birch seedlings. *Oecologia* 51, 343–350.

Osunkoya, O.O., Ash, J.E., Graham, A.W. and Hopkins, M.S. (1993) Growth of tree seedlings in tropical rain forests of north Queensland, Australia. *Journal of Tropical Ecology* 9, 1–18.

Osunkoya, O.O., Ash, J.E., Hopkins, M.S. and Graham, A.W. (1994) Influence of seed size and seedling ecological attributes on shade-tolerance of rain forest tree species in northern Queensland. *Journal of Ecology* 82, 149–163.

Peat, H.J. and Fitter, A.H. (1994) Comparative analyses of ecological characteristics of British angiosperms. *Biological Review of the Cambridge Philosophical Society* 69, 95–115.

Philippi, T. and Seger, J. (1989) Hedging one's evolutionary bets, revisited. *Trends in Ecology and Evolution* 4, 41–44.

Piper, J.K. (1986) Germination and growth of bird-dispersed plants: effects of seed size and light on seedling vigour and biomass allocation. *American Journal of Botany* 73, 959–965.

Price, M.V. and Joyner, J.W. (1997) What resources are available to desert granivores: seed rain or soil seed bank? *Ecology* 78, 764–773.

Primack, R.B. (1987) Relationships among flowers, fruit and seeds. *Annual Review of Ecology and Systematics* 18, 409–430.

Rabinowitz, D. (1978) Abundance and diaspore weight in rare and common prairie grasses. *Oecologia* 37, 213–219.

Raven, J.A. (1999) The minimum size of seeds and spores in relation to the ontogeny of homoiohydric plants. *Functional Ecology* 13, 5–14.

Reader, R.J. (1991) Relationship between seedling emergence and species frequency on a gradient of ground cover density in an abandoned pasture. *Canadian Journal of Botany* 69, 1397–1401.

Reader, R.J. (1993) Control of seedling emergence by ground cover and seed predation in relation to seed size for some old-field species. *Journal of Ecology* 81, 169–175.

Reader, R.J. (1997) Potential effects of granivores on old field succession. *Canadian Journal of Botany – Revue Canadienne de Botanique* 75, 2224–2227.

Rees, M. (1993) Trade-offs among dispersal strategies in British plants. *Nature* 366, 150–152.

Rees, M. (1995) Community structure in sand dune annuals: is seed weight a key quantity? *Journal of Ecology* 83, 857–863.

Rees, M. (1996) Evolutionary ecology of seed dormancy and seed size. *Philosophical Transactions of the Royal Society of London – Series B* 351, 1299–1308.

Rees, M. and Westoby, M. (1997) Game-theoretical evolution of seed mass in multi-species ecological models. *Oikos* 78, 116–126.

Reich, P.B., Tjoelker, M.G., Walters, M.B., Vanderklein, D.W. and Bushena, C. (1998) Close association of relative growth rate, leaf and root morphology, seed mass and shade tolerance in seedlings of nine boreal tree species grown in high and low light. *Functional Ecology* 12, 327–338.

Rincón, E. and Huante, P. (1993) Growth response of tropical deciduous tree seedlings to contrasting light conditions. *Trees* 7, 202–207.

Rydin, H. and Borgegard, S. (1991) Plant characteristics over a century of primary succession on islands: Lake Hjalmaren. *Ecology* 72, 1089–1101.

Ryser, P. (1993) Influences of neighbouring plants on seedling establishment in limestone grassland. *Journal of Vegetation Science* 4, 195–202.

Salisbury, E.J. (1942) *The Reproductive Capacity of Plants*. G. Bell and Sons, London.

Salisbury, E.J. (1975) Seed size and mass in relation to environment. *Proceedings of the Royal Society, London B* 186, 83–88.

Saverimuttu, T. and Westoby, M. (1996) Seedling longevity under deep shade in relation to seed size. *Journal of Ecology* 84, 681–689.

Schimpf, D.J. (1977) Seed weight of *Amaranthus retroflexus* in relation to moisture and length of growing season. *Ecology* 58, 450–453.

Seiwa, K. and Kikuzawa, K. (1991) Phenology of tree seedlings in relation to seed size. *Canadian Journal of Botany* 69, 532–538.

Seiwa, K. and Kikuzawa, K. (1996) Importance of seed size for the establishment of seedlings of five deciduous broad-leaved tree species. *Vegetatio* 123, 51–64.

Shipley, B. and Dion, J. (1992) The allometry of seed production in herbaceous angiosperms. *The American Naturalist* 139, 467–483.

Shipley, B. and Peters, R.H. (1990) The allometry of seed weight and seedling relative growth rate. *Functional Ecology* 4, 523–529.

Shipley, B., Keddy, P.A., Moore, D.R.J. and Lemkt, K. (1989) Regeneration and establishment strategies of emergent macrophytes. *Journal of Ecology* 77, 1093–1110.

Silvertown, J.W. (1981) Seed size, lifespan and germination date as co-adapted features of plant life history. *The American Naturalist* 118, 860–864.

Smith, C.C. and Fretwell, S.D. (1974) The optimal balance between size and number of offspring. *The American Naturalist* 108, 499–506.

Sorenson, F.C. and Miles, R.S. (1978) Cone and seed weight relationships in Douglas-fir from western and central Oregon. *Ecology* 59, 641–644.

Spence, J.R. (1990) Seed rain in grassland, herbfield, snowback and fellfield in the alpine zone, Craigieburn range, South Island, New Zealand. *New Zealand Journal of Botany* 28, 439–450.

Stanton, M.L. (1984) Seed variation in wild radish: effect of seed size on components of seedling and adult fitness. *Ecology* 65, 1105–1112.

Stebbins, G.L. (1976) Seed and seedling ecology in annual legumes I. A comparison of seed size and seedling development in some annual species. *Oecologia Plantarum* 11, 321–331.

Stock, W.D., Pate, J.S. and Delfs, J. (1990) Influence of seed size and quality on seedling development under low nutrient conditions in five Australian and South African members of the Proteaceae. *Journal of Ecology* 78, 1005–1020.

Stromberg, J.C. and Patten, D.T. (1990) Variation in seed size of a southwestern riparian tree, Arizona walnut (*Juglans major*). *The American Midland Naturalist* 124, 269–277.

Sydes, C. and Grime, J.P. (1981) Effects of tree leaf litter on herbaceous vegetation in deciduous woodland I. Field investigations. *Journal of Ecology* 69, 237–248.

Sydes, C.L. and Grime, J.P. (1984) A comparative study of root development using a simulated rock crevice. *Journal of Ecology* 72, 937–946.

Telenius, A. and Torstensson, P. (1991) Seed wings in relation to seed size in the genus *Spergularia*. *Oikos* 61, 216–222.

ter Steege, H., Bokdam, C., Boland, M., Dobbelsteen, J. and Verburg, I. (1994) The effects of man-made gaps on germination, early survival and morphology of *Chlorocardium rodiei* seedlings. *Journal of Tropical Ecology* 10, 245–260.

Thompson, K. (1984) Why biennials are not as few as they ought to be. *The American Naturalist* 123, 854–861.

Thompson, K. (1987) Seeds and seed banks. *New Phytologist* 106 (Suppl.), 23–34.

Thompson, K. (1990) Genome size, seed size and germination temperature in herbaceous angiosperms. *Evolutionary Trends in Plants* 4, 113–116.

Thompson, K. and Rabinowitz, D. (1989) Do big plants have big seeds? *The American Naturalist* 133, 722–728.

Thompson, K., Band, S.R. and Hodgson, J.G. (1993) Seed size and shape predict persistence in soil. *Functional Ecology* 7, 236–241.

Thompson, K., Green, A. and Jewels, A.M. (1994) Seeds in soil and worm casts from a neutral grassland. *Functional Ecology* 8, 29–35.

Tripathi, R.S. and Khan, M.L. (1990) Effects of seed weight and microsite characteristics on germination and seedling fitness in two species of *Quercus* in a subtropical wet hill forest. *Oikos* 57, 289–296.

van Tooren, B.F. (1988) The fate of seeds after dispersal in chalk grassland: the role of the bryophyte layer. *Oikos* 53, 41–48.

Vaughton, G. and Ramsey, M. (1998) Sources and consequences of seed mass variation in *Banksia marginata* (Proteaceae). *Journal of Ecology* 86, 563–573.

Vazquez-Yanes, S.C. and Orozco-Segovia, A. (1992) Effects of litter from a tropical rainforest on tree seed germination and establishment under controlled conditions. *Tree Physiology* 11, 391–400.

Venable, D.L. and Brown, J.S. (1988) The selective interactions of dispersal, dormancy and seed size as adaptations for reducing risks in variable environments. *The American Naturalist* 131, 360–384.

Walters, M.B. and Reich, P.B. (2000) Seed size, nitrogen supply and growth rate affect tree seedling survival in deep shade. *Ecology* 81, 1887–1901.

Westoby, M. (1998) A leaf–height–seed (LHS) plant ecology strategy scheme. *Plant and Soil* 199, 213–227.

Westoby, M., Rice, B. and Howell, J. (1990) Seed size and plant growth form as factors in dispersal spectra. *Ecology* 71, 1307–1315.

Westoby, M., Jurado, E. and Leishman, M.R. (1992) Comparative evolutionary ecology of seed size. *Trends in Ecology and Evolution* 7, 368–372.

Westoby, M., Leishman, M.R. and Lord, J. (1995a) On misinterpreting the 'phylogenetic correction'. *Journal of Ecology* 83, 531–534.

Westoby, M., Leishman, M.R. and Lord, J. (1995b) Further remarks on phylogenetic correction. *Journal of Ecology* 83, 727–729.

Westoby, M., Leishman, M.R. and Lord, J. (1995c) Issues of interpretation after relating comparative data sets to phylogeny. *Journal of Ecology* 83, 892–893.

Westoby, M., Leishman, M.R. and Lord, J. (1996) Comparative ecology of seed size and dispersal. *Philosophical Transactions of the Royal Society, London, Series B* 351, 1309–1318.

Willson, M.F. (1983) *Plant Reproductive Ecology*. John Wiley & Sons, New York, USA.

Winn, A.A. (1988) Ecological and evolutionary consequences of seed size in *Prunella vulgaris*. *Ecology* 69, 1537–1544.

Wood, D.M. and Morris, W.F. (1990) Ecological constraints to seedling establishment on the pumice plains, Mount St Helens, Washington. *American Journal of Botany* 77, 1411–1418.

Wright, I.J. and Westoby, M. (1999) Differences in seedling growth behaviour among species: trait correlations across species, and trait shifts along nutrient compared to rainfall gradients. *Journal of Ecology* 87, 85–97.

Wright, I.J., Clifford, H.T., Kidson, R., Reed, M.L., Rice, B.L. and Westoby, M. (2000) A survey of seed and seedling characters in 1744 Australian dicotyledon species: cross-species trait correlations and correlated trait-shifts within evolutionary lineages. *Biological Journal of the Linnean Society* 69, 521–547.

Wulff, R.D. (1986) Seed size variation in *Desmodium paniculatum* II. Effects on seedling growth and physiological performance. *Journal of Ecology* 74, 99–114.

Zhang, J. and Maun, M.A. (1991) Establishment and growth of *Panicum virgatum* L. seedlings on a Lake Erie sand dune. *Bulletin of the Torrey Botanical Club* 118, 141–153.

Chapter 3

Maternal Effects on Seeds During Development

Yitzchak Gutterman

The Jacob Blaustein Institute for Desert Research and Department of Life Sciences,
Ben-Gurion University of the Negev, Israel

Introduction

In most plant species, the seeds vary in their degree of germinability between and within populations and between and within individuals. Some of this variation may be of genetic origin, but much of it is known to be phenotypic. That is, it is caused by the local conditions under which the seeds matured. These conditions consist of a combination of the microenvironment experienced by the seed due to its position on the parent plant and the abiotic environment of the plant (e.g. the ambient temperature, day length, water availability, etc.).

In different plant species, maternal factors, such as the position of the inflorescence on the mother plants or the position of the seeds in the inflorescence or in the fruit, can markedly influence the germinability of seeds (Evenari, 1963; Koller and Roth, 1964; Datta *et al.*, 1970; Evenari *et al.*, 1977; Thomas *et al.*, 1979; Jacobsohn and Globerson, 1980; Gutterman, 1980/81a, b, 1990a, 1993, 1994a, b, 1996b; Grey and Thomas, 1982). The age of the mother plant during flower induction (Kigel *et al.*, 1979) or seed maturation (Gutterman and Evenari, 1972; Do Cao *et al.*, 1978; Gutterman, 1978a) and, in the case of certain grasses, even the order of the caryopsis that the mother plant originated from can

all have an influence on seed germinability (Datta *et al.*, 1972a).

There are numerous cases recorded of seed germinability being modified by environmental factors operating during development and maturation. Examples include day length (Lona, 1947; Jacques, 1957, 1968; Koller, 1962; Cumming, 1963; Wentland, 1965; Evenari *et al.*, 1966; Gutterman, 1969, 1973, 1974, 1978a, b, 1982a, 1985, 1992b, 1993, 1994b, 1996a, b; Karssen, 1970; Gutterman and Evenari, 1972; Gutterman and Porath, 1975; Pourrat and Jacques, 1975); temperature (Juntila, 1973; Heide *et al.*, 1976); parental photothermal environment (Datta *et al.*, 1972b; Wurzburger and Koller, 1976; Kigel *et al.*, 1977); light quality (Cumming, 1963; McCullough and Shropshire, 1970; Gutterman, 1974, 1992b; Gutterman and Porath, 1975; Jacobsohn and Globerson, 1980; Cresswell and Grime, 1981); and altitude (Dorne, 1981). Achenes of *Lactuca serriola* (which mature during summer and autumn), as well as summer- and winter-maturing seeds on the same mother plants of some biseasonal-flowering perennial shrubs of the *Aizoaceae*, have been found to differ in germinability (Gutterman, 1991, 1992a). Fenner (1991, 1992) reviews environmental effects on seed size, chemical composition and germination.

The maturation of different seeds with different germinability on one mother plant has a very important ecological advantage, especially under extreme habitats, such as deserts (Gutterman, 1980/81b, 1982a, 1983, 1985, 1992a, b, 1993, 1994a, b, c, 1998a; Roach and Wulff, 1987; Fenner, 1991). In such areas, the time (day or night) of the first rains of the season that cause germination may affect the emergence of plants with seeds that require light or dark for germination (Gutterman, 1996a, b). The date, the amount and the distribution of rain, and the temperatures during and following these rains may be completely unpredictable. For example, in the Negev Highlands of Israel, the date of appearance of a rainfall that causes germination could range from mid-November to the end of February (3.5 months). In one season, one to three rains could cause germination (Gutterman, 1982a) when, in most years, the long, hot, dry summer starts towards the beginning of May (Evenari et al., 1982; Gutterman, 1993, 1998a). In many plant species occurring in the Irano-Turanian and Saharo-Arabian phytogeographical regions, including large areas of the Negev and Judaean Deserts, the phenotypic plasticity of seed germination decreases the risk to species survival by increasing the diversity of seed germination (Gutterman, 1993, 1997, 1998a, b). In Pteranthus dichotomus (Caryophyllaceae) (Evenari, 1963), even seeds from the same inflorescence may germinate in different years. A similar phenomenon of intermittent germination is also found in the germination of some other plant species, including weeds, that survive under unpredictable conditions (Harper, 1977; Gutterman, 1985; Cavers, 1995; Baskin and Baskin, 1998). In many plant species, different maternal and environmental factors may increase the phenotypic diversity. This ensures that, during or after a rainfall, only a portion of the total seed bank of a certain plant species will germinate.

It has been found that, in many plant species, the fate of the next generation or generations, as far as the germination of seeds is concerned, is dependent, at least to a certain degree, on the maturation conditions of the seeds when they are still on the mother plant (Datta et al., 1972a, b).

Seed position affecting seed germination

Position in capsules, pods and fruit

Seed position in different organs on the mother plants can affect seed colour, size, morphology and germination in many plant species. Even within a single capsule, the position of a seed may influence its germinability. Mesembryanthemum nodiflorum L. (Aizoaceae) is an annual desert plant originating in South Africa, with very wide phytogeographical distribution, including the Saharo-Arabian, Mediterranean and Siberian regions. Plants collected from populations in the Judaean Desert near the Dead Sea were studied. This area is in the Saharo-Arabian phytogeographical region, receiving an average rainfall of about 100 mm year^{-1}. The M. nodiflorum fruit, which contains about 60 seeds, is a dry capsule and the seeds are dispersed by rain. When the fruit is wetted, the terminal seeds are shed first (after 15 min), followed by the middle and lower seeds, which are shed after 200 and 320 min, respectively. The terminal seeds germinate much more readily than those lower in the fruit. Germination was 61, 5.5 and 1% in the terminal, middle and lower seeds, respectively, after 8 years of storage. These differences persist indefinitely (Fig. 3.1) and can be demonstrated in laboratory-stored seeds even after 28 years. In addition, the terminal seeds have an endogenous annual cycle of dormancy, resulting in high levels of germination in winter and spring and low levels in summer and autumn (Gutterman, 1990a, b, 1994a). The combination of the position effect and the annual rhythm of germination ensures that seed dispersal and germination are spread over time and occur at the right season (Gutterman, 1980/81b).

In Medicago spp. (Fabaceae), the spiral pods are multiseeded dispersal units.

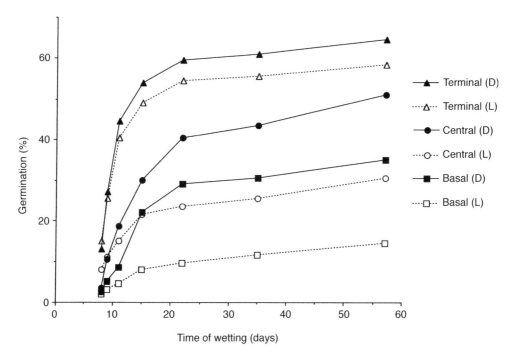

Fig. 3.1. *Mesembryanthemum nodiflorum* seed germination according to their position in their capsules – terminal, central and basal groups of seeds – after 19 years of dry storage and up to 60 days of wetting at 15°C in dark and light (from Gutterman, 1990b).

Seed weight and impermeability to water decrease from the calyx to the stylar end (Kirchner and Andrew, 1971; McComb and Andrews, 1974). In *Platystemon californicus* (*Papaveraceae*), the seeds that mature in the carpels of the fruit are much more dormant than seeds formed in the central chamber (Hannan, 1980). In *Cakile edentula* var. *lacustris* (*Brassicaceae*), the pods are divided into two segments. At sub- or supra-optimal temperatures of 25/5 or 15/5°C, in the dark, seeds of the lower segment germinate to higher percentages than seeds of the higher segment (Maun and Payne, 1989; Baskin and Baskin, 1998).

Effect of position of capsules or fruit

In some species, there are often marked differences between seeds from different capsules on the same plant. For example, on one individual of the South African shrub *Glottiphyllum linguiforme* (*Aizoaceae*) occurring in the Karoo Desert (Herre, 1971), there are central and peripheral capsules that are different in size, number of loculi valves and number of seeds. In Petri dishes, seeds from capsules from the centre, incubated at 25°C for 18 days, germinated to very low percentages, in the light and in the dark. In the same time, under the same conditions, the seeds from the peripheral capsules reached about 80% germination. When seeds from central capsules that had matured during the previous 3 years of the experiment were placed in wet soil, they did not germinate. However, approximately 20% germination was counted in seeds from the peripheral capsules that were placed under the same conditions. The capsules in the central part of the plant are much bigger and contain approximately 200 seeds, whereas the peripheral capsules are smaller and contain approximately 125 seeds. The peripheral capsules are easily separated from the mother plant and it is possible that they act as a dispersal unit, which could be dispersed by wind or floods. The central capsules remain below the canopy in the central part of the shrub, covered by a hillock that forms below the shrub. These capsules may provide the long-living seeds of the local seed bank, supplying seedlings to replace the dead mother plant (Gutterman, 1990a).

Neotorularia torulosa (= *Torularia torulosa*) (*Brassicaceae*) is common in the Irano-Turanian phytogeographical region. It develops two types of pods, according to the position of the flowers in the inflorescence. On the upper part of the inflorescence, the yellow pods are less lignified and the seeds are dispersed after a light touch to the pod or by wind during the summer following seed maturation. The dark brown pods that develop from the lower flowers of the inflorescence are lignified and the seeds are dispersed only after periods of wetting by rain. The seeds from the yellow pods of the upper part germinate faster and to higher percentages than those from the lignified pods (Gutterman, 1998b).

Heteromorphism (the bearing of seeds of different sizes, shapes or colours) is found in a number of species. In some cases, two types are produced (dimorphism), in others, three (trimorphism). A case of the former is seen in *Salicornia europaea* (*Chenopodiaceae*). This species is one of the highly salt-tolerant annual halophyte pioneers that occupy dried-up saline marshes. It flowers in groups of three, with the middle flower situated above two laterals (Zohary, 1966). The single seed of the median flower is larger and heavier than the single seed produced by the lateral flowers. The large seeds germinate to about 90% after 59 days of wetting and the small seeds germinate to only 50% in the same time. It was also found that, after 6 weeks of stratification and 1 week of wetting, large seeds germinate to 74% in light and to 53% in the dark. In contrast, the small seeds germinate to 30% in light and only 16% in the dark. The recovery of seeds after 56 days in 5% NaCl, when wetted by distilled water for 42 days, was 91%

germination for the large seeds and only 16% for small seeds. The small seeds appear to be much less salt-tolerant than the large ones. Germination percentages in NaCl concentrations were higher for large seeds than for small ones (Ungar, 1979; Philipupillai and Ungar, 1984).

In *Salsola komarovii* (*Chenopodiaceae*), the fruits at the distal position have longer wings and faster after-ripening than the fruits at proximal positions (Yamaguchi *et al.*, 1990). In *Halothamnus hierochutnicus* (= *Aellenia autrani*) (*Chenopodiaceae*), fruits that mature at the distal parts of the branches have narrow wings and produce green seeds that are non-dormant. Fruits at the basal position have thick and wide wings and their seeds are yellow and dormant. Fruits producing green or yellow seeds have also been found in *Salsola volkensii* (Negbi and Tamari, 1963; Werker and Many, 1974; Baskin and Baskin, 1998).

Atriplex dimorphostegia (*Chenopodiaceae*) is an Irano-Turanian and Saharo-Arabian annual desert plant of the sandy and/or saline areas (Zohary, 1966). Two types of dispersal units are formed on the same branch: flat or humped. The humped type appears on the distal ends of the branches and the flat type below them. The flat type appears and matures earlier. When the fruit is separated from the dispersal unit, the one seed of the flat type germinates to 20% at 20°C in light, compared with only 6% for the humped type. In the dark, 68 and 38%, respectively, germinate (Koller, 1954, 1957; Koller and Negbi, 1966). Dimorphic dispersal units are also formed in *Atriplex* species (*Chenopodiaceae*), such as *A. rosea* (Kadman, 1954), *A. semibaccata semibaccata* and *A. holocarpa* (= *A. spongiosa holocarpa*) F. Mueller (= *A. spongiosa*), as well as in *A. inflata* in Australia (Beadle, 1952), in which the fruits also differ.

In *Spergularia diandra* (*Caryophyllaceae*), three different seed types are found. The capsules that develop from the first flowers that terminate the main stem contain black seeds, which have low dormancy. They are the heaviest seeds that develop on this plant. On the hairy genotype, the hairs on the seeds are straight, ending in a round knob. The flowers on lateral branches of the main stem develop capsules that contain brown seeds, which are lighter than the black seeds and have a higher percentage of dormancy. The last seeds to mature are those in the capsules of the flowers that appear when the plant is in the process of senescence. These seeds are yellow and are lighter than the black and brown seeds. They have the highest percentage of dormancy and their seed-coat hairs are 'cobra-shaped'. The yellow seeds make up most of the long-term perennial seed bank of these species (Gutterman, 1994b, 1996b; Fig. 3.2).

Hedypnois cretica Dum.-Courset (= *H. rhagadioloides*; (*Asteraceae*), which is a winter annual composite that inhabits Mediterranean and desert areas of Israel, produces three different diaspore morphs: (i) the smallest (1.03 mg) inner achenes, which have a pappus and the highest percentage (77–86%) of germination at 15°C in light; (ii) the larger (2.14 mg) outer achenes, with the lowest germination (41–46%); and (iii) the largest (4.48 mg) marginal epappose achenes, with germination of 42–51% (Kigel, 1992). In *Dimorphotheca polyptera* (*Asteraceae*), found in the Namaqualand desert of South Africa, the three types of diaspores also enable the species to spread germination in time and location (Beneke, 1991).

Effect of seed position in the inflorescence

In *Pteranthus dichotomus*, a Saharo-Arabian annual plant, the whole inflorescence is the dispersal unit. The seed position on the inflorescence influences seed germination. In good conditions, the plant produces dispersal units containing seven pseudocarps, each consisting of a one-seeded fruit, in three orders. The first order contains one pseudocarp, the second order two pseudocarps and the third order four pseudocarps, which are the terminals. In unfavourable conditions, only one or two orders develop. The one to seven seeds are protected for many years in their ligni-

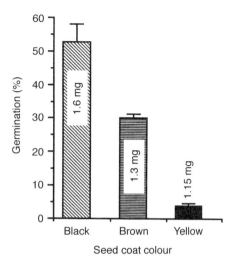

Fig. 3.2. Comparison of germination after 9 days of wetting (average % ± SE of 4 × 50) of black, brown and yellow hairy *Spergularia diandra* seeds harvested on 27 June 1989 from a natural population near Sede Boker. Seeds were wetted on 2 May 1994 at 15°C in the dark for 6 days. The average weight (mg) of 100 seeds is marked. (From Gutterman, 1994b.)

fied dispersal units and usually one seed germinates per year, depending on its position in the capsule. The terminals of each dispersal unit always germinate better than the subterminals. In dispersal units of one pseudocarp, the seed germinates as well as the terminals of order III (Evenari, 1963; Evenari *et al.*, 1982).

Aegilops geniculata (= *A. ovata*) (*Poaceae*) is common in the Mediterranean phytogeographical region. The spike, composed of two to four spikelets, is the dispersal unit (Fig. 3.3). In a dispersal unit spike with three spikelets, each of the lower spikelets contains two caryopses (grains), a_1, a_2 and b_1, b_2. The terminal spikelet contains only one caryopsis, c. Germinability, time of emergence and subsequent time to flowering all vary between spikelets. Even caryopses from the same spikelet differ in this way. Thus, when caryopses were sown under short days or long days, a_1 and b_1 caryopses gave higher germination and emerged and flowered more quickly than caryopses a_2 and b_2. Plants originating from caryopsis c had the lowest germination after the longest time

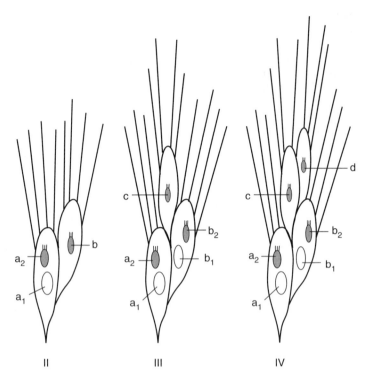

Fig. 3.3. Schematic drawing of the three types of spikes (II, III, IV) of *Aegilops geniculata*. Position and number of caryopses (a_1– d) shown in the different spikes containing II, III or IV spikelets. (Adapted from Datta *et al.*, 1970.)

and also had the longest time to flowering. The position of the caryopses in the dispersal unit has an influence on their germination and the plant development for more than one generation (Datta *et al.*, 1972a; Table 3.1).

When lettuce, *Lactuca sativa*, achenes were imbibed in leachate from hulls of *A. geniculata* spikelets a, b and c of a three-spikelet dispersal unit (25.6 g hulls 300 ml^{-1} water), an inhibitory effect by the leachate was observed. The inhibitory effect is highest in hulls from c spikelets and lowest in hulls from a spikelets (Datta *et al.*, 1970). Under dark conditions, the inhibition is much smaller. The main inhibitor was found to be monoepoxylignanolid (MEL), which inhibits *L. sativa* cv. 'Great Lakes' in incandescent light but not in the dark (Lavie *et al.*, 1974; Gutterman *et al.*, 1980). The different caryopses of a dispersal unit are thus different in size, weight, colour and germinability. There is

also a difference in the inhibitory effect of the hulls of the different spikelets of the dispersal unit. All of these components are involved in the heteroblasty of the caryopses of this plant. This ensures the spread of germination of the caryopses of each spike in time. Only one or two caryopses out of any one spike were observed to germinate in one season in the field. Similar results were observed in *Aegilops kotschyi* (Wurzburger and Koller, 1976). Differing germinability has also been found in caryopses of *Aegilops neglecta*, *Aegilops triuncialis*, *Agrostis curtisii*, *Avenula marginata* and *Pseudarrhenatherum longifolium* (all Poaceae) (Gonzalez-Rabanal *et al.*, 1994). In the spikes of all of these grasses, the lower caryopsis is larger than the upper one and less dormant.

The position of a seed in an individual bur or capitulum can influence its germinability. For example, in *Xanthium canadense* (= *X. strumarium* var. *canadense*)

Table 3.1. Position effect on average weight (mg) and germination (% ± SE) after 24 h in light at 15°C on *Aegilops ovata* caryopses harvested from plants originating from a_1, b_2 and c caryopses and grown under 18 h long days at day/night temperatures of 15/10°C and 28/22°C (from Datta et al., 1972b).

Three-spikelet (A, B and C) dispersal unit and position of caryopses (a_1–c)	Order of caryopses from which mother plant developed	Order of caryopses collected from mother plant	Average weight of caryopses (mg)		Germination (%)	
			15/10°C	28/22°C	15/10°C	28/22°C
	a_1	a_1	20.6 ± 0.7	13.9 ± 0.5	84.4	100.0
		b_2	9.5 ± 0.8	6.7 ± 0.3	10.0	60.0
		c	6.1 ± 0.6	3.0 ± 0.3	8.5	63.1
	b_2	a_1	22.9 ± 0.5	12.9 ± 0.4	55.0	100.0
		b_2	9.2 ± 0.2	7.1 ± 0.2	2.3	85.3
		c	3.8 ± 0.4	3.0 ± 0.2	0	90.0
	c	a_1	27.3 ± 1.9	14.7 ± 1.7	21.2	100.0
		b_2	13.0 ± 0.3	6.9 ± 0.8	0	76.0
		c	4.2 ± 0.7	3.5 ± 4.2	0	86.7

(*Asteraceae*), the upper seed of two in the dispersal unit germinates before the lower one (Crocker, 1906). However, in *X. strumarium*, after 12 weeks of cold stratification, the two seeds germinated together in 18% of the dispersal units or burs (Baskin and Baskin, 1998). In *Trifolium subterraneum* (*Fabaceae*), the larger seed in the bur germinates before the small one. In the three-seeded spikelet burs of *Cenchrus longispinus* (*Poaceae*), the seed in the central spikelet is largest and comes out of dormancy after dry storage much earlier than the smaller seeds of the lateral spikelets (Baskin and Baskin, 1998).

In *Asteriscus hierochunticus* (= *A. pygmaeus*) (*Asteraceae*), there are mechanisms that delay achene dispersal and spread dispersal and germination over time. The capitula are closed when dry and open when wet. During some rain events, a few of the peripheral achenes are dispersed (Fahn, 1947). Only disconnected achenes germinate (Koller and Negbi, 1966). The percentage germination of achenes from the peripheral whorls is much higher than that of achenes from the sub-peripheral whorls. Each year, some of the achenes are disconnected and dispersed by rain. The aerial seed bank of this desert annual can remain protected in the capitula of the lignified dead mother plant for many years (Gutterman and Ginott, 1994).

Where plants have hermaphroditic flowers and female-only flowers in the same inflorescence, their seeds may differ in germinability. For example, *Parietaria judaica* (= *P. diffusa*) (*Urticaceae*), a perennial herb found in shady habitats in the Mediterranean and Irano-Turanian regions, has an inflorescence of this type. The central female flower opens first and the hermaphroditic flowers 2–4 days later. Seeds from female flowers were found to have higher germinability than seeds from the hermaphroditic flowers. The same is true of their longevity: after maturation or 1–2 years of storage, the seeds from female flowers germinate earlier than those from hermaphroditic flowers, and the seeds from female flowers are mainly heterozygotic. The seedlings from these seeds are more

resistant in unpredictable conditions than the seedlings from hermaphroditic flowers. Furthermore, the dispersal units of the seeds originating from the female flowers are more hairy and can be dispersed further. This correlates with the fact that the seedlings do not compete well with the adult plants. Seeds from female flowers produce plants that grow well, far away from the adult plants, and are more resistant to water stress in comparison with the seeds from the hermaphroditic flowers. The latter are dispersed over only short distances and their seedlings grow well near the adult plant (Roiz, 1989).

In the *Apiaceae*, seeds from different positions on the umbel vary greatly in size and degree of dormancy. In three cultivars of celery, *Apium graveolens*, a difference in the achene ('seed') weight and germination was found, depending on the position of the umbels on which they matured. Achenes were collected from primary (p), secondary (s), tertiary (t) and quarternary (q) umbels (Fig. 3.4). Achenes from the primary umbels are the heaviest but have the lowest percentage germination, in comparison with achenes from the other umbels. The highest percentage germination was observed in achenes from the tertiary or quarternary umbels, depending on the cultivar (Table 3.2; Thomas *et al.*, 1979).

Position of the inflorescence on the plant (amphicarpy)

The position effect of the inflorescence on the plant is most marked in species with amphicarpic fruit. One example is *Gymnarrhena micrantha* (*Asteraceae*), a desert annual plant that has telechoric aerial achenes. The hydrochastic capitulum and pappus are opened by rain, after which the achenes are dispersed by wind (Gutterman, 1990b). The aerial achenes differ in their germination requirements from the subterranean atelechoric achenes, which germinate *in situ* from the dead mother plant (Zohary, 1937). The aerial achenes, with a well-developed pappus, are much smaller (0.37 mg) than the subter-

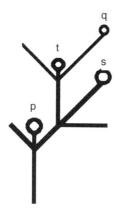

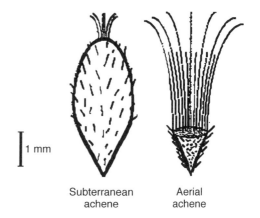

Fig. 3.4. Diagram of the structure of a celery (*Apium graveolens*) inflorescence. p, Primary; s, secondary; t, tertiary; q, quaternary umbels. (From Thomas *et al.*, 1979.)

Fig. 3.5. *Gymnarrrhena micrantha*. Comparison of aerial and subterranean achenes: a schematic drawing (from Gutterman, 1993).

ranean achenes (6.50 mg), which have an undeveloped pappus (Fig. 3.5). At 25°C in the light, final germination of the subterranean achenes was 87% in comparison with 38% for the aerial achenes. In the dark, the subterranean achenes germinated to 30% and the aerial to 4%. In the lower temperatures of 5 or 10°C, germination reached above 90% for both types of achene. However, the seedlings of the subterranean achenes were much more

drought-tolerant than seedlings from the aerial achenes (Koller and Roth, 1964). In some years with above the average amounts of rain and lower temperatures, clusters of seedlings have emerged in the Negev Desert highlands (Evenari and Gutterman, 1976; Loria and Noy-Meir, 1979/80; Evenari *et al.*, 1982; Gutterman, 1993).

Emex spinosa (*Polygonaceae*), a Mediterranean species extending into

Table 3.2. Seed position on umbel, weight (mg) and germination (%) after 21 days at 18°C in light, in three celery (*Apium graveolens*) cultivars. Least significant difference at 5% in parentheses. (From Thomas *et al.*, 1979.)

Cultivars	Umbel position	Mean seed weight (mg)	Germination (%)
'Greensnap'	Primary	0.590	51
	Secondary	0.440	85
	Tertiary	0.386	94
	Quaternary	0.382	80
		(0.069)	(9.8)
'Lathom Blanching'	Primary	0.474	50
	Secondary	0.438	72
	Tertiary	0.380	94
	Quaternary	0.348	82
		(0.069)	(9.2)
'Ely White'	Primary	0.590	59
	Secondary	0.468	62
	Tertiary	0.490	80
	Quaternary	0.520	87
		(0.086)	(7.3)

Saharo-Arabian territories (Zohary, 1966), is another amphicarpic plant. The subterranean propagules, which germinate *in situ* from the dead mother plant, are smooth and much larger (75 mg) than the aerial ones, which are spiny and range from 2 to 24 mg, depending on position. The aerial propagules are dispersed by wind, floods or animals. In this case, the germinability of the aerial propagules is much higher than that of the subterranean ones in all conditions tested in populations that inhabit the Negev Desert (10 versus 0% at 15°C). The leachate of the aerial fruit contains germination inhibitors not found in the subterranean ones. When seeds were transferred for 8 h daily from 15 to 30°C in the dark, the aerial propagules germinated to a much higher percentage 7 days after wetting than the subterranean ones (60 versus 20%) (Evenari *et al.*, 1977). The lower germination of the subterranean atelechoric propagules and their low numbers may be important for preventing competition and ensuring dispersal of germination in time. Usually, only one of the subterranean propagules germinates in one season. The inhibitors that are in the aerial telechoric propagules could have an influence on the amount of rain or washing by floods that is needed before these propagules germinate. These germination inhibitors could act as a rain-gauge or rain clock, ensuring that the germination will take place only after sufficient rainfall has occurred for the establishment of the seedlings. This, in addition to the better germination in the dark and the long time (7 days) of wetting needed for germination, may give the buried propagules a better chance to germinate in more favourable microhabitats, such as depressions and porcupine diggings (Gutterman *et al.*, 1990).

Position effects in the following generation

The position effect can be detected even in the following generation and possibly beyond. An experiment that illustrates this is one in which *Aegilops geniculata* plants were grown from a_1, b_2 and c caryopses and the germination of their grains compared. Germinability of the second-generation grains was markedly influenced by the order of the caryopses from which the mother plant was originally derived. Under cool temperatures (15/10°C), similar to temperatures existing during the growing season of the plant in the natural habitat, large differences were found in germinability between grains from parents derived from caryopses of different orders. However, these were not found when the plants were grown at higher temperatures (28/22°C). It was found that the origin of the mother plant also has an influence on the weight of the different caryopses. It is interesting to note that, in this case at least, the position effect has a very strong influence on the next generation and that its expression is dependent on the environmental conditions under which the second-generation plants are grown (Table 3.1; Datta *et al.*, 1972b).

Age effects

The age of the mother plant can affect the germination of its seeds. For example, in *Amaranthus retroflexus* L. (*Amaranthaceae*), a widespread weed of summer crops (Zohary, 1966), germinability declines with the age of the parent plant at the time of flower induction (Kigel *et al.*, 1979). In *Oldenlandia corymbosa* (*Rubiaceae*), less dormant seeds develop on younger plants in comparison with those from older plants (Do Cao *et al.*, 1978). When these plants were grown under 16 h day lengths, the seeds that matured in July germinated to 80–90%. However, seeds that had matured between August and October germinated to only 1–15% (Attims, 1972). Senescence of the mother plant can affect seed germination too. Seeds of *Trigonella arabica* and *Ononis sicula* (Gutterman 1980/81a) that mature under long days, when the plant has started to dry out at the end of the season, have incomplete seed-coats, which are green or brown. The imbibition and germination of these seeds is much faster than in the typical yellow seeds matured under long days in younger plants (see section below) (Gutterman, 1993).

In *Spergularia diandra*, the yellow seeds that develop on the plants at the senescence stage have the highest dormancy and are the lightest in weight, in comparison with the black or brown seeds that develop on these plants earlier (see section above) (Fig. 3.2; Gutterman, 1994b, 1996b). *S. marina* seeds matured in July germinated in the greenhouse to 4–8% and seeds that matured in August germinated to 80–82% (Okusanya and Ungar, 1983). However, in cases where seeds are collected from plants grown under natural conditions, it is difficult to distinguish between true age effects and the effect of the changed seasonal environment under which the later seeds develop.

Environmental effects

Effect of maturation under natural conditions

L. serriola (*Asteraceae*), a widespread annual (or biennial), is a long-day plant for flowering (Gutterman *et al.*, 1975). Ripe achenes, which were collected each month separately from July to October 1989 from a natural plant population near Sede Boker, differed in germinability when tested in October and December 1989 and January, March and May 1990. A difference was observed in each month of harvest as well as the period of storage (Fig. 3.6). The plants started to produce seeds from the first capitula in July, when the photoperiodic day length is the longest (15 h), and terminated in October, when the day length is much shorter (12 h), when the majority of the leaves are in senescence (Gutterman, 1992a).

In biseasonal-flowering perennial shrubs, such as *Cheiridopsis aurea* (*Aizoaceae*) from South Africa, which were grown in the Negev Desert highlands, large differences in seed germinability according to the season of seed maturation were found in natural day length and temperatures. Seeds matured in winter, when temperatures are mild and days are short (the shortest being *c.* 11 h on 21 December), or in summer, when temperatures are high and days are long (the longest being *c.* 15 h on 21 June), seem to differ in germination when this occurs in the following growing season (Gutterman, 1991).

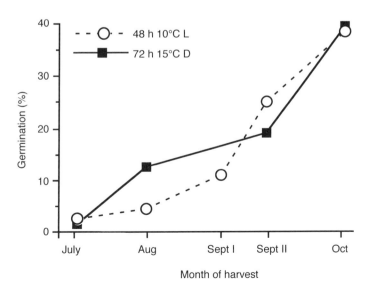

Fig. 3.6. Germination (%) of *Lactuca serriola* achenes after wetting for 48 h in light or 72 h in dark. Achenes harvested from July to October 1989, and tested on 21 January 1990 (after Gutterman, 1992a).

The germinability of *Chenopodium bonus-henricus* seeds that were collected from plants of natural populations from altitudes of 600 m differed markedly from that of seeds collected from plants from altitudes of 2600 m. The higher the altitude, the thicker the seed-coat, the higher the polyphenol content in the seed and the lower the germination. Plants transferred from one altitude to another matured seeds that were typical of the altitude to which they were transplanted. The polyphenols that accumulated in the thicker seed-coats of seeds matured at the higher altitude inhibit germination. It is possible that increase in the polyphenol content is due to increased visible radiation and the lower temperature at the higher elevation (Dorne, 1981).

Day length during seed development

Long- or short-day effect

Day length and other environmental factors may contribute to the phenotypic plasticity and diversity of seed germination in many plant species. The germinability of seeds of many species is affected by day length during seed development and maturation. In some plant species, short days result in higher germinability. For example, when seeds of *Chenopodium album* (*Chenopodiaceae*), a pluriregional plant and a common weed in irrigated crops (Zohary, 1966; Holm *et al.*, 1977), were matured under 8 h days from flower-bud formation, the germination was higher in comparison with seeds matured under 18 h days, when tested either in the light or in the dark. Alternating diurnal temperatures (22/12°C) during seed development also resulted in higher germination than in seeds developed under constant temperatures (22/22°C) (Table 3.3; Karssen, 1970).

In *Ononis sicula* Guss. (*Fabaceae*), an annual plant in the southern Mediterranean and Near East regions (Zohary, 1972), treatments of the mother plants during seed maturation affect the germinability of the seeds, through changes in the development of the seed-coat and its surface structure.

Table 3.3. The effect on seed germinability (% ± SE) of *Chenopodium album* of a daily temperature shift (22/12°C) or a continuous temperature (22°C) during the growth of the mother plant under 18 h (LL) or transferred from 18 h to 8 h days at flower-bud formation (LS). Germination was tested in continuous light or continuous dark at 23°C. (Adapted from Karssen, 1970.)

Temperature (°C)	Photoperiodic conditions	Light	Dark
22/12	LS	100	96
	LL	90	42
22/22	LS	85	61
	LL	71	22

Certain day-length treatments modify the seed-coat permeability to water, seed-coat resistance to fungus and seed longevity. Under long days (14.5 to 20 h), yellow seeds were produced with well-developed seed-coats. Under short days (8 to 11 h), there developed brown seeds with undeveloped seed-coats and/or green seeds with intermediate seed-coat structure and water permeability and higher germinability than the long-day seeds (Gutterman and Evenari, 1972; Gutterman and Heydecker, 1973; Gutterman, 1973, 1993).

In *Trigonella arabica* (*Fabaceae*), an annual desert plant with a Saharo-Arabian geographical distribution, yellow seeds or yellow seeds with green spots are developed during maturation under long days and green seeds or brown seeds are developed during maturation under short days. The seed-coat structure is well developed in the less permeable yellow seeds, in comparison with the less-developed seed-coat of the brown seeds, which also have the most permeable seed-coats (as in *O. sicula*) (Gutterman, 1978c). These day-length influences on the seed-coat are dependent on the last 8 days of maturation, when the fruit has reached its final size but is still green. At this stage it was shown that when seeds mature under 8 h days on plants in which half the fruits are covered with aluminium foil, seeds from both covered and uncovered fruits germinated to 100% after 3 weeks of wetting. In contrast, the seeds matured on plants under 15 h days, from both covered and uncovered fruit, swelled only 29 and 10%, respectively. This indicates that the effect of the different day lengths is not due to action directly on the developing seeds, but is transmitted to the seeds from the vegetative parts of the mother plant (Gutterman, 1998a).

Seeds of *O. sicula* or *T. arabica* that mature on a young plant under long days have a well-developed seed-coat and are referred to as 'hard seeds' (impermeable to water). These are possibly the long-term seeds of the seed bank in the soil. The seeds that mature on the same mother plants during the short days have seed-coats that are more water permeable and are possibly the short-term seeds of the seed bank (Gutterman and Evenari, 1972; Gutterman, 1973, 1978c, 1993).

Quantitative short-day effect

Portulaca oleracea (*Portulacaceae*) is an annual plant of the warm-temperate regions of the world (Zohary, 1966) and is listed as one of the eight most common weeds on earth (Holm *et al.*, 1977). Seeds matured on plants grown under 8 h days germinate to a higher percentage than seeds from plants grown under 16 h days. This day-length effect is a quantitative short-day effect, and the critical time is the last 8 days of seed maturation. When plants were transferred from 16 h days to either 13 h or 8 h days during the last 8 days of maturation, the seeds increased their germinability. The shorter the day length, the higher the germination (Fig. 3.7; Gutterman, 1974). This also applied in the case of *Chenopodium polyspermum*, an annual plant of wasteland and cultivated ground. The shorter the day length (from 24 h to 10 h), the higher the germination and the heavier the seeds (Table 3.4). The seed-coat thickness of seeds matured under 8 h days is 20 μm, in comparison with 46 μm of seeds matured under 24 h days. Scarification of the seed-coat brings the seeds to 100% germination (Jacques, 1968). This was also seen in *C. album* (Karssen, 1970).

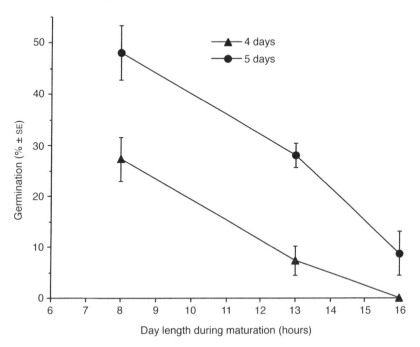

Fig. 3.7. The influence of the change of day length during the last 8 days of maturation of *Portulaca oleracea* L. seeds, from 16 h to either 13h or 8 h days on germination at 40°C in the dark with 5 min of white light, once in 24 h. The first illumination was given 1.5 h after wetting (four batches of 50 seeds each per treatment). Germination (% ± SE) after 4 and 5 days. (From Gutterman, 1974.)

Table 3.4. The influences of 10, 12, 14 and 24 h day-length treatments of *Chenopodium polyspermum* plants during seed maturation on seed weight and seed germinability (from Jacques, 1968).

	Day length (h)			
	10	12	14	24
Weight (mg) of 100 seeds	63	54	45	23
% Germination	80	26	6	0

Day length also affects seed germination in plants with soft fruit. The tomato, *Lycopersicum esculentum* (*Solanaceae*), is a natural day-length plant for flowering but the day length has been found to affect the seed germination. Under 6 h days in over-ripe fruit, 38% of seeds were found to germinate in the fruit in comparison with only 0.2% that germinated when the plant was under a day length of 13 h. None of the seeds were found to germinate in fruit of plants under 20 h days. Tomato seeds harvested from plants that were under 6 h days germinated to a much higher percentage in comparison with seeds from plants grown under 20 h days. The differences in germination were found to be even greater when the fruit was covered with aluminium foil. Almost the same results were seen as in *T. arabica* (Gutterman, 1978c, 1998a). The day length during fruit maturation also has an influence on the inhibitory effect of the juice of the tomato fruit. When lettuce achenes were imbibed in half-strength tomato juice from fruit matured under 6 h days, the percentage of germination of the lettuce achenes was much higher than that in juice from fruit matured under 20 h days (58 versus 8%, respectively). Moreover, these soft fruits respond to day length when they are separated from the mother plant and there is an effect of the day length on the germinability of the seeds. In this situation, the isolated fruit responds in the same way as it does on the mother plant. In postharvested tomato fruit that were exposed to 6, 9 or 17 h of day length during fruit ripening and seed maturation, the ethylene amounts released were found to be affected quantitavely by the day length. The longer the day length, the higher the quantities of ethylene that were released from the fruit, even after a single photoperiod. This phenomenon lasted for a number of days until the fruits were fully ripe. The application of ethephon (which releases ethylene) to the tomato fruits during storage under the different photoperiods increased the effect of day length on ethylene release. These treatments also have an effect on the germinability of the seeds harvested from the fruits treated with ethephon under different day-length regimes (Gutterman, 1978b).

In cucumbers, *Cucumis sativus* (*Cucurbitaceae*), postharvested fruit stored for 15 days under different photoperiodic regimes were found to have an influence on the germination of their seeds even after the seeds had been kept for 270 days in dry storage. The seeds were wetted at 20°C in dark and the germination, after 24 h of wetting, was 84% for seeds from postharvested ripe fruit kept at 8 h days and only 12.5% for seeds from fruit held under long days. Injection of ethephon into the soft fruit of cucumbers affects the germination response to day-length treatments in the opposite direction: only 2% for seeds from fruit kept at 8 h days and 79% from fruit held under long days (Gutterman and Porath, 1975; Gutterman, 1978b). It is well known from other plant species that hormone application to plants during maturation has an influence on seed germinability (Zeevaart, 1966; Jackson, 1968; Ingram and Browning, 1979; Baskin and Baskin, 1998). Specific examples are *Avena fatua* (Black and Naylor, 1959), *Lactuca scariola* (Gutterman *et al.*, 1975), *Phaseolus vulgaris* (Felippe and Dale, 1968) and *Salsola komarovii* (Takeno and Yamaguchi, 1991).

Quantitative long-day effect

In contrast to all of the above cases, there are some species in which short days result in lower germinability. *Cucumis prophetarum* is a perennial desert plant distributed in the East Saharo-Arabian phytogeographical region. Seeds from postharvested soft fruit of *C. prophetarum* that had been stored under day lengths of 8, 11, 13, 15 and 18 h were tested for germination. After 7 days of wetting at 20°C in light, the longer the storage day length of the fruit from which the seeds had originated, from day length of 8 to 15 h, the higher their germination percentages (Gutterman, 1992b).

Polypogon monspeliensis (*Poaceae*) plants were grown under six different day lengths, outdoors as well as in the greenhouse, during vegetative growth and seed maturation. In seeds harvested both outdoors and from greenhouse conditions, the longer the day length, the higher the percentage of germination at 25°C in continuous light (Table 3.5; Gutterman, 1982a). Similarly, after 7 days of wetting at 10°C in the dark, *Schismus arabicus* (*Poaceae*) 'seeds' that had developed and matured under constant day lengths of 8, 12 or 18 h showed a facultative long-day response in natural outdoor conditions at Sede Boker. This response was less pronounced in greenhouse conditions (Gutterman, 1996a). *S. diandra* black hairy seeds that developed and matured under day lengths of 8, 12 or 18 h also showed a quantitative long-day response for germination after 47 days of wetting (Gutterman, 1994b).

Carrichtera annua (*Brassicaceae*), an annual desert plant from the Saharo-Arabian geographical region (Zohary, 1966) has a day-length-independent response for flowering (Evenari and Gutterman, 1966). However, seeds matured on plants under 20 h days have much better germinability

(51.5%) in comparison with seeds from 8 h days (7% germination) when germinated at 25°C in light (Gutterman, 1973). Potted plants of this species were transferred from natural day length of 13 h to either 8 h or 20 h days, after the green fruit that had achieved the final size were covered with aluminium foil. The seeds that matured in the covered fruit at the three day lengths, 8, 13 and 20 h, were compared for their germination at 25°C in the light 144 h after wetting. Again, the longer the day length, the higher the percentage germination (Gutterman, 1978b). From these results, it is possible to speculate that *C. annua* seeds matured under long days in the late spring are the seeds that will germinate first. The seeds that are matured under short days in the early winter are the seeds that will remain in the seed bank for an extended period. Since the fruit was covered, the photoperiodic effect on seeds must be mediated via the leaves and green stems of the mother plants, as was shown in *T. arabica* (Gutterman, 1978c) and *L. esculentum* (Gutterman 1978a).

The critical time of day length during seed maturation

In *O. sicula*, *T. arabica* (Gutterman and Evenari, 1972; Gutterman, 1978c), *Lactuca sativa* (Table 3.6; Gutterman, 1973), *Portulaca oleracea* (Fig. 3.7; Gutterman, 1974) and *C. annua* (Gutterman, 1978b,

Table 3.5. The influence of different day lengths during growth and seed maturation, under greenhouse or outdoor conditions, on seed germinability of *Polypogon monspeliensis*. The photoperiodic treatments started from 20–24 December 1979 until the harvesting of the seeds in March 1980. The seeds were wetted at 25°C in light in October 1980 and the results were observed after 3 days of wetting. (From Gutterman, 1982a.)

Day length during growth of mother plant and seed maturation (h)	Seeds from greenhouse plants: germination (%)	Seeds from outdoor plants: germination (%)
9.0	0.0	1.5
11.0	0.0	9.5
12.0	17.5	64.0
13.5	38.0	98.5
15.0	60.5	97.0
18.0	90.5	98.0
Control (natural day length)	93.0	91.5

Table 3.6. Photoperiodic treatments given to the plants of *Lactuca sativa* 'Grand Rapids 517' and the germinability of their seeds (% ± SE), at 26°C in the dark with short (5 min) illumination of white light. There were 200 (4 × 50) seeds in each treatment. The germination experiment began immediately after harvest. (From Gutterman, 1973.)

Photoperiodic conditions during growth of mother plants	Germination (%) after 2 days	Germination (%) after 11 days
SD, then LD (80)[a]	0	4 ± 1.4
LD, then SD (80)	29.5 ± 2.6	32 ± 3.2
SD, then LD (12)	5 ± 1.2	8 ± 1.4
LD, then SD (12)	13.5 ± 3.4	18 ± 3.4
Continuous LD	5 ± 0.6	6 ± 0.8
Continuous SD	16.5 ± 2.2	24 ± 2.5

[a] Number of days under the photoperiodic conditions before harvest.
SD, short day, 8 h; LD, long day, 16 h.

1982a), the critical time for the day-length effects is during the last stages of seed maturation: that is, from the time when the fruit reaches its final size but is still green, to full maturation 7–14 days later (Gutterman, 1978c). In soft fruit, such as cucumbers and tomatoes, 5–15 days are also sufficient for the day-length effect (Gutterman and Porath, 1975; Gutterman, 1978a). The importance of the final period of seed maturation in determining germinability is seen in lettuce. *L. sativa* plants were transferred from long days to short days or vice versa 12 days before the harvest of the achenes. Controls were maintained at either constant long or constant short days. Seeds from plants grown under short or long days in the last 12 days of ripening (having been transferred from the other day length) behaved like those grown under continuous short or long days, respectively (Table 3.6). In *Chenopodium album*, the transfer from long to short days at flower-bud formation increased germination in comparison with that obtained from seeds of plants kept under continuous long days (Karssen, 1970). In *C. polyspermum*, the 8 days after flower-bud formation is the critical time (Pourrat and Jacques, 1975).

Influences of light quality during maturation on seed germination

Light quality during seed maturation can influence germinability. For example, in *Arabidopsis thaliana* (*Brassicaceae*), seeds matured under white light with a high red/far-red (R/FR) ratio have a higher dark germination than seeds from plants grown under light with a low R/FR ratio (McCullough and Shropshire, 1970). Seeds matured under incandescent light (high R-absorbing phytochrome (P_r)) require light for germination but not seeds matured under fluorescent light (high in FR-absorbing phytochrome (P_{fr})). Immature seeds are sensitive to light quality up to 1 day before full seed maturation (Hayes and Klein, 1974).

The quality of the light received at particular times in the 24 h cycle is crucial in determining seed germinability – for example, in *Portulaca oleracea*, which has a quantitative response to day length during maturation affecting germination (Fig. 3.7; Gutterman, 1974).

It seems that the light treatments during *P. oleracea* plant development and seed maturation have different influences on the number of leaves at the time of the appearance of the first flower-bud and the germination percentages of the seeds harvested from these plants. These different responses of flowering and seed germination are even more pronounced after treatment of 8 h R or FR light given before or after the 8 h dark (D) period, in comparison with 8 h of white light following the 8 h of daylight. The 8 h R or FR following the 8 h of daylight has a 'short-day effect' on the number of leaves at flower-bud appearance and the 8 h R or FR following the 8 h D has a pronounced 'long-day effect' on the number of leaves at flower-bud appearance. But all of these four treatments have a 'short-day effect' as far as the germination percentages are concerned. This is different from the treatment of 8 h white light, which has a 'long-day effect' for both flowering and germination (Gutterman, 1974). It is interesting to note that the photoperiodic germination response was found in other plant species: *Carrichtera annua*, *Lycopersicum esculentum* and *Cucumis sativus*. In these species there is no photoperiodic effect on flowering but there is an influence on seed germinability. From all the above, it would seem that the mechanisms affecting flowering are different from the mechanisms affecting seed germinability, at least in the plant species mentioned.

If mature and turgid fruit of *Cucumis prophetarum* and *C. sativus* are stored in continuous FR light, the dark germination of their seeds will be reduced because most of their photoreversible phytochrome has been converted to the P_r form. At this stage, the seeds require light to germinate. If such fruit are kept under R light, germination is very high, due to the conversion of the photoreversible phytochrome to the P_{fr} form. The total photoreversible phytochrome has been shown to be much higher in seeds after the storage of the fruit under FR light in comparison with seeds that were separated from fruit stored under R light (Table 3.7). However, after the seeds were exposed to a period of dry storage, these differences in the dark germination completely disappeared (Gutterman and

Porath, 1975). In *Chenopodium album*, the R light effect during seed maturation disappears 4 months after maturation (Karssen, 1970).

Postharvested turgid fruit of *Cucumis prophetarum* were stored in the laboratory for 9 days under continuous R, FR or dark conditions. The seeds were separated and germinated at 20°C in continuous light. Seeds originating from fruits stored in FR light, which contained higher amounts of photoreversible phytochrome (Table 3.7), germinated to high percentages 7 days after wetting, in comparison with seeds from fruit stored under R light, which contained lower amounts of photoreversible phytochrome. The seeds from fruit stored in the dark, which contained high photoreversible phytochrome, reached the highest germination percentages (Gutterman, 1992b, 1993).

Leaf canopies have been found to inhibit germination of matured light-sensitive seeds (Black, 1969; Vander Veen, 1970; Gorski, 1975; King, 1975; Fenner 1980a, b). Only 1 h under leaf-transmitted light is required to inhibit germination in the dark of detached *Bidens pilosa* (*Asteraceae*) seeds (Fenner, 1980b). During seed maturation on the mother plant in different plant species, there is a relationship between the chlorophyll concentrations that surround the developing seeds during the different stages of seed maturation and dehydration. In seeds matured entirely surrounded by green maternal tissues, most of their phytochrome will be arrested in the inactive P_r

Table 3.7. Amounts of photoreversible phytochrome, the state of phytochrome in the seeds in the post-harvested fruit and germination (% ± SE) (at 25°C) of *Cucumis prophetarum* (after 50 h) and *C. sativus* seeds (after 170 h), influenced by fruit storage under different light conditions (at 23–25°C) (adapted from Gutterman and Porath, 1975).

Species	Light conditions to the harvested fruits	Photoreversible phytochrome in the seeds (Δ OD) $\times 10^{-4}$	Far-red-absorbing phytochrome in seeds (% P_{fr})	Seed germination (%) in the dark
C. prophetarum	Red	8.6 ± 0.4	93.5	100.0
	Far red	20.0 ± 0.7	0.0	0.0
C. sativus	Red	24.0 ± 2.1	53.3	91.9 ± 4.3
	Far red	35.1 ± 2.0	0.0	27.5 ± 4.3
	Dark	36.4 ± 1.8	8.5	22.5 ± 2.5
	Sunlight	28.8 ± 2.0	47.2	100.0

OD, optical density.

form and therefore these seeds will require a light stimulus for germination in the dark (Cresswell and Grime, 1981).

Water stress during maturation affecting seed germination

Desiccation during maturation enhances germinability. For example, the green premature seeds of *Hirschfeldia incana* (*Brassicaceae*) did not germinate when wetted 4–6 weeks after anthesis (WAA). However, when seeds were taken off the parent plant 4–6 WAA and dried for 2 weeks at room temperature, they remained green but germinated to 91% when wetted in the light at 26°C (Evenari, 1965). Similarly, immature developing seeds of soybean (*Glycine max*) will not germinate when wetted unless they are previously desiccated (Adams *et al.*, 1983). However, in soybean plants exposed to drought stress levels during seed fill, the greater the number of stress days, the lower the standard germination percentage (Dornbos *et al.*, 1989). Drought during seed maturation may affect seed germinability by changing the properties of the maternal tissue surrounding the seed. Benech Arnold *et al.* (1992) showed that the increased germination of seeds of *Sorghum halepense* subjected to drought during maturation was due to a modification of the glumes rather than of the caryopses themselves.

Immature developing seeds of castor bean (*Ricinus communis*) and *Phaseolus vulgaris* that were removed from the capsules and wetted did not germinate. When such seeds were removed and stored at relatively high humidity, their water content slightly declined and they germinated when immersed in water (Bewley *et al.*, 1989). Water stress during stages of seed maturation may cause it to switch from the seed-developing system to the seed-germinating system (Kermode *et al.*, 1986). This switch involves changes in protein patterns (Lalonde and Bewley, 1986) and messenger RNA (mRNA) (Bewley *et al.*, 1989). Desiccation also induces changes in mRNA population within the endosperm of *R.*

communis (*Euphorbiaceae*) (Kermode *et al.*, 1989).

Out of 13 plant species examined in the literature by Baskin and Baskin (1998), water stress during seed development decreased dormancy in seven species and increased it in six. There seems to be no consistent pattern distinguishing species with physiological dormancy from those with physical dormancy.

Temperatures during maturation affecting seed germination

In different plant species, sometimes even small differences in temperature during plant development and seed maturation can have an influence on the germinability of the seeds produced by these plants. For example, germination of *Amaranthus retroflexus* seeds was higher when matured under temperatures of 27/22°C than under 22/17°C. Similarly, in *Aegilops ovata* (*A. geniculata*), maturation temperatures of 28/22°C produced caryopses that were more germinative than those grown at 15/10°C (Datta *et al.*, 1972a; Table 3.1). In both these cases, the higher temperature resulted in seeds with higher germinability. The seeds produced under warmer conditions were lighter than those which developed at lower temperatures.

In some cases, there is an inverse relationship between maturation temperature and germinability. For example, in soybeans, Keigley and Mullen (1986) found that the more accumulated days of high temperatures (32/28°C) after flowering, the lower the germination in comparison with maturation and seed fill under temperatures of 27/22°C. The germination of *Chenopodium album* seed was lower after maturation at 22/22°C than at 22/12°C (Karssen, 1970). The question of the effect of temperatures during seed development on their subsequent germination needs more detailed study. In this connection, Plett and Larter (1986) show the importance of testing germination over a range of temperatures when investigating the effect of maturation temperature. The maturation temperature resulting in the highest germination depended on the germination temperature at which the seeds were tested.

Baskin and Baskin (1998) have summarized the responses when mother plants produce seeds under different temperatures. They found that an increase in temperatures during the time of seed development has a preconditioning effect on at least four plant species to decrease seed dormancy. Under controlled temperatures, many plant species that grow under higher temperatures produce seeds with higher germinability. Fenner (1991) lists 15 cases where high temperatures during maturation result in lower dormancy. However, three plant species, grown in higher controlled temperatures produce seeds with increased dormancy. In still other plant species, the time of exposure to varying temperatures during seed development has an important effect on seed germination requirements.

Mineral nutrition

Baskin and Baskin (1998) have summarized 24 plant species in which different mineral nutrients affect seed germination as preconditioning during seed development. As a general rule, the addition of nutrient fertilizers (notably nitrogen) to parent plants decreases dormancy in the seeds. Fenner (1991) cites a number of such cases. The physiological mechanism involved is unknown.

Conclusions

From the examples given in this chapter, it seems that, in various plant species, seeds with different germinability develop on the same mother plant and on plants of the same species growing in different environments; maternal position and environmental factors cause these differences by their influence on plant development and seed maturation. At least in some species, it was shown that the last 5–15 days of seed maturation is the critical time.

The genotypic influences of a plant species ensure that the adaptation of the plant to its habitat conditions is such that germination is likely to occur in the right season and in the right place. The phenotypic influences, including environmental and maternal effects, during maturation on seed germination ensure that, even under optimal conditions, only a portion of the seed population will germinate in one rain event or in one season. It was observed in the Negev Desert highlands that only one, or at most two, seedlings appear in one season from a dispersal unit of the desert plant *Pteranthus dichotomus* containing seven pseudocarps (seeds). The same phenomenon was observed in the Mediterranean plant *Aegilops geniculata* (= *A. ovata*) in its natural habitat. Out of five or six caryopses of a dispersal unit, only one or two seedlings appear in one season. In *Ononis sicula* and *Trigonella arabica*, the brown seeds with the undeveloped seed-coats will swell and some of these will germinate in the following season, along with a small portion of the green seeds. The yellow seeds with the well-developed seed-coats will germinate much later.

The importance of this heteroblasty for the survival of species by dispersing the germination in time is obvious. The main question is whether there is a general biochemical pathway at the relevant stage of maturation which is affected by the maternal position and/or environmental factors resulting in differences in seed germinability, or, in the different plant species, are there different biochemical pathways that are affected by different maternal position and environmental factors? Whether the first or the second possibility is the correct one, the biochemical events that are involved in these phenomena are still not known. At this stage, we can only speculate, as has already been done in the past (Gutterman, 1980/81b, 1982b, 1985, 1993), that it is possible that during seed maturation the different factors affect the accumulation of different relative amounts of materials that are involved later on in the germinability of the seeds, which could react through three main pathways:

1. They could lead to the development of seed-coats with different degrees of impermeability, according to the day length and age effects, as was observed in some species of the *Fabaceae*: *O. sicula* (Gutterman, 1973) and *T. arabica* (Gutterman, 1978c).

2. These materials may result in an accumulation of germination inhibitors in the fruit, as was found in tomato, and/or in the embryos and hulls, as was found in *A. geniculata* (Datta *et al.*, 1972a, b). There could be an accumulation of germination inhibitors, such as polyphenols, in the seed-coats that increase with the higher altitude, as was found in *Chenopodium bonus-henricus* (Dorne, 1981). Is there an accumulation of different materials in the embryo, according to the day length during maturation, that are involved later on in the germination process and result in seed germinability, as was found in *Carrichtera annua*? Such materials could include germination inhibitors.

3. Are part of these materials hormones, such as ethylene? In tomatoes different rates of ethylene are released from maturing fruits, depending upon the length of even one night (Gutterman, 1978a, 1982a). In *Cucumis sativus*, an additional amount of ethylene changes the levels of germination of seeds matured under short days (SD) or long days (LD) to the opposite direction (Gutterman and Porath, 1975). It is possible that, at different elevations, enzymes are activated to produce polyphenols, such as those that accumulate in seed-coats of *Chenopodium bonus-henricus*, and inhibit the germination (Dorne, 1981). Enzymes and other materials that accumulate in the embryo in different relative amounts, according to the environmental conditions during seed maturation, possibly affect the germinability.

In some species of the *Papilionaceae*, the day length affects the water permeability of the seed-coat according to the day length during the last stage of seed maturation. It was also shown that the 'day-length effects' are transferred from the leaves and affect the development of the seed-coat in covered fruit. So far, the biochemistry of this process and the material or materials that are transferred from the leaves to the seeds and affect the process that leads to the degree of seed-coat development have not yet been identified.

Another very interesting point is the fact that, in some plant species, the effect of the environmental factors, such as day length, on flowering differs from the effect of the same treatment on seed germination. Therefore, it is possible that the regulation of flowering and the regulation of seed germination involve two different biochemical pathways. For instance, in *O. sicula* and *T. arabica*, the LD treatment accelerates flowering but the SD treatment increases the seed germinability. In some neutral-day plants for flowering, the germinability of the seeds is accelerated by SD, as in *Lycopersicum esculentum* and *Cucumis sativus*, or by LD, as in *Carrichtera annua*.

Are the observed responses by seeds to maternal environmental conditions necessarily adaptive or merely random in their action? At least in some of the plant species inhabiting the most extreme and unpredictable desert conditions, the heteroblasty contributes very strongly to the survival of the species. Heteroblasty prevents catastrophes of mass germination after a relatively heavy rain, followed by a long dry period, which would cause all the seedlings to die out. In some other plant species inhabiting deserts, there are also other mechanisms that regulate the dispersal and germination in time (Gutterman, 1993, 1994a, 1998a). In many plant species found in deserts, there are serotinous aerial seed banks (Baskin and Baskin, 1998; van Rheede van Oudtshoorn and van Rooyen, 1999). Another such mechanism is the seed dispersal by rain, as studied in *Blepharis* spp. (Gutterman *et al.*, 1967). The number of seeds that are released is regulated in such a way that in one rain event only a part of the seed bank stored on the dead mother plant will be released and germinate. A similar phenomenon is seen in *Asteriscus pygmaeus* (Fahn, 1947; Koller and Negbi, 1966; Gutterman and Ginott, 1994).

There are a number of major areas of research that need to be carried out on the effects of maternal position and environmental conditions during seed development on germination and dormancy, such as the following:

- Which material or materials are transferred from the leaves to the seeds and affect the germinability, and what are the physiological processes involved?
- What are the physiological processes involved in the accumulation of different materials by the seed during development, including possibly germination accelerators or inhibitors? For example, are there interrelations between the aerial and subterranean capitula, which affect the germinability of the seeds in the amphicarpic plants *Emex spinosa* (Evenari *et al.*, 1977) and *Gymnarrhena micrantha*? It is interesting to note that, under SD in the greenhouse or outdoors, *G. micrantha* plants developed both aerial and subterranean capitula, but under LD in the greenhouse only the aerials developed. Under LD outdoors, only the subterraneans developed (Evenari and Gutterman, 1966). Are there also some interactions between the individual seeds within dispersal units, e.g. in *Aegilops geniculata* and *Pteranthus dichotomus*? In spikelets a and b of *A. geniculata*, the terminal caryopses, a_2 and b_2, are inhibited in their germination in comparison with the lower caryopses, a_1 and b_1 (Datta *et al.*, 1970, 1972a, b). In *P. dichotomus*, the opposite was found: the terminals germinate the best and the lowest seeds germinate the poorest. The upper order changes the germinability of the seeds of the lower order (Evenari, 1963). Similarly, in the capsule of *Mesembryanthemum nodiflorum*, the terminal seeds germinate the best and the seeds from the lowest group germinate the poorest (Fig. 3.1; Gutterman, 1980/81a, 1993, 1994a).
- What are the materials involved in the different relative levels of germination of *Portulaca oleracea* (Gutterman, 1974), *Cheiridopsis* spp., *Juttadinteria* (Gutterman, 1990a) and *Lactuca serriola* seeds (Gutterman, 1992a) under different

temperatures during wetting? A detailed kinetic study is necessary for a better understanding of these changes in the levels of germination. For example, *P. oleracea* seeds matured under different environmental conditions and transferred during the germination process from low to high temperatures germinated to different percentages at different temperatures. In each temperature, the seed population reaches another level of germination and the relative amounts of germination of the different treatments differed from one temperature to another. The relative effect of parental light-quality treatments on germination depends on the temperature at which the seeds are tested (Gutterman, 1974).

- In some plant species the photoreversible phytochrome is arrested in the P_r form, which causes the seeds to require light for germination. Does this phenomenon depend on the chlorophyll content of the maternal tissue surrounding the developing seeds during dehydration (Cresswell and Grime, 1981)? In which other plants do the seeds that mature under red light and have a high percentage of P_{fr} also require light for germination after a short time of storage (Karssen, 1970; Gutterman and Porath, 1975)?

There are about 16 plant species in which it was found that the seeds that mature early in the growing season are heavier than seeds that mature late (Baskin and Baskin, 1998). The opposite was found in *Atriplex heterosperma* (Frankton and Bassett, 1968). Seed size is one of the results of the position effect on the mother plant during maturation.

As summarized in great detail by Baskin and Baskin (1998), there are 79 species in which seed size (and in many cases also germination) has been observed to be affected by maternal position and environmental factors to which the mother plant has been exposed during seed maturation. There is no doubt, that in most species, individual plants produce seeds that vary phenotypically in their germinability. The ecological effect is presumed to be catastrophe avoidance by spreading the risk of mass mortality among seedlings.

References

Adams, C.A., Fjerstad, M.C. and Rinne, R.W. (1983) Characteristics of soybean seed maturation: necessity for slow dehydration. *Crop Science* 23, 265–267.

Attims, Y. (1972) Influence de l'âge physiologique de la plante mère sur la dormance des graines d'*Oldenlandia corymbosa* L. (Rubiacées). *Comptes Rendus de l'Académie des Sciences Paris, Série D* 275, 1613–1616.

Baskin, C.C. and Baskin, J.M. (1998) *Seeds – Ecology, Biogeography, and Evolution of Dormancy and Germination.* Academic Press, San Diego, 666 pp.

Beadle, N.C.W. (1952) Studies of halophytes. I. The germination of the seed and establishment of the seedlings of five species of *Atriplex* in Australia. *Ecology* 33, 49–62.

Benech Arnold, R.L., Fenner, M. and Edwards, P.J. (1992) Changes in dormancy levels in *Sorghum halepense* (L.) Pers. seeds induced by water stress during seed development. *Functional Ecology* 6, 596–605.

Beneke, K. (1991) Fruit polymorphism in ephemeral species of Namaqualand. MSc thesis, University of Pretoria, South Africa.

Bewley, J.D., Kermode, A.R. and Misra, S. (1989) Desiccation and minimal drying treatments of seeds of Castor Bean and *Phaseolus vulgaris* which terminate development and promote germination cause changes in protein and messenger RNA synthesis. *Annals of Botany* 63, 3–17.

Black, M. (1969) Light-controlled germination of seeds. *Society of Experimental Biology Symposium* 23, 193–217.

Black, M. and Naylor, J.M. (1959) Prevention of onset of seed dormancy by gibberellic acid. *Nature* 184, 468–469.

Cavers, P.B. (1995) Seed banks: memory in soil. *Canadian Journal of Soil Science* 75, 11–13.

Cresswell, E.G. and Grime, J.P. (1981) Induction of a light requirement during seed development and its ecological consequences. *Nature* 291, 583–585.

Crocker, W. (1906) Role of seed coats in delayed germination. *Botanical Gazette* 42, 265–291.

Cumming, B.G. (1963) The dependence of germination on photoperiod, light quality and temperature in *Chenopodium* ssp. *Canadian Journal of Botany* 41, 1211–1223.

Datta, S.C., Evenari, M. and Gutterman, Y. (1970) The heteroblasty of *Aegilops ovata* L. *Israel Journal of Botany* 19, 463–483.

Datta, S.C., Evenari, M. and Gutterman, Y. (1972a) Photoperiodic and temperature responses of plants derived from the various heteroblastic caryopses of *Aegilops ovata* L. *Journal of the Indian Botanical Society* 50A, 546–559.

Datta, S.C., Gutterman, Y. and Evenari, M. (1972b) The influence of the origin of the mother plant on yield and germination of their caryopses in *Aegilops ovata* L. *Planta* 105, 155–164.

Do Cao, T., Attims, Y., Corbineau, F. and Côme, D. (1978) Germination des grains produits par les plantes de deux lignées d'*Oldenlandia corymbosa* L. (Rubiacées) cultivées dans des conditions contrôlées. *Physiologie Végétale* 16, 521–531.

Dornbos, D.L., Jr, Mullen, R.E. and Shibles, R.M. (1989) Drought stress effects during seed fill on soybean seed germination and vigor. *Crop Science* 29, 476–480.

Dorne, C.J. (1981) Variation in seed germination inhibition of *Chenopodium bonus-henricus* in relation to altitude of plant growth. *Canadian Journal of Botany* 59, 1893–1901.

Evenari, M. (1963) Zur Keimungsökologie zweier Wüstenpflanzen. *Mitteilungen der Floristisch-soziologischen Arbeitsgemeinschaft* 10, 70–81.

Evenari, M. (1965) Physiology of seed dormancy, after-ripening and germination. *Proceedings of International Seed Testing Association* 30, 49–71.

Evenari, M. and Gutterman, Y. (1966) The photoperiodic response of some desert plants. *Zeitschrift für Pflanzenphysiologia* 54, 7–27.

Evenari, M. and Gutterman, Y. (1976) Observations on the secondary succession of three plant communities in the Negev desert, Israel. I. *Artemisietum herbae albae.* In: Jacques, R. (ed.) *Hommage au Prof. P. Chouard. Études de Biologie Végétale.* CNRS, Gif sur Yvette, Paris, pp. 57–86.

Evenari, M., Koller, D. and Gutterman, Y. (1966) Effects of the environment of the mother plants on the germination by control of seed-coat permeability to water in *Ononis sicula* Guss. *Australian Journal of Biological Science* 19, 1007–1016.

Evenari, M., Kadouri, A. and Gutterman, Y. (1977) Eco-physiological investigations on the amphicarpy of *Emex spinosa* (L.) Campd. *Flora* 166, 223–238.

Evenari, M., Shanan, L. and Tadmor, N. (1982) *The Negev: the Challenge of a Desert.* 2nd edn. Harvard University Press, Cambridge, Massachusetts, 438 pp.

Fahn, A. (1947) Physico-anatomical investigations in the dispersal apparatus of some fruits. *Palestine Journal of Botany* 4, 136–145.

Felippe, G.M. and Dale, J.E. (1968) Effects of CCC and gibberellic acid on the progeny of treated plants. *Planta* 80, 344–348.

Fenner, M. (1980a) The inhibition of germination of *Bidens pilosa* seeds by leaf canopy shade in some natural vegetation types. *New Phytologist* 84, 95–101.

Fenner, M. (1980b) The induction of a light requirement in *Bidens pilosa* seeds by leaf canopy shade. *New Phytologist* 84, 103–106.

Fenner, M. (1991) The effects of the parent environment on seed germinability. *Seed Science Research* 1, 75–84.

Fenner, M. (1992) Environmental influences on seed size and composition. *Horticultural Reviews* 13, 183–213.

Frankton, C. and Bassett, I.J. (1968) The genus *Atriplex* (Chenopodiaceae) in Canada. I. Three introduced species: *A. heterosperma*, *A. oblongifolia*, and *A. hortensis*. *Canadian Journal of Botany* 46, 1309–1313.

Gonzalez-Rabanal, R., Casal, M. and Trabaud, L. (1994) Effects of high temperatures, ash and seed position in the inflorescences on the germination of three Spanish grasses. *Journal of Vegetation Science* 5, 389–394.

Gorski, T. (1975) Germination of seeds in the shadow of plants. *Physiologia Plantarum* 34, 342–346.

Grey, D. and Thomas, T.H. (1982) Seed germination and seedling emergence as influenced by the position of development of the seed on, and chemical applications to, the parent plant. In: Khan, A.A. (ed.) *The Physiology and Biochemistry of Seed Development, Dormancy and Germination.* Elsevier, New York, pp. 81–110.

Gutterman, Y. (1969) The photoperiodic response of some plants and the effect of the environment of the mother plants on the germination of their seeds. PhD thesis, The Hebrew University, Jerusalem (Hebrew with English summary).

Gutterman, Y. (1973) Differences in the progeny due to daylength and hormonal treatment of the mother plant, In: Heydecker, W. (ed.) *Seed Ecology.* Butterworth, London, pp. 59–80.

Gutterman, Y. (1974) The influence of the photoperiodic regime and red/far-red light treatments of *Portulaca oleracea* L. plants on the germinability of their seeds. *Oecologia* 17, 27–38.

Gutterman, Y. (1978a) Germinability of seeds as a function of the maternal environments. *Acta Horticulturae* 83, 49–55.

Gutterman, Y. (1978b) Influence of environmental conditions and hormonal treatment of the mother plants during seed maturation on the germination of their seeds. In: Malik, C.P. (ed.) *Advances in Plant Reproductive Physiology.* Kalyani Publishers, New Delhi, pp. 288–294.

Gutterman, Y. (1978c) Seed coat permeability as a function of photoperiodical treatment of the mother plants during seed maturation in the desert annual plant *Trigonella arabica* Del. *Journal of Arid Environments* 1, 141–144.

Gutterman, Y. (1980/81a) Annual rhythm and position effect in the germinability of *Mesembryanthemum nodiflorum. Israel Journal of Botany* 29, 93–97.

Gutterman, Y. (1980/81b) Review: influences on seed germinability: phenotypic maternal effects during seed maturation, In: Mayer, A.M. (ed.) *Control Mechanisms in Seed Germination. Israel Journal of Botany* 29, 105–117.

Gutterman, Y. (1982a) Phenotypic maternal effect of photoperiod on seed germination. In: Khan, A.A. (ed.) *The Physiology and Biochemistry of Seed Development, Dormancy and Germination.* Elsevier Biomedical Press, Amsterdam, pp. 67–79.

Gutterman, Y. (1982b) Survival mechanisms of desert winter annual plants in the Negev Highlands of Israel. In: Mann, H.S. (ed.) *Scientific Reviews on Arid Zone Research.* Scientific Publishers, Jodhpur, India, pp. 249–283.

Gutterman, Y. (1983) Mass germination of plants under desert conditions: effects of environmental factors during seed maturation, dispersal, germination and establishment of desert annual and perennial plants in the Negev Highlands, Israel. In: Shuval, H.I. (ed.) *Developments in Ecology and Environmental Quality.* Balaban ISS, Rehovot/Philadelphia, pp. 1–10.

Gutterman, Y. (1985) Flowering, seed development, and the influences during seed maturation on seed germination of annual weeds. In: Duke, S.O. (ed.) *Weed Physiology*, Vol. I. CRC Press, Baco Raton, Florida, pp. 1–25.

Gutterman, Y. (1990a) Do the germination mechanisms differ in plants originating in deserts receiving winter or summer rain? *Israel Journal of Botany* 39, 355–372.

Gutterman, Y. (1990b) Seed dispersal by rain (ombrohydrochory) in some of the flowering desert plants in the deserts of Israel and the Sinai Peninsula. *Mitteilungen aus dem Institut fur Allgemeine Botanik Hamburg* 23b, 841–852.

Gutterman, Y. (1991) Comparative germination study on seeds matured during winter or summer of some bi-seasonal flowering perennial desert plants from the Aizoaceae. *Journal of Arid Environments* 21, 283–291.

Gutterman, Y. (1992a) Maturation dates affecting the germinability of *Lactuca serriola* L. achenes collected from a natural population in the Negev Desert highlands: germination under constant temperatures. *Journal of Arid Environments* 22, 353–362.

Gutterman, Y. (1992b) Influences of daylength and red or far red light during the storage of post harvested ripe *Cucumis prophetarum* fruit, on the light germination of the seeds. *Journal of Arid Environments* 23, 443–449.

Gutterman, Y. (1993) *Seed Germination in Desert Plants. Adaptations of Desert Organisms*. Springer, Berlin, 253 pp.

Gutterman, Y. (1994a) Long-term seed position influences on seed germinability of the desert annual, *Mesembryanthemum nodiflorum* L. *Israel Journal of Plant Sciences* 42, 197–205.

Gutterman, Y. (1994b) In memoriam – Michael Evenari and his desert: seed dispersal and germination strategies of *Spergularia diandra* compared with some other desert annual plants inhabiting the Negev Desert of Israel. *Israel Journal of Plant Sciences* 42, 261–274.

Gutterman, Y. (1994c) Germinability under natural temperatures of *Lactuca serriola* L. achenes matured and collected on different dates from a natural population in the Negev Desert highlands. *Journal of Arid Environments* 28, 117–127.

Gutterman, Y. (1996a) Effect of day length during plant development and caryopsis maturation on flowering and germination, in addition to temperature during dry storage and light during wetting, of *Schismus arabicus* (Poaceae) in the Negev Desert, Israel. *Journal of Arid Environments* 33, 439–448.

Gutterman, Y. (1996b) Environmental influences during seed maturation, and storage affecting germinability in *Spergularia diandra* genotypes inhabiting the Negev Desert, Israel. *Journal of Arid Environments* 34, 313–323.

Gutterman, Y. (1997) Genotypic, phenotypic and opportunistic germination strategies of some common desert annuals compared with plants with other seed dispersal and germination strategies. In: Ellis, R.H., Black, M., Murdoch, A.J. and Hong, T.O. (eds) *Basic and Applied Aspects of Seed Biology. Proceedings of 5th Workshop on Seeds, Reading, UK, 1995*. Kluwer Academic Publishers, Dordrecht, pp. 611–622.

Gutterman, Y. (1998a) Ecological strategies of desert annual plants. In: Ambasht, R.S. (ed.) *Modern Trends in Ecology and Environment*. Backhuys Publishers, Leiden, pp. 203–231.

Gutterman, Y. (1998b) Ecophysiological genotypic and phenotypic strategies affecting seed 'readiness to germinate' in plants occurring in deserts. In: Taylor, A.G. and Huang, X.-L. (eds) *Progress in Seed Research: Proceedings of the 2nd International Conference on Seed Science and Technology*. Communication Services of the New York State Agricultural Experiment Station, Geneva, New York, pp. 10–19.

Gutterman, Y. and Evenari, M. (1972) The influence of day length on seed coat colour, an index of water permeability of the desert annual *Ononis sicula* Guss. *Journal of Ecology* 60, 713–719.

Gutterman, Y. and Ginott, S. (1994) The long-term protected 'seed bank' in the dry inflorescence, the mechanism of achenes (seeds) dispersal by rain (ombrohydrochory) and the germination of the annual desert plant *Asteriscus pygmaeus*. *Journal of Arid Environments* 26, 149–163.

Gutterman, Y. and Heydecker, W. (1973) Studies of the surfaces of desert plant seeds. I. Effect of day length upon maturation on the seed coat of *Ononis sicula* Guss. *Annals of Botany* 37, 1049–1050.

Gutterman, Y. and Porath, D. (1975) Influences of photoperiodism and light treatments during fruits storage on the phytochrome and on the germination of *Cucumis prophetarum* L. and *Cucumis sativus* L. seeds. *Oecologia* 18, 37–45.

Gutterman, Y., Witztum, A. and Evenari, M. (1967) Seed dispersal and germination in *Blepharis persica* (Burm.) Kuntze. *Israel Journal of Botany* 16, 213–234.

Gutterman, Y., Thomas, T.H. and Heydecker, W. (1975) Effect on the progeny of applying different day length and hormone treatments to parent plants of *Lactuca scariola*. *Physiologia Plantarum* 34, 30–38.

Gutterman, Y., Evenari, M., Cooper, R., Levy, E.C. and Lavie, D. (1980) Germination inhibition activity of a naturally occurring lignin from *Aegilops ovata* L. in green and infrared light. *Experientia* 26, 662–663.

Gutterman, Y., Golan, T. and Garsani, M. (1990) Porcupine diggings as a unique ecological system in a desert environment. *Oecologia* 85, 122–127.

Hannan, G.L. (1980) Heteromericarp and dual seed germination modes in *Platystemon californicus* (Papaveraceae). *Madrono* 27, 163–170.

Harper, J.L. (1977) *Population Biology of Plants*. Academic Press, New York, 892 pp.

Hayes, R.G. and Klein, W.H. (1974) Spectral quality influence of light during development of *Arabidopsis thaliana* plants in regulating seed germination. *Plant Cell Physiology* 15, 643–653.

Heide, O.M., Juntila, O. and Samuelsen, R.T. (1976) Seed germination and bolting in red beet as affected by parent plant environment. *Physiologia Plantarum* 36, 343–349.

Herre, H. (1971) *The Genera of Mesembryanthemaceae*. Tafelberg Uitgewers Beperk, Cape Town, 316 pp.

Holm, L.G., Plucknett, D.L., Pancho, J.V. and Herberger, J.P. (1977) *The World's Worst Weeds: Distribution and Biology*. University Press of Hawaii, Honolulu, 609 pp.

Ingram, T.J. and Browning, G. (1979) Influence of photoperiod on seed development in genetic line of peas G_2 and its relation to changes in endogenous gibberellins measured by combined gas chromatography and mass spectrometry. *Planta* 146, 423–432.

Jackson, G.A.D. (1968) Hormonal control of fruit development, seed dormancy and germination with particular reference to *Rosa*. *Plant Growth Regulators* 31, 127–156.

Jacobsohn, R. and Globerson, D. (1980) *Daucus carota* (carrot) seed quality. I. Effects of seed size on germination, emergence and plant growth under subtropical conditions. II. The importance of the primary umbel in carrot seed production. In: Hebblethwaite, P.D. (ed.) *Seed Production*. Butterworth, London, pp. 637–646.

Jacques, R. (1957) Quelques données sur le photoperiodisme de *Chenopodium polyspermum* L., influence sur la germination des graines. In: *Colloque International sur le Photo-thermo Periodisme*. Publication 34, Serie B, IUBS, Parma, pp. 125–130.

Jacques, R. (1968) Action de la lumière par l'intermédiaire du phytochrome sur la germination, la croissance et le développement de *Chenopodium polyspermum* L. *Physiologica Végétale* 6, 137–164.

Juntila, O. (1973) Seed and embryo germination in *Syringa vulgaris* and *S. reflexa* as affected by temperature during seed development. *Physiologia Plantarum* 29, 264–268.

Kadman, A. (1954) Germination of some summer-annuals. MSc thesis, Hebrew University of Jerusalem (Hebrew with English summary).

Karssen, C.M. (1970) The light promoted germination of the seeds of *Chenopodium album* L. III. Effect of the photoperiod during growth and development of the plants on the dormancy of the produced seeds. *Acta Botanica Neerlandica* 19, 81–94.

Keigley, P.J. and Mullen, R.E. (1986) Changes in soybean seed quality from high temperatures during seed fill and maturation. *Crop Science* 26, 1212–1216.

Kermode, A.R., Bewley, J.D., Dasgupta, J. and Misra, S. (1986) The transition from seed development to germination: a key role for desiccation? *HortScience* 21, 1113–1118.

Kermode, A.R., Pramanik, S.K. and Bewley, J.D. (1989) The role of maturation drying in the transition from seed development to germination. VI. Desiccation-induced changes in messenger RNA populations within the endosperm of *Ricinus communis* L. seeds. *Journal of Experimental Botany* 40(210), 33–41.

Kigel, J. (1992) Diaspore heteromorphism and germination in populations of the ephemeral *Hedypnois rhagadioloides* (L.) F.W. Schmidt (Asteraceae) inhabiting a geographic range of increasing aridity. *Acta Oecologia* 13, 45–53.

Kigel, J., Ofir, M. and Koller, D. (1977) Control of the germination responses of *Amaranthus retroflexus* L. seeds by their parental photothermal environment. *Journal of Experimental Botany* 28, 1125.

Kigel, J., Gibly, A. and Negbi, M. (1979) Seed germination in *Amaranthus retroflexus* L. as affected by the photoperiod and age during flower induction of the parent plants. *Journal of Experimental Botany* 30, 997–1002.

King, T.J. (1975) Inhibition of seed germination under leaf canopies in *Arenaria serphyllifolia*, *Veronica arvensis* and *Cerastium holosteoides*. *New Phytologist* 75, 87–90.

Kirchner, R. and Andrew, W.D. (1971) Effects of various treatments on hardening and softening of seeds in pods of barrel medic (*Medicago trunculata*). *Australian Journal of Experimental Agriculture and Animal Husbandry* 11, 536–540.

Koller, D. (1954) Germination regulating mechanisms in desert seeds. PhD thesis, Hebrew University of Jerusalem (Hebrew with English summary).

Koller, D. (1957) Germination-regulating mechanisms in some desert seeds. IV. *Atriplex dimorphostegia* Kar. et Kir. *Ecology* 38, 1–13.

Koller, D. (1962) Preconditioning of germination in lettuce at time of fruit ripening. *American Journal of Botany* 49, 841–844.

Koller, D. and Negbi, M. (1966) *Germination of Seeds of Desert Plants*. Report to the USDA, Hebrew University of Jerusalem, Jerusalem, pp. 1–180.

Koller, D. and Roth, N. (1964) Studies on the ecological and physiological significance of amphicarpy in *Gymnarrhena micrantha* (Compositae). *American Journal of Botany* 51, 26–35.

Lalonde, L. and Bewley, J.D. (1986) Patterns of protein synthesis during the germination of pea axes, and the effects of an interrupting desiccation period. *Planta* 167, 504–510.

Lavie, D., Levy, E.C., Cohen, A., Evenari, M. and Gutterman, Y. (1974) New germination inhibitor from *Aegilops ovata* L. *Nature* 249, 388.

Lona, F. (1947) L'influenza della condizioni ambientali, durante l'embriogenesi, sulla caratteristiche del seme e della pianta che ne deriva. In: *Lavori di Botanica*, Special volume on the occasion of the 70th birthday of Prof. Gola (Italian), pp. 313–352.

Loria, M. and Noy-Meir, I. (1979/80) Dynamics of some annual populations in a desert loess plain. *Israel Journal of Botany* 28, 211–225.

McComb, J.A. and Andrews, R. (1974) Sequential softening of hard seeds in burrs of annual medics. *Australian Journal of Experimental Agriculture and Animal Husbandry* 14, 68–75.

McCullough, J.M. and Shropshire, W., Jr (1970) Physiological predetermination of germination responses in *Arabidopsis thaliana* (L.) Heynh. *Plant Cell Physiology* 11, 139–148.

Maun, P.A. and Payne, A.M. (1989) Fruit and seed polymorphism and its relation to seedling growth in the genus *Cakile*. *Canadian Journal of Botany* 67, 2743–2750.

Negbi, M. and Tamari, B. (1963) Germination of chlorophyllous and achlorophyllous seeds of *Salsola volkensis* and *Aellenia autrani*. *Israel Journal of Botany* 12, 124–135.

Okusanya, O.T. and Ungar, I.A. (1983) The effects of time of seed production on the germination response of *Spergularia marina*. *Physiologia Plantarum* 59, 335–342.

Philipupillai, J. and Ungar, I.A. (1984) The effect of seed dimorphism on the germination and survival of *Salicornia europaea* L. populations. *American Journal of Botany* 71, 542–549.

Plett, S. and Larter, E.N. (1986) Influence of maturation temperature and stage of kernel development in sprouting tolerance of wheat and triticale. *Crop Science* 26, 804–807.

Pourrat, Y. and Jacques, R. (1975) The influence of photoperiodic conditions received by the mother plant on morphological and physiological characteristics of *Chenopodium polyspermum* L. seeds. *Plant Science Letters* 4, 273–279.

Roach, D.A. and Wulff, R.D. (1987) Maternal effects in plants. *Annual Review of Ecology and Systematics* 18, 209–235.

Roiz, L. (1989) Sexual strategies in some gynodioecious and gynomonoecious plants. PhD thesis, Tel-Aviv University, Tel-Aviv.

Takeno, K. and Yamaguchi, H. (1991) Diversity in seed germination behavior in relation to heterocarpy in *Salsola komarovii* Iljin. *Botanical Magazine of Tokyo* 104, 207–215.

Thomas, T.H., Biddington, N.L. and O'Toole, D.F. (1979) Relationship between position on the parent plant and dormancy characteristics of seed of three cultivars of celery (*Apium graveolens*). *Physiologia Plantarum* 45, 492–496.

Ungar, I.A. (1979) Seed dimorphism in *Salicornia europaea* L. *Botanical Gazette* 140, 102–108.

Vander Veen, R. (1970) The importance of the red–far red antagonism in photoblastic seeds. *Acta Botanica Neerlandica* 19, 809–812.

van Rheede van Oudtshoorn, K. and van Rooyen, M.W. (1999) *Dispersal Biology of Desert Plants: Adaptations of Desert Organisms*. Springer, Berlin, 242 pp.

Wentland, M.J. (1965) The effect of photoperiod on the seed dormancy of *Chenopodium album*. PhD thesis, University of Wisconsin-Madison.

Werker, E. and Many, T. (1974) Heterocarpy and its ontogeny in *Aellenia autrani* (Post) Zoh: light and electron-microscope study. *Israel Journal of Botany* 23, 132–144.

Wurzburger, J. and Koller, D. (1976) Differential effects of the parental photothermal environment on development of dormancy in caryopses of *Aegilops kotschyi*. *Journal of Experimental Botany* 27, 43–48.

Yamaguchi, H., Ichihara, K., Takeno, K., Hori, Y. and Saito, T. (1990) Diversities in morphological characteristics and seed germination behavior in fruits of *Salsola komarovii* Iljin. *Botanical Magazine of Tokyo* 103, 177–190.

Zeevaart, J.A.D. (1966) Reduction of the gibberellin content of pharbitis seeds by CCC and after-effects in the progeny. *Plant Physiology* 41, 856–862.

Zohary, M. (1937) Die verbreitungsökologischen Verhältnisse der Flora Palästinas. *Beihefte zum Botanischen Zentralblatt* 56(I), 1–155.

Zohary, M. (1966) *Flora Palestina*, Part I – *Text*. Israel Academy of Sciences and Humanities, Jerusalem, 364 pp.

Zohary, M. (1972) *Flora Palestina*, Part II – *Text*. Israel Academy of Sciences and Humanities, Jerusalem, 489 pp.

Chapter 4
The Ecology of Seed Dispersal

Mary F. Willson[1] and Anna Traveset[2]

[1]5230 Terrace Place, Juneau, Alaska, USA; [2]Institut Mediterrani d'Estudis Avançats (CSIC–UIB), Esporles, Balearic Islands, Spain

Introduction

Seed dispersal has long been a topic of interest to naturalists, but it has not been until the last three decades that the ecology of dispersal has received much rigorous scientific attention. Many theoretical and empirical advances have recently been made, although important lacunae in our understanding still need to be filled before dispersal ecology becomes a coherent body of knowledge. A major goal of this chapter is to review the existing literature on seed dispersal, highlighting these essential but missing kinds of information.

This review is divided into two sections, the first dealing with the evolution of dispersal mechanisms and the second with the consequences of dispersal at population and community levels. Dispersal can occur in both space and time, but only the former will be treated here, except where some relationship between the two axes is known (Venable and Brown, 1988; Leck *et al.*, 1989; Eriksson and Ehrlén, 1998a; see also Murdoch and Ellis, Chapter 8, this volume).

The seed shadow

The spatial distribution of dispersed seeds around their source is called a 'seed shadow' (Janzen, 1971). The term is most commonly (and perhaps properly) used in reference to the postdispersal distribution of seed around the maternal parent, but it can also be used to refer to the distribution of seeds around a source composed of multiple parents. Although the unit of dispersal may be technically a fruit or a group of fruits and, thus, the generic label should be diaspore or propagule, we shall continue to use the more euphonious term of seed shadow, with the intent of conceptually including all diaspores of any morphological and genetic derivation.

Seed shadows exist in two horizontal dimensions for most seed plants (a salient exception is found in epiphytic plants, in which the third, vertical, dimension is presumably well developed). The shape of the seed distribution in the horizontal plane may be asymmetric with respect to the source (if, for example, the wind carries most seeds in a downwind direction). This simple descriptor of shape is generally augmented by information on seed density, such that the seed shadow resembles a topographic map, with peaks of high seed density. Two factors, then, can be used to describe seed shadows – the relationship of seed numbers or density to distance from the source and the directionality with respect to the source.

Although directionality is clearly

significant for many ecological questions, it is common to discuss seed distributions chiefly in terms of distance from the source. Conventional wisdom describes the seed number/distance relationship as leptokurtic (with a higher peak and longer tail than a normal distribution); from the peak outward, seed numbers are generally considered to decrease monotonically, fitting a negative exponential curve (or sometimes a negative power function: Okube and Levin, 1989; Fig. 4.1). Most measured seed shadows conform to this expectation (Fig. 4.2; Willson, 1993a). The nature of the source (single or multiple individuals) can influence the location of the peak of the curve, and the use of seed density rather than numbers can change the overall shape of the curve, including the proximity of the peak to the source (Peart, 1985; Greene and Johnson, 1996).

Deviations from the conventional seed shadow shape can result from patchiness of habitat structure (Hoppes, 1988; Debussche and Lepart, 1992; Debussche and Isenmann, 1994; Kollmann and Pirl, 1995; Aguiar and Sala, 1997) and other

ecological factors, including behaviour patterns of the seed vectors that lead to nucleation processes (Willson and Crome, 1989; McClanahan and Wolfe, 1993; Verdú and García-Fayos, 1996, 1998; Julliot, 1997). For species with polymorphic seeds (i.e. with and without dispersal devices, or with two or more different kinds of dispersal devices), the shape of the seed shadow for each seed type may differ, such that the combined seed shadow for a given parent may have a very unconventional shape.

Many factors can alter the location of the peak and the slope and shape of the tail of the seed shadow for particular species and individuals (e.g. Rabinowitz and Rapp, 1981; Johnson, 1988; Debussche and Lepart, 1992; Verdú and García-Fayos, 1996, 1998; Julliot, 1997). Some factors are, from the plant's perspective, strictly environmental and thus outside the control of the plant (e.g. the strength and direction of the wind, the social behaviour of animal dispersal agents, the patterns of rainfall and relative humidity). Other factors, such as plant height, have a strong environmental component (Greene and Johnson, 1996)

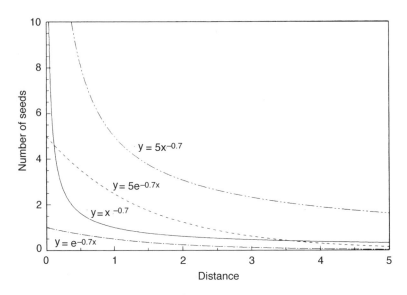

Fig. 4.1. Idealized curves commonly used to describe the distribution of seeds at increasing distances (arbitrary units) from the seed source. Real seed shadows often peak at some distance from the source; in that case, the curves refer to the part of the seed distribution from the peak outward. For a given set of coefficients, the negative exponential curve ($y = ax^{-mx}$) drops less steeply than the negative power function ($y = ax^{-m}$); it is converted to a straight line on a semilogarithmic scale. (From Okube and Levin, 1989.)

but may also have a genetic component. Still others are probably controlled both by environment and the genetic constitution of the parent plant, the balance depending on the species and circumstances (e.g. fruit size, seed size, ease of dehiscence or abscission). All such factors can contribute to variation in the size and shape of the seed shadow among species and among conspecific individuals.

Few data yet exist, either from the tropics or from the temperate zone, to compare seed shadows generated by different dispersal modes (but see Gorchov *et al.*, 1993; Portnoy and Willson, 1993; Willson, 1993a). Even less is known about how the

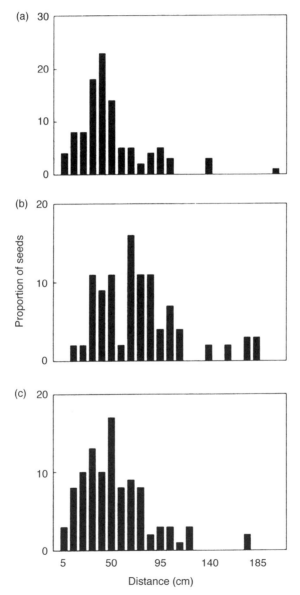

Fig. 4.2. Seed shadows of three individuals of *Lithospermum caroliniense*. The tails of all three seed shadows fit a negative exponential curve, but the slope of that curve (on a semilog plot) varies from −1.47 to −1.72, and the location of the peak differs. (From Westelaken and Maun, 1985.)

loss or the addition of a dispersal agent alters the seed shadow of a plant. Moreover, we cannot yet make any generalization about the relative ecological importance of different portions of the seed shadow for the evolutionary ecology of plants.

In order to understand the ecological and evolutionary consequences of variation in the length and shape of the seed shadows, we need to experimentally modify seed distributions and monitor the fitness of recruits in different parts of the distribution, for a variety of species and circumstances (Portnoy and Willson, 1993). The tail of the distribution may be at least as important as the modal portion of the curve, although little theoretical effort has been devoted to this question (but see, for instance, Cain et al., 1998). Propagules in the distribution's tail potentially spread the parental genes more widely, and plant traits that affect the behaviour of such tails can be subject to selection. The examination of 68 data sets has shown a lack of association between tail shape and dispersal mode, suggesting that, in most circumstances, selection for tail behaviour contributes little to the evolution of the dispersal mode itself (Portnoy and Willson, 1993).

The evolution of dispersal

Why are seeds dispersed?

If the dispersal of offspring increases the fitness of a parent, we should expect that dispersed offspring survive and reproduce better than undispersed offspring, either because they avoid detrimental conditions near the parent or because they reach better conditions farther away (which amounts to the same thing, from a different perspective). If seeds fall directly beneath the canopy of a parent, the physical separation hardly constitutes real dispersal, but fallen seeds are normally treated as part of the seed shadow, for purposes of comparing seed fates. Van der Pijl (1982) actually treats simple seed fall as a separate mode

of dispersal for species that have no evident special means. The principal factors that favour dispersal are avoidance of natural enemies or sibling interactions and the probability of finding a physically suitable establishment site.

1. Some natural enemies of seeds and seedlings respond to density and/or distance from the parent (or other conspecific). Pathogens, postdispersal seed predators, parasites and herbivores often concentrate their activities where their resources are common, and more distant seeds/seedlings may survive better than those close to the parent (e.g. Howe and Smallwood, 1982; Augspurger, 1983a, b; Augspurger and Kelly, 1984; Howe, 1993; Peres et al., 1997; Hulme, 1998); the magnitude of this effect varies among species (Augspurger, 1984; Howe, 1993). The impact of such consumers depends, in part, on their specificity to genotype or to species of resource – some attackers may specialize in the offspring of particular parents or in particular taxa. The density or proximity of other genotypes or species of seed would have little impact on the availability of suitable resources for such specialists, and thus the effect of density- and distance-responsive enemies must often vary with their specificity.

The ability of natural enemies to depress seed and seedling density also depends, of course, on other factors limiting their abundance and activity. These will vary among consumers, and even among populations of the same species of consumer. Thus, although numerous cases of density- and distance-responsive attackers have been reported, we do not yet have a general picture of which plant species are subject to such attacks and in what circumstances (e.g. habitat, season, geographical region, adult densities).

2. Because the seeds and seedlings of any one parent are genetically related (at least half-sibs), they are subject, potentially, to sibling competition. Conventional wisdom suggests that sib competition may often be more severe than competition with non-sib conspecifics, because their patterns of resource use are probably more similar (see references in Ellstrand and Antonovics, 1985; see also McCall *et al.*, 1989). However, in a number of cases, sib competition does not have a detectably different outcome from that of non-sib competition, and sibs may even profit, at some early stages of the life history, from the proximity of genetic relatives (Smith, 1977; Williams *et al.*, 1983; Willson *et al.*, 1987; Kelley, 1989; McCall *et al.*, 1989). Furthermore, self-fertilized seeds disperse less well than outcrossed seeds in *Impatiens capensis* and *Amphicarpaea bracteata* (Schmitt and Ehrhardt, 1987; Trapp, 1988), which is the opposite of what would be expected if dispersal were an adaptation to reduce sib competition. Also, if sib competition were critical, genetically variable offspring should have higher fitness than clonal progeny where sib densities are high, but this was not the case in *Anthoxanthum odoratum* (Kelley *et al.*, 1988).

Another potential disadvantage of high sibling densities is the possibility of inbreeding when (and if) the offspring reach adulthood (Ghiselin, 1974). The relative effects of extreme inbreeding and outcrossing on the genetic variance of offspring are not predictable in any simple way, because they depend on many aspects of the genetic system, as well as past episodes of inbreeding (McCall *et al.*, 1989). As a result, the degree of offspring similarity and the intensity of potential sib competition are likewise difficult to predict. A degree of inbreeding may be advantageous under certain circumstances (e.g. Shields, 1982; Jarne and Charlesworth, 1993). Furthermore, even if inbreeding is disadvantageous, there are other ways for plants to reduce inbreeding (e.g. through changes in the floral biology and mating system), so it is difficult to assess the importance of inbreeding avoidance as a factor that selects for offspring dispersal. Moreover, dispersal has its own costs (e.g. Cohen and Motro, 1989), which may outweigh the costs of some inbreeding (Waser *et al.*, 1986); the benefit/cost ratio is likely to vary among species (e.g. Augspurger, 1986).

Attacks by parasites and pathogens may be more devastating when sibs grow in close proximity to each other (see Alexander and Holt (1998) for a recent review on the interaction between plant competition and disease). Just as conspecifics growing in a monoculture are often more heavily hit by pests than they are when growing in a mixed stand, so also are the genetic monocultures of closely related individuals sometimes more heavily hit by certain kinds of pests than stands of mixed parentage (e.g. Parker, 1985; Burdon, 1987). To the extent that a species is subject to such attacks on particular genetic lineages, there may be selection for dispersal, which lowers the concentration of any one lineage in a given area (Fig. 4.3).
3. Some species have special physical requirements for germination and establishment that are met only in scattered locations, such as fallen logs, tree-fall gaps or badger mounds. In the absence of well-developed dormancies and the ability to wait for suitable conditions to arrive, selection may favour dispersal in order to increase the probability of finding the necessary kind of location (Platt and Weis, 1985; Reid, 1989; Sargent, 1995). Even with good dormancy mechanisms, dispersal should enhance the probability that a waiting seed will eventually encounter the right conditions for establishment. Theoretically, dispersal generally enhances the likelihood that at least some offspring reach appropriate sites (Hamilton and May, 1977; Comins *et al.*, 1980). These theoretical expectations need to be examined for a variety of particular cases. If the seed shadow is adapted to the distribution of safe sites, species with widespread seed shadows should have more far-flung establishment sites than species with restricted seed shadows (Green, 1983; see also Geritz *et al.*, 1984; Horvitz and Le Corff, 1993).

What portion of the seed shadow is most effective in yielding successful offspring and how does this vary with species and conditions? The peak portion of the seed shadow, where the most seeds are deposited, often receives the most attention from ecologists. Although the peak may be

Fig. 4.3. A 'bear garden' – seedlings of *Ribes bracteosum* growing from a faecal deposit of an Alaskan brown bear. The small squares in the grid beside the garden are 3 cm on a side. Both sib and non-sib competition must be intense, predation by rodents can be severe and survivorship may ultimately be low. Some bear deposits contain seeds of two or three species, and germination may occur a month to a year or more after deposition. (Photo by J. Zasada.)

ecologically important as a source of food for seed predators and other consumers, we seldom know if it is the most important in terms of parental fitness. In part because consumers or intense competition can level the peak and in part because rare events can be ecologically important, it seems essential that the tail of the distribution should receive more attention in the future. Moreover, we need to know if there are any patterns (among species, for instance) in the relative importance of the distributional tail. A number of studies have shown that seed and seedling survival increases with distance from the parent (see above), or that more distantly dispersing seeds are more likely to reach good sites (Platt and Weis, 1985). However, a few studies have shown that seeds at the end of the seed shadow often do poorly (Horvitz and Schemske, 1986; Augspurger and Kitajima, 1992) and some authors have argued (without experimental documentation, however) that selection may actually oppose dispersal in some species (Zohary, in van der Pijl, 1982; but see Ellner and Schmida, 1981). Recent studies by Russell and Schupp (1998, and references therein)

show that patterns of initial seed-fall density are more affected by distance from a seed source than by the physical structure of the microhabitat, at least for wind-dispersed plants. On the other hand, investigations by Donohue (1997, 1998), who has decoupled the fitness effects of density and distance from the 'home site', have shown that selection on dispersion patterns is likely to be through density rather than distance effects. This author has also investigated the maternal environment effects on seed dispersal within an evolutionary context. Although maternal characters (e.g. fruit and seed traits, architectural traits, plant size, fruit production) are known to influence seed dispersal, only a few studies address the evolution of dispersal within the context of maternal character evolution (see also Thiede and Augspurger, 1996).

How are seeds dispersed?

If dispersal is advantageous, we would expect to find that diaspores have adaptations that enhance dispersal (Ridley, 1930; van der Pijl, 1982). Morphological devices that enhance dispersal are often quite readily evident and interpretable (Kerner, 1898; Ridley, 1930; van der Pijl, 1982), although some dispersal-enhancing traits (such as buoyancy of seeds dispersed by water) are less immediately obvious. Wind-borne diaspores often have wings or plumes that increase air resistance and slow the rate of fall. Diaspores carried by animal consumers commonly have edible appendages or coverings that are consumed by animals that later eject the seeds (Fig. 4.4); in some cases, the seeds themselves are harvested and either eaten (killed) or cached and sometimes abandoned by the harvester (Sork, 1983; Price and Jenkins, 1986). Other animal-dispersed diaspores travel by means of hooks or sticky coatings that adhere to the exteriors of the animal vectors. Some plants disperse their offspring ballistically, by the explosive opening of the fruits or the springing of a trip-lever. Seeds of certain plant species combine two

or even three modes of dispersal (Westoby and Rice, 1981; Clifford and Monteith, 1989; Stamp and Lucas, 1990; Aronne and Wilcock, 1994; Traveset and Willson, 1997), as in some *Viola* (ballistic plus ants), *Disporum, Rhamnus, Myrtus, Smilax* (birds plus ants) and *Petalostigma pubescens* (birds plus ballistic plus ants); a great number of species are dispersed by both birds and mammals (e.g. Herrera, 1989b; Willson, 1993b; Traveset and Willson, 1997).

The dispersal potential of the different modes of dispersal varies greatly. Both wind and vertebrates can potentially carry seeds far from the parent plant, but ants and ballistic mechanisms typically generate shorter seed shadows. A preliminary survey (Willson, 1993a) for herbaceous species indicates that peak and maximum dispersal distances are greater, and the slope of the tail of the seed distribution is less steep, for wind and ballistic dispersal

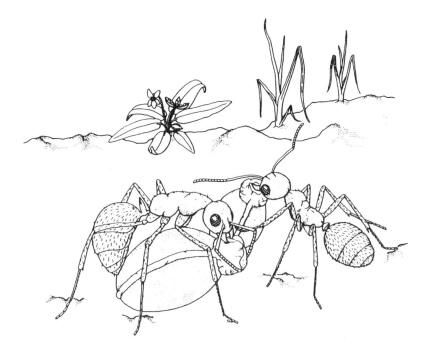

Fig. 4.4. Ants (*Formica podzolica*) picking up seeds of *Viola nuttallii*. The seed bears an attractive and edible appendage. Ants carry the entire seed back to their nest, eat the appendage and discard the seed. Dispersal of seeds by ants is very common in some floras, but the advantage of ant dispersal may vary greatly among species or regions (e.g. escape from predators or other destructive agents, or deposition in an especially favourable site for germination and growth). (From Beattie, 1985, p. 74.)

than for species with no special devices on the diaspore. Maximum distances are greater for wind-dispersed than for ballistically dispersed seeds of herbs, although peak distances and slopes are similar. Such results indicate that, on average, dispersal devices seem to work. But variation around the averages was great, and both sampling methods and environmental conditions of dispersal affect the outcome. A much better database is needed to make good comparisons of the seed shadows produced in different ways.

Of course, there is also great variation in the dispersal potential within each general mode of dispersal. The relative size of seed and wing or plume can have enormous effects on the seed shadow of wind-borne seeds (Augspurger and Franson, 1987; Sacchi, 1987; Benkmann, 1995). The size and chemistry of the edible appendage on ant-dispersed seeds may influence the rate of seed removal (Gunther and Lanza, 1989; Gorb and Gorb, 1995; Mark and Olesen, 1996), the array of dispersing ant species and the eventual fate of the seeds. Dispersal of fleshy fruits by ground-foraging vertebrates generates different seed shadows from those produced by flying fruit-eaters. Nuts favoured by scatter-hoarding squirrels may be spread more widely than less favoured species of nuts (Stapanian and Smith, 1984), but jays carry acorns much further than squirrels carry any nuts (Johnson and Webb, 1989). Seeds dispersed by frugivorous lizards also show different patterns of deposition from those dispersed by mammals (Traveset, 1995). By contrasting dispersal syndromes in a family (*Marantaceae*) of tropical understorey herbs, Horvitz and Le Corff (1993) found that bird-dispersed species went further than ant-dispersed species; however, dispersion patterns did not vary among types of dispersal, most species having a clumped spatial patterning.

For vertebrate-dispersed species, it has been hypothesized that plants can exert some kind of control over seed shadows produced by frugivores, by specific laxative and/or constipative chemicals in the fruit pulp, which affect seed retention time in the dispersers' guts (Murray *et al.*, 1994; Cipollini and Levey, 1997; Wahaj *et al.*, 1998). In turn, seed retention time inside the disperser, together with other factors (reviewed in Traveset, 1998), can affect the germinability, the rate of germination, or both, in certain species.

Although many species exhibit the morphological devices that are presumed to enhance dispersal, large numbers of species lack any evident dispersal device (Ridley, 1930; Willson *et al.*, 1990a; Willson, 1993a; Cain *et al.*, 1998). The seeds of some of these species are so small that they are easily wind-borne without special devices (e.g. orchids). Others have small, hard seeds that are consumed by herbivorous vertebrates along with the foliage and dispersed after passing through the animal's gut, but whether or not this constitutes an 'evolutionary design' can be debated (Janzen, 1984; Collins and Uno, 1985; Dinerstein, 1989). Still other species have small, round seeds that are shaken out of the fruit when the plant is stirred by a passing breeze or animal (e.g. *Papaver*); this is considered to be a special mechanism for dispersal by some (van der Pijl, 1982). Yet many species remain that lack any apparent device for dispersal in space and that also appear to have little capacity for dispersal in time (Willson, 1993a). Although some of these species may be discovered to be dispersed by one of the major modes, it is reasonable to ask why so many species seem to lack dispersal devices. How do such species achieve effective dispersal? Or is dispersal less advantageous in these species?

Variation in dispersal devices is often evident, both within species and among closely related species. Examples of within-species variation are provided by many *Asteraceae* (seeds with and without a device for wind dispersal) and amphicarpic species, which have both above-ground seeds, which may be dispersed by any of the normal vectors, and below-ground seeds, which may not disperse at all or may be harvested and cached by rodents (van der Pijl, 1982). Examples of variation among related species are many,

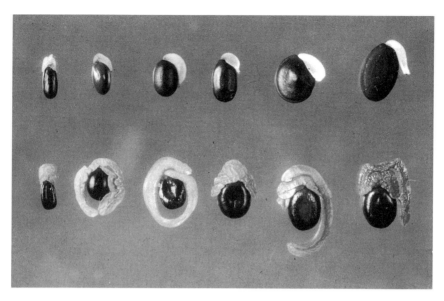

Fig. 4.5. Variations on a theme – *Acacia* seeds apparently adapted for dispersal by ants (above) and by birds (below). The food reward in the appendage on the seed is larger and usually more colourful (red, orange, yellow versus white) in bird-dispersed species. (Photo courtesy of D.J. O'Dowd.)

one of the most striking being found in the *Acacia* genus, which is widespread through Latin America, Australia, Asia and Africa. Most Australian acacias exhibit morphological adaptations for dispersal by ants or birds (Fig. 4.5); the American species are dispersed by birds or large mammals (O'Dowd and Gill, 1986); the African species are dispersed by large mammals and, reportedly, by wind (Coe and Coe, 1987). Different populations of *Acacia ligulata* in Australia seem to be adapted to dispersal by ants or birds (Davidson and Morton, 1984). Even if this enormous genus is eventually split into several new genera, the basic observation of diversity of dispersal within a set of related species remains valid. In the genus *Pinus*, seed mass is associated with the mode of dispersal; pine seeds weighing less than about 100 mg are wind-dispersed, while heavier seeds usually have adaptations for bird dispersal (Benkman, 1995). In lineages where long-distance seed dispersal predominates, clonal propagation has evolved on frequent occasions, possibly to make genet fitness less dependent on local dispersal by seed (Eriksson, 1992).

The mode(s) of dispersal of any plant species must reflect many different pressures and constraints. Because natural selection must work with existing variation and many plants have very long generation times, there are inevitable phylogenetic constraints. Entire families or genera sometimes exhibit only slight variations on a single mode of dispersal. However, extensive variation within families (e.g. *Liliaceae*), genera (e.g. *Acacia, Pinus*) and species (e.g. *Spergularia marina, Heterotheca latifolia*) demonstrates that such constraints are neither universal nor totally confining.

Constraints on the evolution of diaspores also emerge from the many, sometimes potentially conflicting, selection pressures that impinge on diaspore design (e.g. Ellner and Schmida, 1981; Benkman *et al.*, 1984; Westoby *et al.*, 1991; Armstrong and Westoby, 1993; Leishman and Westoby, 1994a, b; Kelly, 1995; Winn and Miller, 1995; see also Leishman *et al.*, Chapter 2, this volume). For example, seeds must be endowed with adequate resources to accomplish germination and establishment. If these processes must take place in sites where the seedling can initially generate few resources on its own or in sites where intra- or interspecific competition is severe, there can be selection to increase the energy reserves within the seed, often resulting in a larger seed (e.g. Foster and Janson, 1985; Winn and Miller, 1995). Large seeds are generally harder to disperse than small ones – they need larger animals, stronger winds or more powerful propulsion (Kelly, 1995). Thus, selection for large seed size may bring with it selection for altered dispersal devices or may constrain the array of efficacious dispersal agents. Very large seeds cannot go far ballistically or by adhering to animal exteriors, they cannot be carried by small animals (such as ants) and they need very large wings to be successfully wind-dispersed. One option may be dispersal by vertebrates (Willson *et al.*, 1990a), and larger seeds generally require larger vertebrates to carry them (Foster and Janson, 1985; Wheelwright, 1985; Hammond and Brown, 1995). Even the chemical composition of seeds is associated with the dispersal mode; wind- and animal-dispersed species generally have greater proportions of fat and less of protein and carbohydrates than passively dispersed seeds (Lokesha *et al.*, 1992).

Furthermore, seeds must be protected against the physical environment and from natural enemies (see Chambers and MacMahon, 1994), and the demands of protection may sometimes interfere with dispersal by certain means. Also, fruits are often photosynthetic; selection to enhance the photosynthetic capacity of the fruit could affect fruit size, colour, shape and other design features that influence dispersal. In addition, a long style can increase the intensity of competition among male gametophytes, but it also affects dispersal distance in the ballistically dispersed *Geranium* (M.F. Willson and J. Ågren, unpublished). The timing of dispersal affects the probability and pattern of dispersal and the susceptibility to enemies, and hence can affect the selection pressures on diaspore morphology. The physiological costs of the various modes of

dispersal are generally unknown, but they constitute potential constraints on the evolution of dispersal devices.

Plant size and growth form show some correlations with dispersal mode, which may affect the evolution of dispersal traits (Thompson and Rabinowitz, 1989; Willson *et al.*, 1990a). For instance, few plants that are dispersed by ants or externally on animals are tall in stature. One reason may be that tall plants typically have large crowns, and ants commonly carry seeds for relatively short distances, such that the seeds of large-crowned, tall plants would seldom be carried beyond the crown of the parent and dispersal would be relatively ineffective. Ant dispersal of small acacia trees in Australia may be the exception that proves the rule and thus worthy of special study. Given that a species has become tall, the range of efficacious modes of dispersal may be limited. Most ballistically dispersed plants are small in stature, but at least a few trees use this mode. Dispersal by wind is frequent among species that are relatively tall within their respective habitats (e.g. trees in forests, tall forbs in fields). Although some understorey plants are also dispersed by wind, the common inference is that relatively short stature often renders wind dispersal less advantageous than other modes. Stature and growth form are sometimes correlated with seed size (Foster and Janson, 1985; Willson *et al.*, 1990a; but see Kelly, 1995), which may thus influence dispersal mode indirectly (through growth form), as well as more directly.

Whatever the array of constraints on diaspore evolution may be, it is also necessary to ascertain the occurrence and magnitude of selection on dispersal traits. At least two fundamental approaches are useful.

First, studies explicitly designed to measure selection on dispersal traits are essential. Seemingly small differences in the design of dispersal devices can have profound effects on dispersal ability: on aerodynamic performance by wind-dispersed species (Augspurger and Franson, 1987; Matlack, 1987; Sacchi, 1987); on capacity for attachment for diaspores car ried externally by animals (Bullock and Primack, 1977; Carlquist and Pauly, 1985; Sorensen, 1986); on foraging preferences of avian fruit consumers (Howe and vande Kerckhove, 1980; Wheelwright, 1985; Whelan and Willson, 1994; Traveset *et al.*, 1995; Loiselle *et al.*, 1996; Rey *et al.*, 1997; Traveset and Willson, 1998); and on the average distance of seed dispersal in a ballistically dispersed herb (M.F. Willson and J. Ågren, unpublished). Furthermore, individual plants often differ in the allocation of resources to dispersal devices on the diaspore (Willson *et al.*, 1990b; Jordano, 1995a), although the extent to which such differences are inheritable is seldom established (but see Wheelwright, 1993). Individual variation in seed shadows has been documented for a few species (Augspurger, 1983a; McCanny and Cavers, 1987; Thiede and Augspurger, 1996; Donohue, 1997, 1998). Such information needs to be brought together, so that we know the extent and pattern of individual variation in dispersal devices: how the variation affects the seed shadows of the respective parents; how the seed shadow affects parental fitness; and how these relationships vary among species and conditions. Several studies have shown that selection by avian dispersal agents may be relatively weak (e.g. Manasse and Howe, 1983; Herrera, 1984c, 1987, 1988; Jordano, 1987, 1993, 1994, 1995a, b; Guitián *et al.*, 1992; Willson and Whelan, 1993; Traveset, 1994; Whelan and Willson, 1994), although, in most cases, fitness is indexed by removal rates rather than by the eventual pattern of offspring dispersion.

Secondly, a less direct but still useful approach is to document patterns of variation in the array of dispersal modes present in plant communities (i.e. the dispersal spectra of those communities). Examination of the pattern can help generate hypotheses about the relative advantage of different dispersal modes in different regions and habitats. A few patterns have begun to emerge, but we seldom know how general they are. One consistent trend is that a high proportion of species in tropical wet forest is dispersed by vertebrate consumers (see references in Willson *et al.*, 1989), although there are biogeographical differences in the strength of the trend (Karr, 1976; Snow, 1981; Fleming *et al.*, 1987). In temperate zones, forests commonly have more vertebrate-dispersed species than other habitats, and the frequency of fleshy-fruited species is especially high in certain southern-hemisphere forests (Willson *et al.*, 1990a). Vertebrate dispersal apparently increases on moister sites, on fertile soils (in Australia) and in floras dominated by shrubs and/or trees (Willson *et al.*, 1990a). The causal factors for such patterns are not yet clear. In contrast, when comparing the seed dispersal spectra of five different types of communities on the Iberian Peninsula (potential woodland, forest fringe, scrubland, nitrophile communities and montane communities) between Mediterranean and Eurosiberian regions, no significant differences are found for any type of community, although biotic dispersal appears to be consistently more prevalent at mature stages of succession (Guitián and Sánchez, 1992).

A conspicuous and well-documented observation is the extraordinarily high frequency of ant-dispersed species in Australia and South Africa, particularly in sclerophyll vegetation on infertile soils. Several hypotheses (reviewed in Westoby *et al.*, 1991) have been proposed to explain such patterns, although few have been tested thoroughly (Hughes *et al.*, 1993). Seed size, the cost of dispersal structures and the availability of dispersal agents are among the potentially important factors. The availability of potassium and nitrogen in the soil may be limiting for the production of fleshy fruits and elaiosomes, respectively (Hughes *et al.*, 1993). The importance of seed deposition in nutrient-enriched ant mounds is debated (e.g. Bond and Stock, 1989) and may, indeed, vary from place to place. Seed burial may also protect seeds from fire or from surface-foraging seed predators (Bennett and Krebs, 1987). See Stiles (Chapter 5, this volume) for further discussion of myrmecochory.

External dispersal on vertebrates is common in riparian zones in arid parts of

Southern Africa and in disturbed and grazed habitats (Sorensen, 1986; Milton *et al.*, 1990; Willson *et al.*, 1990a; Fischer *et al.*, 1996). This pattern may reflect, in part, the level of activity of terrestrial mammals in such areas (i.e. the availability of dispersal agents).

If dispersal spectra are constructed with some measure of abundance (number of stems, percentage cover) instead of species, quite different trends may appear. For example, ants disperse 29% of the herbaceous species in a North American deciduous forest, but ant-dispersed species constitute 50–60% of the stems (Handel *et al.*, 1981). However, such comparisons are, at present, even rarer than those based on species composition (Frantzen and Bouman, 1989; Willson *et al.*, 1990a; Guitián and Sánchez, 1992). The construction of comparative dispersal spectra based on both species counts and abundances for many vegetation types in diverse regions would be heuristically productive.

When are seeds dispersed?

Less seems to be known about the evolutionary ecology of the 'when' of dispersal than of the 'why' and 'how'. Numerous ecological factors may contribute to dispersal phenology. Ideally, seed maturation and dispersal would be timed to match the seasonal availability of good dispersal agents (where required) and the availability of good germination conditions (for seeds lacking dormancy). Constraints on the ideal may derive from selection to avoid seed predators or to shift the flowering time, as well as the length of time required for fruit maturation. In addition, there is environmental variation in the time of fruit maturation and the timing of vector activity in a given area with concomitant differences in rates and quality of dispersal.

A few general patterns in dispersal phenology have been described. Wind-dispersed neotropical trees often mature their seeds during the dry season, when trade winds are strong and trees are leafless (Foster, 1982; Morellato and Leitao, 1996).

This contrasts with the production of fleshy or dry fruits throughout the year (De Lampe *et al.*, 1992). Fleshy-fruited plants in the north temperate zone commonly produce mature fruit crops in late summer and autumn, when avian frugivores are usually abundant, but, a little further south, more fruit maturation occurs in winter, when flocks of wintering migrant birds are foraging (Thompson and Willson, 1979; Willson and Thompson, 1982; Herrera, 1984a, 1995; Skeate, 1987; Snow and Snow, 1988). In contrast, ant-dispersed plants in central North America generally mature their seeds in early summer, at a time when avian frugivores are relatively few but ants are very active; given that they bloom in early spring, if they held their fruits until autumn, their low stature would keep the fruits inconspicuous to birds, beneath the foliage of other plants, and the maturing fruits would be exposed to predators all summer long (Thompson, 1981).

As these patterns of phenology in relation to disperser availability have emerged, evidence has appeared that they may not be entirely interpretable as adaptations to dispersal. Fruiting patterns in western Europe tend to match bird phenology at the community level, because abundant species with strictly northern distributions fruit earlier than those with strictly southern distributions, but wide-ranging plant species show no latitudinal shift in fruiting times, as would be expected if their fruiting seasons were adapted to disperser phenology (Fuentes, 1991; French, 1992; Willson and Whelan, 1993). Marked annual variation in the seasonal timing of fruit maturation of temperate plants indicates that events during earlier phases of reproduction can have a large impact on fruit timing and serves as a reminder that compromises may be required between the timing of flowering and the timing of fruit production (see Fenner, 1998).

Eriksson and Ehrlén (1998a) have examined structural and nutritional features of fleshy fruits of temperate plants in relation to phenology, finding that some secondary compounds containing nitrogen decrease during the season, i.e. are more abundant in early- than in late-fruiting species. Whether this pattern is adaptive remains an open question. In contrast, no phenological trends in lipid or carbohydrate contents were found. The adaptive value of secondary compounds is controversial. Cipollini and Levey (1997, 1998) suggested that they probably have a specific function in the fruits, postulating different adaptive hypotheses, whereas Ehrlén and Eriksson (1993) and Eriksson and Ehrlén (1998b) argue that the distribution patterns of secondary compounds in plant tissues do not call for any adaptive explanation of their presence in the fruits. Another pattern that appears when comparing early- with late-ripening species is in seed number per fruit, which decreases seasonally (Eriksson and Ehrlén, 1998a); assuming trade-offs of numbers and size, this is attributed to developmental constraints imposed by a demand for a long developmental time in large seeds or, alternatively, to a 'better' dissemination of small seeds early in the season.

Future investigation must be directed to unravelling both the ecological causes of seasonal patterns of dispersal and the consequences of variation in dispersal phenology. We know very little about how much variation occurs among individuals in the timing of dispersal and whether this is the result of genetic or site differences in conditions controlling fruit maturation (but see, for instance, Heideman, 1989; see also review in van Schaik *et al.*, 1993). We know even less about the possibility that differences in timing of seed maturation result in accompanying differences in seed dispersal and offspring success. From the perspective of animal dispersal agents, variation in timing and abundance of fruit production can have great effects on consumers, resulting in massive population movements (e.g. van der Wall and Balda, 1977; Leighton and Leighton, 1983; Ostfeld *et al.*, 1996; Selas, 1997; Hansson, 1998) or catastrophic mortality (see below), and such major effects on the consumer community are likely to have reciprocal, lingering effects on seed dispersal for several years.

For certain, usually long-lived, species, fruits are produced only once every few years in an unpredictable pattern. This phenomenon, called 'masting', has received much attention from ecologists and from evolutionary biologists (Kelly, 1994, and references therein). Three types of masting are distinguished: (i) strict masting, when the population reproduces synchronously and the distribution of seed crop size among years is bimodal (bamboos, *Strobilanthes*); (ii) normal masting, when synchrony is poor and there are many overlapping cohorts (e.g. imperfectly synchronized monocarps and many polycarps; *Fagus, Quercus, Pinus*); and (iii) putative masting, where variation in seed output is due only to environmental variation, without any evolutionary significance. A number of hypotheses (reviewed in Kelly, 1994) have been postulated to explain this phenomenon, the most widely accepted being related to economy of scale (i.e. larger reproductive efforts are more efficient in terms of successful pollination or seed production and survival) (e.g. Sork, 1993; Tapper, 1996; Kelly and Sullivan, 1997; Shibata *et al.*, 1998). The great variation in fruit production from year to year can have strong effects not only on the recruitment of the plant population itself (e.g. Schupp, 1990; Jones *et al.*, 1994; Crawley and Long, 1995; Shibata and Nakashizuka, 1995; Forget, 1997a), but also on the animal populations that consume their seeds (Ostfeld *et al.*, 1996, and references therein; Selas, 1997; Hansson, 1998; see also Crawley, Chapter 7, this volume).

The usefulness and ecological significance of the masting concept have been questioned by Herrera *et al.* (1998), who argue that a critical re-examination of patterns of annual variability in seed production is necessary, as most species apparently fall along broad continua of interannual variability in seed production, with no indication of multimodality (Kelly, 1994). In reviewing almost 300 data sets, Herrera *et al.* (1998) failed to identify distinct groups of species with contrasting levels of annual variability in seed output, although most polycarpic plants had alternating supra-annual schedules, consisting of either high- or low-reproduction years. Annual variability in seed production appears to be weakly associated with pollination mode (wind versus animal pollination). In contrast, animal-dispersed species are less variable than those dispersed by either inanimate means or animals that usually act as seed predators (Herrera *et al.*, 1998). These associations certainly deserve further investigation.

Ecological consequences of dispersal

Population structure

Dispersal of offspring away from the natal site is one way that genes move through a population or into new populations. Movement of genes also occurs at pollination in outcrossing species. Paternally transmitted genes in outcrossing species move twice in each seed generation, once during pollination and again during seed dispersal; maternally transmitted genes in outcrossers and all genes in self-fertilized seeds move only once, during seed dispersal. Thus, in any one seed generation, paternally transmitted genes are likely to move farther from their source than maternally transmitted genes. Lloyd (1982) used this observation as a factor favouring the evolution of cosexuality in seed plants. He also noted that the difference in paternal and maternal gene movement is less in plants with very effective seed dispersal (e.g. by birds), which might facilitate the evolution of dioecism in bird-dispersed plants.

Gene movement is often limited within a population, such that many plant populations consist of genetic 'neighbourhoods' of more or less related individuals (Levin, 1981; Gibson and Wheelwright, 1995). Dispersal by caching animals (Furnier *et al.*, 1987) or by animals that use sleeping sites (Julliot, 1997) or that are attracted to infected hosts (e.g. mistletoe dispersers: Larson, 1996) can lead to clusters of related plants within populations, even when the seeds have been carried

long distances. Microdifferentiation of local populations can occur on a very small spatial scale, in response to localized selection and/or very restricted gene flow (e.g. Schemske, 1984; Turkington and Aarssen, 1984; Parker, 1985; Berg and Hamrick, 1995; Linhart and Grant, 1996; Nagy, 1997; Nagy and Rice, 1997). Thus, the dispersal pattern of seeds contributes to the genetic structure of populations and to the potential for both genetic drift and responses to natural selection. Although some correlations of dispersal mode with the degree of local differentiation have emerged, these correlations are not very tight, and other factors must also contribute to observed genetic structuring (Hamrick and Loveless, 1986; Hamrick *et al.*, 1993; and see review in Linhart and Grant, 1996; Schnabel *et al.*, 1998).

On the other hand, the at least occasional passage of genes out of a local neighbourhood or between conspecific populations is important in maintaining the genetic diversity of the recipient population and presumably slows the rate of population differentiation (e.g. Slatkin, 1987; Hamrick *et al.*, 1993; Linhart and Grant, 1996). To the extent that outcrossing is advantageous, the most effective outbreeding in populations with neighbourhood structure will occur when genes pass from one neighbourhood to another. Thus, in neighbourhood-structured populations, the 'best' outcrossing is rare, by definition. When neighbourhoods reflect ecotypic differentiation to local conditions, however, the 'best' outcrossing may occur between individuals that are not too close together and yet not too far apart. The concept of 'optimal outcrossing' has been controversial, and the extent to which selection may favour a degree of inbreeding is still unclear (Shields, 1982; Waddington, 1983; Jarne and Charlesworth, 1993, and references therein). Seed dispersal patterns have a clear potential to affect the level of outcrossing achieved.

The demographic and evolutionary consequences of seed dispersal began to receive attention only a few years ago (Houle, 1992, 1995, 1998; Herrera *et al.*,

1994; Horvitz and Schemske, 1994, 1995; Jordano and Herrera, 1995; Schupp, 1995; Schupp and Fuentes, 1995; Shibata and Nakashizuka, 1995; Kollman and Schill, 1996; Forget and Sabatier, 1997; Valverde and Silvertown, 1997; Carlton and Bazzaz, 1998; Dalling *et al.*, 1998). The scarcity of information available on the causes and consequences of spatial patterns of dispersal at a variety of scales from seeds to new adults is certainly a major gap in our knowledge on the ecology of seed dispersal. Most of the studies that consider the multistaged nature of recruitment find no strong and consistent relationships between seed and seedling spatial patterns of abundance. The causes of this 'uncoupling' are mainly attributed to the spatiotemporal variation in the relative importance of mortality factors (e.g. predation, pathogens, competition) for seeds and seedlings (Houle, 1992, 1995, 1998). Seed–seedling conflicts occur, for instance, in those microhabitats where the probability of seed survival is low but seedling survival is high (Jordano and Herrera, 1995). These conflicts probably play an important role in structuring many natural systems, as they appear to be rather common (Schupp, 1995). Plant population dynamics in patchy environments depends not only on stage-specific survival and growth in different patches, but also on the degree of discordance in patch suitabilities across stages (Kollmann and Pirl, 1995; Schupp, 1995; Schupp and Fuentes, 1995; Kollmann and Schill, 1996; Aguiar and Sala, 1997; Forget, 1997b; Russell and Schupp, 1998). Such discordance can have major impacts on both the quantity and the spatial patterning of recruits. Furthermore, site suitability may not be independent of seed arrival (due to density-dependent mortality factors).

In the case of animal-dispersed plants, the influence of frugivorous animals depends on the extent of coupling of the different stages in the recruitment process, which can vary among sites and populations. Factors acting at the end of the recruitment process can potentially 'screen off' the effects acting at the beginning,

making less predictable the demographic consequences of seed dispersal (Herrera *et al.*, 1994; Schupp, 1995). Intra- and inter-population variation in the composition of disperser assemblages visiting a plant species has been little documented (Snow and Snow, 1988; Reid, 1989; Guitián *et al.*, 1992; Jordano, 1994; Traveset, 1994; Loiselle and Blake, 1999), despite its potential demographic importance. Different species of frugivores generate characteristic seed shadows, depending on foraging behaviour, seed retention times, patterns of fruit selection and response to the vegetation structure (Herrera, 1995; Rey, 1995). In order to evaluate the effect of seed vectors on plant demography we need to know the disperser effectiveness, i.e. the proportion of the seed crop dispersed by a particular species (Schupp, 1993), and to examine how suitable the microsite is where seeds are deposited for germination and establishment. As effectiveness is difficult to estimate in the field, Bustamante and Canals (1995) have proposed a model to estimate it indirectly.

Colonization and plant community structure

Dispersal mode is one factor that affects the ability of a plant species to colonize a new area, especially one at some distance from the seed source. Long-distance dispersal capacity is poorly developed in ballistic and ant-dispersed species and much better developed in wind- and vertebrate-dispersed species. Wind and birds account for the arrival of most species in an isolated cloud forest in Colombia (Sugden, 1982). But wind dispersal is insufficient to result in frequent colonization of extremely distant islands, where many colonists arrive inside avian guts or stuck to the feathers (good numbers also arrive, without special devices, in the mud on birds' feet, and some come on ocean currents (Carlquist, 1974)). Birds may have been responsible for post-Pleistocene colonization of habitat islands on mountain-tops in western North America by conifers (Wells, 1983). Many colonists in the island flora of the Great

Lakes are bird-dispersed, and a similar proportion may travel by water (Morton and Hogg, 1989). Likewise, a study by Whittaker and Jones (1994) showed that 30% of the flora of Krakatoa island has arrived and expanded, since the volcanic eruption in 1883, by endozoochory (specifically by birds and bats). Thus, the composition of island floras reflects, in part, the dispersal ability of potential colonists. The initial colonization of debris avalanches after a volcanic eruption on Mount St Helens, Washington, was accomplished primarily by wind-dispersed species, although colonization was independent of distance to the source area (Dale, 1989). The ability of a species to establish a new population at unoccupied sites is a critical feature in the maintenance of biological diversity. Current habitat fragmentation creates barriers to dispersal, however, impeding the natural dispersal of some species out of their range in response to global climate change (Primack and Miao, 1992).

Harper (1977) modified the original model of van der Plank (1960) and suggested that patterns of colonization may differ as a function of the shape of the seed shadow. If the slope of the regression of seed number versus distance (on a log–log scale) is steeper than -2, Harper proposed that colonization would frequently occur by 'fronts' of invasion, in which phalanxes of colonizers gradually invade new areas relatively close to the seed source. But, if the slope is less steep, colonization may occur chiefly by far-flung outposts of establishment. There is a weak association of dispersal mode with steepness of the log–log slope of the seed shadow tail (Portnoy and Willson, 1993). However, many other factors also affect colonization patterns (e.g. postdispersal seed predation, germination requirements, conditions for dispersal).

After colonization has occurred, the spatial distribution of the colonizers may persist for decades or centuries, with repercussions for the establishment of subsequent colonists (e.g. Yarranton and Morrison, 1974). The presence of small trees and shrubs in an old field or pasture often increases the deposition of bird- or bat-dispersed seeds beneath them (see references in Willson, 1991; Debussche and Isenmann, 1994; Verdú and García-Fayos, 1996, 1998) and decreases the deposition of wind-dispersed seeds (Willson and Crome, 1989). Clusters of individuals of fleshy-fruited species often persist even after the initial perch tree has died. On the other hand, the early colonizers may inhibit further colonization if they establish themselves so densely that few other plants can grow beneath them. Thus, some aspects of the spatial patterning of plant succession can be related to dispersal. In the Mediterranean region, dispersal of fleshy-fruited plants by birds appears to be unimportant for plant dynamics in open herbaceous communities and in dense forests, but it is crucial when woody patches appear with succession in the open communities or when grassy patches appear in the forest (Debussche and Isenmann, 1994; see also Kollmann, 1995). In desert playas of western North America, what seems to be limiting the initiation of primary succession is not seed dispersal but the low rates of seed entrapment in these habitats (Fort and Richards, 1998).

Plant dispersal and animal communities

Plant propagules (i.e. the dispersing phase of the life history) are critical food resources for a vast number of animal species. Legions of insect species have specialized to a life of seed predation (both pre- and postdispersal), and some of the prodigious radiation of insects is associated with these specializations. Whole taxonomic groups of birds and mammals also use seeds as central resources. In turn, these predators have exerted selection pressures on plants to develop and diversify chemical and structural defences.

Fleshy-fruited plants engage in mutualisms with their dispersal agents; these relationships are quite generalized, very ancient, extremely widespread and extraordinarily frequent in certain communities (see references in Willson *et al.*, 1989, 1990a; Willson, 1993b). Many vertebrate populations rely on fleshy fruits as food for migration, breeding and winter maintenance. Fruit resources are thought to be crucial in sustaining certain vertebrate populations in some tropical areas (e.g. Terborgh, 1986; Gautier-Hion and Michaloud, 1989; Julliot, 1997). Heavy use of fruit resources may account for part of the great diversity of tropical vertebrates (Karr, 1980) and may have been related to the radiation of certain tropical bird families (Snow, 1981). In turn, the biotic dispersal of seeds seems to have contributed to some extent to angiosperm diversification (Tiffney and Mazer, 1995, and references therein; but see also Ricklefs and Renner, 1994).

Non-mutualistic animals also exploit mutualistic interactions and effectively become parasites on the mutualistic system. Both vertebrate and invertebrate consumers (plus fungi and microbes) capitalize on fleshy fruits, without dispersing the seeds. Some insects have become fruit-pulp specialists to the extent that the radiation of certain families (e.g. Tephritidae: Bush, 1966) is associated with this kind of parasitism. The effect of invertebrate parasitism of fruit pulp on potential vertebrate dispersal agents varies. Although microbial and fungal infestations generally depress effective dispersal (Knoch *et al.*, 1993), infestation by insect larvae can either increase or decrease fruit consumption by birds, depending on the bird species (Willson and Whelan, 1990; Traveset *et al.*, 1995, and references therein). In addition, the foraging of frugivores may decrease the abundance and change the distribution of insect frugivores, with reciprocal, often beneficial, effects on plant reproduction (Herrera, 1984b, 1989b; but see Traveset, 1992, 1993). In sum, dispersal mutualisms between plants and animals have had prodigious and ramifying effects on the animal community.

Dependence on animals for seed transport means that the plants are susceptible to dispersal failure when their seed vectors become rare or extinct. Disruption of this mutualism can have serious consequences for the maintenance of the plant populations. Loss of native seed-dispersing ants from certain habitats in South Africa means poor seed dispersal and low seed survival and may lead to the extinction of many rare and endemic plants (Bond and Slingsby, 1984; Bond, 1994). Extinction of the dodo on Mauritius has probably affected the population structure of the tree whose seeds it dispersed (Temple, 1977; but see also Owadally, 1979; Temple, 1979; Witmer and Cheke, 1991). When a plant has many dispersal agents (as is true for many small-fruited, vertebrate-dispersed species in North America, for instance), the loss of one species of vector may have minor consequences for plant population biology. However, as both temperate and equatorial forests continue to be decimated, the remnant stands are losing many of their dispersers, with potentially severe consequences for their continued survival. Evidence is growing from some of the South Pacific Islands (Cox *et al.*, 1991) and from Chatham Island (Given, 1995) that disappearance of the main dispersers of some plant species deeply alters their reproductive success. Likewise, the extinction of the Pleistocene megafauna may have left many tropical trees with only a few substitute dispersers (e.g. Hallwachs, 1986; but see Howe, 1985), although the population consequences of the historical change cannot be examined.

Phylogenetic patterns in dispersal

Dispersal modes often differ greatly within taxonomic units, and a single mode may arise independently many times (e.g. wind dispersal in the legumes: Augspurger, 1989). It seems likely that some morphological transformations are more easily made than others. For instance, a plume for wind dispersal may be converted to a hook for dispersal on vertebrate exteriors (e.g. *Anemone, sensu lato*) or vice versa. The loss of a wing contributed to a change from wind to bird dispersal in *Pinus* but was accompanied by changes in cone structure and seed size as well; nevertheless, the shift between these two modes of dispersal has occurred several times (see Strauss and Doerksen, 1990; Tomback and Linhart, 1990; Benkman, 1995). A change from bird to ant dispersal in *Acacia* necessitated chiefly a shift in the size and colour of the food body attached to the seed (O'Dowd and Gill, 1986; see Fig. 4.5), but a similar shift in *Trillium* occasioned a seemingly more complex change, from a fleshy fruit enclosing several unappendaged seeds to a dry fruit with elaiosome-bearing seeds (Berg, 1958). We need phylogenetic analyses of whole families or genera with respect to dispersal modes to determine: (i) how often dispersal mode has changed within a taxon; (ii) what the directions of the change are; (iii) what kinds of changes are most common; and (iv) for wide-ranging taxa, how the biogeographical history of different regions influences the evolution of diaspore traits. The answers to such questions, in conjunction with ecological data, will contribute to our understanding of community dispersal spectra, patterns of selection on dispersal devices and other aspects of population and community biology related to dispersal. A synthesis of modern systematics and evolutionary ecology is a powerful tool in elucidating questions about diaspore adaptation and phylogenetic radiation (e.g. Wanntorp *et al.*, 1990; Bremer and Eriksson, 1992; Ricklefs and Renner, 1994; Tiffney and Mazer, 1995).

Conclusion

The study of the dispersal of plants has advanced relatively fast in the last decade, as essential elements of the evolutionary and ecological causes and consequences of dispersal have been examined. The link between seed dispersal and its demographic and genetic consequences is one major gap that still needs to be filled, although some recent studies are already paying attention to it. Much remains to be discovered yet in terms of geographical and habitat patterns, as well as the dynamics of colonization, population differentiation and plant/animal interactions. Dispersal ecology is a rapidly developing field that still offers a wealth of investigative opportunity at levels ranging from good natural history to sophisticated modelling and conceptual synthesis.

Acknowledgements

C.K. Augspurger and J.N. Thompson graciously provided constructive comments on the manuscript written for the first edition of this book.

References

Aguiar, M.R. and Sala, O.E. (1997) Seed distribution constrains the dynamics of the Patagonian steppe. *Ecology* 78, 93–100.

Alexander, H.M. and Holt, R.D. (1998) The interaction between plant competition and disease. *Perspectives in Plant Ecology, Evolution and Systematics* 1, 206–220.

Armstrong, D.P. and Westoby, M. (1993) Seedlings from large seeds tolerate defoliation better: a test using phylogenetically independent contrasts. *Ecology* 74, 1092–1100.

Aronne, C. and Wilcock, C.C. (1994) First evidence of myrmecochory in fleshy–fruited shrubs of the Mediterranean region. *New Phytologist* 127, 781–788.

Augspurger, C.K. (1983a) Seed dispersal of the tropical tree, *Platypodium elegans*, and the escape of its seedlings from fungal pathogens. *Journal of Ecology* 71, 759–771.

Augspurger, C.K. (1983b) Offspring recruitment around tropical trees: changes in cohort distance with time. *Oikos* 40, 189–196.

Augspurger, C.K. (1984) Seedling survival of tropical tree species: interactions of dispersal distance, light-gaps, and pathogens. *Ecology* 65, 1705–1712.

Augspurger, C.K. (1986) Morphology and dispersal potential of wind-dispersed diaspores of neotropical trees. *American Journal of Botany* 73, 353–363.

Augspurger, C.K. (1989) Morphology and aerodynamics of wind-dispersed legumes. *Advances in Legume Biology* 29, 451–466.

Augspurger, C.K. and Franson, S.E. (1987) Wind dispersal of artificial fruits varying in mass, area, and morphology. *Ecology* 68, 27–42.

Augspurger, C.K. and Kelly, C.K. (1984) Pathogen mortality of tropical tree seedlings: experimental studies of the effects of dispersal distance, seedling density, and light conditions. *Oecologia* 61, 211–217.

Augspurger, C.K. and Kitajima, K. (1992) Experimental studies of seedling recruitment from contrasting seed distributions. *Ecology* 73, 1270–1284.

Beattie (1985) *The Evolutionary Ecology of Ant–Plant Mutualisms.* Cambridge University Press, Cambridge.

Benkman, C.W. (1995) Wind dispersal capacity of pine seeds and the evolution of different seed dispersal modes in pines. *Oikos* 73, 221–224.

Benkman, C.W., Balda, R.P. and Smith, C.C. (1984) Adaptations for seed dispersal and the compromises due to seed predation in limber pine. *Ecology* 65, 632–642.

Bennett, A. and Krebs, J. (1987) Seed dispersal by ants. *Trends in Ecology and Evolution* 2, 291–292.

Berg, E.E. and Hamrick, J.L. (1995) Fine-scale genetic structure of a turkey oak forest. *Evolution* 49, 110–120.

Berg, R.Y. (1958) Seed dispersal, morphology, and phylogeny of *Trillium. Skrifter utgitt av det Norske Videnskaps-Akademi i Oslo 1. Matematisk-Naturuidenskapelig Klasse* 1, 1–36.

Bond, W.J. (1994) Do mutualisms matter? Assessing the impact of pollinator and disperser disruption on plant extinction. *Philosophical Transactions of the Royal Society of London* 344, 83–90.

Bond, W.J. and Slingsby, P. (1984) Collapse of an ant-plant mutualism: the Argentine ant (*Iridomyrmex humilis*) and myrmecochorous Proteaceae. *Ecology* 65, 1031–1037.

Bond, W.J. and Stock, W.D. (1989) The costs of leaving home: ants disperse myrmecochorous seeds to low nutrient sites. *Oecologia* 81, 412–417.

Bremer, B. and Eriksson, O. (1992) Evolution of fruit characteristics and dispersal modes in the tropical family Rubiaceae. *Biological Journal of the Linnean Society* 47, 79–95.

Bullock, S.H. and Primack, R.B. (1977) Comparative experimental study of seed dispersal on animals. *Ecology* 58, 681–686.

Burdon, J.J. (1987) *Diseases and Plant Population Biology.* Cambridge University Press, Cambridge.

Bush, G.L. (1966) The taxonomy, cytology, and evolution of the genus *Rhagoletis* in North America north of Mexico. *Bulletin of the Museum of Comparative Zoology* 134, 431–562.

Bustamante, R.O. and Canals, L.M. (1995) Dispersal quality in plants: how to measure efficiency and effectiveness of a seed disperser. *Oikos* 73, 133–136.

Cain, M.L., Damman, H. and Muir, A. (1998) Seed dispersal and the Holocene migration of woodland herbs. *Ecological Monographs* 68, 325–347.

Carlquist, S. (1974) *Island Biology.* Columbia University Press, New York.

Carlquist, S. and Pauly, Q. (1985) Experimental studies on epizoochorous dispersal in California plants. *Aliso* 11, 167–177.

Carlton, G.C. and Bazzaz, F.A. (1998) Regeneration of three sympatric birch species on experimental hurricane blowdown microsites. *Ecological Monographs* 68, 99–120.

Chambers, J.C. and MacMahon, J.A. (1994) A day in the life of a seed: movements and fates of seeds and their implications for natural and managed systems. *Annual Review of Ecology and Systematics* 25, 263–292.

Cipollini, M.L. and Levey, D.J. (1997) Secondary metabolites of fleshy vertebrate-dispersed fruits: adaptive hypotheses and implications for seed dispersal. *American Naturalist* 150, 346–372.

Cipollini, M.L. and Levey, D.J. (1998) Secondary metabolites as traits of ripe fleshy fruits: a response to Eriksson and Ehrlén. *American Naturalist* 152, 908–911.

Clifford, H.T. and Monteith, G.B. (1989) A three phase seed dispersal mechanism in Australian quinine bush (*Petalostigma pubescens* Domin). *Biotropica* 21, 284–286.

Coe, M. and Coe, C. (1987) Large herbivores, acacia trees and bruchid beetles. *South African Journal of Science* 83, 624–635.

Cohen, D. and Motro, U. (1989) More on optimal rates of dispersal: taking into account the cost of the dispersal mechanism. *American Naturalist* 134, 659–663.

Collins, S.L. and Uno, C.E. (1985) Seed predation, seed dispersal, and disturbances in grassland: a comment. *American Naturalist* 125, 866–872.

Comins, H.N., Hamilton, W.D. and May, R.M. (1980) Evolutionarily stable dispersal strategies. *Journal of Theoretical Biology* 82, 205–230.

Cox, P.A., Elmquist, T., Pierson, E.D. and Rainey, W.E. (1991) Flying foxes as strong interactors in South Pacific island ecosystems: a conservation hypothesis. *Conservation Biology* 5, 448–454.

Crawley, M.J. and Long, C.R. (1995) Alternate bearing, predator satiation and seedling recruitment in *Quercus robur* L. *Journal of Ecology* 83, 683–696.

Dale, V.H. (1989) Wind dispersed seeds and plant recovery on the Mount St. Helens debris avalanche. *Canadian Journal of Botany* 67, 1434–1441.

Dalling, J.W., Hubbell, S.P. and Silvera, K. (1998) Seed dispersal, seedling establishment and gap partitioning among tropical pioneer trees. *Journal of Ecology* 86, 674–689.

Davidson, D.W. and Morton, S.R. (1984) Dispersal adaptations of some *Acacia* species in the Australian arid zone. *Ecology* 65, 1038–1051.

Debussche, M. and Isenmann, P. (1994) Bird-dispersed seed rain and seedling establishment in patchy Mediterranean vegetation. *Oikos* 69, 414–426.

Debussche, M. and Lepart, J. (1992) Establishment of woody plants in Mediterranean old fields: opportunity in space and time. *Landscape Ecology* 6, 133–145.

De Lampe, M.G., Bergeron, Y., McNeil, R. and Leduc, A. (1992) Seasonal flowering and fruiting patterns in tropical semiarid vegetation of north-eastern Venezuela. *Biotropica* 24, 64–76.

Dinerstein, E. (1989) The foliage-as-fruit hypothesis and the feeding behavior of south Asian ungulates. *Biotropica* 21, 214–218.

Donohue, K. (1997) Seed dispersal in *Cakile edentula* var. *lacustris*: decoupling the fitness effects of density and distance from the home site. *Oecologia* 110, 520–527.

Donohue, K. (1998) Maternal determinants of seed dispersal in *Cakile edentula*: fruit, plant, and site traits. *Ecology* 79, 2771–2788.

Ehrlén, J. and Eriksson, O. (1993) Toxicity in fleshy fruits – a non-adaptive trait? *Oikos* 66, 107–113.

Ellner, S. and Schmida, A. (1981) Why are adaptations for long-range seed dispersal rare in desert plants? *Oecologia* 51, 133–144.

Ellstrand, N.C. and Antonovics, J. (1985) Experimental studies of the evolutionary significance of sexual reproduction. II. A test of the density-dependent selection hypothesis. *Evolution* 39, 657–666.

Eriksson, O. (1992) Evolution of seed dispersal and recruitment in clonal plants. *Oikos* 63, 439–448.

Eriksson, O. and Ehrlén, J. (1998a) Phenological adaptations in fleshy vertebrate-dispersed fruits of temperate plants. *Oikos* 82, 617–621.

Eriksson, O. and Ehrlén, J. (1998b) Secondary metabolites in fleshy fruits: are adaptive explanations needed? *American Naturalist* 152, 905–907.

Fenner, M. (1998) The phenology of growth and reproduction in plants. *Perspectives in Plant Ecology, Evolution and Systematics* 1, 78–91.

Fischer, S.F., Poschlod, P. and Beinlich, B. (1996) Experimental studies on the dispersal of plants and animals on sheep in calcareous grasslands. *Journal of Applied Ecology* 33, 1206–1222.

Fleming, T.H., Breitwisch, R. and Whitesides, G.H. (1987) Patterns of vertebrate frugivore diversity. *Annual Review of Ecology and Systematics* 18, 91–109.

Forget, P.M. (1997a) Ten-year seedling dynamics in *Vouacapoua americana* in French Guiana: a hypothesis. *Biotropica* 29, 124–126.

Forget, P.M. (1997b) Effect of microhabitat on seed fate and seedling performance in two rodent-dispersed tree species in rainforest in French Guiana. *Journal of Ecology* 85, 693–703.

Forget, P.M. and Sabatier, D. (1997) Dynamics of the seedling shadow of a frugivore-dispersed tree species in French Guiana. *Journal of Tropical Ecology* 13, 767–773.

Fort, K.P. and Richards, J.H. (1998) Does seed dispersal limit initiation of primary succession in desert playas? *American Journal of Botany* 85, 1722–1731.

Foster, R.B. (1982) The seasonal rhythm of fruitfall on Barro Colorado Island. In: Leigh, E.G., Rand, A.S. and Windsor, D.M. (eds) *The Ecology of a Tropical Forest*. Smithsonian, Washington, DC, pp. 151–172.

Foster, S.A. and Janson, C.H. (1985) The relationship between seed size and establishment conditions in tropical woody plants. *Ecology* 66, 773–780.

Frantzen, N.M.L.H.F. and Bouman, F. (1989) Dispersal and growth form patterns of some zonal paramo vegetation types. *Acta Botanica Neerlandica* 38, 449–465.

French, K. (1992) Phenology of fleshy fruits in a wet sclerophyll forest in southeastern Australia: are birds an important influence? *Oecologia* 90, 366–373.

Fuentes, M. (1991) Latitudinal and elevational variation in fruiting phenology among western European bird-dispersed plants. *Ecography* 15, 177–183.

Furnier, G.R., Knowles, P., Clyde, M.A. and Dancik, B.P. (1987) Effects of avian seed dispersal on the genetic structure of whitebark pine populations. *Evolution* 41, 607–612.

Gautier-Hion, A. and Michaloud, G. (1989) Are figs always keystone resources for tropical frugivorous vertebrates? A test in Gabon. *Ecology* 70, 1826–1833.

Geritz, S.A.H., de Jong, T.J. and Klinkhamer, P.G.L. (1984) The efficacy of dispersal in relation to safe site area and seed production. *Oecologia* 62, 219–221.

Ghiselin, M. (1974) *The Economy of Nature and the Evolution of Sex*. University of California Press, Berkeley, California.

Gibson, J.P. and Wheelwright, N.T. (1995) Genetic structure in a population of a tropical tree *Ocotea tenera* (Lauraceae): influence of avian seed dispersal. *Oecologia* 103, 49–54.

Given, D.R. (1995) Biological diversity and the maintenance of mutualisms. In: Vitousek, P.M., Loope, L.L. and Adsersen, H. (eds) *Islands: Biological Diversity and Ecosystem Function*. Springer Verlag, Berlin, pp. 149–162.

Gorb, S.N. and Gorb, E.V. (1995) Removal rates of seeds of five myrmecochorous plants by the ant *Formica polyctena* (Hymenoptera: Formicidae). *Oikos* 73, 367–374.

Gorchov, D.L., Cornejo, F., Ascorra, C. and Jaramillo, M. (1993) The role of seed dispersal in the natural regeneration of rainforest after strip-cutting in the Peruvian Amazon. *Vegetatio* 107/108, 339–349.

Green, D.S. (1983) The efficacy of dispersal in relation to safe site density. *Oecologia* 56, 356–358.

Greene, D.F. and Johnson, E.A. (1996) Wind dispersal of seeds from a forest into a clearing. *Ecology* 77, 595–609.

Guitián, J. and Sánchez, J.M. (1992) Seed dispersal spectra of plant communities in the Iberian Peninsula. *Vegetatio* 98, 157–164.

Guitián, J., Fuentes, M., Bermejo, T. and López, B. (1992) Spatial variation in the interactions between *Prunus mahaleb* and frugivorous birds. *Oikos* 63, 125–130.

Gunther, R.W. and Lanza, J. (1989) Variation in attractiveness of *Trillium* diaspores to a seed-dispersing ant. *American Midland Naturalist* 122, 321–328.

Hallwachs, W. (1986) Agoutis (*Dasyprocta punctata*): the inheritors of guapinol (*Hymenaea courbaril*: Leguminosae). In: Estrada, A. and Fleming, T.H. (eds) *Frugivores and Seed Dispersal*. Junk, Dordrecht, pp. 285–304.

Hamilton, W.D. and May, R.M. (1977) Dispersal in stable habitats. *Nature* 269, 578–581.

Hammond, D.S. and Brown, V.K. (1995) Seed size of woody plants in relation to disturbance, dispersal, soil type in wet neotropical forests. *Ecology* 76, 2544–2561.

Hamrick, J.L. and Loveless, M.D. (1986) The influence of seed dispersal mechanisms on the genetic structure of plant populations. In: Estrada, A. and Fleming, T.H. (eds) *Frugivores and Seed Dispersal*. Junk, Dordrecht, pp. 211–223.

Hamrick, J.L., Murawski, D.A. and Nason, J.D. (1993) The influence of seed dispersal mechanisms on the genetic structure of tropical tree populations. *Vegetatio* 107/108, 281–297.

Handel, S.N., Fisch, S.B. and Schatz, G.E. (1981) Ants disperse a majority of herbs in a mesic forest community in New York State. *Bulletin of the Torrey Botanical Club* 108, 430–437.

Hansson, L. (1998) Mast seeding and population dynamics of rodents: one factor is not enough. *Oikos* 82, 591–594.

Harper, J.L. (1977) *Population Biology of Plants*. Academic Press, London.

Heideman, P.D. (1989) Temporal and spatial variation in the phenology of flowering and fruiting in a tropical rainforest. *Journal of Ecology* 77, 1059–1079.

Herrera, C.M. (1984a) A study of avian frugivores, bird-dispersed plants, and their interaction in Mediterranean scrublands. *Ecological Monographs* 54, 1–23.

Herrera, C.M. (1984b) Avian interference with insect frugivory: an exploration into the plant–bird–fruit pest evolutionary triad. *Oikos* 42, 203–210.

Herrera, C.M. (1984c) Seed dispersal and fitness determinants in wild rose: combined effects of hawthorn, birds, mice and browsing ungulates. *Oecologia* 63, 386–393.

Herrera, C.M. (1987) Vertebrate-dispersed plants of the Iberian peninsula: a study of fruit characteristics. *Ecological Monographs* 57, 305–331.

Herrera, C.M. (1988) The fruiting ecology of *Osyris quadripartita*: individual variation and evolutionary potential. *Ecology* 69, 233–249.

Herrera, C.M. (1989a) Vertebrate frugivores and their interaction with invertebrate fruit predators: supporting evidence from a Costa Rican dry forest. *Oikos* 54, 185–188.

Herrera, C.M. (1989b) Frugivory and seed dispersal by carnivorous mammals and associated fruit characteristics in undisturbed Mediterranean habitats. *Oikos* 55, 250–262.

Herrera, C.M. (1995) Plant–vertebrate seed dispersal systems in the Mediterranean: ecological, evolutionary, and historical determinants. *Annual Review of Ecology and Systematics* 26, 705–727.

Herrera, C.M., Jordano, P., López–Soria, L. and Amat, J.A. (1994) Dispersal ecology of a mast-fruiting, bird-dispersed tree: bridging frugivory activity and seedling establishment. *Ecological Monographs* 64, 315–344.

Herrera, C.M., Jordano, P., Guitián, J. and Traveset, A. (1998) Annual variability in seed production by woody plants and the masting concept: reassessment of principles and relationship to pollination and seed dispersal. *American Naturalist* 152, 576–594.

Hoppes, W.G. (1988) Seedfall pattern of several species of bird-dispersed plants in an Illinois woodland. *Ecology* 69, 320–329.

Horvitz, C.C. and Le Corff, J. (1993) Spatial scale and dispersion pattern of ant- and bird-dispersed herbs in two tropical lowland rainforests. *Vegetatio* 107/108, 351–362.

Horvitz, C.C. and Schemske, D.W. (1986). Seed dispersal and environmental heterogeneity in a neotropical herb: a model of population and patch dynamics. In: Estrada, A. and Fleming, T.H. (eds) *Frugivores and Seed Dispersal*. Junk, Dordrecht, pp. 169–186.

Horvitz, C.C. and Schemske, D.W. (1994) Effects of dispersers, gaps, and predators on dormancy and seedling emergence in a tropical herb. *Ecology* 75, 1949–1958.

Horvitz, C.C. and Schemske, D.W. (1995) Spatiotemporal variation in demographic transitions of a tropical understory herb: projection matrix analysis. *Ecological Monographs* 65, 155–192.

Houle, G. (1992) Spatial relationship between seed and seedling abundance and mortality in a deciduous forest of north-eastern North America. *Journal of Ecology* 80, 99–108.

Houle, G. (1995) Seed dispersal and seedling recruitment: the missing link(s). *Ecoscience* 2, 238–244.

Houle, G. (1998) Seed dispersal and seedling recruiment of *Betula alleghaniensis*: spatial inconsistency in time. *Ecology* 79, 807–818.

Howe, H.F. (1985) Gomphothere fruits: a critique. *American Naturalist* 125, 853–865.

Howe, H.F. (1993) Aspects of variation in a neotropical seed dispersal system. *Vegetatio* 107/108, 149–162.

Howe, H.F. and Smallwood, J. (1982) Ecology of seed dispersal. *Annual Review of Ecology and Systematics* 13, 201–228.

Howe, H.F. and vande Kerckhove, G.A. (1980) Nutmeg dispersal by tropical birds. *Science* 210, 925–927.

Hughes, L., Westoby, M. and Johnson, A.D. (1993) Nutrient costs of vertebrate- and ant-dispersed fruits. *Functional Ecology* 7, 54–62.

Hulme, P.E. (1998) Post-dispersal seed predation: consequences for plant demography and evolution. *Perspectives in Plant Ecology, Evolution and Systematics* 1, 32–46.

Janzen, D.H. (1971) Seed predation by animals. *Annual Review of Ecology and Systematics* 2, 465–492.

Janzen, D.H. (1984) Dispersal of small seeds by big herbivores: foliage is the fruit. *American Naturalist* 123, 338–353.

Jarne, P. and Charlesworth, D. (1993) The evolution of the selfing rate in functionally hermaphrodite plants and animals. *Annual Review of Ecology and Systematics* 24, 441–466.

Johnson, W.C. (1988) Estimating dispersibility of *Acer, Fraxinus* and *Tilia* in fragmented landscapes from patterns of seedling establishment. *Landscape Ecology* 1, 175–187.

Johnson, W.C. and Webb, T. (1989) The role of blue jays (*Cyanocitta cristata* L.) in the postglacial dispersal of fagaceous trees in eastern North America. *Journal of Biogeography* 16, 561–571.

Jones, R.H., Sharitz, R.R., Dixon, P.M. and Segal, D.S. (1994) Woody plant regeneration in four floodplain forests. *Ecological Monographs* 64, 345–367.

Jordano, P. (1987) Avian fruit removal: effects of fruit variation, crop size, and insect damage. *Ecology* 68, 1711–1723.

Jordano, P. (1993) Geographical ecology and variation of plant–seed disperser interactions: southern Spanish junipers and frugivorous thrushes. *Vegetatio* 107/108, 85–104.

Jordano, P. (1994) Spatial and temporal variation in the avian–frugivore assemblage of *Prunus mahaleb*: patterns and consequences. *Oikos* 71, 479–491.

Jordano, P. (1995a) Frugivore-mediated selection on fruit and seed size: birds and St. Lucie's cherries, *Prunus mahaleb. Ecology* 76, 2627–2639.

Jordano, P. (1995b) Angiosperm fleshy fruits and seed dispersers: a comparative analysis of adaptation and constraints in plant–animal interactions. *American Naturalist* 145, 163–191.

Jordano, P. and Herrera, C.M. (1995) Shuffling the offspring: uncoupling and spatial discordance of multiple stages in vertebrate seed dispersal. *Ecoscience* 2, 230–237.

Julliot, C. (1997) Impact of seed dispersal by red howler monkeys *Alouatta seniculus* on the seedling population in the understorey of tropical rainforest. *Journal of Ecology* 85, 431–440.

Karr, J.R. (1976) Within- and between-habitat avian diversity in African and neotropical lowland habitats. *Ecological Monographs* 46, 457–481.

Karr, J.R. (1980) Geographical variation in the avifaunas of tropical forest undergrowth. *Auk* 97, 283–298.

Kelley, S.E. (1989) Experimental studies of the evolutionary significance of sexual reproduction. VI. A greenhouse test of the sib competition hypothesis. *Evolution* 43, 1066–1074.

Kelley, S.E., Antonovics, J. and Schmitt, J. (1988) A test of the short-term advantage of sexual reproduction. *Nature* 331, 714–716.

Kelly, C.K. (1995) Seed size in tropical trees: a comparative study of factors affecting seed size in Peruvian angiosperms. *Oecologia* 102, 377–388.

Kelly, D. (1994) The evolutionary ecology of mast seeding. *Trends in Ecology and Evolution* 9, 465–470.

Kelly, D. and Sullivan, J.J. (1997) Quantifying the benefits of mast seeding on predator satiation and wind pollination in *Chionochloa pallens* (Poaceae). *Oikos* 78, 143–150.

Kerner, A. (1898) *The Natural History of Plants: Their Form, Growth, Reproduction and Distribution* (English translation by F.W. Oliver), 2 vols. Holt, New York.

Knoch, T.R., Faeth, S.H. and Arnott, D.L. (1993) Endophytic fungi alter foraging and dispersal by desert seed-harvesting ants. *Oecologia* 95, 470–473.

Kollmann, J. (1995) Regeneration window for fleshy-fruited plants during scrub development on abandoned grassland. *Ecoscience* 2, 213–222.

Kollmann, J. and Pirl, M. (1995) Spatial pattern of seed rain of fleshy-fruited plants in a scrubland grassland transition. *Acta Oecologica* 16, 313–329.

Kollmann, J. and Schill, H.P. (1996) Spatial patterns of dispersal, seed predation and germination during colonization of abandoned grassland by *Quercus petraea* and *Corylus avellana*. *Vegetatio* 125, 193–205.

Larson, D.L. (1996) Seed dispersal by specialist versus generalist foragers: the plant's perspective. *Oikos* 76, 113–120.

Leck, M.A., Parker, V.T. and Simpson, R.L. (eds) (1989) *Ecology of Soil Seed Banks*. Academic Press, London.

Leighton, M. and Leighton, D.R. (1983) Vertebrate responses to fruiting seasonality within a Bornean rainforest. In: Sutton, S.L., Whitmore, T.C. and Chadwick, A.C. (eds) *Tropical Rain Forest Ecology and Management*. Blackwell Scientific Publications, Oxford, pp. 181–196.

Leishman, M.R. and Westoby, M. (1994a) Hypotheses on seed size: tests using the semiarid flora of western New South Wales, Australia. *American Naturalist* 143, 890–906.

Leishman, M.R. and Westoby, M. (1994b) The role of large seed size in shaded conditions: experimental evidence. *Functional Ecology* 8, 205–214.

Levin, D.A. (1981) Dispersal versus gene flow in plants. *Annals of the Missouri Botanical Garden* 68, 233–253.

Linhart, Y.B. and Grant, M.C. (1996) Evolutionary significance of local genetic differentiation in plants. *Annual Review of Ecology and Systematics* 27, 237–277.

Lloyd, D.G. (1982) Selection of combined versus separate sexes in seed plants. *American Naturalist* 120, 571–585.

Loiselle, B.A. and Blake, J.G. (1999) Dispersal of melastome seeds by fruit-eating birds of tropical forest understory. *Ecology* 80, 330–336.

Loiselle, B.A., Ribbens, E. and Vargas, O. (1996) Spatial and temporal variation of seed rain in a tropical lowland wet forest. *Biotropica* 28, 82–95.

Lokesha, R., Hegde, S.G., Shaanker, R.U. and Ganeshaiah, K.N. (1992) Dispersal mode as a selective force in shaping the chemical composition of seeds. *American Naturalist* 140, 520–525.

McCall, C., Mitchell-Olds, T. and Waller, D.M. (1989) Fitness consequences of outcrossing in *Impatiens capensis*: tests of the frequency-dependent and sib-competititon models. *Evolution* 43, 1075–1084.

McCanny, S.J. and Cavers, P.B. (1987) The escape hypothesis: a test involving a temperate annual grass. *Oikos* 49, 67–76.

McClanahan, T.R. and Wolfe, R.W. (1993) Accelerating forest succession in a fragmented landscape. *Conservation Biology* 7, 279–288.

Manasse, R.S. and Howe, H.F. (1983) Competition for dispersal agents among tropical trees: influences of neighbors. *Oecologia* 59, 185–190.

Mark, S. and Olesen, J.M. (1996) Importance of elaiosome size to removal of ant-dispersed seeds. *Oecologia* 107, 95–101.

Matlack, G.R. (1987) Diaspore size, shape, and fall behavior in wind-dispersed plant species. *American Journal of Botany* 74, 1150–1160.

Milton, S.J., Siegfried, W.R. and Dean, W.R.J. (1990) The distribution of epizoochoric plant spectres: a clue to the prehistoric use of arid Karoo rangelands by large herbivores. *Journal of Biogeography* 17, 25–34.

Morellato, P.C. and Leitao, H.F. (1996) Reproductive phenology of climbers in a south-eastern Brazilian forest. *Biotropica* 28, 180–191.

Morton, J.K. and Hogg, E.M. (1989) Biogeography of island floras in the Great Lakes. II. Plant dispersal. *Canadian Journal of Botany* 67, 1803–1820.

Murray, K.G., Russell, S., Picone, C.M., Winnett-Murray, K., Sherwood, W. and Kuhlmann, M.L. (1994) Fruit laxatives and seed passage rates in frugivores: consequences for plant reproductive success. *Ecology* 75, 989–994.

Nagy, E.S. (1997) Frequency-dependent seed production and hybridization rates: implications for gene flow between locally adapted plant populations. *Evolution* 51, 703–714.

Nagy, E.S. and Rice, K.J. (1997) Local adaptation in two subspecies of an annual plant: implications for migration and gene flow. *Evolution* 51, 1079–1089.

O'Dowd, D.J. and Gill, A.M. (1986) Seed dispersal syndromes in Australian *Acacia*. In: Murray, D.R. (ed.) *Seed Dispersal*. Academic Press, Sydney, pp 87–121.

Okube, A. and Levin, S.A. (1989) A theoretical framework for data analysis of wind dispersal of seeds and pollen. *Ecology* 70, 329–338.

Ostfeld, R.S., Jones, C.G. and Wolff, J.O. (1996) Of mice and mast: ecological connections in eastern deciduous forests. *Bioscience* 46, 323–330.

Owadally, A.W. (1979) The dodo and the tambalacoque tree. *Science* 203, 1363–1364.

Parker, M.A. (1985) Local population differentiation for compatibility in an annual legume and its host-specific fungal pathogen. *Evolution* 39, 713–723.

Peart, D.R. (1985) The quantitative representation of seed and pollen dispersal. *Ecology* 66, 1081–1083.

Peres, C.A., Schiesari, L.C. and Dias-Leme, C.L. (1997) Vertebrate predation of Brazil-nuts (*Bertholletia excelsa*, Lecythidaceae), an agouti-dispersed Amazonian seed crop: a test of the escape hypothesis. *Journal of Tropical Ecology* 13, 69–79.

Platt, W.J. and Weis, I.M. (1985) An experimental study of competition among fugitive prairie plants. *Ecology* 66, 708–720.

Portnoy, S. and Willson, M.F. (1993) Seed dispersal curves: behavior of the tail of the distribution. *Evolutionary Ecology* 7, 25–44.

Price, M.V. and Jenkins, S.H. (1986) Rodents as seed consumers and dispersers. In: Murray, D.R. (ed.) *Seed Dispersal*. Academic Press, Sydney, pp. 191–235.

Primack, R.B. and Miao, S.L. (1992) Dispersal can limit local plant distribution. *Conservation Biology* 6, 513–519.

Rabinowitz, D. and Rapp, J.K. (1981) Dispersal abilities of seven sparse and common grasses from a Missouri prairie. *American Journal of Botany* 68, 616–624.

Reid, N. (1989) Dispersal of mistletoes by honeyeaters and flowerpeckers: components of seed dispersal quality. *Ecology* 70, 137–145.

Rey, P.J. (1995) Spatio–temporal variation in fruit and frugivorous bird abundance in olive orchards. *Ecology* 76, 1625–1635.

Rey, P.J., Gutiérrez, J.E., Alcántara, J. and Valera, F. (1997) Fruit size in wild olives: implications for avian seed dispersal. *Functional Ecology* 11, 611–618.

Ricklefs, R.E. and Renner, S.S. (1994) Species richness within families of flowering plants. *Evolution* 48, 1619–1636.

Ridley, H.N. (1930) *The Dispersal of Plants throughout the World*. Reeve, Ashford, Kent.

Russell, S.K. and Schupp, E.W. (1998) Effects of microhabitat patchiness on patterns of seed predation of *Cercocarpus ledifolius* (Rosaceae). *Oikos* 81, 434–443.

Sacchi, C.F. (1987) Variability in dispersal ability of common milkweed, *Asclepias syriaca*, seeds. *Oikos* 49, 191–198.

Sargent, S. (1995) Seed fate in a tropical mistletoe: the importance of host twig size. *Functional Ecology* 9, 197–204.

Schemske, D.W. (1984) Population structure and local selection in *Impatiens pallida* (Balsaminaceae), a selfing annual. *Evolution* 38, 817–832.

Schmitt, J. and Ehrhardt, D.W. (1987) A test of the sib-competition hypothesis for outcrossing advantage in *Impatiens capensis*. *Evolution* 41, 579–590.

Schnabel, A., Nason, J.D. and Hamrick, J.L. (1998) Understanding the population genetic structure of *Gleditsia triacanthos* L.: seed dispersal and variation in female reproductive success. *Molecular Ecology* 7, 819–832.

Schupp, E.W. (1990) Annual variation in seedfall, postdispersal predation, and recruitment of a neotropical tree. *Ecology* 71, 504–515.

Schupp, E.W. (1993) Quantity, quality and the effectiveness of seed dispersal by animals. *Vegetatio* 107/108, 13–29.

Schupp, E.W. (1995) Seed–seedling conflicts, habitat choice, and patterns of plant recruitment. *American Journal of Botany* 82, 399–409.

Schupp, E.W. and Fuentes, M. (1995) Spatial patterns of seed dispersal and the unification of plant population ecology. *Ecoscience* 2, 267–275.

Selas, V. (1997) Cyclic population fluctuations of herbivores as an effect of cyclic seed cropping of plants: the mast depression hypothesis. *Oikos* 80, 257–268.

Shibata, M. and Nakashizuka, T. (1995) Seed and seedling demography of four co-occurring *Carpinus* species in a temperate deciduous forest. *Ecology* 76, 1099–1108.

Shibata, M., Tanaka, H. and Nakashizuca, T. (1998) Causes and consequences of mast seed production of four co–occurring *Carpinus* species in Japan. *Ecology* 79, 54–64.

Shields, W.M. (1982) *Philopatry, Inbreeding, and the Evolution of Sex*. State University of New York, Albany.

Skeate, S.T. (1987) Interactions between birds and fruits in a northern Florida hammock community. *Ecology* 68, 297–309.

Slatkin, M. (1987) Gene flow and the geographic structure of natural populations. *Science* 236, 787–792.

Smith, A.P. (1977) Albinism in relation to competition in bamboo *Phyllostachys bambusoides*. *Nature* 266, 527–529.

Snow, B. and Snow, D. (1988) *Birds and Berries.* Poyser, Calton, UK.

Snow, D.W. (1981) Tropical frugivorous birds and their food plants: a world survey. *Biotropica* 13, 1–14.

Sorensen, A.E. (1986) Seed dispersal by adhesion. *Annual Review of Ecology and Systematics* 17, 443–463.

Sork, V.L. (1983) Mammalian seed dispersal of pignut hickory during three fruiting seasons. *Ecology* 64, 1049–1056.

Sork, V.L. (1993) Evolutionary ecology of mast-seeding in temperate and tropical oaks (*Quercus* spp.). *Vegetatio* 107/108, 528–541.

Stamp, N.E. and Lucas, J.R. (1990) Spatial patterns and dispersal distances of explosively dispersing plants in Florida sandhill vegetation. *Journal of Ecology* 78, 589–600.

Stapanian, M.A. and Smith, C.C. (1984) Density-dependent survival of scatterhoarded nuts: an experimental approach. *Ecology* 65, 1387–1396.

Strauss, S.H. and Doerksen, A.H. (1990) Restriction fragment analysis of pine phylogeny. *Evolution* 44, 1081–1096.

Sugden, A.M. (1982) Long-distance dispersal, isolation, and the cloud forest flora of the Serrania de Macuira, Guajira, Colombia. *Biotropica* 14, 208–219.

Tapper, P. (1996) Long-term patterns of mast fruiting in *Fraxinus excelsior. Ecology* 77, 2567–2572.

Temple, S.A. (1977) Plant–animal mutualism: coevolution with dodo leads to near extinction of plant. *Science* 197, 885–886.

Temple, S.A. (1979) The dodo and the tambalacoque tree. *Science* 203, 1364.

Terborgh, J. (1986) Community aspects of frugivory in tropical forests. In: Estrada, A. and Fleming, T.H. (eds) *Frugivores and Seed Dispersal.* Junk, Dordrecht, pp. 371–384.

Thiede, D.A. and Augspurger, C.K. (1996) Intraspecific variation in seed dispersion of *Lepidium campestre* (Brassicaceae). *American Journal of Botany* 83, 856–866.

Thompson, J.N. (1981) Elaiosomes and fleshy fruits: phenology and selection pressures for ant-dispersed seeds. *American Naturalist* 117, 104–108.

Thompson, J.N. and Willson, M.F. (1979) Evolution of temperate fruit/bird interactions: phenological strategies. *Evolution* 33, 973–982.

Thompson, K. and Rabinowitz, D. (1989) Do big plants have big seeds? *American Naturalist* 133, 722–728.

Tiffney, B.H. and Mazer, S.J. (1995) Angiosperm growth habit, dispersal and diversification reconsidered. *Evolutionary Ecology* 9, 93–117.

Tomback, D.F. and Linhart, Y.B. (1990) The evolution of bird-dispersed pines. *Evolutionary Ecology* 4, 185–219.

Trapp, E.J. (1988) Dispersal of heteromorphic seeds in *Amphicarpaea bracteata* (Fabaceae). *American Journal of Botany* 75, 1535–1539.

Traveset, A. (1992) Effect of vertebrate frugivores on bruchid beetles that prey on *Acacia farnesiana* seeds. *Oikos* 63, 200–206.

Traveset, A. (1993) Weak interactions between avian and insect frugivores: the case of *Pistacia terebinthus* L. (Anacardiaceae). *Vegetatio* 107/108, 191–203.

Traveset, A. (1994) Influence of type of avian frugivory on the fitness of *Pistacia terebinthus* L. *Evolutionary Ecology* 8, 618–627.

Traveset, A. (1995) Seed dispersal of *Cneorum tricoccon* L. (Cneoraceae) by lizards and mammals in the Balearic Islands. *Botanical Journal of the Linnean Society* 117, 221–232.

Traveset, A. (1998) Effect of seed passage through vertebrate frugivores' guts on germination: a review. *Perspective in Plant Ecology, Evolution and Systematics* 1, 151–190.

Traveset, A. and Willson, M.F. (1997) Effect of birds and bears on seed germination of fleshy-fruited plants in temperate rainforests of southeast Alaska. *Oikos* 80, 89–95.

Traveset, A. and Willson, M.F. (1998) Ecology of the fruit-colour polymorphism in *Rubus spectabilis. Evolutionary Ecology* 12, 331–345.

Traveset, A., Willson, M.F. and Gaither, J.C., Jr (1995) Avoidance by birds of insect-infested fruits of *Vaccinium ovalifolium. Oikos* 73, 381–386.

Turkington, R. and Aarssen, L.W. (1984) Local-scale differentiation as a result of competitive interactions. In: Dirzo, R. and Sarukhan, J. (eds) *Perspectives in Plant Population Ecology.* Sinauer, Sunderland, Massachusetts, pp. 107–127.

Valverde, T. and Silvertown, J. (1997) An integrated model of demography, patch dynamics and seed dispersal in a woodland herb, *Primula vulgaris. Oikos* 80, 67–77.

van der Pijl, L. (1982) *Principles of Dispersal in Higher Plants,* 3rd edn. Springer-Verlag, Berlin.

van der Plank, J.E. (1960) Analysis of epidemics. *Plant Pathology* 3, 229–289.

vander Wall, S.B. and Balda, R.P. (1977) Coadaptations of the Clark's nutcracker and the pinyon pine for efficient seed harvest and dispersal. *Ecological Monographs* 47, 89–111.

van Schaik, C.P., Terborgh, J.W. and Wright, S.J. (1993) The phenology of tropical forests: adaptive significance and consequences for primary consumers. *Annual Review of Ecology and Systematics* 24, 353–377.

Venable, D.L. and Brown, J.S. (1988) The selective interactions of dispersal, dormancy, and seed size as adaptations for reducing risk in variable environments. *American Naturalist* 131, 360–384.

Verdú, M. and García–Fayos, P. (1996) Nucleation processes in a Mediterranean bird-dispersed plant. *Functional Ecology* 10, 275–280.

Verdú, M. and García-Fayos, P. (1998) Old-field colonization of *Daphne gnidium*: seedling distribution and spatial dependence at different scales. *Journal of Vegetation Science* 9, 713–718.

Waddington, K.D. (1983) Pollen flow and optimal outcrossing distance. *American Naturalist* 122, 147–151.

Wahaj, S.A., Levey, D.J., Sanders, A.K. and Cipollini, M.L. (1998) Control of gut retention time by secondary metabolites in ripe *Solanum* fruits. *Ecology* 79, 2309–2319.

Wanntorp, H.E., Brooks, D.R., Nilsson, T., Nylin, S., Ronquist, F., Stearns, S.C. and Wedell, N. (1990) Phylogenetic approaches in ecology. *Oikos* 57, 119–132.

Waser, P.M., Austad, S.N. and Keane, B. (1986) When should animals tolerate inbreeding? *American Naturalist* 128, 529–537.

Wells, F.V. (1983) Paleobiogeography of montane islands in the Great Basin since the last glaciopluvial. *Ecological Monographs* 53, 341–382.

Westelaken, I.L. and Maun, M.A. (1985) Spatial pattern and seed dispersal of *Lithospermum caroliniense* on Lake Huron sand dunes. *Canadian Journal of Botany* 63, 125–132.

Westoby, M. and Rice, B. (1981) A note on combining two methods of dispersal-for-distance. *Australian Journal of Ecology* 6, 189–192.

Westoby, M., French, K., Hughes, L., Rice, B. and Rodgerson, L. (1991) Why do more plant species use ants for dispersal on infertile compared with fertile soils? *Australian Journal of Ecology* 16, 445–456.

Wheelwright, N.T. (1985) Fruit size, gape width, and the diets of fruit-eating birds. *Ecology* 66, 808–818.

Wheelwright, N.T. (1993) Fruit size in a tropical tree species: variation, preference by birds, and heritability. *Vegetatio* 107/108, 163–174.

Whelan, C.J. and Willson, M.F. (1994) Fruit choice in migrating North American birds: field and aviary experiments. *Oikos* 71, 137–151.

Whittaker, R.J. and Jones, S.H. (1994) The role of frugivorous bats and birds in the rebuilding of a tropical forest ecosystem, Krakatau, Indonesia. *Journal of Biogeography* 21, 245–258.

Williams, C.G., Bridgewater, F.E. and Lambeth, C.C. (1983) Performance of single family versus mixed family plantation blocks of loblolly pine. *Proceedings of the Southern Forest Tree Improvement Conference* 17, 194–201.

Willson, M.F. (1991) Dispersal of seed by frugivorous animals in temperate forests. *Revista Chilena de Historia Natural* 64, 537–554.

Willson, M.F. (1993a) Dispersal mode, seed shadows, and colonization patterns. *Vegetatio* 107/108, 261–280.

Willson, M.F. (1993b) Mammals as seed-dispersal mutualists in North America. *Oikos* 67, 159–176.

Willson, M.F. and Crome, F.H.J. (1989) Patterns of seed rain at the edge of a tropical Queensland rainforest. *Journal of Tropical Ecology* 5, 301–308.

Willson, M.F. and Thompson, J.N. (1982) Phenology and ecology of color in bird-dispersed fruits, or why some fruits are red when they are 'green'. *Canadian Journal of Botany* 60, 701–713.

Willson, M.F. and Whelan, C.J. (1990) The evolution of fruit color in fleshy-fruited plants. *American Naturalist* 136, 790–809.

Willson, M.F. and Whelan, C.J. (1993) Variation of dispersal phenology in a bird-dispersed shrub, *Cornus drummondii. Ecological Monographs* 63, 151–172.

Willson, M.F., Hoppes, W.C., Goldman, D.A., Thomas, P.A., Katusic-Malmborg, P.L. and Bothwell, J.L. (1987) Sibling competition in plants: an experimental study. *American Naturalist* 129, 304–311.

Willson, M.F., Irvine, A.K. and Walsh, N.G. (1989) Vertebrate dispersal syndromes in some Australian and New Zealand plant communities, with geographical comparisons. *Biotropica* 21, 133–147.

Willson, M.F., Rice, B. and Westoby, M. (1990a) Seed dispersal spectra: comparison of temperate plant communities. *Journal of Vegetation Science* 1, 547–562.

Willson, M.F., Michaels, H.J., Bertin, R.I., Benner, B., Rice, S., Lee, T.D. and Hartgerink, A.P. (1990b) Intraspecific variation in seed packaging. *American Midland Naturalist* 123, 179–185.

Winn, A.A. and Miller, T.E. (1995) Effect of density on magnitude of directional selection on seed mass and emergence time in *Plantago wrightiana* Dcne.(Plantaginaceae). *Oecologia* 103, 365–370.

Witmer, M.C. and Cheke, A.S. (1991) The dodo and the tambalocoque tree: an obligate mutualism reconsidered. *Oikos* 61, 133–137.

Yarranton, G.A. and Morrison, R.G. (1974) Spatial dynamics of a primary succession: nucleation. *Journal of Ecology* 62, 417–428.

Chapter 5
Animals as Seed Dispersers

Edmund W. Stiles

Department of Biological Sciences, Rutgers University, Piscataway, New Jersey, USA

Animals: vectors for seed movement

Seed plants for the most part are 'rooted' to one spot and have limited ability for self-propulsion. This intimate attachment to the soil poses interesting challenges, as successful colonization of new sites is dependent upon the arrival of seeds. There are significant advantages, in the currency of genes passed into the next generation, for plants bearing traits that increase the probability of successful dispersal. Seeds falling beneath the parent plant are faced with competition for resources with their parent, higher levels of density-dependent seed predation and higher densities of competing siblings, with the associated epidemiological problems associated with high densities, such as fungal or viral transmission among individuals.

Plants have evolved diverse arrays of adaptations that result in the movement of seeds away from their parents. Movement of wind and water provides predictable physical forces selecting for many morphological and phenological adaptations that facilitate seed dispersal; but the greatest diversity of adaptations found in the diaspores of plants are those that facilitate seed movement by animals. Adaptations of plant diaspores have evolved in response to the morphology and physiology of animals as well as the behavioural choices made by animals.

The primary consideration in the relationship between plant seed and animal dispersal agent is that animals are mobile. Sessile animals are of little use in these interactions. Beyond this basic premise, the movement of seeds by animals is dependent upon the diverse array of animal morphologies and behaviours.

In mobile animals, movement patterns may transport seeds thousands of kilometres with transcontinental or transoceanic migrant birds (Proctor, 1968) or millimetres in the guts of earthworms (Ridley, 1930). Habitat selection by animals will dictate the specificity of sites of seed arrival at potential colonization locations.

The vast majority of animals that disperse seeds are either vertebrates or ants. Among the vertebrates, the birds are probably the most important seed dispersers, as determined by numbers of successful propagules disseminated, followed by mammals, fishes, reptiles and amphibians. For invertebrates, ants are the only major group of seed dispersers, with small numbers of seeds moved by molluscs and annelids.

In this chapter, I shall consider the process of seed acquisition by animals (or animal acquisition by seeds), the nature of seed treatment by the animals, animal mobility and seed deposition patterns, and finally the diversity of animals that disperse seeds.

Seed acquisition

Animals may acquire seeds either actively, through the process of selecting different seeds or fruits, or passively, as hitchhikers attached to fur or feathers or consumed incidentally with other foods.

Passive – external

Characteristics of external morphology, such as hair or feathers, determine the probability of seed attachment to the bodies of animals, the strength of attachment and the time before the seed is dislodged.

Large numbers of seeds are moved by passive attachment to the fur of mammals. This relationship exists among terrestrial mammals and plants bearing seeds within the height range contacted by these species. As mammals walk through vegetation, seeds with a diverse array of different sizes and distributions of hooks are dislodged from the parent plants and attached to the fur. Different plant parts may aid the attachment, including branchlets of the inflorescence, armed bracts, glumes of grasses, adhesive perianth lobes, adhesive calyces, adhesive corollas, hooked styles, spiny and hooked fruits, hooked hairs and bulbils (Ridley, 1930). The subsequent pattern of dissemination of these seeds is dependent on a number of factors associated with both seed and animal morphology and on animal behaviour (Sorensen, 1986). Where seeds have different sizes, numbers and distributions of hooks, mammals have great differences in lengths and density of hairs to which the seeds may attach (Agnew and Flux, 1970; Bullock and Primack, 1977). The seed distribution pattern (or seed shadow) for a plant/mammal combination is dependent on the nature of the hook/hair interaction. Some seeds, such as those of *Arctium* or *Xanthium*, become strongly tangled in longer animal fur and can only be removed with great effort, while other seeds, such as those of *Geum* or *Bidens*, may be dislodged with a light brush of vegetation. The length of time a seed remains attached to a mammal is also

a function of the individual animal behaviour. Seeds not only become attached by animal movement through vegetation, but can also be dislodged by subsequent movement. How these species move through the vegetation influences the seed shadows of these passively dispersed plants. One result is a non-random distribution of seeds along pathways frequented by mammals. Mammals also differ in their grooming behaviour. The frequency and locations of grooming may influence the nature of the seed shadow in these passively dispersed species as well (Agnew and Flux, 1970). Almost 200 species from 28 plant families were found in the wool of sheep and had been successfully moved around the world by human commerce (Ridley, 1930).

Adhesive achenes and viscid seeds are moved by the activities of terrestrial and aquatic birds and mammals in a fashion similar to that involved in the case of hooked seeds. Seeds may adhere to feathers and even snail shells with a sticky substance produced by the seeds. The dwarf mistletoes (*Arceuthobium*) are parasitic plants of the coniferous forests of the western USA. Nicholls *et al.* (1984) identified 18 bird species and five mammal species that carried the sticky seeds of the dwarf mistletoe. Seeds are expelled at initial velocities of 27 m s^{-1} for distances up to 16.5 m (Ostry *et al.*, 1983). Birds and mammals contact the seeds when foraging in mistletoe plants and subsequently groom off the sticky seeds. Ridley (1930) recounts examples of the viscid fruits of the genus *Pisonia*, found in tropical regions worldwide. For one extreme example with *Pisonia aculeata* from Uganda, C.B. Ussher found 'a bird lying helpless on the ground covered with the fruit of this creeper. Its feathers were all stuck together, and it was unable to fly.' Bird movement through the trees and shrubs brings them in contact with sticky fruits.

Active – external

A number of birds and mammals select seeds for food but transport the seeds and often store them for a time before eating them. These species are the cachers. Seeds store well and can provide resources for birds and mammals over periods when seeds are not being produced by plants. These seeds are not passed through the rigours of the digestive system before storage, and those that escape from subsequent detection by a consuming animal have often been placed in a superior site for germination (Tomback, 1982, 1983, 1986; Stapanian and Smith, 1984, 1986). Some corvids, including Clark's nutcracker (Kamil and Balda, 1985) and grey jays (Bunch and Tomback, 1986), find stored seeds using a highly developed spatial memory. Squirrels combine a keen spatial memory with an equally keen sense of smell to locate stored seeds (Stapanian and Smith, 1984, 1986). Seed selection by birds or mammals is influenced by seed weight (Stiles and Dobi, 1987).

Passive – internal

Some seeds are transported by animals ingesting them as an incidental part of some other food they are eating. The best examples of this come from the seeds dispersed through the grazing activities of large herbivores (Janzen, 1982c). Quinn *et al.* (1994) tested Janzen's 'foliage is the fruit' hypothesis with buffalo grass (*Buchloe dactyloides*), demonstrating both the high quality of the foliage and the positive effects on germination and seedling growth for seeds passed through cattle. Dinerstein (1989), however, examined over 40,000 deer pellets from four species of Asian deer and over 1000 kg of dung from Indian bison and greater one-horned rhinoceros. He found no obvious examples supporting the 'foliage is the fruit' hypothesis, but he found that conventional frugivory by large ungulates was important in the South Asian flood-plains.

Active – internal

Both seeds and the fleshy parts of fruits are used by animals for food. Feeding behav-

iour and foraging choices also generate patterns of association between seeds and particular animal species. Characteristics of internal morphology may influence whether a seed survives the potentially hostile environment of the mouth and digestive systems of animals.

Animals select food items based on a complex set of criteria based on food availability, food quality and the perceived need for food. Selection of fruits or seeds by animals is based also on the location of fruits or seeds by the animals, morphological constraints, involving fruit or seed size or structure, and nutritional demands and avoidance of toxins by individual foragers.

Fruit and seed location

Animals depend on different senses for the location of fruits or seeds. Species with colour vision, such as birds, primates, tortoises and squirrels, employ colour as a primary cue for finding food, and the colours of fruits or associated plant parts have been identified as cues used by birds for location of food. In addition to the ripe colour of the fruits, fruits may go through a two-stage colour change or associated plant parts may develop contrasting colours, called preripening fruit flags (Stiles, 1982) or bicoloured fruits (Willson and Thompson, 1982), prior to fruit ripening, advertising the imminent presence of ripe fruits. In autumn in the temperate zone, some species change leaf colour early, providing a long-distance signal for spatially naïve, migrant frugivores, advertising the potential presence of fruit (Stiles, 1982).

The importance of the many different colours of fruit, both within and between species, remains somewhat obscure (Turcek, 1963; Knight and Siegfried, 1983; Wheelwright and Janson, 1985; Willson *et al.*, 1990), but the importance of certain colours is suggested by the non-random distribution of black and red fruits eaten by birds and the dominance of yellow and green in fruits eaten by animals not having colour vision. Seeds are also located by sight in birds, but, for most interactions, selection has been for crypsis, reducing the encounter rates with the seeds for foraging birds (Cook *et al.*, 1971).

Mammals use a powerful sense of smell to locate seeds under many circumstances in temperate forests (Calahane, 1942; Howard *et al.*, 1968; Drickamer, 1970, 1976), in grasslands (Gibson *et al.*, 1990) and in deserts (Davidson *et al.*, 1984). Many nocturnal mammals locate fruits using their sense of smell. Musty odours are produced by fruits, which can then be located by foraging mammals (van der Pijl, 1982; Rieger and Jakob, 1988). Reptiles, including tortoises and iguanas, also locate fruits by smell. Box tortoises are often attracted to fallen crops of mulberry (*Morus* sp.) or ripening strawberries (*Fragaria* sp.), apparently using both sight and smell as cues (Klimstra and Newsome, 1960). Little work has been done on the importance of smell in the location of fruits by birds, but evidence from starlings, cedar waxwings and tree swallows (Clark, 1991) indicates that some birds may use smell in fruit choice more frequently than thought previously.

Ants are attracted to seeds by oil-rich elaiosomes attached to the seeds. They transport seeds to their nests, usually relatively short distances – an average of 75 cm in one study (Culver and Beattie, 1978). Seeds are stored in ant nests or the elaiosome is cut off and typically fed to larvae. The seed is discarded in a refuse pile near the nest, which is nutrient-enriched by ant faeces and the carcasses of prey taken by the ants. Five hypotheses predicting the evolutionary advantages of seed dispersal by ants are: (i) avoidance of competition (Handel, 1978); (ii) avoidance of fire (Berg, 1975; Bond and Slingsby, 1984); (iii) increased dispersal distance (Westoby and Rice, 1981); (iv) increased nutrient availability (Beattie and Culver, 1983; but see Rice and Westoby, 1986); and (v) avoidance of seed predation under the parent (O'Dowd and Hay, 1980; Heithaus, 1981). Some support exists for all five hypotheses, and it is probable that all are important in different interactions.

Even sound is used infrequently for location of fruits or seeds. Agoutis in dense tropical forest will respond positively to the sound of falling palm fruits, and fishes congregate under ripening fig trees, presumably responding to the sound of the figs hitting the water.

Birds may change their preference ranking for fruits of different species depending on the position of the fruits relative to their perch (Denslow and Moermond, 1982; Moermond and Denslow, 1983, 1985). Bird morphology constrains the ease of access. For example, perch-feeding *Tangara larvata* can reach farther below a perch to pick fruit than can aerial-feeding *Manacus candei*.

Seed treatment

Once fruit or seed are located, the process of dealing with the diaspore may result in from 0 to 100% death of the seeds.

Morphological constraints

The relationship between animal size and seed size is a basic consideration, as some seeds that are dispersed by animals are larger than some animals that disperse seeds. Ants do not disperse acorns. Some seeds are carried and dropped, as in the case of many ant-dispersed (Beattie, 1985) and bat-dispersed seeds (Fleming, 1981). Most animal-dispersed seeds, however, are passed through at least a portion of the digestive tract. To pass unharmed through a dispersal agent, a seed must be able to fit into the mouth and throat of the animal.

One of the primary constraints that has been examined, primarily with birds, has been the relationship between fruit or seed diameter and the gape width of the consumers. Small birds do not disperse the seeds of large-seeded species. Larger seeds are generally not available to smaller animals, whereas small-seeded fruits are generally available for most frugivores, as has been illustrated for the fruits of lower montane forests at Monteverde, Costa Rica

(Wheelwright, 1985). Large seeds are usually borne by the plants in single-seeded fruits. S.J. Manzer and N.T. Wheelwright (unpublished) have proposed that among members of the family *Lauraceae* larger fruits may be relatively more elongated under selection for smaller minimum diameter. This would increase the diversity of frugivores able to eat the fruits with relatively large seed mass.

At the other end of the spectrum, many fruits have very small seeds. Usually these are borne as multiple-seeded fruits, such as strawberries (*Fragaria*), blackberries (*Rubus*) and mulberries (*Morus*) in the temperate zone and *Miconia* and *Ficus* in the tropics. Seeds from these fruits are more likely to be defecated than regurgitated and are more likely to escape mammalian predation.

Placement of seeds or fruits on a plant may restrict access to some animals. This is especially important for many bird-disseminated fruits that bear seeds within the size that is attractive to seed-eating mammals. Seeds of certain plants in eastern North America that are dispersed by birds and mammals are placed so that they are available not only to seed-dispersing mammals, but also to the primary mammalian seed predators of the region, mice of the genus *Peromyscus*. These fruits bear seeds that are either very small or very large, decreasing the profitability of foraging seed predators. Other fruits with intermediate-sized seeds are located on the ends of racemes or high in the trees and are better protected from foraging mice. Many birds, such as parrots and finches, are primarily seed predators, but some seeds may pass unharmed through their digestive tracts. An extreme example is that described by Roessler (1936), where seven of 40,025 seeds germinated successfully after passage through house finches.

Mammalian teeth represent an equally formidable rite of passage for many seeds. Very small, hard seeds, such as those of blueberries (*Vaccinium*) or strawberries (*Fragaria*), are eaten with fruit pulp and defecated intact, even by the smallest seed predators. Rodents are voracious predators

of intermediate and larger seeds. Successful dispersal of propagules is accomplished in some cases by multiple embryos in concrescent seeds, as in *Arctostaphylos*. The predator consumes some of the embryos within the seed, leaving others to germinate. The same result is accomplished in *Sterculia* (Janzen, 1972), where seed predators drop pods with only some seeds eaten to remove numerous urticating hairs that get stuck in soft mouth-parts as they feed.

Internal treatment of seeds is influenced by the physical and chemical processing that takes place in the gut. This is additionally affected by the time a seed spends in the gut before being deposited by the animal. Most highly frugivorous species have short guts and seed passage is very rapid (McKey, 1975). Highly frugivorous species of Old World flycatchers (Muscicapidae) process fruits and defecate seeds faster than less frugivorous species (29 ± 9 versus 64 ± 23 min: Herrera, 1984). The phainopepla (*Phainopepla nitens*) passes mistletoe seeds in 29 min (range 12–45 min: Walsberg, 1975). Fukui (1996) has pointed out the potential conflicts between fruiting plants and frugivores associated with seed retention time in the gut and the number of seeds deposited in a single faecal pellet.

Seed predators usually have longer guts, with associated grinding action of teeth in mammals and the gizzard in many bird species. Turkeys can crush pecans (*Carya illinoensis*) in about 1 h in their gizzard, but it takes more than 30 h to crush shag-bark hickory nuts (*Carya ovata*) (Schorger, 1960), and some seeds are hard enough to pass in viable condition even through this formidable grinding mill (Temple, 1977). Some species gain in percentage germination following passage through an animal gut (Krefting and Roe, 1949; Temple, 1977).

Seed retention in animal guts for longer periods of time may even induce germination while in the gut, and death of the seed follows. Janzen (1981, 1982a) found that horses kill a substantial fraction of the *Enterolobium* seeds that they ingest, resulting in only 9–56% of ingested seeds being defecated alive. Box tortoises likewise retain seeds for long periods in the gut, increasing germination during gut passage and subsequent seed death (E.W. Stiles, unpublished). Deposition of seeds with faecal material may provide a nitrogen source that increases nutrients for early seedling growth, although large numbers of seeds may be removed from the dung by seed predators (Janzen, 1982b, c).

Aspects of fruit or seed structure other than size or placement may also impose restrictions. Many fruit protect both pulp and seeds in indehiscent capsules. Only animals with strong teeth or bills may reach the pulp or seeds. In Papua New Guinea, large (> 12 mm diameter) fruits are either structurally unprotected and taken mostly by fruit-pigeons and bowerbirds, or structurally protected, usually within a capsule, and taken by birds of paradise. Birds of paradise use complex food-handling techniques for removing fruits from the capsules (Pratt and Stiles, 1985). Primate fruits are also often in indehiscent capsules, an adaptation that may reduce access to other frugivores. For a full discussion of frugivory, see Jordano (Chapter 6, this volume).

Nutrients and toxins

Animals select fruit or seeds based in part on the nutritional rewards available or the secondary chemicals that reduce palatability and increase toxicity of foods. Both the quality and quantity of carbohydrates, fats and protein may influence preferences of frugivores. Fruits vary appreciably in their content of all three of these nutrient groups.

Fruits differ in the amounts of glucose, fructose and sucrose present in their pulp. Some birds can preferentially select fruits with as little as 3% higher sugar content (Levey, 1987). Physiological differences among species may have strong influences on fruit choice. For example, American robins lack the enzyme sucrase and may be influenced in their choice of fruits by the proportion of disaccharides and monosaccharides in fruit pulp (Martínez del Rio et al., 1988; Martínez del Rio and Restrepo, 1993). Pulp of fruits in New Jersey, USA, ingested by both birds and mammals is higher in sugars than that of fruits eaten primarily by birds (Stiles, 1980; White, 1989).

Fruits and seeds differ in both the amounts and types of lipids present. In fruit pulp, lipids may be in the form of lower-molecular-weight fatty acids up to high-molecular-weight waxes. These lipids may be assimilated differently by different species, resulting in different seed-dispersing associations. For example, few birds or mammals can metabolize the high-molecular-weight waxes found in bayberries (Myrica sp.), and most frugivorous birds within the range of these species of plants along the east coast of the USA rarely take these fruits. The yellow-rumped warbler and the tree swallow can digest the waxes and are commonly associated with shrubs of these species. In fact, these two species are the most northerly-wintering species of their respective taxa, Parulinae and Hirundinidae, apparently surviving periods of low insect abundance by subsisting on the waxes of these fruits (Place and Stiles, 1992).

During autumn bird migration in the eastern deciduous forests of the USA, high-lipid fruits are the most highly preferred species. Grey catbirds selected higher-lipid fruits in 36 of 42 paired preference tests for fruits in New Jersey (27 with $P < 0.05$), and American robins selected higher-lipid fruits in 28 of 42 paired trials (19 with $P < 0.05$) (E.W. Stiles, unpublished). In both these species, the probability of selecting the higher-lipid fruit increased with the difference in lipid content between the pair of fruits offered.

Elaiosomes on seeds differ in the amount of lipid available for ants; some are quite rich, while others are relatively poor (Keeler, 1989). Beattie (1985) indicates that some lipids in elaiosomes may be essential for ant nutrition, but the specificity of this requirement is still in need of further tests.

Animal mobility and seed deposition

The vast majority of seeds produced by a plant end up near the parent. The dispersion patterns of seeds leaving the parent for animal-dispersed seeds depend on the type of dispersal agent (Janzen, 1970). Seed deposition patterns are an interaction between seed retention by the animal and animal morphology and behaviour. From an individual source plant, the distribution of seed densities with increasing distance, or the seed shadow, shows a sharp decline. This is, first and foremost, because the area into which the seed can be dropped increases in relation to the area of the circle with expanding radius around the parent plant (Pratt, 1983); secondly, because the retention times of seeds are usually not great, often only a few minutes; and, thirdly, because home ranges of animals restrict movement.

Although precise seed shadows of individual plants are difficult to obtain, some have been attempted (Smith, 1975). Howe and Smallwood (1982) identified three hypotheses that explore the different advantages to seed dispersal and the relative importance of different shapes of seed shadows generated by different animal species: (i) the 'escape hypothesis', which would select for a seed shadow that avoids seed and seedling competition or mortality near the parent; (ii) the 'colonization hypothesis', which places seeds in newly disturbed areas; and (iii) the 'directed dispersal hypothesis', which deposits seeds in microsites with superior probabilities of supporting a successful propagule. See Willson and Traveset (Chapter 4, this volume) for a discussion of the advantages of dispersal. These competing hypotheses are not mutually exclusive and the relative importance of each is still uncertain.

Several characteristics of animal morphology and behaviour affect these dispersion patterns. The size of the animal relative to the seed may influence how the animal treats the seed. Seeds may be ingested and defecated or they may be separated from the pulp in the foregut and regurgitated. Animals gain food-processing efficiency by regurgitating rather than defecating seeds, as regurgitation reduces the time seeds spend in animal guts. White (1989) found that three species of birds (veery, 34 g, grey catbird, 38 g, and wood thrush, 55 g) regurgitated significantly more for six plants species with seeds 10–15 mm in diameter than the larger American robins (78 g). For smaller seeds of *Phytolacca americana* (< 8 mm), no regurgitation was found and, for *Lindera benzoin* and *Sassafras albidum* (> 16 mm), most seeds were regurgitated by all species. For individual species of birds eating many species of fruits, seed size influences the probability of regurgitation or defecation. For 77 North American species with fleshy fruits, seed circumference ranged from 1.90 to 20.15 mm, but the estimated smallest seed regurgitated by American robins is 8 mm and the largest defecated seed is 18 mm. Seed treatment and hence dispersion patterns differ among species of animals consuming the same fruits, and this may depend on the interaction between frugivore gut morphology and seed size (Martínez del Rio and Restrepo, 1993).

Whereas many birds swallow fruits whole, others mash fruits in their bills, ingesting the pulp but often dropping the seeds as they eat the fruit (Levey, 1987). This is especially true for tanagers (Moermond, 1983) and emberizid finches (Johnson *et al.*, 1985), and fruit consumption by these species may select for small seed size (Levey, 1986). Fruits consumed by mammals may also be selected for small seed size to escape both mastication by larger mammals and seed predation by mice, since these fruits are available to these terrestrial consumers. Larger mammals deposit these small seeds in large piles, which may result in severe intraspecific competition among the germinating seeds or in secondary dispersal (see below).

Seeds are not distributed evenly over the environment by the animals that eat the fruits. Not only are there smaller piles of seeds from individual defecations, but also points of intensive activity, such as roosts or nests, may yield large numbers of deposited seeds. Bat roosts of *Corollia perspicillata* deposit over 90% of the seeds of *Piper* (*Piperaceae*) they consume in piles under the roost (Fleming, 1981). Bats will carry fruits back to specific roosting sites and drop larger seeds there, as well as defecating smaller seeds of ingested species. Many species of ants carry seeds to their nests, where they cut off the lipid-rich elaiosome from the seed and discard the seeds in refuse piles adjacent to their nest (Beattie, 1985). Not only are such concentrations of seeds disadvantageous because of the potential for strong competitive interactions with other plants, but they also increase the probability of fungal attacks on the seeds, encouraged by the large amounts of faeces and fruit parts. Oilbirds deposit large numbers of seeds in their nesting caves, where probability of seed survival is zero (Snow, 1962).

Patchy distribution of seeds in the environment may also take place in subtler ways. In open fields, birds often perch on the tallest available perch site. They frequently defecate or regurgitate seeds either while on the perch or on taking off, yielding a high density of seeds, or recruitment focus, around the perch (McDonnell and Stiles, 1983; McDonnell, 1986; McClanahan and Wolfe, 1993). Similar non-random distributions are found after strip-cutting in the Peruvian Amazon (Gorchov *et al.*, 1993). Individual fruiting plants also serve as recruitment foci for seeds of species fruiting at the same time. Most species of frugivores eat many species of fruits, and seeds of species fruiting at the same time of year are deposited more frequently under other fruiting individuals of the same or different species. This may result in patterns of seed movement that depend on the distribution of species of plants fruiting at the same time of year (Pratt, 1983; E.W. Stiles, unpublished).

Patchy distributions of seeds are also created through site selection by scatter hoarders and larder hoarders. Scatter hoarders, such as squirrels, bury seeds individually or in small groups, eventually returning to retrieve the seeds. Although

scatter hoarders are extremely good at finding the seeds they store (they usually find more than 85%), a few seeds are not recovered. The seeds are often buried near the surface in ideal locations for successful germinations. In a study of bear oak (*Quercus ilicifolia*) in New York, acorns that were protected from dispersers and predators but left on the surface of the soil did not survive. All of the seeds appeared to have died from desiccation (Unnash, 1990). Larder hoarders hide seeds in groups. For wood rats (*Neotoma*), red squirrels (*Tamiasciurus*) and kangaroo-rats (*Dipodomys*), seeds usually have a fate similar to those in the oilbird caves. If they are not consumed, they are in locations not acceptable for seed germination and survival.

Animal seed dispersers

Most birds disperse seeds in some way or other. A great number of species actually consume the pulp of fruits or the seeds themselves for food and move around varying numbers of seeds in the process. Dominant among species that eat fruit and deposit seeds unharmed are members of the orders of perching birds (Passeriformes), pigeons (Columbiformes), woodpeckers (Piciformes) and cassowaries (Casuariiformes). Other important seed-dispersing groups of birds are spread throughout the other orders, including the hornbills, motmots, trogons, oilbirds and turacos. Species that are primarily seed predators but whose actions result in seed movement in some situations are also found within the perching birds and pigeons, as well as the parrots (Psittaciformes), gallinaceous birds (Galliformes), tinamous (Tinamiformes) and rheas (Rheiformes).

Other species move seeds stuck in mud on their feet. These include ducks and geese (Anseriformes), herons (Ciconiiformes), cranes (Gruiformes), shorebirds (Charadriiformes), loons (Gaviiformes) and grebes (Podocipediformes). Birds of prey, including hawks (Falconiformes) and owls (Strigiformes), are often secondary dispersers, moving the seeds already in the guts of prey species. This leaves only a few orders of birds without representative seed dispersers, including penguins (Sphenisciformes), pelagic tubenoses (Procellariiformes) and swifts and hummingbirds (Apodiformes).

Birds vary markedly in the extent of their frugivory. A few species have been noted for their high level of frugivory. The white-cheeked cotinga (*Ampelion stresemanni*) was observed to feed only on the fruit of the genus *Tristerix*, an orange-flowered mistletoe of the high Andes (Snow, 1982). Individually regurgitated seeds were wiped from the bill on to the limbs and Snow never saw one fall to the ground. Earlier examples of mistletoe specialization have also been reported, including the Australian flowerpeckers (*Dicaeum* spp.) (Docters van Leeuwen, 1954), the phainopepla in the south-western USA (Walsberg, 1975) and the cedar waxwing in Florida (Skeate, 1985). Another close relationship occurs among the waxy fruits of bayberries, wax myrtles (*Myrica*) and poison ivy (*Toxicodendron*) and the yellow-rumped warbler, tree swallow and some woodpeckers, especially the northern flicker. These birds are able to metabolize the waxy coatings on these fruits rapidly and efficiently. These high-energy waxes are unavailable to most frugivores, but the ability of the yellow-rumped warblers and tree swallows to feed on these fruits during times of low insect abundance enable them to maintain the most northerly wintering ranges of any member of the wood warblers and swallows (Place and Stiles, 1992).

Variation in the extent of frugivory is seasonal in the temperate zones, with many frugivorous species consuming fruit in autumn and winter and animal food in spring and summer. American robins consume 79% animal matter in the spring and 81% fruit in the autumn (Martin *et al.*, 1951; Wheelwright, 1986). Some migrant species, such as the eastern kingbird and chestnut-sided warbler (Greenberg, 1981), rarely eat fruits on the breeding grounds but become highly frugivorous during the winter in the tropics. The changing demands for nutrients available in fruit versus animals seems to fuel the change in diet during different periods in the annual cycle for these species.

Mammals too are represented by major groups of fruit- and seed-eating species. Major frugivorous groups include the primates (Primates), the bats (Chiroptera), carnivores (Carnivora), elephants (Proboscidea) and some marsupials (Marsupialia). Of the mammals, some of the bats are the most heavily frugivorous, although two large families, the flying foxes (Pteropidae) of the Old World and the American leaf-nosed bats (Phyllostomatidae), are the major fruit-eating groups.

Most of the primates include fruit as an important part of their diets, and dogs (Canidae), bears (Ursidae), racoons (Procyonidae) and civets (Viverridae) are the primary fruit-eating families among the carnivores. Elephants eat fruits as well as many other types of vegetation. Frugivorous marsupials are concentrated in the opossums (Didelphidae) and phalangers (Phalangeridae).

Seed-eating species are found primarily within the rodents (Rodentia), although both odd-toed ungulates (Perissodactyla) and even-toed ungulates (Artiodactyla) are heavy consumers of seeds, often as a by-product of eating the leafy portion of the plant (Janzen, 1982c). These two orders of mammals have been central to the hypothesis that, for many grasses, as well as members of the *Leguminosae*, *Scrophulariaceae* and *Amaranthaceae*, the foliage serves as the resource attracting seed dispersers to the plants, much as the fruit pulp does in fleshy-fruited species (Quinn *et al.*, 1994). Large families within the rodents, including the squirrels (Sciuridae), heteromyid rats (Heteromyidae), cricetid rats (Cricetidae) and murid rats (Muridae), are all major seed predators, as well as dispersers of seeds through their scatter-hoarding and larder-hoarding behaviours.

Da Silva and co-workers (1989) demonstrated that the neotropical treefrog, *Hyla truncata*, is a frugivorous seed disperser. They found fruits and seeds of *Anthurium harrisii* (*Araceae*), *Erythroxylum ovalifolium* (*Erythroxylaceae*) and two unidentified species in the guts of the frogs, and demonstrated that the frogs would eat the fruits and defecate viable seeds of *Anthurium* in the laboratory. The other five groups have larger numbers of seed-dispersing species.

Fruit- and seed-eating habits of fishes are as yet poorly studied. The excellent work of Goulding (1980) in the seasonally flooded forests of the Amazon basin found that seven species within the family Characidae and members of three catfish families, the Pimelodidae, Doradidae and Auchenipteridae, were primarily seed predators but also served as seed dispersal agents for a variety of plants. The large characins of the genera *Colossoma* and *Brycon* were especially important.

Very few reptiles are important seed dispersal agents. These include large herbivorous lizards (*Iguana* spp.), which consume fruits and leaves as adults, and the more terrestrial tortoises. Aquatic turtles are highly carnivorous, but the terrestrial forms, including box tortoises (*Terrapene*) (Klimstra and Newsome, 1960), gopher tortoises (*Gopherus*) (Woodbury and Hardy, 1948) and the Galapagos tortoise (*Testudo elephantropus*) (Rick and Bowman, 1961), as well as other high-shelled species, are omnivorous in diet. They include fruits as important parts of their food and pass viable seeds.

The only non-vertebrate group to disperse appreciable numbers of seeds is the ants. However, ants are the most numerous and widely dispersed group of animals on Earth. They range from the tree line in the Arctic to the southern tips of Africa and South America and are found on virtually every oceanic island. At any moment, there are at least 10^{15} living ants on Earth (Wilson, 1971). Seed-dispersing ants fall into two groups: primary dispersers, which remove seeds from plants, and secondary dispersers, which remove seeds from the faeces of other animals, although the distinction between the two groups lies only in the stage of dispersal in which they participate.

Known primary seed-dispersing ants belong to four subfamilies of the Formicidae. These are the Formicinae, Myrmicinae, Ponerinae and Dolichoderinae. Common, widespread genera, such as *Pheidole* and *Crematogaster*, occur throughout the tropics, and dispersing species are found in virtually all ant communities (Beattie, 1985). As elaiosomes vary among seeds, ants vary in their response to them. This variation is not only among species of ants, but also among populations of the same species and even within a population on different days (Beattie and Culver, 1981).

Most bird defecations in a Costa Rican forest were found to be visited by ants within a few hours. Small seeds were taken to nests close by (< 1 m) in the leaf litter and most were consumed. Some seeds, however, are placed, intact, on refuse piles, often within twig nests. This placement may afford additional protection for these seeds (Byrne and Levey, 1993; Levey and Byrne, 1993).

Conclusion

For many species of plants, animals provide the means for the critical mobile stage in the plant's life history. The disseminated seeds may become part of the seed bank, preserving genetic variation for contribution to future populations, or they may germinate soon after they are deposited. Many of the seeds will die either before or shortly after germination, succumbing to poor physical conditions of the deposition site or aggressive herbivores, but others will find themselves in superior sites for establishment and will contribute their genes to the production of more seeds. We have much to learn about the factors that influence this interaction among seeds and seed dispersers, but there is little doubt concerning the importance of these interactions for the evolution of many characteristics of both the plants and the animals.

References

Agnew, A.D.Q. and Flux, J.E.C. (1970) Plant dispersal by hares (*Lepus capensis* L.) in Kenya. *Ecology* 51, 735–737.

Beattie, A.J. (1985) *The Evolutionary Ecology of Ant–Plant Mutualisms*. Cambridge University Press, Cambridge.

Beattie, A.J. and Culver, D.C. (1981) The guild of myrmecochores in the herbaceous flora of West Virginia forests. *Ecology* 62, 107–115.

Beattie, A.J. and Culver, D.C. (1983) The nest chemistry of two seed-dispersing ant species. *Oecologia* 56, 99–103.

Berg, R.Y. (1975) Myrmecochorous plants in Australia and their dispersal by ants. *Australian Journal of Botany* 23, 475–488.

Bond, W.J. and Slingsby, P. (1984) Collapse of an ant–plant mutualism: the Argentine ant (*Iridomyrmex humilis*) and myrmecochorous Proteaceae. *Ecology* 65, 1031–1037.

Bullock, S.H. and Primack, R.B. (1977) Comparative experimental study of seed dispersal on animals. *Ecology* 58, 681–686.

Bunch, K.G. and Tomback, D.F. (1986) Bolus recovery by gray jays: an experimental analysis. *Animal Behavior* 34, 754–762.

Byrne, M.M. and Levey, D.J. (1993) Removal of seeds from frugivore defecations by ants in a Costa Rican rainforest. *Vegetatio* 108, 363–374.

Calahane, V.H. (1942) Catching and recovery of food by the western fox squirrel. *Journal of Wildlife Management* 6, 338–352.

Clark, L. (1991) Odor detection thresholds in tree swallows and cedar waxwings. *Auk* 108, 177–180.

Cook, A.D., Atsatt, P.R. and Simon, C.A. (1971) Doves and dove weed: multiple defense against avian predation. *Bioscience* 21, 277–281.

Culver, D.C. and Beattie, A.J. (1978) Myrmecochory in *Viola*: dynamics of seed–ant interactions in some West Virginia species. *Journal of Ecology* 66, 53–72.

Da Silva, H.R., De Britto-Pereira, M.C. and Caramaschi, U. (1989) Frugivory and seed dispersal by *Hyla truncata*, a neotropical treefrog. *Copeia* 3, 781–783.

Davidson, D.W., Inouye, R.S. and Brown, J.H. (1984) Granivory in a desert ecosystem: experimental evidence for indirect facilitation of ants by rodents. *Ecology* 65, 1780–1786.

Denslow, J.S. and Moermond, T.C. (1982) The effect of accessibility on rates of fruit removal from tropical shrubs: an experimental study. *Oecologia* 54, 170–176.

Dinerstein, E. (1989) The foliage-as-fruit hypothesis and the feeding behavior of South Asian ungulates. *Biotropica* 21, 214–218.

Docters van Leeuwen, W.M. (1954) On the biology of some Loranthaceae and the role birds play in their life-history. *Beaufortia* 4, 105–208.

Drickamer, C.C. (1970) Seed preferences in wild caught *P. maniculatus bairdi* and *P. leucopus noveboracensis*. *Journal of Mammology* 57, 191–194.

Drickamer, C.C. (1976) Hypothesis linking food habits and habitat selection in *Peromyscus*. *Journal of Mammology* 57, 763–766.

Fleming, T.H. (1981) Fecundity, fruiting pattern, and seed dispersal in *Piper amalago* (Piperaceae), a bat dispersed tropical shrub. *Oecologia* 51, 42–46.

Fukui, A. (1996) Retention time of seeds in bird guts: costs and benefits for fruiting plants and frugivorous birds. *Plant Species Biology* 11, 141–147.

Gibson, D.J., Freeman, C.C. and Hubert, L.C. (1990) Effects of small mammal and invertebrate herbivory on plant species richness and abundance in tallgrass prairie. *Oecologia* 84, 169–175.

Gorchov, D.L., Cornejo, F., Ascorra, C. and Jaramillo, M. (1993) The role of seed dispersal in the natural regeneration of rainforest after strip-cutting in the Peruvian Amazon. *Vegetatio* 107/108, 339–349.

Goulding, M. (1980) *The Fishes and the Forest*. University of California Press, Berkeley, California.

Greenberg, R. (1981) Frugivory in some migrant tropical forest wood warblers. *Biotropica* 13, 215–223.

Handel, S.N. (1978) The competitive relationship of three woodland sedges, and its bearing on the evolution of ant dispersal of *Carex pedunculata*. *Evolution* 32, 151–163.

Heithaus, E.R. (1981) Seed propagation by rodents on three ant-dispersal plants. *Ecology* 62, 136–145.

Herrera, C.M. (1984) Adaptation to frugivory of Mediterranean avian seed dispersers. *Ecology* 65, 609–617.

Howard, W.E., Marsh, R.E. and Cole, R.E. (1968) Food detection by deer mice using olfactory rather than visual cues. *Animal Behavior* 16, 13–17.

Howe, H.F. and Smallwood, J. (1982) Ecology of seed dispersal. *Annual Review of Ecology and Systematics* 13, 201–228.

Janzen, D.H. (1970) Herbivores and the number of tree species in tropical forests. *American Naturalist* 104, 501–528.

Janzen, D.H. (1972) Escape in space by *Sterculia apetala* seeds from the bug *Dysdercus fasciatus* in a Costa Rican deciduous forest. *Ecology* 53, 350–361.

Janzen, D.H. (1981) *Enterolobium cyclocarpum* seed passage rates in cows and horses, Costa Rican Pleistocene seed dispersal agents. *Ecology* 62, 593–601.

Janzen, D.H. (1982a) Differential seed survival and passage rates in cows and horses, surrogate Pleistocene dispersal agents. *Oikos* 38, 150–156.

Janzen, D.H. (1982b) Removal of seeds from dung by tropical rodents: influence of habitat and amount of dung. *Ecology* 63, 1887–1900.

Janzen, D.H. (1982c) Dispersal of small seeds by big herbivores: foliage is the fruit. *American Naturalist* 123, 338–353.

Johnson, R.A., Willson, M.F., Thompson, J.N. and Bertin, R.I. (1985) Nutritional values of wild fruits and consumption by migrant frugivorous birds. *Ecology* 66, 819–827.

Kamil, A.C. and Balda, R.P. (1985) Cache recovery and spatial memory in Clark's nutcrackers (*Nucifraga columbiana*). *Journal of Experimental Psychology, Animal Behavior Proceedings* 11, 95–111.

Keeler, K.H. (1989) Ant–plant interactions. In: Abramson, W.G. (ed.) *Plant–Animal Interactions*. McGraw-Hill, New York, pp. 207–242.

Klimstra, W.D. and Newsome, F. (1960) Some observations on the food coactions of the common box turtle *Terrapene carolina*. *Ecology* 41, 639–647.

Knight, R.S. and Siegfried, W.R. (1983) Inter-relationships between type, size and colour of fruits and dispersal in southern African trees. *Oecologia* 56, 405–412.

Krefting, L.W. and Roe, E.I. (1949) The role of some birds and mammals in seed germination. *Ecological Monographs* 19, 269–286.

Levey, D.J. (1986) Methods of seed processing by birds and seed deposition patterns. In: Estrada, A. and Fleming, T.H. (eds) *Frugivores and Seed Dispersal*. Junk, Dordrecht, pp. 147–158.

Levey, D.J. (1987) Seed size and fruit-handling techniques of avian frugivores. *American Naturalist* 129, 471–485.

Levey, D.J. and Byrne, M.M. (1993) Complex ant–plant interactions: rainforest ants as secondary dispersers and postdispersal seed predators. *Ecology* 74, 1802–1812.

McClanahan, T.R. and Wolfe, R.W. (1993) Accelerating forest succession in a fragmented landscape: the role of birds and perches. *Conservation Biology* 7, 279–288.

McDonnell, M.J. (1986) Old field vegetation height and the dispersal pattern of bird disseminated wood plants. *Bulletin of the Torrey Botanical Club* 113, 6–11.

McDonnell, M.J. and Stiles, E.W. (1983) The structural complexity of old field vegetation and the recruitment of bird-dispersed plant species. *Oecologia* 56, 109–116.

McKey, D. (1975) The ecology of coevolved seed dispersal systems. In: Gilbert, L.E. and Raven, P.H. (eds) *Coevolution of Animals and Plants*. University of Texas Press, Austin, pp. 159–191.

Martin, A.C., Zim, H.S. and Nelson, A.L. (1951) *American Wildlife and Plants*. McGraw-Hill, New York.

Martínez del Rio, C. and Restrepo, C. (1993) Ecological and behavioral consequences of digestion in frugivorous animals. In: Fleming, T.H. and Estrada, A. (eds) *Frugivory and Seed Dispersal: Ecological and Evolutionary Aspects*. Kluwer Academic Publisher, Dordrecht, pp. 205–216.

Martínez del Rio, C., Stevens, B.R., Daneke, D.E. and Andreadis, P.T. (1988) Physiological correlations of preferences and aversion for sugars in three species of birds. *Physiological Zoology* 61, 222–229.

Moermond, T.C. (1983) Suction-drinking in tanagers and its relation to fruit handling. *Ibis* 125, 545–549.

Moermond, T.C. and Denslow, J.S. (1983) Fruit choice in neotropical birds: effects of fruit type and accessibility on selectivity. *Journal of Animal Ecology* 52, 407–420.

Moermond, T.C. and Denslow, J.S. (1985) Neotropical frugivores: patterns of behavior, morphology, and nutrition with consequences for fruit selection. In: Buckley, P.A., Foster, M.S., Morton, E.S., Ridgely, R.S. and Buckley, N.G. (eds) *Neotropical Ornithology*. American Ornithologists' Union Monographs, Allen Press, Lawrence, Kansas, pp. 865–897.

Nicholls, T.H., Hawksworth, F.G. and Merrill, L.M. (1984) Animal vectors of dwarf mistletoe, with special reference to *Arceuthobium americanum* on lodgepole pine. In: *Biology of Dwarf Mistletoes. Proceeding of the Symposium GTR RM-111, Fort Collins, Colorado*. pp. 102–110.

O'Dowd, D.J. and Hay, M.E. (1980) Mutualism between harvester ants and a desert ephemeral: seed escape from rodents. *Ecology* 61, 531–540.

Ostry, M.E., Nicholls, T.H. and French, D.W. (1983) *Animal Vectors of Eastern Dwarf Mistletoe of Black Spruce*. Research Paper NC-232, North Central Forest Experiment Station, Forest Service, US Department of Agriculture, St Paul, Minnesota, 16 pp.

Place, A.R. and Stiles, E.W. (1992) Living off the wax of the land: bayberries and yellow-rumped warblers. *Auk* 109, 334–345.

Pratt, T.K. (1983) Seed dispersal in a montane forest in Papua New Guinea. Unpublished PhD thesis, Rutgers University, New Brunswick, New Jersey.

Pratt, T.K. and Stiles, E.W. (1985) The influence of fruit size and structure on composition of frugivore assemblages in New Guinea. *Biotropica* 17, 314–321.

Proctor, V.W. (1968) Long-distance of seeds by retention in digestive tract of birds. *Science* 166, 321–322.

Quinn, J.A., Mowrey, D.P., Emanuele, S.M. and Whalley, R.D.B. (1994) The 'foliage is the fruit' hypothesis: *Buchloe dactyloides* (Poaceae) and the shortgrass prairie of North America. *American Journal of Botany* 81, 1545–1554.

Rice, B. and Westoby, M. (1986) Evidence against the hypothesis that ant-dispersed seeds reach nutrient-enriched microsites. *Ecology* 67, 1270–1274.

Rick, C.M. and Bowman, R.I. (1961) Galapagos tomatoes and tortoises. *Evolution* 15, 407–417.

Ridley, H.N. (1930) *The Dispersal of Plants throughout the World.* L. Reeve, Ashford, Kent.

Rieger, J.F. and Jakob, E.M. (1988) The use of olfaction in food location by frugivorous bats. *Biotropica* 20, 161–164.

Roessler, E.S. (1936) Viability of weed seeds after ingestion by California linnets. *Condor* 38, 62–65.

Schorger, A.W. (1960) The crushing of *Carya* nuts in the gizzard of the turkey. *Auk* 77, 337–340.

Skeate, S.T. (1985) Mutualistic interactions between birds and fruits in a northern Florida hammock community. Unpublished PhD thesis, University of Florida, Gainesville, Florida.

Smith, A.J. (1975) Invasion and ecesis of bird-disseminated woody plants in a temperate forest sere. *Ecology* 56, 19–34.

Snow, D.W. (1962) The natural history of the oilbird, *Steatornis caripensis*, in Trinidad, WI. II. Population, breeding ecology, and food. *Zoologica* 47, 199–221.

Snow, D.W. (1982) *The Cotingas.* British Museum (Natural History), London.

Sorensen, A.E. (1986) Seed dispersal by adhesion. *Annual Review of Ecology and Systematics* 17, 443–463.

Stapanian, M.A. and Smith, C.C. (1984) Density-dependent survival of scatterhoarded nuts: an experimental approach. *Ecology* 65, 1387–1396.

Stapanian, M.A. and Smith, C.C. (1986) How fox squirrels influence the invasion of prairies by nut-bearing trees. *Journal of Mammology* 67, 326–332.

Stiles, E.W. (1980) Patterns of fruit presentation and seed dispersal in bird-disseminated woody plants in the eastern deciduous forest. *American Naturalist* 116, 670–688.

Stiles, E.W. (1982) Fruit flags: two hypotheses. *American Naturalist* 120, 500–509.

Stiles, E.W. and Dobi, E.T. (1987) Scatterhoarding of horse chestnuts by eastern gray squirrels. *Bulletin of the New Jersey Academy of Science* 32, 1–3.

Temple, S.A. (1977) Plant–animal mutualism: coevolution with dodo leads to near extinction of plant. *Science* 197, 885–886.

Tomback, D.F. (1982) Dispersal of whitebark pine seeds by Clark's nutcrackers: a mutualism hypothesis. *Journal of Animal Ecology* 51, 451–467.

Tomback, D.F. (1983) Nutcrackers and pines: coevolution or coadaptation? In: Nitecki, N.H. (ed.) *Coevolution.* University of Chicago Press, Chicago, pp. 179–223.

Tomback, D.F. (1986) Post-file regeneration of Krummholz whitebark pine: a consequence of nutcracker seed caching. *Madrono* 33, 100–110.

Turcek, F.J. (1963) Color preference in fruit and seed-eating birds. *Proceedings of the International Ornithological Congress* 13, 285–292.

Unnash, R.S. (1990) Seed predation and the limits to the recruitment in two species of pine barren oak. Unpublished PhD thesis, State University of New York at Stony Brook, New York.

van der Pijl, L. (1982) *Principles of Dispersal in Higher Plants*, 3rd edn. Springer-Verlag, New York.

Walsberg, G.E. (1975) Digestive adaptations of *Phainopepla nitens* associated with the eating of mistletoe berries. *Condor* 77, 169–174.

Westoby, M. and Rice, B. (1981) A note on combining two methods of dispersal-for-distance. *Australian Journal of Ecology* 6, 23–27.

Wheelwright, N.T. (1985) Fruit size, gape width, and the diets of fruit-eating birds. *Ecology* 66, 808–818.

Wheelwright, N.T. (1986) The diet of American robins: an analysis of US Biological Survey records. *Auk* 103, 710–725.

Wheelwright, N.T. and Janson, C.H. (1985) Colors of fruit displays of bird-dispersed plants in two tropical forests. *American Naturalist* 126, 777–799.

White, D.W. (1989) North American bird-dispersed fruits: ecological and adaptive significance of nutritional and structural traits. Unpublished PhD thesis, Rutgers University, New Brunswick, New Jersey.

Willson, M.F. and Thompson, J.N. (1982) Phenology and ecology of color in bird-dispersed fruits, or why some fruits are red when they are 'green'. *Canadian Journal of Botany* 60, 701–713.

Willson, M.F., Graff, D.A. and Whelan, C.J. (1990) Color preferences of frugivorous birds in relation to the colors of fleshy fruits. *Condor* 92, 545–555.

Wilson, E.O. (1971) *The Insect Societies*. Belknap Press, Cambridge, Massachusetts.

Woodbury, A.M. and Hardy, R. (1948) Studies of the desert tortoise, *Gopherus agassizii*. *Ecological Monographs* 18, 145–200.

Chapter 6
Fruits and Frugivory

Pedro Jordano

Estación Biológica de Doñana, CSIC, Apdo. 1056, Sevilla, Spain

Introduction

The pulp of fleshy fruits, with the soft, edible, nutritive tissues surrounding the seeds, is a primary food resource for many frugivorous animals, notably mammals and birds, but also reptiles (Howe, 1986). These animals either regurgitate, defecate, spit out or otherwise drop undamaged seeds away from the parent plants; they are the seed dispersers that establish a dynamic link between the fruiting plant and the seed–seedling bank in natural communities. Therefore, frugivory is a central process in plant populations where natural regeneration is strongly dependent upon seed dissemination by animals.

Early conceptual contributions to the study of frugivory emphasized dichotomies in frugivory patterns and fruit characteristics that presumably originated by co-evolved interactions (Snow, 1971; McKey, 1975; Howe and Estabrook, 1977; Howe, 1993). Fruits with pulps of a high energetic content and nutritive value surrounding a single large seed would be one extreme of specialization by interacting with specialized frugivores providing high-quality dispersal; fruits with succulent, watery, carbohydrate-rich pulps occupy the other extreme by having their numerous small seeds dispersed by opportunist frugivores. Subsequent work during the last two decades has centred around this seminal paradigm and there is a wealth of information about patterns of frugivory in particular taxa, variation in fruit charateristics and detailed descriptions of plant/frugivore interactions for particular plant species or communities (for recent reviews, see Howe, 1984, 1993; Estrada and Fleming, 1986; Herrera, 1995; Corlett, 1998). However, studies of frugivory have rarely been linked conceptually with demographic patterns in the plant population; also, the evolutionary consequences of frugivore choices, fruit processing and movement patterns have seldom been examined in an explicit evolutionary context, where fitness differentials in plant populations are measured and associated with individual variation in dispersal-related traits. Frugivory and dispersal influence the evolution of plant traits through effects on population processes, but predictive frameworks that link frugivory patterns, associated differences in seed/seedling mortality and differential reproductive success with demographic patterns in natural plant populations are very scarce (Howe, 1989; Jordano and Herrera, 1995; Schupp and Fuentes, 1995; Wenny and Levey, 1998).

Recent reviews of seed dispersal and frugivore ecology show that, for most frugivores, fleshy fruits are a non-exclusive food resource, which is supplemented with

© CAB *International* 2000. *Seeds: The Ecology of Regeneration in Plant Communities, 2nd edition* (ed. M. Fenner)

animal prey, vegetative plant parts, seeds, etc. (Hladik, 1981; Moermond and Denslow, 1985; Fleming, 1986; Howe, 1986; Willson, 1986; Corlett, 1998). Very few vertebrates rely totally on fruit food, but many species are 'partial' frugivores, which consume other prey together with various amounts of fruit; dietary habits among these species range from sporadic fruit consumption to almost totally frugivorous diets. For example, only 17 families of birds (15.6%) can be considered as strictly frugivorous, but at least 21 families (19.3%) consume a mixed diet with a large proportion of fruits and a minor contribution of animal prey; and 23 families (21.1%) mix, in roughly equal proportions, fruits and other material in their diets (see Snow, 1981). Total frugivory among mammals is non-existent. Among bats, only pteropodids (Old World bats) and phyllostomids (New World fruit-bats) can be considered largely frugivorous (Gardner, 1977; Marshall, 1983; Fleming, 1986), supplementing fruit food with insects (Courts, 1998) and/or leaves (Kunz and Diaz, 1995). Fruit is the most widely used type of food among primates, found in the diets of 91% of the species examined to date (Harding, 1981; Hladik, 1981), and certain frugivorous forest ungulates, such as brocket deer (*Mazama* spp.) and African cephalophines (*Cephalophus* spp.), can include up to 85% of fruit material in their diet (Dubost, 1984; Bodmer, 1989a, 1990). However, partially frugivorous mammals include opossums, phalangers, kangaroos, lemurs, lorises, apes, foxes, bears, elephants, horses and other ungulates (Harding, 1981; Janzen, 1983; Howe, 1986). Finally, among reptiles, tortoises, lizards and iguanids can have an important role as seed dispersers, even with infrequent and non-obligate frugivory (Barquín and Wildpret, 1975; Losos and Greene, 1988).

Frugivorous animals, relying sporadically or obligately on fruits for food, have a central role in demography and plant community evolution because: (i) their interaction with plants takes place at the final stage of each plant reproductive episode, having a potential to 'screen off' or nullify previous effects of the pollination and fruit growth phases (Herrera, 1988a; Jordano, 1989); (ii) by directing the early spatial distribution of the seeds, i.e. the 'seed shadow' (Janzen *et al.*, 1976), they provide a template over which future spacing patterns of adult plants will build up; and (iii) seed deposition patterns by frugivores directly affect patterns of early seed survival and seedling establishment (Howe *et al.*, 1985; Katusic-Malmborg and Willson, 1988; Schupp, 1988; Willson, 1988; Herrera *et al.*, 1994).

The purpose of this chapter is to dissect this fleshy-fruit/frugivore interface, which brings up both characteristics of the fruits as 'prey items', which must be sought, handled and efficiently processed, and the ability of frugivores to perform these tasks, with consequences for the plants themselves. Throughout the chapter, any mention of fruits will be with reference to fleshy fruits, loosely defined to include any structure enclosing seeds surrounded by a fleshy, edible, pulp layer (Howe and Smallwood, 1982). Most references to frugivorous animals will be to birds, primates, ungulates and bats that behave as seed dispersers. The first section of the chapter describes fruits as prey items from the perspective of the foraging animal, and examines their characteristics, temporal and spatial patterns of availability and intrinsic traits, such as design and nutritive value. The second part reviews frugivore traits that influence fruit choice, fruit and seed processing and foraging movements that have implications for seed deposition patterns.

Fruit production and availability

Fleshy fruits are, for the organisms consuming them, discrete food items available in an extremely diverse array of spatial and temporal configurations. The various characteristics (Table 6.1) include those that define their spatial distribution and the temporal patterns of availability, both seasonally and between years, and their food value as prey that must be processed as

Table 6.1. Summary of major characteristics of fleshy fruits as food resources for frugivorous vertebrates.

A. Availability characteristics
    Marked seasonal changes in abundance
    Non-renewable in the short term
    Strong between-year changes in availability for certain species
    Heterogeneous spatial distribution: highly clumped; local superabundance; few species available at
    the same particular location

B. 'Intrinsic' characteristics as prey items
    High water content
    Strong imbalance between energetic and protein components
    Presence of voluminous mass of indigestible material (seeds)
    Presence of secondary metabolites

discrete items. Availability characteristics influence overall abundance of frugivores in particular habitat patches, their foraging movements and important aspects of their annual cycles. Intrinsic features determine fruit and seed processing and, consequently, how the seeds reach the ground. Both groups of traits ultimately influence seed deposition patterns, because they determine the movement patterns of frugivores foraging for fruits in relation to the mosaic of habitat patches.

Production and abundance of fruits

Variation among communities in the frequency of endozoochorous seed dispersal is broadly associated with variation in precipitation and moisture (Gentry, 1982), and a latitudinal gradient is also evident. Vertebrate seed dispersal is very common among woody plants in neotropical (70–94% of woody species), Australian (82–88%) and African rainforests (approximately 80%) (Table 6.2). Mediterranean scrubland and some tropical dry and humid forests and woodlands usually range between 50 and 70%; temperate coniferous and broad-leaved forests vary within 30–40% of animal-dispersed woody species. Frugivory and endozoochorous seed dispersal are virtually absent or unimportant in grasslands, extreme deserts, alpine vegetation and certain types of scrublands on nutrient-poor sites.

Table 6.2. Percentages of woody species adapted for endozoochorous seed dispersal by vertebrates in different vegetation types.

Vegetation type	Mean (Range)	References[a]
Temperate coniferous forest	41.8 (33.3–56.5)	1–4
Temperate deciduous forest	35.4 (9.5–53.8)	1–5
Savannah woodland	41.2 –	6
Mediterranean scrubland (Spain)	56.1 (47.1–64.3)	7, 8
Mediterranean scrubland (Chile)	41.9 (20.0–55.1)	9
Mediterranean scrubland (California)	34.4 (16.7–43.3)	9
Mediterranean scrubland (Australia)	22.5 (10.0–50.0)	9–11
Neotropical dry forest	46.2 (27.0–58.7)	12–14
New Zealand lowland forest	64.0	15
Subtropical humid forest	69.4 (65.2–73.5)	16, 17
Neotropical and palaeotropical humid forest	74.7 (62.1–82.1)	5, 18–22
Tropical rainforest	89.5 (70.0–93.5)	5, 22–24

[a] References: 1, Johnson and Landers (1978); 2, Marks and Harcombe (1981); 3, Schlesinger (1978); 4, Franklin *et al.* (1979); 5, Howe and Smallwood (1982) and references therein; 6, Poupon and Bille (1974); 7, Herrera (1984b); 8, Jordano (1984); 9, Hoffmann and Armesto (1995); 10, Milewski (1982); 11, Milewski and Bond (1982); 12, Gentry (1982); 13, Frankie *et al.* (1974b); 14, Daubenmire (1972); 15, Burrows (1994); 16, Frost (1980); 17, Boojh and Ramakrishnan (1981); 18, Charles-Dominique *et al.* (1981); 19, Alexandre (1980); 20, Lieberman (1982); 21, Tanner (1982); 22, Willson *et al.* (1989) and references therein; 23, Putz (1979); 24, Janson (1983).

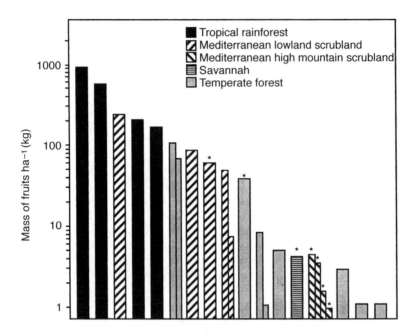

Fig. 6.1. Total production (per unit area) of fleshy fruits in different plant communities (placed in order of decreasing magnitude of production). Bars with asterisks indicate biomass figures as wet mass, all others are dry mass. Double bars indicate data for two localities in a single study. References: Leigh (1975); Johnson and Landers (1978); Baird (1980); Stransky and Halls (1980); Charles-Dominique *et al.* (1981); Hladik (1981); Sorensen (1981); Guitián (1984); Herrera, (1984b, c); Jordano (1984, and unpublished data).

This range of variation is also exemplified when considering between-community variation in production of fleshy fruits, both in numbers and biomass. Overall levels of fruit production in particular habitats are strongly associated with the relative importance of zoochory as an adaptation for the dispersal of seeds (Fig. 6.1), but rigorous estimation of absolute abundance is subject to numerous potential biases (Blake *et al.*, 1990; Chapman *et al.*, 1992b, 1994; Zhang and Wang, 1995). Fruit production in temperate forests of the northern hemisphere is always below 10^5 fruits ha^{-1}, representing less than 10 kg ha^{-1} (dry mass). Mediterranean scrublands have productions similar to those of some tropical forests, in general around 80 kg (dry mass) ha^{-1}, but fruit density might reach more than 1.4×10^6 fruits ha^{-1} in good crop years (Herrera, 1984b; Jordano, 1985); however, high-elevation Mediterranean scrublands have productions more similar to

those of temperate forests (Fig. 6.1). Tropical rainforests range widely in production, usually between 180 and approximately 1000 kg ha^{-1} (dry mass). For additional data, see Blake *et al.* (1990).

Extreme between-year variations in the production of fleshy fruits have been found (e.g. Davies, 1976; Foster, 1982; Jordano, 1985; Herrera, 1988c, 1998), but a direct, causal relationship between these fluctuations and frugivore numbers has rarely been documented. In most instances, studies with long-term data are lacking and inferences about causal associations due to the plant/frugivore interaction are unwarranted or are established without a proper evaluation of the influence of external variables (e.g. climate, food resource levels outside the study area, etc.). Between-year variations in availability of fruits, paralleled or not by variations in frugivore numbers, add an important stochastic component to plant/frugivore interactions, and long-term data are needed to begin a realistic assessment of their demographic implications (Herrera, 1998).

Seasonality

The overall production figures outlined above illustrate broad patterns of variation in fruit abundance but mask actual availability for frugivores, which frequently face seasonal and annual shortages of this food resource. Figure 6.2 summarizes variation in the phenology of ripe fruit availability in six major community types. In general, fruiting peaks occur during periods of low photosynthetic activity or after periods of high rates of reserve accumulation towards the end of the growing season (French, 1992; see review by Fenner, 1998). Fruiting peaks occur at the end of the dry seasons, matching generalized increases in precipitation, and these trends are evident even without shifting the graphs to compensate for latitudinal differences. Unimodal fruiting peaks of the highly seasonal forests are not replicated in the very humid rainforests, where several peaks of different importance occur as a result of both variations in rainfall intensity within the rainy season and delays in the phenological responses of different growth forms (Frankie *et al.*, 1974a; Croat, 1978; Opler *et al.*, 1980). Several authors point out the absence of significant flowering and fruiting seasonality in certain rainforests of South-East Asia (Koelmeyer, 1959; Putz, 1979) and Colombia (Hilty, 1980). Seasonality in the number of plant species bearing ripe fruits decreases from temperate to tropical forests, largely as a result of the increase in the average duration of the fruiting phenophase (although the seasonal pattern can be strikingly similar in some cases; see Fig. 6.2). Average duration of period of ripe fruit availability for a given species is always less than 1.5 months (mean = 0.6–1.3 months) in temperate forests and always more than 4 months (mean = 4.3–5.8 months) in tropical forests (Herrera, 1984c; see also references in Table 6.2). Lowland Mediterranean scrublands

(Herrera, 1984c; Jordano, 1984) have intermediate averages of 2.2–4.0 months. It would be interesting to know if these consistent patterns of variation reflect similar environmental influences or if, as evidenced for the flowering seasons of temperate forest plants, they are largely attributable to phylogenetic affinities (Kochmer and Handel, 1986; Fenner, 1998).

These differences in the seasonal patterns of fruit availability between the tropics and temperate zones define important differences in frugivory patterns. Temperate frugivory is a strongly seasonal phenomenon among migrant birds (Thompson and Willson, 1978; Stiles, 1980; Herrera, 1982, 1998; Jordano, 1985; Wheelwright, 1986, 1988; Willson, 1986; Snow and Snow, 1988; Noma and Yumoto, 1997; Parrish, 1997) and mammal species, such as carnivores (Debussche and Isenmann, 1989; Herrera, 1989) or warm-temperate pteropodid bats (Funakoshi *et al.*, 1993), which show marked seasonal shifts in diet composition. Tropical frugivores usually exploit fruit food during the whole year, but important seasonal dietary shifts also take place (e.g. Snow, 1962a, b, c; Decoux, 1976; Hilty, 1977; Worthington, 1982; Terborgh, 1983; Leighton and Leighton, 1984; Sourd and Gauthier-Hion, 1986; Fleming, 1988; Erard *et al.*, 1989; Rogers *et al.*, 1990; Williamson *et al.*, 1990; Conklin-Brittain *et al.*, 1998; Wrangham *et al.*, 1998).

Seasonality of fruit availability causes dietary shifts by frugivorous animals, which 'track' the changes in the fruit supply (Loiselle and Blake, 1991). For whole-year resident frugivores, this type of resource tracking involves the sequential consumption of a great variety of fruit species, with a major effect on nutrient dietary balance and nutrient intake (Witmer and van Soest, 1998; Wrangham *et al.*, 1998). Important aspects of the annual cycles of frugivores, such as reproduction, breeding, migratory movements, etc., are associated with seasonal fruiting peaks. However, in most cases, a direct causal link between both cyclic phenomena cannot be established. The long-term studies by

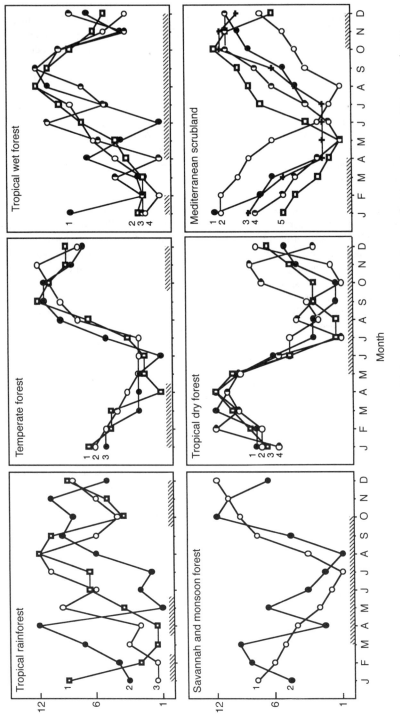

Fig. 6.2. Seasonality patterns in availability of ripe fleshy fruits in several habitat types. Months have been ranked (scores from 1 to 12 in vertical axis) according to proportion of woody species with ripe fruit available. The shaded bars on the abscissa depict the rainy seasons. References: Tropical rainforest, Davis (1945)[1, 3], Hilty (1980)[2]. Temperate forest, Halls (1973)[1]; Sorensen (1981)[2]; Guitián (1984)[3]. Tropical wet forest, Frankie et al. (1974b)[1]; Crome (1975)[2]; Alexandre (1980)[3]; Medway (1972)[4]. Savannah and monsoon forest, Poupon and Bille (1974)[1]; Boojh and Ramakrishnan (1981)[2]. Tropical dry forest, Daubenmire (1972)[1]; Frankie et al. (1974a)[2]; Morel and Morel (1972)[3]; Lieberman (1982)[4]. Mediterranean scrubland, Herrera (1984c)[1, 3]; Mooney et al. (1977), California[2], Chile[4], Jordano (1984)[5].

Crome (1975) and Innis (1989) in the rainforests of Queensland (Australia) clearly show that seasonal patterns of abundance of certain fruit-pigeons are strongly associated with the seasonal patterns of fruit ripening. Similarly, Leighton and Leighton (1984) found a good correlation between local densities of major frugivorous vertebrates (fruit-pigeons, hornbills, primates and ungulates) and fruit abundance in a Bornean rainforest; regional migration, nomadism, exploitation of aseasonal fruit types (e.g. *Ficus*) or alternate food resources were means of escaping seasonal fruit scarcity in time and space (see also Whitney and Smith (1998) for African *Ceratogymna* hornbills). Wheelwright (1983) describes marked shifts in habitat selection by resplendent quetzals, which track the seasonal sequence of ripe fruit availability among Lauraceae. Migratory or nomadic movements among Megachiroptera (Marshall, 1983) can be associated with changes in the fruit supply. Also, the annual cycle of frugivorous bird abundance in Mediterranean scrubland has been found to track closely the abundance and biomass cycle of ripe fruits (Jordano, 1985). On the other hand, Reid (1990) showed no clear relation between the seasonal abundance patterns of the mistletoe bird (*Dicaeum hirundinaceum*) and its preferred fruit, *Amyema quadang* (Loranthaceae), in Australia. The breeding seasons of certain tropical frugivorous birds (e.g. Snow, 1962a, b; Worthington, 1982), bats (Marshall, 1983; Fleming, 1988) and primates (e.g. Terborgh, 1983) all match local maxima of ripe fruit availability. Loiselle and Blake (1991) found that frugivorous birds bred when the fruit supply was low but, after the breeding season, moved into areas where fruit was more abundant. Seasonal use of fruits as an alternative food resource for temperate passerines is probably the major impelling influence on the evolution of long-distance migratory movements in the Nearctic and Palaearctic (Levey and Stiles, 1992).

The evidence outlined by these studies suggests that seasonal fruiting patterns can have a great effect on the annual cycles of most frugivores (van Schaik *et al.*, 1993). Frugivorous animals, on the other hand, probably have a negligible effect on shaping the abundance patterns of fleshy fruits in time. Thus, for western European bird-dispersed plants, Fuentes (1992) found parallel seasonal trends in bird abundance and the number and biomass of fruits, but not in the proportion of species with ripe fruit; frugivores might favour the seasonal displacement of fruit availability by positive demographic effects on particular plant species fruiting when birds are most abundant. Major patterns of convergence in community-level fruiting patterns strongly support the findings of previous studies showing: (i) a complex role of climate (alternation of drought and rainfall seasons) in shaping the fruiting curves at a community level in relation to flowering and leafing activity (Janzen, 1967; Borchert, 1983; Gautier-Hion *et al.*, 1985a; Hopkins and Graham, 1989); (ii) a prominent role of germination requirements at the start of the rainy season (Garwood, 1983); (iii) phylogenetic constraints in the timing and duration of the fruiting phenophase (Kochmer and Handel, 1986; Gorchov, 1990); (iv) the effect of physiological constraints derived from the integration of flowering, fruit growth, ripening and seed dispersal phases of the reproductive cycle (Primack, 1987; Fenner, 1998); and (v) potential effects of frugivores in shaping fruit availability patterns but not the fruiting phenophase itself (Debussche and Isenmann, 1992; Fuentes, 1992).

Spatial distribution

Relative to other food resources, such as animal prey (e.g. insects), fruits are extremely aggregated in space, usually in relatively isolated patches, with high local abundance. In addition to the intrinsic spacing patterns of the adult trees, which determine the spacing patterns of the fruits themselves, the spatial distribution of fruits as food resources for foraging animals is constrained by two major factors: (i) successional characteristics of the patch; and (ii) relative frequency of fruit-bearing trees in the patch. Fruit abundance increases in gaps and secondary growth of temperate forests (Thompson and Willson, 1978; Willson et al., 1982; Martin, 1985), and fruiting individuals of a given species usually bear larger crops when growing in open sites rather than the forest interior (Piper, 1986a; Denslow, 1987). Work in tropical rainforest (De Foresta et al., 1984; Levey, 1988a, b; Murray, 1988; Restrepo and Gómez, 1998) showed that patchiness in fruit availability is predictably associated with tree-fall gaps and other disturbances. Individual plants growing in Costa Rican tree-fall gaps produced more fruit over a longer period of time than conspecifics growing in intact forest understorey; the diversity of fruiting plants also increased in gaps (Levey, 1988b, 1990).

The same pattern exists in temperate forests, where mature stands are dominated by *Quercus*, *Fagus* and *Acer* species, among others, and fleshy-fruited shrubs and treelets are characteristic of early successional stages and forest gaps (Marks, 1974; Smith, 1975; Kollmann and Poschlod, 1997). Forest gaps of temperate forest are sites of increased local concentration of fruits (Sherburne, 1972; Sorensen, 1981; Blake and Hoppes, 1986; Martin and Karr, 1986). For example, Blake and Hoppes (1986) found an average fruit abundance at the start of the fruiting season (September) of approximately 50 fruits 80 m^{-2} in Illinois forest gaps versus approximately five fruits 80 m^{-2} in forest interior plots. Among the reasons for these trends in both tropical and temperate forests are: (i) increased abundance of individual plants in gaps; (ii) increased diversity of fleshy-fruit-producing species; and (iii) increased crop sizes among individuals growing in gaps.

In Mediterranean shrubland, however, pioneer, successional species with dry fruits and capsules are progressively substituted by endozoochorous species, which eventually dominate the late successional stands (Bullock, 1978; Houssard et al., 1980; Debussche et al., 1982; Herrera,

1984d). For example, average cover of fleshy-fruited species in southern Spanish Mediterranean, lowland shrubland, mature stands (Jordano, 1984) is 96.88% and it is 62.00% in open, successional stands.

Two additional sources of local patchiness in fruit availability have seldom been considered. First, abundance will be influenced by the frequent association of dioecism with production of fleshy fruits (Givnish, 1980; Donoghue, 1989). In Mediterranean shrubland, the relative cover of female individuals can vary on local patches between 20 and 95%, and increasing local abundance of male, non-fruiting plants is associated with decreased fruit availability (Jordano, 1984). This factor is probably irrelevant as a source of patchiness in fruit abundance in temperate forests, but might prove to be important in tropical habitats, where dioecism is relatively frequent. Secondly, fleshy-fruiting plants are frequently associated with particular patches below the closed canopy of taller trees, probably because of increased recruitment in these foci as a result of increased seed rain beneath trees (McDonnell and Stiles, 1983; Tester et al., 1987; Hoppes, 1988; Izhaki et al., 1991; Holl, 1998). Bat roosts, nests of frugivorous birds, fruiting plants where frugivores defend feeding territories, traditional perches for sexual displays and latrines of certain 'carnivore' mammals are among the many types of sites that create recruitment foci, with seed density orders of magnitude greater than in sites elsewhere in the forest (Lieberman and Lieberman, 1980; Stiles and White, 1986; Dinerstein and Wemmer, 1988; Théry and Larpin, 1993; Fragoso, 1997; Kinnaird, 1998). In addition, seed rain of fleshy-fruited species is significantly higher beneath female, fruit-bearing, plants compared with male plants of dioecious species (Herrera et al., 1994), a result of preferential foraging by fruit-seeking frugivores. All these processes generate predictable spatial patterns of fruit availability, which, in turn, influence the pattern of patch use by foraging frugivores.

Fruit characteristics

Fruits are particulate foods, which frugivorous animals usually harvest, handle and swallow as individual items. Relevant traits of fleshy fruits, from the perspective of the foraging animal, include design (e.g. size, number and size of seeds, mass of pulp relative to fruit mass), nutrient content (relative amounts of lipids, protein, carbohydrates and minerals per unit mass of fruit processed) and secondary metabolites (Table 6.1B). These traits influence the overall, intrinsic profitability of fruits, by determining both the total amount of pulp ingested per fruit handled and the nutrient concentration of the ingesta (Herrera, 1981a), but the profitability of a given fruit should be examined in the context of an interaction with a particular frugivore species (Martínez del Rio and Restrepo, 1993).

Fruit size and design

The ability to handle, swallow and process a given fruit efficiently depends on fruit size relative to body size of the frugivorous animal, particularly the gape width and mouth size. These types of constraints are similar to those found among gape-limited predators seeking particulate food and, from the plant perspective, they restrict the potential range and diversity of frugivores and dispersers (Pratt and Stiles, 1985; Wheelwright, 1985). Consumption of extremely large-seeded fruits (e.g. family Lauraceae, Palmae, etc.) by frugivorous birds is largely confined to large-bodied species (toucans, trogons, bellbirds: Wheelwright, 1985; see also Pratt, 1984) or terrestrial species (trumpeter (Psophia crepitans): Erard and Sabatier, 1988; cassowary (Casuarius casuarius): Pratt, 1983; Stocker and Irvine, 1983). Bonaccorso (1979) reported a significant positive relationship between body-mass variation among individual phyllostomid bats of three species and the mass of individual fruits taken. Extremely large seeds (> 3 cm length) have been reported to be dispersed exclusively by large mammals (apes and elephants: Tutin et al., 1991; Chapman et al., 1992a).

The maximum and mean diameter of fruit species included in the diets of Costa Rican birds is positively correlated with gape width, and the number of bird species feeding on the fruits of a particular species of Lauraceae was inversely correlated with fruit diameter (Wheelwright, 1985). Reduced species richness of avian frugivores visiting large-fruited species was also reported by Green (1993) in subtropical Australian rainforest. Lambert (1989a, b) found that seven species of frugivorous pigeons in Malaysia fed on at least 22 Ficus species, and a positive relation exists between body size and mean fig diameter of the species consumed. Fig size choice by different bird species was influenced by body size, in spite of the fact that the structure of the syconium enables exploitation by birds of all sizes (Jordano, 1983; Lambert, 1989a). In turn, gape width strongly limited the size and variety of fruits included in the diet of six warbler species (genus Sylvia) in southern Spain (Jordano, 1987b). The average fruit size consumed (calculated by weighting the fruit diameter of each fruit species by the relative consumption) was positively correlated with gape width (Fig. 6.3a; but see Johnson et al. (1985) for North American migrant birds). In addition, the average percentage of fruits dropped during short feeding bouts decreased in the larger species with wider gape (Fig. 6.3b), indicating increasingly larger handling costs for smaller species. Snow and Snow (1988) reported a similar decrease in fruit-handling success with fruit diameter/bill width ratios greater than 1.0. Rey and Gutiérrez (1996) reported that blackcaps switch from swallowing whole wild olive fruits to fruit pecking in the olive orchards, where seeds are twice as large; as a result, only 4.9% of faecal samples from orchards contained seeds, but 58.1% of those from the wild contained wild olive seeds. In a more exhaustive set of experiments with several Mediterranean passerine species, Rey et al. (1997) showed that fruit size

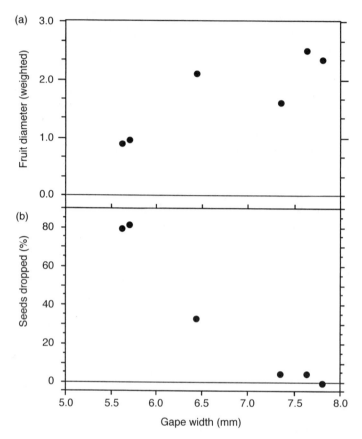

Fig. 6.3. Relationship between mean gape width (width of the bill measured at the commissures) of six species of *Sylvia* warblers and mean fruit size in the diet. (a) fruit size of each plant species consumed weighted by the frequency of consumption and, (b) the mean percentage of fruits which are dropped during feeding sequences at *Prunus mahaleb* trees, a species with average fruit diameter of 8.4 mm. Data from Jordano (1987b), Jordano and Schupp (2000). Dots, in order of increasing gape width, indicate *S. conspicillata, S. cantillans, S. melanocephala, S. atricapilla, S. communis* and *S. borin.*

determined a shift from swallowing to pecking, as pecking frequency increased with the enlargement of the fruit size; all the species showed increased fruit-handling failure rate when trying to swallow increasingly large fruits. These trends reflect the increase in handling cost associated with picking, seizing and positioning in the bill of increasingly larger fruits, but the main effect of fruit size on handling success, especially in drupes and other single-seeded fruits, is due to seed size and not to fruit size.

Few studies have concentrated, however, on intraspecific comparisons of fruit removal as related to fruit size variation among individual plants. Bonaccorso (1979) reported strong selectivity by individual bats of figs of *Ficus insipida* differing in size, which suggests strong fruit size selection limited by aerodynamic constraints on fruit transport on the wing. Howe (1983) reported that an average of 62% of variation in seed removal of *Virola surinamensis* by birds was accounted for by the aril : seed ratio of individual trees; 78% variation in seed size of this species is among individual crops (Howe and Richter, 1982). Significant correlations are frequently obtained between seed dispersal efficiency (the percentage of the seed crop dispersed) and both fruit and seed size, although the sign most probably varies as a result of the degree of gape limitation of the particular set of frugivores interacting with a plant species (Herrera, 1988a; White and Stiles, 1991; Sallabanks, 1992; Herrera *et al.*, 1994; Jordano, 1995b).

The potential selective pattern on fruit seediness differs with seed size and seed packaging, and complex allocation patterns to flesh, seed endocarp and seed content exist in fleshy fruits (Lee *et al.*, 1991). For multiseeded fruits, the fraction of total fruit mass allocated to seeds increases with seed number, and frugivores are expected to select few-seeded fruits (Herrera, 1981b). In drupes and other single-seeded fruits, seed burden per unit pulp mass increases with increasing fruit size, and frugivores are expected to select small fruits, especially if gape-limited (Snow and Snow, 1988; Jordano, 1995b; Rey *et al.*, 1997). Future studies should bridge the gap in our knowledge of the demographic effect of these types of selective pressures on the plant populations by considering simultaneously the effect of fruit size and seed size on germination and early seedling vigour and survival.

Allocating many small seeds within a given fruit increases the potential diversity of dispersers by allowing small frugivores to ingest pulp pieces and seeds. Levey (1987) found that the percentage of seeds dropped during feeding trials with several tanager (Thraupidae) species in captivity increased as a function of seed size; birds consistently dropped more than 60% of seeds that were greater than 2.0 mm in length. These birds are 'mashers', which crush all fruits in their bills; the largest seeds are worked to the edge of the bill and dropped and the smallest seeds are swallowed along with pulp pieces. In contrast, manakins (Pipridae) are 'gulpers', which swallow the whole fruits and defecate all seeds up to the 10 mm threshold imposed by their gape width; however, the percentage of fruits taken by manakins decreased as seed size increased. See Rey and Gutiérrez (1996) for a similar example of switching between 'gulper' and 'masher' behaviour.

The same trend is also exhibited by other taxonomic groups. The smallest species of African forest frugivorous ungulates of genus *Cephalophus* (*C. monticola*, 4.9 kg) take no fruit above 3 cm diameter and the largest (*C. sylvicultor*) consumes fruit up to 6 cm in diameter (Dubost, 1984). Similar size-related constraints have been found in bats (Fleming, 1986) and primates (Hylander, 1975; Terborgh, 1983; Corlett and Lucas, 1990; Tutin *et al.*, 1996; Kaplin and Moermond, 1998). For example, seed size strongly influences whether seeds are swallowed, spat out or dropped *in situ* by long-tailed macaques (*Macaca fascicularis*); seeds of most species with individual seeds less than 4.0 mm width are swallowed (Corlett and Lucas, 1990; see also Gautier-Hion, 1984). Kaplin and Moermond (1998) report that most seeds > 10 mm are dropped by *Cercopithecus* monkeys, but variability in behaviour as seed predators or legitimate dispersers was observed. In summary, all this evidence indicates that small frugivores are limited in the largest fruit they can efficiently handle and process and, on the other hand, increase in fruit size generally limits the range of potential seed dispersers to the largest frugivores. Both assertions are especially true for drupes or other single-seeded fruits, and have important implications for the resulting seed dispersal pattern, the evolution of fruit and seed shape and their biogeographical patterns (Mack, 1993). Thus, evidence of negative allometry in the development of large-fruited species (e.g. *Lauraceae*) has been interpreted as an adaptation to gape-limited avian frugivores (Mazer and Wheelwright, 1993; but see Herrera, 1992).

As stated by Wheelwright (1985), fruit size alone does not explain the wide variability in the number of frugivore species feeding at different plant species that have fruits of the same size. Studies examining interspecific trends in fruit structural characteristics have also found that overall size provides the main source of functional variation in fruits relative to the types of frugivores consuming them, but additional important traits were the number of seeds per fruit, the mass of each seed and the mass of pulp per seed (Janson, 1983; Wheelwright et al., 1984; Gautier-Hion et al., 1985b; O'Dowd and Gill, 1986; Debussche et al., 1987; Herrera, 1987; Debussche, 1988). However, only fruit size among another 15 fruit traits examined by Jordano (1995a; see Appendix to this chapter) was associated with a major type of seed disperser when accounting for phylogenetic affinities in a comparative analysis of a large data set of angiosperms.

Nutrient content of the pulp

Comparative studies of the nutrient content of fleshy fruits have revealed that most variation in components can be explained by a few major patterns of covariation that have a major correlate with phylogeny, especially at the family and genus level (Jordano, 1995a). Herrera (1987) found, by means of factor analysis, that 46.5% of the variance in nutrient content among 111 species of the Iberian Peninsula was accounted for by the strong negative correlation between lipid and non-structural carbohydrate (NSC) content; three additional factors accounted for 51.1% of variance. Therefore, rather than the succulence continuum suggested by some authors, pulp composition patterns included: high lipid–low NSC–low fibre; low lipid–high NSC–low fibre; and medium lipid–medium NSC–high fibre. Variation in protein and water content was independent of these pulp types. Similar patterns have been described by other authors (Wheelwright et al., 1984; Gautier-Hion et al., 1985b; Johnson et al., 1985; O'Dowd and Gill, 1986; Debussche et al., 1987; Jordano, 1995a) and are probably caused by the great variation in lipid content among angiosperm fruit pulps relative to other constituents and its strong inverse correlation with carbohydrate content.

The pulp of fruits has been considered repeatedly as deficient in certain nutrients, especially nitrogen and protein (Snow, 1971; Morton, 1973; White, 1974; Berthold, 1977; Thomas, 1984). Relative to other dietary items usually consumed by verte- brate frugivores (Table 6.3; Appendix to this chapter), the fruit pulp shows the highest concentration of soluble carbohydrates and the lowest relative amount of protein. Lipid content is relatively high but shows extreme interspecific variation. The importance of the mineral fraction is relatively constant among food types, but the content of particular cations is very variable (Nagy and Milton, 1979; Piper, 1986b; Herrera, 1987; Pannell and Koziol, 1987). Fruits are extremely poor in protein in comparison with leaves and insects. However, their energetic value in terms of soluble carbohydrates and lipids exceeds that of any other food type (Table 6.3). Therefore, the combination of traits that best characterizes the fruit pulp nutritive content is the excess of digestible energy relative to protein, the high water content and the extreme deficiency in some compounds relative to others (i.e. imbalance between components).

The Appendix to this chapter summarizes most of the information available at present on the nutrient content of the pulp of the main angiosperm families dispersed by vertebrate frugivores. Detailed reports for local or regional floras include, among others: Hladik et al. (1971); Sherburne (1972); White (1974); Crome (1975); Frost (1980); Stiles (1980); Viljoen (1983); Wheelwright et al. (1984); Johnson et al. (1985); O'Dowd and Gill (1986); Piper (1986b); Debussche et al. (1987); Herrera (1987); Fleming (1988); Snow and Snow (1988); Eriksson and Ehrlén (1991); Hughes et al. (1993); Corlett (1996); Witmer (1996); Heiduck (1997); Ko et al. (1998).

In the case of frugivorous birds, virtually nothing is known about the protein demand in natural conditions, although recent efforts have been made to understand the nutritional limitations of fruits (Sorensen, 1984; Karasov and Levey, 1990; Martínez del Rio and Karasov, 1990; Levey and Grajal, 1991; Levey and Duke, 1992; Witmer, 1996, 1998a; Witmer and van Soest, 1998). Information available, mostly from domestic, granivorous species, indicates that a diet with 4–8% protein (wet mass) is necessary for maintenance (several authors cited in Moermond and Denslow,

Table 6.3. Summary of nutrient contents of different food types consumed by vertebrate frugivores. Figures are mean and range of % of each component relative to dry mass. Data for seeds refer to wet mass.

Food type	Water	Protein	Lipids	Non-structural carbohydrates	Minerals
Insects[1]	63.7 (56.8–70.4)	68.3 (59.9–75.9)	16.8 (9.4–21.2)	14.9 (0.5–20.0)	8.9 (3.1–19.0)
Seeds[2]	11 (4–12)	11 (6–14)	4 (0.3–9)	69 (61–73)	2.2 (1.1–5.3)
Neotropical fruits[3]	71.3 (38.0–95.2)	7.8 (1.2–24.5)	18.5 (0.7–63.9)	67.8 (5.6–98.3)	5.6 (1.3–19.4)
Mediterranean fruits[4]	69.9 (36.9–90.1)	6.4 (2.5–27.7)	9.0 (3.7–58.8)	80.1 (33.2–93.7)	4.6 (1.1–13.1)
Mature leaves[5]	59.4 (46.2–76.2)	12.6 (7.1–26.1)	3.3 (0.7–10.7)	6.9 (1.9–14.7)	4.9 (1.5–11.3)
Young leaves[5]	71.9 (54.0–82.3)	18.2 (7.8–36.3)	3.2 (0.7–6.3)	15.4 (1.8–32.7)	5.0 (3.4–7.5)

References: [1], White (1974); [2], Jenkins (1969) cited in Moermond and Denslow (1985); [3], see references in Appendix; [4], Herrera (1987); [5], Hladik (1978); Oates (1978); Oates *et al.* (1980); Waterman *et al.* (1980).

1985), by providing a daily consumption of 0.43 g N kg$^{-0.75}$ day^{-1} (Robbins, 1983). Considering that the high amount of water in the pulp of fleshy fruits acts as a 'solvent' of the included nutrients, most fruits contain amounts of protein, relative to dry mass of pulp, within the limits adequate for maintenance. Thus, average protein content for a sample of angiosperm fleshy fruits (Appendix to this chapter) is 6.12 ± 4.47% (mean ± SD, $n = 477$ species), ranging between 0.1 and 27.7%.

These nutrient levels are adequate if the fruit supply in nature is not limiting, but this is an infrequent situation (Foster, 1977; Witmer, 1996, 1998a). Dinerstein (1986) found that the protein content of the fruits consumed by frugivorous bats (*Artibeus, Sturnira*) in Costa Rican cloud forest (mean = 6.7% protein, dry mass) was apparently sufficient to sustain the protein demands of lactating females; otherwise females could be depending on previously accumulated protein reserves. The data available regarding *Carollia perspicillata* (Herbst, 1986; Fleming, 1988) indicate that dietary mixing of a protein-rich fruit, such as *Piper* spp. (*Piperaceae*) and an energy-rich fruit, such as *Cecropia peltata* (*Cecropiaceae*), adequately balanced the daily net energy and nitrogen requirements. In contrast to these phyllostomid bats, totally frugivorous pteropodid bats relying on low-quality *Ficus* fruit food (less than 4.0% protein, dry mass) obtain sufficient protein by overingesting energy from fruits, but are unable to supplement this diet with animal prey (Thomas, 1984). In other pteropodids (*Rousettus*), Korine *et al.* (1996) reported a positive nitrogen balance on a totally fruit diet due to exceptionally low nitrogen demands (55% lower than expected from allometry), apparently as an adaptation to periods of low fruit availability. Overingestion of energy to meet the protein needs has been reported for the totally frugivorous oilbird *Steatornis caripensis* (Steatornithidae) (White, 1974). Early findings by Berthold (1976) that lipids and protein in fruits were insufficient for maintenance and migratory fat deposition by warblers (*Sylvia* spp.) have been challenged by the experiments of Simons and Bairlein (1990) demonstrating significant body mass gain by *Sylvia borin* when fed on a totally frugivorous diet, although additional work has confirmed loss of body mass and nitrogen on diets of sugary fruits for some species (Izhaki and Safriel, 1989; Witmer, 1996, 1998a; Witmer and van Soest, 1998). Several studies reveal a positive nitrogen balance of specialized frugivorous birds, such as phainopeplas or waxwings, when feeding on fruits with a protein content greater than 7.0% dry mass (Walsberg, 1975; Berthold and Moggingen, 1976; Studier et al., 1988; Witmer, 1998a).

Therefore, the poor value of fruits as a unique food largely results from the internal imbalance of major nutritive components relative to others – basically the extreme protein and nitrogen deficiency relative to energy content. Thus, it is paradoxical that certain neotropical fruits, qualified as highly nutritious, had calorie : protein ratios greater than 1500 (Moermond and Denslow, 1985), when others, considered as poor (Rubiaceae, Melastomataceae), had ratios more similar to those of insects. The main effect of these types of relative deficiencies for frugivorous animals is that the assimilation of a particular nutrient can be limited by the impossibility of processing enough food material to obtain it, and not by the scarcity of the nutrient itself. That is, the effect is due to a digestive bottleneck (Kenward and Sibly, 1977; Sibly, 1981). Consumption of minor amounts of animal prey provides the necessary nitrogen input to escape the constraint imposed by the overingestion of energy, as demonstrated by field studies of phyllostomid bats and frugivorous warblers (Fleming, 1988; Jordano, 1988; see also Bowen et al., 1995).

Direct interaction among different components present in the pulp, such as secondary metabolites, can limit nutrient digestibility and assimilation (Herrera, 1981a; Izhaki and Safriel, 1989; Mack, 1990; Cipollini and Levey, 1992, 1997). The presence of tannins, together with alkaloids and saponins, is particularly frequent among Mediterranean species (Jordano, 1988, and references therein). The presence of tannins in the pulp may cause lower assimilation of proteins and damage the digestive epithelium (Hudson et al., 1971; Swain, 1979). Experiments by Sherburne (1972) demonstrate that other types of secondary compounds, such as glycosides or alkaloids, have a direct effect on frugivore foraging by preventing feeding or drastically reducing the palatability of unripe fruits. However, little is known about the effects of metabolites that act like tannins and phenols, reducing the assimilation efficiency (Izhaki and Safriel, 1989; Mack, 1990; Cipollini and Levey, 1997).

Finally, the content in the fruit pulp of cations and microelements, such as calcium, phosphorus, iron, manganese and zinc, is frequently below the requirements of frugivorous birds, and situations of negative balance in wild birds have been reported (Studier et al., 1988). These types of effects should be controlled in experiments assessing the nutritional limitation of fruit food for frugivores.

Frugivory

Frugivory appears to be a feeding mode that is open to many types of organisms. No special adaptations, such as deep beaks or special digestive processing of the ingesta, are necessary to consume fruit, but certain morphological, anatomical and physiological characteristics determine an animal's ability to rely extensively on fruit food. The purpose of this section is to review patterns of anatomical and physiological variation associated with exclusive or extensive frugivory.

At least three basic types of frugivory can be defined, relative to their potential consequences for seed dispersal. First, legitimate dispersers swallow whole fruits and defecate or regurgitate seeds intact. Secondly, pulp consumers tear off pulp pieces while the fruit is attached to its peduncle, or they mandibulate the fruits and ingest only the pulp by working the seed(s) out. Finally, seed predators may extract seeds from fruits, discard the pulp, crack the seed and ingest its contents or can swallow whole fruits and digest both pulp and seeds. From the plant's perspective, these categories define a wide gradient of seed dispersal 'quality' (Snow, 1971; McKey, 1975; Howe, 1993; Schupp, 1993), from frugivores that deliver seeds unharmed (dispersers) to those that destroy seeds (granivores), with no clear-cut limits between them (Jordano and Schupp, 2000). Single traits, such as body size, wing form or bill width, are not satisfactory predictors of frugivory intensity or the type of frugivorous behaviour shown by a species, and simultaneous consideration of a number of traits is needed. Herrera (1984a) found that a multiple discriminant analysis of body mass and six ratios describing bill shape accurately predicted the assignment of Mediterranean scrubland birds to three frugivory types. Seed dispersers showed larger body size and flatter and wider bills than non-frugivores and pulp/seed consumers. Consumers of pulp that discarded the seeds beneath the plants (finches, emberizids and parids) were characterized by smaller size, deeper beaks and narrower gapes. Non-frugivores showed more slender bills than the other two groups. Actually, species of seed dispersers, pulp/seed consumers and non-frugivores occupy a continuum along the discriminant function, emphasizing the absence of clear limits between categories.

Whether a given frugivore behaves as a seed disperser, pulp consumer or seed predator in a particular interaction with plants is not only dependent on frugivore ecomorphology and behaviour, but also on fruit characteristics (especially seed size) of the plants in the specific situation. Detailed descriptions of these categories and associated behavioural patterns are given by, among others: Hladik and Hladik (1967); Hladik (1981); Janzen (1981a, b, c, 1982); Fleming (1982); Herrera (1984c); Moermond and Denslow (1985); Levey (1986, 1987); Bonaccorso and Gush (1987); Snow and Snow (1988); Bodmer (1989a); Corlett and Lucas (1990); Green (1993); Corlett (1998); Jordano and Schupp (2000). It is apparent from these studies that the different types of frugivory are present in all groups of vertebrate frugivores, but in markedly different proportions.

Anatomical characteristics of frugivores

Frugivore size and form

Body mass is a major determinant of intensity of frugivory. The relative importance of fruit in the diet of Mediterranean passerines is strongly correlated with body mass (Herrera, 1984a, 1995; Jordano, 1984, 1987c). Smaller birds, such as those in genera *Phylloscopus*, *Saxicola*, *Hippolais* and *Acrocephalus*, only sporadically consume fruits. Fruit makes up 30–70% of diet volume among medium-sized *Phoenicurus*, *Luscinia*, the smaller *Sylvia* warblers and *Erithacus* and always more than 80% in the larger species (*Sylvia atricapilla*, *S. borin*, *Turdus* spp., *Cyanopica cyanus* and *Sturnus* spp.). Katusic-Malmborg and Willson (1988) found a similar relationship for eastern North American frugivorous birds, but Willson (1986) found no consistent differences in body size between frugivores and non-frugivores in a number of habitats in this region.

Body size affects frugivory intensity by limiting the maximum number of fruits that can be swallowed or otherwise processed in feeding bouts (e.g. during short visits to plants) and the maximum amount of pulp mass that can be maintained within the gut, since gut capacity is strongly correlated with body mass. Thus, average number of fruits ingested per feeding visit to *Prunus mahaleb* plants is 1.5 for *Phoenicurus ochruros* (16.0 g), 9.0 for *Turdus viscivorus* (107.5 g), and 21.0 for *Columba palumbus* (460.0 g) (Jordano and Schupp, 2000). The number of fruits consumed per visit by frugivorous birds has been found to be strongly correlated with body mass in a number of studies (Fig. 6.4). Therefore, body size alone sets an upper limit to the potential maximum number of seeds that a given frugivore can disperse after a feeding bout. Note that sporadic visits by large frugivores can have a far greater effect on crop removal than consistent visitation by small frugivores, but the net result on seed dispersal also depends on differences in postforaging movements between small and large frugivores (Schupp, 1993).

Body size differs markedly among species showing different types of frugivory, and influences fruit and seed handling prior to ingestion or immediately after it. Usually, small species tend to be

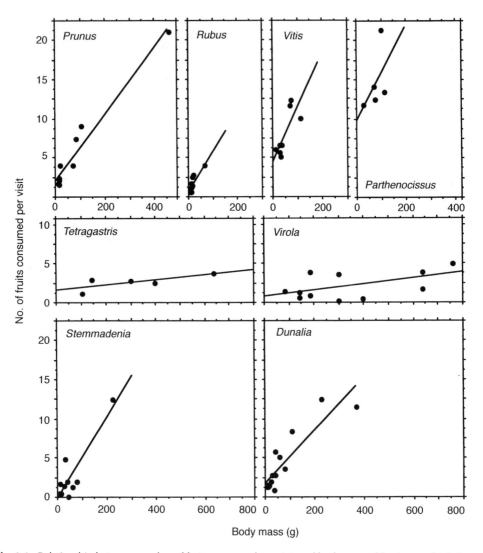

Fig. 6.4. Relationship between number of fruits consumed per visit and body mass of frugivorous birds in different plant species. Data from Jordano (1982) (*Rubus ulmifolius*); Howe and Vande Kerckhove (1981) (*Virola surinamensis*); Howe (1980) (*Tetragastris panamensis*); McDiarmid *et al.* (1977) (*Stemmadenia donnell-smithii*); Cruz (1981) (*Dunalia arborescens*); P. Jordano (unpublished data) (*Prunus mahaleb*) and Katusic-Malmborg and Willson (1988) (*Vitis vulpina* and *Parthenocissus quinquefolia*).

pulp consumers rather than legitimate dispersers, mostly though their inability to handle fruits efficiently and swallow them intact. Thus, fruit and seed swallowing among frugivorous primates is restricted to large hominoids and cebids (Corlett and Lucas, 1990); smaller species either spit out seeds (some cercopithecines; but see Kaplin and Moermond, 1998) or consume only pulp and discard seeds (Terborgh, 1983), although some small species, such as *Saguinus*, can swallow very large seeds (Garber, 1986).

The use by frugivores of different foraging manoeuvres to reach fruits on plants is constrained by external morphology and body proportions, which can be considered in most cases as preadaptations to other forms of prey use. Fitzpatrick (1980) showed that fruit use among tyrannid flycatchers is restricted to three groups of genera, with generalist foraging modes and fruit-feeding techniques that reflect the typical insect-foraging manoeuvres. Among Mediterranean frugivorous birds, the relative importance of fruits in the diet is significantly larger for foliage-gleaning species than for those with more specialized or stereotyped means of prey capture, such as sallyers, flycatchers and trunk foragers (Jordano, 1981). Therefore, it is reasonable to conclude that the ecomorphological configuration of a species is a preadaptation limiting feeding on fruit food, especially for those partial frugivores that consume other prey types; functional and behavioural predisposition, rather than specific adaptations, is expected (Herrera, 1984a; but see Moermond and Denslow, 1985).

Differences in fruit capture modes among frugivores show strong ecomorphological correlations, especially with wing morphology, bill form or dental characteristics and locomotory morphology (Hylander, 1975; Karr and James, 1975; Moermond and Denslow, 1985; Moermond *et al.*, 1986; Bonaccorso and Gush, 1987; Levey, 1987; Snow and Snow, 1988; Corlett and Lucas, 1990). Fleming (1988) reported relatively more elongated wings and higher wing loadings (g cm^{-2} of wing surface) among plant-visiting phyllostomid bats, which are more able to perform rapid, straight flights and hovering than insectivorous or carnivorous species. Frugivorous bats are quite conservative in the way they reach fruits, major differences being found in fruit handling and postforaging movements. The ecomorphological patterns that define the patterns of habitat selection among groups of these species (canopy-dwelling stenodermines and ground-storey carollines and glossophagines) strongly influence frugivory patterns, fruit selectivity and fruit-foraging behaviour (Bonaccorso and Gush, 1987; Fleming, 1988; see also Marshall and McWilliam (1982) and Marshall (1983) for information on Old World pteropodids).

Among frugivorous birds, fruits may be taken from a perch or on the wing (Herrera and Jordano, 1981; Moermond and Denslow, 1985; Foster, 1987; Snow and Snow, 1988; Jordano and Schupp, 2000). Ground-foraging frugivorous birds are larger and rarely use branches (Erard and Sabatier, 1988), but some perching species also forage for fruits on the ground (e.g. *Turdus* spp.; Snow and Snow, 1988). The description that follows relies heavily on detailed accounts and experiments reported by Denslow and Moermond (1982); Levey *et al.* (1984); Santana and Milligan (1984); Moermond and Denslow (1985); Levey (1986, 1987); Moermond *et al.* (1986); Foster (1987); Snow and Snow (1988); Green (1993); and Jordano and Schupp (2000). In addition to reaching from a perch, Moermond and Denslow (1985) describe four distinct flight manoeuvres by which birds pluck fruits: hovering, the method used by manakins, flycatchers and small tanagers; stalling, used by trogons and similar to hovering; and swooping and stalling, involving a continuous movement from perch to perch plucking the fruit on the way, which is the method used by most cotingids; and taking fruit from perches by picking, reaching and hanging. The first two manoeuvres are the two most commonly used, but those species that take most fruit on the wing are unable to reach well from a perch.

From the plant's perspective, the patterns described above have important implications for seed dispersal. These studies demonstrated that consistent choices between fruit species are made by foraging birds, based on accessibility restrictions that set different foraging costs, depending on anatomical characteristics. Consequences for seed dispersal are important, because small changes in accessibility override preferences for particular fruits; hence non-preferred fruits are consumed when accessibility to preferred fruits decreases. Other things being equal, decreasing fruit accessibility to legitimate dispersers would increase fruit retention time on branches and increase the probability of resulting damage or consumption by non-disperser frugivores (Denslow and Moermond, 1982; Jordano, 1987a). The ability to access and pick fruits of a given species by different frugivores varies, depending on the positions of the fruits within the infructescence or their locations relative to the nearest perch (and the thickness of that perch). In turn, differences in feeding techniques may influence dietary diversity by affecting which specific types of fruit displays are accessible. For example, frugivorous birds that take fruit on the wing show lower diet diversity and are more selective than species that pick fruits from perches (Wheelwright, 1983; Levey *et al.*, 1984; Wheelwright *et al.*, 1984; Moermond *et al.*, 1986). An ecomorphologically diverse array of visitors might result in a more thorough removal of the crop if different species predominantly take fruits from different positions in the canopy differing in accessibility to their foraging mode (Kantak, 1979; Herrera and Jordano, 1981; Santana and Milligan, 1984; Jordano and Schupp, 2000). In addition, if microhabitat selection is related to ecomorphological variation, individual trees differing in their relative position within a given habitat can differ markedly in the particular frugivore assemblage visiting the tree (see, for example, Manasse and Howe, 1983; Traveset, 1994).

Once the fruit is plucked, differences in dental characteristics, mouth size and bill shape among frugivores have important consequences for external seed treatment and seed dispersal. Two basic handling modes, gulping and mashing, originally described for frugivorous birds (Levey, 1987), can probably be expanded to accommodate fruit handling behaviour by most vertebrate frugivores. For example, phyllostomid bats (*Artibeus* spp.) take single bites out of fruits (*Ficus* spp.), slowly masticating the pulp and then pressing the food bolus against the palate with the tongue; thus, they squeeze the juice and expectorate the pulp along with seeds (Morrison, 1980; Bonaccorso and Gush, 1987). In contrast, *Carollia* spp. masticate the pulp and swallow it along with the seeds and discard the fruit skin (Bonaccorso and Gush, 1987; Fleming, 1988). Both behaviours are functionally similar to mashing, but the consequences

for the plant depend on frugivore movement after fruit plucking. Many ungulates swallow whole fruits and defecate seeds (Alexandre, 1978; Merz, 1981; Short, 1981; Lieberman *et al.*, 1987; Dinerstein and Wemmer, 1988; Bodmer, 1989b; Sukumar, 1990; Chapman *et al.*, 1992a; Fragoso, 1997) and others spit out seeds (Janzen, 1981c, 1982). Seed spitting is a common behaviour among primates, especially cercopithecines, which use cheek pouches to store food and later spit out the seeds, but whether a particular seed is defecated, spat out or destroyed is strongly dependent upon seed size and fruit structure (Corlett and Lucas, 1990; Tutin *et al.*, 1996; Kaplin and Moermond, 1998). New World apes (ceboids) and Old World hominoids apparently swallow and defecate most seeds intact (Hladik and Hladik, 1967; Hladik *et al.*, 1971; Hladik, 1981; Garber, 1986; Idani, 1986; Janson *et al.*, 1986; Rogers *et al.*, 1990; Tutin *et al.*, 1991, 1996; Wrangham *et al.*, 1994; Corlett, 1998), but some species mash fruits or tear off pulp pieces and can spit out or destroy seeds (Howe, 1980; Terborgh, 1983). Colobines and some cercopithecines destroy most seeds they consume (McKey *et al.*, 1981; Davies *et al.*, 1988), but at least some *Cercopithecus* can disperse relatively large seeds by dropping or defecating them unharmed (Kaplin and Moermond, 1998).

In summary, frugivore ecomorphology *per se* determines, from the plant perspective, the position of each frugivore species along a gradient ranging between zero and 1.0 survival probability for the seeds after interaction; and the main result of the studies discussed above is that vertebrate frugivore ecomorphologies are not distributed at random over this gradient.

Digestion of fruits

The bizarre digestive structures of some specialized frugivorous birds were documented long ago by ornithologists (Forbes, 1880; Wetmore, 1914; Wood, 1924; Desselberger, 1931; Cadow, 1933; Docters van Leeuwen, 1954; Walsberg, 1975;

Decoux, 1976). Typically, in birds, an oesophagus, which may or may not be dilated into a crop, is continued in a stomach, with a glandular proventriculus and a muscular ventriculus or gizzard. Common traits of modified digestive systems of frugivorous birds (Fig. 6.5; also including *Ducula* and *Ptilinopus* pigeons (Cadow, 1933) are: (i) absence or extreme reduction and simplification of the crop and/or proventriculus; (ii) presence of a thin-walled, non-muscular gizzard; (iii) lateral position of the simplified gizzard as a 'diverticulum' and an almost direct continuation of the oesophagus into the duodenum; and (iv) short intestines relative to body size. Despite the absence of a distinct crop, some specialized frugivorous birds, such as waxwings, can store fruits in the distensible oesophagus (Levey and Duke, 1992). This ability to store fruits oral to the gizzard somewhat offsets the problem of process-rate limitation, by allowing ingestion of two meals of fruit in a single foraging bout. Frugivorous bats also show a

typical stomachal structure, where the oesophagus leads into a cardiac vestibule and the rest of the stomach is an elongated tube, with a conspicuous, large, fundic caecum (Bhide, 1980, and references therein; see also Fleming, 1988).

Extreme diversification is also found in the anatomy of the digestive tract among non-volant, mammalian frugivores (Langer, 1986). Apart from ruminant artiodactyls, which consume fleshy fruits only sporadically (Bodmer, 1990), the digestive processing by non-ruminant frugivores differs chiefly between foregut and hindgut fermenters. To my knowledge, no comparative assessment has been made of the differential consequences for seed survival within the gut between these two types of digestive strategies (but see Bodmer, 1989a) and what fruit or seed traits, if any, are consistently associated with safe seed delivery by these frugivorous mammals. However, it is well known that fore-stomach fermenters usually crack seeds before ingestion (e.g. some colobine monkeys and

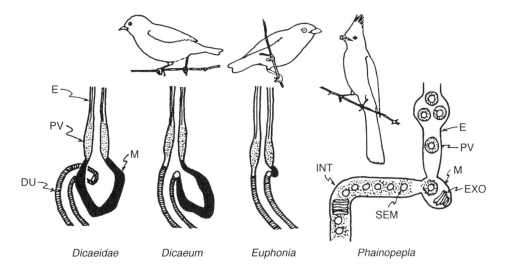

Fig. 6.5. Schematic representation of several types of proventriculus and gizzard configurations in specialized frugivorous birds. Left, arrangement of a relatively differentiated 'normal' muscular gizzard (M) stomach and associated oesophagus (E), proventriculus (PV) and duodenum (DU) in insectivorous Dicaeidae (after Desselberger, 1931). Note the normal approximation of the cardiac and pyloric ends of the stomach similar to most birds. Extreme simplification of the gizzard, with thinner walls and lack of hard epithelium and location of the gizzard as a lateral diverticulum along the oesophagus–duodenum axis is characteristic of frugivorous dicaeids (*Dicaeum*) and *Euphonia* tanagers (Forbes, 1880). Right: arrangement in phainopeplas (*Phainopepla nitens*), with schematic view of ingested fruits, exocarps (EXO) being accumulated in the simplified gizzard and seeds (SEM) passing to the small intestine (INT) (after Walsberg, 1975).

peccaries) and some hindgut fermenters also destroy most seeds they ingest (e.g. tapirs and suids: Janzen, 1981a; Corlett, 1998).

These digestive patterns are perhaps extreme examples of specialization not found in partial frugivores. Pulliainen *et al.* (1981) examined the digestive systems of three European granivorous birds and three seed dispersers and found no difference, except for *Bombycilla garrulus*, which is a specialized frugivore (Berthold and Moggingen, 1976; Voronov and Voronov, 1978), which showed the largest liver mass. Eriksson and Nummi (1982) reported higher liver activity and detoxification ability in *B. garrulus* relative to granivorous and omnivorous species. However, Herrera (1984a) showed no significant differences in relative mass of gizzard, liver and relative intestine length among avian seed dispersers, pulp/seed predators and non-frugivores (for additional data, see Magnan, 1912; Cvitanic, 1970). The largest livers were found among muscicapid warblers and would have preadapted them to frugivory by enabling efficient detoxification

of the secondary metabolites present in the pulp. In addition, a closer examination of variation in frugivory among six *Sylvia* warblers (Jordano, 1987b) revealed that most variation in fruit consumption across species was accountable by considering only external morphology. Functional modulation of gut morphology allowing constant digesta retention and extraction efficiency usually require prolonged time periods and do not seem alternatives open to frugivores, which frequently face local and short-term changes in fruit supply (Karasov, 1996; McWilliams and Karasov, 1998). Therefore, rather than elaborate morphological transformations, one finds more functional compensatory modulations to digest a soft, dilute food with low nutrient density that has a large energy content relative to protein (Herrera, 1984a; Moermond and Denslow, 1985; Karasov and Levey, 1990; Afik and Karasov, 1995; Karasov, 1996).

There are marked functional differences among different diet types from the perspective of the digestion process (Table 6.4). Ruminant diets are characteristically

Table 6.4. Some characteristics of ruminant, carnivore and frugivore diets from the perspective of digestive physiology (modified after Morris and Rogers, 1983).

Characteristics	Ruminant diets	Animal prey	Frugivore diets
Nature of diet	Structural and photosynthetic parts of plants	Animal tissue	Fruit pulp
Digestibility	Cell-wall components are refractory to mammalian enzymes	Readily digested by mammalian and avian enzymes	Readily digested, but presence of indigestible seeds
Food passage through the gut	Very slow	Slow	Very rapid
Organic matter digestibility (%)	Most forages < 65	> 85	c. 60–80
Presence of natural toxins	Generalized	None in species normally eaten	Generalized
Proximate constituents of the diet:			
Lipids	Low	High	Variable–low
Protein	Low (generally)	Very high	Very low
Non-structural carbohydrates	Low	Very low	Very high
Structural carbohydrates	Very high	Absent	Variable–low

high in structural hexose and pentose polymers, which require special pregastric microbial digestion, which, in addition, detoxifies many secondary plant substances (Morris and Rogers, 1983). In contrast with this slow digestion process, the digestive processing of the fruit pulp is much more rapid and more similar to digestion of vegetative plant parts by non-ruminant herbivores. In general, both forage and fruit diets show much lower digestibilities than diets based on animal prey. In addition, a sizeable fraction of the fruit food mass ingested by frugivores (the seeds) is actually indigestible and causes gut displacement (Levey and Grajal, 1991; Witmer, 1998b). Herbivore diets, and fruits are no exception, pose a frequent problem by creating digestive bottlenecks (Kenward and Sibly, 1977), which prevent frugivores from increasing fruit intake to compensate for low fruit quality. The energy requirements can be adequately met, but the food-processing rate is too slow to meet the demand for micronutrients and nitrogen, which are deficient in the fruit pulp, and an alternative source is needed (Foster, 1978; Moermond and Denslow, 1985).

Frugivores, as monogastric herbivores, base their feeding on rapid processing of their poor-quality food and maximization of ingestion rate. They thus appear to be process-rate-limited, because ingestion rate is limited by the processing of the previous meal (Sorensen, 1984; Worthington, 1989; Levey and Grajal, 1991; Levey and Duke, 1992). Throughput rate – the rate of flow of digesta past a specified point in the gut – is a function of both gut capacity (intestine length) and food retention time (Sibly, 1981; Hume, 1989; Levey and Grajal, 1991). Rapid processing of separate pulp and seed fractions, rapid passage of seeds, partial emptying of the rectal contents, rectal antiperistalsis and nutrient uptake in the rectum are all characteristics of the digestive process of frugivores to cope with nutrient-poor fruit pulp (Levey and Duke, 1992). For frugivores that defecate seeds, high throughput rates of indigestible seeds must be achieved, with minimum costs for pulp digestion and assimilation. Karasov

and Levey (1990) have demonstrated that this cost exists as a lower digestive efficiency, due to the absence of compensatory high rates of digestive nutrient transport, among frugivores (but see Witmer, 1998b). In consequence, an important functional adaptation among strong frugivores would be a relatively large gut (e.g. long intestine) and extremely short throughput times; therefore, nutrient assimilation is maximized with high throughput rates. Holding constant the throughput rate, a larger gut allows processing of a greater volume of digesta at the same processing speed.

Among strongly frugivorous vertebrate species, high throughput rates are achieved by extreme shortening of throughput times (e.g. Turcek, 1961; Milton, 1981; Sorensen, 1983; Herrera, 1984a; Levey, 1986, 1987; Jordano, 1987b; Worthington, 1989; Karasov and Levey, 1990; Levey and Grajal, 1991). Seeds are processed much more quickly than pulp, either by rapid regurgitation or by 'selective' processing and defecation (but see Levey and Duke, 1992), indicating that they limit fruit processing by gut displacement and that frugivores void them selectively in order to maximize gut capacity for digestible pulp. Time to regurgitate seeds by frugivorous birds is very rapid, frequently 5–20 min, while throughput times for seed defecation are much longer, usually in the range of 0.3–1.5 h (Levey, 1986; Snow and Snow, 1988; Worthington, 1989; Levey and Grajal, 1991). In some species, such as the phainopeplas (Fig. 6.5), an active mechanism for selective pulp retention is used; but, in most instances, differences in throughput times might be caused by the differences in specific gravity between pulp and seeds.

Relative intestine length is greater among Mediterranean frugivorous *Sylvia* warblers than among non-frugivorous muscicapid warblers (Jordano, 1987b), although gut passage time is shorter in the former. For a sample of Mediterranean scrubland frugivorous passerines, variation across species in the relative importance of fruit in the diet is positively correlated with food throughput rate (Fig. 6.6), indicating that the ability to modulate retention

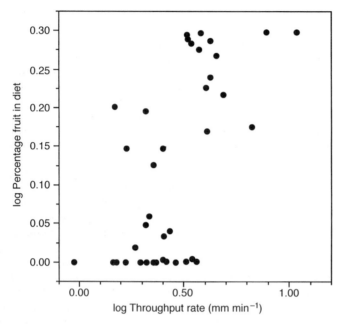

Fig. 6.6. Relationship between relative importance of fruits in the diet of several Mediterranean scrubland passerine birds (as percentage of total diet made up by fruits) and food passage rate (mm min^{-1}), the quotient of intestine length divided by gut passage time. Species ($n = 38$) include *Turdus* spp., *Sylvia* spp., *Phoenicurus phoenicurus*, *Muscicapa striata*, *Ficedula hypoleuca*, *Erithacus rubecula*. Regression fit is $y = 1.539 + 0.041x$ ($r^2 = 0.465$; $F = 8.69$; $P = 0.015$).

time of digesta to achieve a high through-put rate might be important for sustained frugivory. Similarly, McWilliams and Karasov (1998) reported that compensatory modulation of retention time or digesta mixing (and not rate of hydrolysis and absorption) explained the remarkably constant digestive efficiency in waxwings exposed to varied fruit-feeding costs.

Evidence that the size of indigestible seed material limits feeding rates by causing gut displacement and represents an important foraging cost for frugivores mostly comes from observations in captivity (Bonaccorso and Gush, 1987; Levey, 1987; Fleming, 1988; Snow and Snow, 1988; Corlett and Lucas, 1990; Levey and Duke, 1992; but see Witmer, 1998b), which revealed: (i) negative correlations between seed size and the number of seeds ingested per feeding bout; (ii) continuous feeding rates of birds and bats, resulting in at least one ingested seed retained in the gut; (iii) selective throughput times for seeds and pulp; and (iv) immediate consumption of

new fruits after defecation or regurgitation, implying that ingested seeds in the crop limited ingestion of additional fruits. Apparently, however, frugivores might compensate for these costs to achieve adequate intake of basic nutrients (Levey and Duke, 1992; Witmer, 1998b; Witmer and van Soest, 1998). These costs of the internal handling of seed ballast are obviously overcome by frugivorous mashers and spitters, as well as by pulp predators, which manage seeds externally; however, these frugivores have increased handling costs and lower rates of pulp ingestion per fruit handled.

Foraging for fruits and seed transport

Most seed movement away from the parent trees of fleshy-fruited species is a direct consequence of movement patterns by frugivores. Frugivore movements take place on a habitat template with numerous microhabitats, patches, safe sites or other potential 'targets' for seed delivery. These patches differ in potential 'quality' for plant recruitment, measured as the probabilities for early survival of seeds, germination and seedling establishment (Schupp, 1993). From the plant perspective, the potential evolutionary and demographic relevance of the interaction with a particular disperser depends on the number of seeds it moves and how they are delivered over this habitat template, which includes a non-random distribution of patches of variable probability for the establishment and survival of the plant propagules. Therefore, the two main aspects of frugivory that influence the resulting seed dispersal are the seed-processing behaviour (both external and digestive) and the ranging behaviour of the frugivore (Schupp, 1993; Jordano and Schupp, 2000). The former determines the number of seeds that are transported and delivered unharmed, in conditions adequate for germination; the latter defines the potential range of microsites that will intercept delivered seeds. The aim of this final section is to review how the fruit and frugivore characteristics previously considered interact and result in seed deposition patterns with implications for differential seed and seedling survival.

The spatial pattern of seed fall in zoochorous species, i.e. the seed shadow, is a function of the species of frugivore eating the fruit, its movement rates and its seed throughput rates (Hoppes, 1987; Murray, 1988; see also Willson and Traveset, Chapter 4, this volume). Note that two of the factors, namely the species identity and the seed throughput rates, can be expected to remain more or less invariant in their effect on the seed shadow independently of the particular ecological context (e.g. fruit-handling patterns, defecation rates, fruit-capture behaviours and other characteristics of the frugivore). In contrast, movement rates, which depend on movements between foraging locations and the distances between these locations, are much more 'context sensitive' and dependent on the particular ecological situation.

Fruit processing and seed deposition

Fruit processing by frugivores determines how many seeds are delivered to potential safe sites in an unharmed condition. Two important components of fruit processing are the number of fruits handled and the probability that seeds survive the fruit handling by the frugivore. If the number of safe sites increases with distance from parent plants or if the probability of seed and early seedling survival increases with distance, an important component of seed processing will be how fast seeds are delivered after fruit capture.

A typical feeding bout for most frugivores, especially small-sized temperate and tropical birds and phyllostomid bats, includes consumption of one or a few fruits during discrete visits to individual plants that occur along foraging sequences (Herrera and Jordano, 1981; Fleming, 1988; Snow and Snow, 1988; Green, 1993; Sun and Moermond, 1997; Jordano and Schupp, 2000). The resulting pattern of seed delivery will differ markedly between species that process fruits through the digestive tract and defecate seeds and those that process seeds orally by spitting, regurgitating or mashing prior to ingestion. These two general types of seed-processing behaviours are present in most communities and differ in their immediate consequences for seed delivery. I must emphasize here that they do not represent a dichotomy of frugivore strategies but rather a continuum gradient of seed-processing rate (e.g. the number of viable seeds delivered per unit foraging time). Even the same frugivore species can be ranked in different positions along this gradient when interacting with different plant species.

Rapid processing of seeds by frugivores that mash or spit out seeds involves mastication and slow mandibulation of the fruit to separate the pulp from the seeds prior to ingestion, and this usually results in increased risk of seed damage by cracking of the endocarp, excessive mechanical scarification, etc. (Hylander, 1975; Levey, 1987; Corlett and Lucas, 1990). Short-distance delivery of seeds, usually below the parent plant, is the likely result of oral fruit processing, resulting in highly clumped seed distributions, irrespective of how many seeds are dispersed. In addition, low mixing of different seed species is expected, since fruits are processed individually. Frugivores that process fruits orally either expectorate seeds while foraging on the same plant for more fruits (e.g. birds that mash fruits, some neotropical primates) or temporarily exit to nearby perches to process the fruit and then return to the same foraging patch. Highly clumped seed distributions have been reported as a result of the activity of phyllostomid bats, which mash fruits (e.g. *Carollia*) or expectorate a food bolus with seeds (e.g. *Artibeus*) (Bonaccorso and Gush, 1987; Fleming, 1988). The same applies to territorial birds that regurgitate seeds within a close range of the feeding plant or display perches (Pratt and Stiles, 1983; Pratt, 1984; Snow and Snow, 1984; Théry and Larpin, 1993; Kinnaird, 1998; Wenny and Levey, 1998) and tapirs and large primates using recurrent movement patterns (Fragoso, 1997; Julliot, 1997). Clumped seed distributions are not caused by a high number of seeds being processed, since the longer times to handle fruits (birds that regurgitate seeds are an exception) result in slower feeding rates, but are caused by the recurrent use of the same perches and sites for fruit handling, resting, defecation, etc.

In contrast, digestive seed processing involves a longer retention time for seeds and increases the probability that the seed will be moved away from the parent plant. This might result in more scattered seed delivery, unless postforaging movements concentrate seeds at traditional roosts, latrines, pathways, etc. Also, the degree of scattering depends on frugivore size. Blackcaps scatter one to three seeds in single droppings at no particular locations in Mediterranean shrubland (Jordano, 1988; Debussche and Isenmann, 1994), but large ungulates and some primates can concentrate hundreds of seeds in single droppings (Dinerstein and Wemmer, 1988; Fragoso,

1997; Julliot, 1997). The longer retention times of seeds within the gut obviously increase the probability of seed delivery to longer distances. Fruit handling prior to ingestion is minimal, but there is a greater risk of digestive seed damage, especially in frugivores with long retention times, such as ungulates, parrots, some pigeons and terrestrial birds and some finches (Janzen, 1981a, 1982; Gautier-Hion, 1984; Erard and Sabatier, 1988; Murray, 1988; Bodmer, 1989a; Lambert, 1989b). Finally, seed clumping in faeces is strongly dependent on frugivore size (Howe, 1989; White and Stiles, 1990) and this has important implications for seed survival, germination and seedling competition. Few studies, however, have documented how these patterns translate into positive net effects of non-random ('directed') seed dispersal by frugivores (Reid, 1989; Ladley and Kelly, 1996; Wenny and Levey, 1998).

Proximate consequences of seed deposition patterns

Frugivory influences on plant fitness and recruitment do not end with seed delivery. For every dispersal episode, it matters how many and where seeds reach the ground and the particular mix of seed species delivered. There are a number of detailed studies on the ranging behaviour of frugivores and I shall not attempt to consider them in detail here (e.g. Gautier-Hion *et al.*, 1981; Hladik, 1981; Terborgh, 1983; Fleming, 1988; Murray, 1988; Chavez-Ramirez and Slack, 1994; Sun *et al.*, 1997). This is probably the aspect of zoochory that is most 'context-sensitive'. Most of the animal-orientated studies of frugivore movements and ranging behaviour have emphasized the patchy nature of the movements and foraging effort and the influences of external factors, such as seasonality, between-year variations in the fruit supply and numbers of other frugivores, habitat structure and abundance of alternative fruit sources and other food resources. These factors influence the 'where' component of seed deposition patterns, but I wish to concentrate on the 'how' component and point out some recent research and promising directions.

The greater probability of seed mixing for internally processed seeds has far-reaching implications for postdispersal seed and seedling survival, which have only recently been considered in detail in explicit relation to frugivore activity. Bullock (1981) showed that aggregated dispersal of several seeds of *Prunus ilicifolia* in coyote faeces increased seedling survival and that seedlings resulting from clumped dispersal in single droppings showed greater above-ground biomass than spaced seedlings. He reported that grafting between roots was commonly observed among seedlings from a cohort, indicating some direct physiological integration among different genets in such a group of seedlings. Studies by Lieberman and Lieberman (1980); Herrera (1984b, c); Jordano (1988); Loiselle (1990); White and Stiles (1990); Théry and Larpin (1993); Fragoso (1997); and Julliot (1997) strongly support the hypothesis that frugivorous animals can have determinant effects on plant community composition by differentially dispersing particular combinations of seed species. Detailed studies are needed to obtain experimental support for this hypothesis.

Observational evidence indicates that combinations of seed species in the faeces of dispersal agents are not the result of a process of random assortment of the available fruits in the diet, but rather indicate the presence of consistent choice patterns. Preliminary correlative evidence comes from studies of hemiparasitic and parasitic plants, which need highly directed dispersal to particular hosts (Herrera, 1988b; Reid, 1989; Ladley and Kelly, 1996), but a similar effect can be important for vines. Additional evidence has been obtained from detailed studies of individual diet variation in frugivore populations (Jordano, 1988; Loiselle, 1990; White and Stiles, 1990) and seed-rain studies (Stiles and White, 1986). Loiselle (1990) has demonstrated experimentally that specific combinations of dispersed seeds in faeces of

tropical frugivorous birds have a direct influence on seed germination and early seedling vigour and survival.

Studies of germination rates in deposited seeds, early seedling survival and variations in seedling biomass, adequately linked with detailed information of frugivory patterns, such as those described above, are the necessary tools for exploring the potential consequences of the fruit/frugivory interface in plant demography.

Concluding remarks: an agenda for the fruit/frugivory interface

Seed dispersal is a central demographic process in plant populations. The interaction of fruits and frugivores determines the net result of the whole predispersal reproductive phase, being its last step. However, events occurring during this fruit-removal, seed-delivery episode have a direct influence on later-occurring demographic processes, such as germination and early seedling establishment and survival. The studies of fruit–frugivore interactions considered in this chapter have documented what could be designated as the largely 'invariant' fruit and frugivory patterns that characterize each interacting species in the particular scenario where the interaction occurs (e.g. fruit and seed size, design, nutrient configuration, fruiting display, etc.; and body size, ecomorphology, fruit-handling behaviour and digestive process of food, etc.). Description of these patterns has enabled us in the last 25 years to elaborate predictions about the outcomes of particular combinations of characteristics and to test them by evaluating the associated costs in terms of seed losses for the plants or foraging costs for the frugivorous animals.

But we need to translate the effects of these interactions into a demographic and evolutionary context in order to assess the relative contributions of the derived selection pressures in shaping the patterns we are observing. In this context, the net outcomes of the interactions may or may not have evolutionary consequences if their effects are 'screened off' by factors external to the interaction itself. The same can be said for the potential of frugivores to impose 'dispersal limitation' on the recruitment of their food plants (Jordano and Herrera, 1995; Clark et al., 1999). Thus, the outcome of the invariant patterns described above depends, in addition, on 'context-sensitive' effects, which represent a largely stochastic component of the fruit-removal, seed-dispersal phase. Among them, plant spacing patterns, neighbourhood structure, site-specific habitat heterogeneity, density of alternative resources, temporal variations in fruit production and frugivore numbers, etc., produce effects that shape the result of the 'invariant' fruit/frugivore patterns.

A future avenue of research would assess the net demographic outcome of the fruit/frugivory interface by associating probabilities of seed delivery, resulting from a given interaction, with probabilities of seed and seedling survival in different microhabitats (e.g. see Willson and Traveset, Chapter 4, and Crawley, Chapter 7, in this volume). In this way, the relative roles of seed dispersal limitation and recruitment limitation in determining abundance could be gauged (Dalling et al., 1998). The preliminary protocols have been developed (e.g. Heithaus et al., 1982; Herrera, 1988a; Jordano, 1989) for incorporating the consequences of the predispersal events and the deferred consequences for the postdispersal phase (McDonnell and Stiles, 1983; Howe et al., 1985; Fleming, 1988; Katusic-Malmborg and Willson, 1988; Murray, 1988; Schupp, 1988, 1993; Herrera et al., 1994; Jordano and Herrera, 1995; Schupp and Fuentes, 1995; Wenny and Levey, 1998; Clark et al., 1999; Jordano and Schupp, 2000). These studies emphasize the need to estimate the fitness effects of interactions with frugivores for individual plants in natural populations and consider whether the effects of frugivores are offset by events in subsequent stages of recruitment. In addition, it is necessary to

consider how demographic processes (especially seed germination and seedling establishment) are influenced by variation in traits relevant to the plant–frugivore interaction.

In 1591, the Italian painter Giuseppe Arcimboldo finished *Vertumnus*, an oil-painting on wood depicting a portrait of Emperor Rudolf II in a frontal view of head and shoulders. When admired from a distance, this image of Vertumnus, a Roman deity responsible for vegetation and metamorphosis, appears as a neat, brightly coloured and meticulously elaborate picture. On approaching the painting, one discovers that Arcimboldo illustrated at least 34 species of fleshy fruits, which, carefully assembled, served as natural models to produce Vertumnus' image. Grapes, cherries, pears, figs, blackberries, peaches and plums, among many others, serve as the eyes, ears, lips, nose, etc. of this incredible fruit dish. What I admire about this intriguing, funny face is the painter's ability to produce an ordered image from such a chaotic ensemble of fruits and plant parts. I think that the last two decades of research on the fruit–frugivory interface have yielded many fruits, which, like Arcimboldo's model objects, need an elaborate assembly to produce a neat image. The efforts to bridge the consequences of frugivory and seed dispersal with the demographic and evolutionary processes in plant and frugivore populations are a first sketch of that picture.

Acknowledgements

This research was supported by grants PB96–0857 and 1FD97–0743-C03–01 from the Comisión Interministerial de Ciencia y Tecnología (CICYT), the Ministerio de Educación y Ciencia and the European Commission, and also by funds from the Consejería de Educación y Ciencia (Junta de Andalucía).

References

Abrahamson, W.G. and Abrahamson, C.R. (1989) Nutritional quality of animal dispersed fruits in Florida sandridge habitats. *Bulletin of the Torrey Botanical Club* 116, 215–228.

Afik, D. and Karasov, W.H. (1995) The trade-offs between digestion rate and efficiency in warblers and their ecological implications. *Ecology* 76, 2247–2257.

Alexandre, D.Y. (1978) Le rôle disséminateur des éléphants en forêt de Tai, Côte-d'Ivoire. *La Terre et la Vie* 32, 47–72.

Alexandre, D.Y. (1980) Caractère saisonier de la fructification dans une forêt hygrophile de Côte-d'Ivoire. *Revue d'Ecologie (Terre et Vie)* 34, 335–350.

Atramentowicz, M. (1988) La frugivorie opportuniste de trois marsupiaux didelphidés de Guyane. *Revue d'Ecologie (Terre et Vie)* 43, 47–57.

Baird, J.W. (1980) The selection and use of fruit by birds in an eastern forest. *Wilson Bulletin* 92, 63–73.

Barquín, E. and Wildpret, W. (1975) Diseminación de plantas canarias: datos iniciales. *Vieraea* 5, 38–60.

Beehler, B. (1983) Frugivory and polygamy in birds of paradise. *Auk* 100, 1–12.

Berthold, P. (1976) Animalische und vegetabilische Ernährung omnivorer Singvogelarten: Nahrungsbevorzugung, Jahresperiodik der Nahrungswahl, physiologische und ökologische Bedeutung. *Journal für Ornithologie* 117, 145–209.

Berthold, P. (1977) Proteinmangel als Ursache der schädigenden Wirkung rein vegetabilischer Ernhärung omnivorer Singvogelarten. *Journal für Ornithologie* 118, 202–205.

Berthold, P. and Moggingen, S. (1976) Der Seidenschwanz *Bombycilla garrulus* als frugivorer Ernährungsspezialist. *Experientia* 32, 1445.

Bhide, S.A. (1980) Observations on the stomach of the Indian fruit bat, *Rousettus leschenaulti* (Desmarest). *Mammalia* 44, 571–579.

Blake, J.G. and Hoppes, W.G. (1986) Influence of resource abundance on use of tree-fall gaps by birds in an isolated woodlot. *Auk* 103, 328–340.

Blake, J.G., Loiselle, B.A., Moermond, T.C., Levey, D.J. and Denslow, J.S. (1990) Quantifying the abundance of fruits for birds in tropical habitats. *Studies in Avian Biology* 13, 73–79.

Bodmer, R.E. (1989a) Frugivory in Amazonian Artiodactyla: evidence for the evolution of the ruminant stomach. *Journal of Zoology, London* 219, 457–467.

Bodmer, R.E. (1989b) Ungulate biomass in relation to feeding strategy within Amazonian forests. *Oecologia* 81, 547–550.

Bodmer, R.E. (1990) Ungulate frugivores and browser–grazer continuum. *Oikos* 57, 319–325.

Bonaccorso, F.J. (1979) Foraging and reproductive ecology in a Panamanian bat community. *Bulletin of the Florida State Museum* 24, 359–408.

Bonaccorso, F.J. and Gush, T.J. (1987) Feeding behaviour and foraging strategies of captive phyllostomid fruit bats: an experimental study. *Journal of Animal Ecology* 56, 907–920.

Boojh, R. and Ramakrishnan, P.S. (1981) Phenology of trees in a sub-tropical evergreen montane forest in North-east India. *Geo-eco-trop* 5, 189–209.

Borchert, R. (1983) Phenology and control of flowering in tropical trees. *Biotropica* 15, 81–89.

Bowen, S.H., Lutz, E.V. and Ahlgren, M.O. (1995) Dietary protein and energy as determinants of food quality: trophic strategies compared. *Ecology* 76, 899–907.

Bullock, S.H. (1978) Plant abundance and distribution in relation to types of seed dispersal. *Madroño* 25, 104–105.

Bullock, S.H. (1981) Aggregation of *Prunus ilicifolia* (Rosaceae) during dispersal and its effect on survival and growth. *Madroño* 28, 94–95.

Burrows, C.J. (1994) Fruits, seeds, birds and the forests of Banks Peninsula. *New Zealand Natural Sciences* 21, 87–108.

Cadow, A. (1933) Magen und Darm der Fruchttauben. *Journal für Ornithologie* 81, 236–252.

Chapman, L.J., Chapman, C.J. and Wrangham, R.W. (1992a) *Balanites wilsoniana*: elephant dependent dispersal. *Journal of Tropical Ecology* 8, 275–283.

Chapman, C.A., Wrangham, R. and Chapman, L.J. (1992b) Estimators of fruit abundance of tropical trees. *Biotropica* 24, 527–531.

Chapman, C.A., Wrangham, R. and Chapman, L.J. (1994) Indices of habitat-wide fruit abundance in tropical forests. *Biotropica* 26, 160–171.

Charles-Dominique, P., Atramentowicz, M., Charles-Dominique, M., Gerard, H., Hladik, A., Hladik, C.M. and Prevost, M.F. (1981) Les mammifères frugivores arboricoles nocturnes d'une forêt Guyanaise: inter-relations plantes–animaux. *Revue d'Ecologie (Terre et Vie)* 35, 341–435.

Chavez-Ramirez, F. and Slack, R.D. (1994) Effects of avian foraging and post-foraging behavior on seed dispersal patterns of Ashe juniper. *Oikos* 71, 40–46.

Cipollini, M.L. and Levey, D.J. (1992) Relative risks of microbial rot for fleshy fruits: significance with respect to dispersal and selection for secondary defense. *Advances in Ecological Research* 23, 35–91.

Cipollini, M.L. and Levey, D.J. (1997) Secondary metabolites of fleshy vertebrate-dispersed fruits: adaptive hypotheses and implications for seed dispersal. *American Naturalist* 150, 346–372.

Clark, J.S., Beckage, B., Camill, P., Cleveland, B., HilleRisLambers, J., Lichter, J., McLachlan, J., Mohan, J. and Wyckoff, P. (1999) Interpreting recruitment limitation in forests. *American Journal of Botany* 86, 1–16.

Conklin-Brittain, N.L., Wrangham, R.W. and Hunt, K.D. (1998) Dietary responses of chimpanzees and cercopithecines to seasonal variation in fruit abundance. I. Antifeedants. *International Journal of Primatology* 19, 949–970.

Corlett, R.T. (1996) Characteristics of vertebrate-dispersed fruits in Hong Kong. *Journal of Tropical Ecology* 12, 819–833.

Corlett, R.T. (1998) Frugivory and seed dispersal by vertebrates in the Oriental (Indomalayan) region. *Biological Review* 73, 413–448.

Corlett, R.T. and Lucas, P.W. (1990) Alternative seed-handling strategies in primates: seed-spitting by long-tailed macaques (*Macaca fascicularis*). *Oecologia* 82, 166–171.

Courts, S.E. (1998) Dietary strategies of Old World fruit bats (Megachiroptera, Pteropodidae): how do they obtain sufficient protein? *Mammal Review* 28, 185–194.

Croat, T.B. (1978) *Flora of Barro Colorado Island*. Stanford University Press, Stanford.

Crome, F.H.J. (1975) The ecology of fruit pigeons in tropical Northern Queensland. *Australian Wildlife Research* 2, 155–185.

Cruz, A. (1981) Bird activity and seed dispersal of a montane forest tree (*Dunalia arborescens*) in Jamaica. *Biotropica (Suppl.)* 13, 34–44.

Cvitanic, A. (1970) The relationships between intestine and body length and nutrition in several bird species. *Larus* 21–22, 181–190.

Dalling, J.W., Hubbell, S.P. and Silvera, K. (1998) Seed dispersal, seedling establishment and gap partitioning among tropical pioneer trees. *Journal of Ecology* 86, 674–689.

Daubenmire, R. (1972) Phenology and other characteristics of tropical semi-deciduous forest in north-western Costa Rica. *Journal of Ecology* 60, 147–170.

Davies, A.G., Bennett, E.L. and Waterman, P.G. (1988) Food selection by two south-east Asian colobine monkeys (*Prebytis rubicunda* and *Presbytis melalophos*) in relation to plant chemistry. *Biological Journal of the Linnean Society* 34B, 33–56.

Davies, S.J.J.F. (1976) Studies on the flowering season and fruit production of some arid zone shrubs and trees in Western Australia. *Journal of Ecology* 64, 665–687.

Davis, D.E. (1945) The annual cycle of plants, mosquitoes, birds, and mammals in two brazilian forests. *Ecological Monographs* 15, 245–295.

Debussche, M. (1988) La diversité morphologique des fruits charnus en Languedoc méditerranéen: relations avec les caractéristiques biologiques et la distribution des plantes, et avec les disseminateurs. *Acta Oecologica, Oecologia Plantarum* 9, 37–52.

Debussche, M. and Isenmann, P. (1989) Fleshy fruit characters and the choices of bird and mammal seed dispersers in a Mediterranean region. *Oikos* 56, 327–338.

Debussche, M. and Isenmann, P. (1992) A Mediterranean bird disperser assemblage: composition and phenology in relation to fruit availability. *Revue d' Ecologie* 47, 411–432.

Debussche, M. and Isenmann, P. (1994) Bird-dispersed seed rain and seedling establishment in patchy Mediterranean vegetation. *Oikos* 69, 414–426.

Debussche, M., Escarré, J. and Lepart, J. (1982) Ornithochory and plant succession in Mediterranean abandoned orchards. *Vegetatio* 48, 255–266.

Debussche, M., Cortez, J. and Rimbault, I. (1987) Variation in fleshy fruit composition in the Mediterranean region: the importance of ripening season, life-form, fruit type, and geographical distribution. *Oikos* 49, 244–252.

Decoux, J.P. (1976) Régime, comportement alimentaire et régulation écologique du métabolisme chez *Colius striatus*. *La Terre et la Vie* 30, 395–420.

De Foresta, H., Charles-Dominique, P. and Erard, C. (1984) Zoochorie et premières stades de la régénération naturelle après coupe en forêt guyanaise. *Revue d'Ecologie (Terre et Vie)* 39, 369–400.

Denslow, J.S. (1987) Fruit removal rates from aggregated and isolated bushes of the red elderberry, *Sambucus pubens*. *Canadian Journal of Botany* 65, 1229–1235.

Denslow, J.S. and Moermond, T.C. (1982) The effect of fruit accessibility on rates of fruit removal from tropical shrubs: an experimental study. *Oecologia* 54, 170–176.

Desselberger, H. (1931) Der Verdauungskanal der Dicaeiden nach Gestalt und Funktion. *Journal für Ornithologie* 79, 353–374.

Dinerstein, E. (1986) Reproductive ecology of fruit bats and the seasonality of fruit production in a Costa Rican cloud forest. *Biotropica* 18, 307–318.

Dinerstein, E. and Wemmer, C.M. (1988) Fruits rhinoceros eat: dispersal of *Trewia nudiflora* (Euphorbiaceae) in lowland Nepal. *Ecology* 69, 1768–1774.

Docters van Leeuwen, W.M. (1954) On the biology of some Loranthaceae and the role birds play in their life-history. *Beaufortia* 4, 105–208.

Donoghue, M.J. (1989) Phylogenies and the analysis of evolutionary sequences, with examples from seed plants. *Evolution* 43, 1137–1156.

Dowsett-Lemaire, F. (1988) Fruit choice and seed dissemination by birds and mammals in the evergreen forests of upland Malawi. *Revue d'Ecologie (Terre et Vie)* 43, 251–286.

Dubost, G. (1984) Comparison of the diets of frugivorous forest ruminants of Gabon. *Journal of Mammalogy* 65, 298–316.

Erard, C. and Sabatier, D. (1988) Rôle des oiseaux frugivores terrestres dans la dinamique forestière en Guyane française. In: Ouellet, H. (ed.) *Acta XIX Congressus Internationalis Ornithologici.* University of Ottawa Press, Ottawa, Canada, pp. 803–815.

Erard, C., Thery, M. and Sabatier, D. (1989) Régime alimentaire de *Rupicola rupicola* (Cotingidae) en Guyane française. Relations avec la frugivorie et la zoochorie. *Revue d'Ecologie (Terre et Vie)* 44, 47–74.

Eriksson, K. and Nummi, H. (1982) Alcohol accumulation from ingested berries and alcohol metabolism in passerine birds. *Ornis Fennica* 60, 2–9.

Eriksson, O. and Ehrlén, J. (1991) Phenological variation in fruit characteristics in vertebrate-dispersed plants. *Oecologia* 86, 463–470.

Estrada, A. and Fleming, T.H. (eds) (1986) *Frugivores and Seed Dispersal.* Junk, Dordrecht.

Estrada, A., Coates-Estrada, R. and Vázquez-Yanes, C. (1984) Observations on fruiting and dispersers of *Cecropia obtusifolia* at Los Tuxtlas, Mexico. *Biotropica* 16, 315–318.

Fenner, M. (1998) The phenology of growth and reproduction in plants. *Perspectives in Plant Ecology, Evolution and Systematics* 1, 78–91.

Fitzpatrick, J.W. (1980) Foraging behavior of neotropical tyrant flycatchers. *Condor* 82, 43–57.

Fleming, T.H. (1982) Foraging strategies of plant-visiting bats. In: Kunz, T.H. (ed.) *Ecology of Bats.* Plenum Press, New York, pp. 287–325.

Fleming, T.H. (1986) Opportunism versus specialization: the evolution of feeding strategies in frugivorous bats. In: Estrada, A. and Fleming, T.H. (eds) *Frugivores and Seed Dispersal.* Junk, Dordrecht, pp. 105–118.

Fleming, T.H. (1988) *The Short-tailed Fruit Bat: a Study in Plant–Animal Interactions.* University of Chicago Press, Chicago.

Forbes, W.A. (1880) Contributions to the anatomy of passerine birds. Part I. On the structure of the stomach in certain genera of tanagers. *Proceedings of the Zoological Society, London* 188, 143–147.

Foster, M.S. (1977) Ecological and nutritional effects of food scarcity on a tropical frugivorous bird and its fruit source. *Ecology* 58, 73–85.

Foster, M.S. (1978) Total frugivory in tropical passerines: a reappraisal. *Tropical Ecology* 19, 131–154.

Foster, M.S. (1987) Feeding methods and efficiencies of selected frugivorous birds. *Condor* 89, 566–580.

Foster, M.S. and McDiarmid, R.W. (1983) Nutritional value of the aril of *Trichilia cuneata*, a bird-dispersed fruit. *Biotropica* 15, 26–31.

Foster, R.B. (1982) Famine on Barro Colorado Island. In: Leigh, E.G., Rand, E.S. and Windsor, D. (eds) *The Ecology of a Tropical Forest.* Smithsonian Institution Press, Washington, DC, pp. 201–212.

Fragoso, J.M.V. (1997) Tapir-generated seed shadows: scale-dependent patchiness in the Amazon rain forest. *Journal of Ecology* 85, 519–529.

Frankie, G.W., Baker, H.G. and Opler, P.A. (1974a) Comparative phenological studies in tropical wet and dry forests in the lowlands of Costa Rica. *Journal of Ecology* 62, 881–919.

Frankie, G.W., Baker, H.G. and Opler, P.A. (1974b) Tropical plant phenology: applications for studies in community ecology. In: Lieth, H. (ed.) *Phenology and Seasonality Modelling.* Ecological Studies, Vol. 8, Springer Verlag, Berlin, pp. 287–296.

Franklin, J.F., Maeda, T., Ohsumi, Y., Matsui, M., Yagi, H. and Hawk, G.M. (1979) Subalpine coniferous forests of central Houshu, Japan. *Ecological Monographs* 49, 311–334.

French, K. (1992) Phenology of fleshy fruits in a wet sclerophyll forest in Southeastern Australia: are birds an important influence? *Oecologia* 90, 366–373.

Frost, P.G.H. (1980) Fruit–frugivore interactions in a South African coastal dune forest. In: Noring, R. (ed.), *Acta XVII Congressus Internationalis Ornithologici.* Deutsche Ornithologen Gesellschaft, Berlin, pp. 1179–1184.

Fuentes, M. (1992) Latitudinal and elevational variation in fruiting phenology among western european bird-dispersed plants. *Ecography* 15, 177–183.

Funakoshi, K., Watanabe, H. and Kunisaki, T. (1993) Feeding ecology of the northern Ryukyu fruit bat, *Pteropus dasymallus dasymallus*, in a warm-temperate region. *Journal of Zoology* 230, 221–230.

Garber, P.A. (1986) The ecology of seed dispersal in two species of callitrichid primates (*Sanguinus mystax* and *Sanguinus fuscicollis*). *American Journal of Primatology* 10, 155–170.

Gardner, A.L. (1977) Feeding habits. In: Baker, R.J., Knox-Jones, J. and Carter, D.C. (eds) *Biology of Bats in the New World Family Phyllostomatidae*, Part II. Special Publication, Museum Texas Technical University No.13, Austin, Texas, pp. 295–328.

Garwood, N.C. (1983) Seed germination in a seasonal tropical forest in Panama: a community study. *Ecological Monographs* 53, 159–181.

Gautier-Hion, A. (1984) La dissémination des graines par les cercopithecidés forestiers africains. *Revue d'Ecologie (Terre et Vie)* 39, 159–165.

Gautier-Hion, A., Gautier, J.P. and Quris, R. (1981) Forest structure and fruit availability as complementary factors influencing habitat use of monkeys (*Cercopithecus cephus*). *Revue d'Ecologie (Terre et Vie)* 35, 511–536.

Gautier-Hion, A., Duplantier, J.M., Emmons, L., Feer, F., Heckestweiler, P., Moungazi, A., Quris, R. and Sourd, C. (1985a) Coadaptation entre rythmes de fructification et frugivorie en forêt tropicale humide du Gabon: mythe ou réalité? *Revue d'Ecologie (Terre et Vie)* 40, 405–434.

Gautier-Hion, A., Duplantier, J.M., Quris, R., Feer, F., Sourd, C., Decoux, J.P., Dubost, G., Emmons, L., Erard, C. and Hecketsweiler, P. (1985b) Fruit characters as a basis of fruit choice and seed dispersal in a tropical forest vertebrate community. *Oecologia* 65, 324–337.

Gentry, A.H. (1982) Patterns of neotropical plant species diversity. In: Hecht, M.K., Wallace, B. and Prance, G.T. (eds) *Evolutionary Biology*, Vol. 15. Plenum Press, New York, pp. 1–84.

Givnish, T.J. (1980) Ecological constraints on the evolution of breeding systems in seed plants: dioecy and dispersal in gymnosperms. *Evolution* 34, 959–972.

Gorchov, D.L. (1990) Pattern, adaptation, and constraint in fruiting synchrony within vertebrate-dispersed woody plants. *Oikos* 58, 169–180.

Green, R.J. (1993) Avian seed dispersal in and near subtropical rainforests. *Wildlife Research* 20, 535–537.

Guitián, J. (1984) Ecología de una comunidad de Passeriformes en un bosque montano de la Cordillera Cantábrica Occidental. Unpublished PhD thesis, Universidad de Santiago, Santiago.

Halls, L.K. (1973) *Flowering and Fruiting of Southern Browse Species*. Forest Service Research Paper SO-90, US Department of Agriculture, Washington, DC, 10 pp.

Harding, R.S.O. (1981) An order of omnivores: nonhuman primate diets in the wild. In: Harding, R.S.O. and Teleki, G. (eds) *Omnivorous Primates: Gathering and Hunting in Human Evolution*. Columbia University Press, New York, pp. 191–214.

Heiduck, S. (1997) Food choice in masked titi monkeys (*Callicebus personatus melanochir*): selectivity or opportunism? *International Journal of Primatology* 18, 487–502.

Heithaus, E.R., Stashko, E. and Anderson, P.K. (1982) Cumulative effects of plant–animal interactions on seed production by *Bauhinia ungulata*, a neotropical legume. *Ecology* 63, 1294–1302.

Herbst, L.H. (1986) The role of nitrogen from the fruit pulp in the nutrition of the frugivorous bat *Carollia perspicillata*. *Biotropica* 18, 39–44.

Herrera, C.M. (1981a) Are tropical fruits more rewarding to dispersers than temperate ones? *American Naturalist* 118, 896–907.

Herrera, C.M. (1981b) Fruit variation and competition for dispersers in natural populations of *Smilax aspera*. *Oikos* 36, 51–58.

Herrera, C.M. (1982) Seasonal variation in the quality of fruits and diffuse coevolution between plants and avian dispersers. *Ecology* 63, 773–785.

Herrera, C.M. (1984a) Adaptation to frugivory of Mediterranean avian seed dispersers. *Ecology* 65, 609–617.

Herrera, C.M. (1984b) Habitat–consumer interactions in frugivorous birds. In: Cody, M.L. (ed.) *Habitat Selection in Birds*. Academic Press, New York, pp. 341–365.

Herrera, C.M. (1984c) A study of avian frugivores, bird-dispersed plants, and their interaction in Mediterranean scrublands. *Ecological Monographs* 54, 1–23.

Herrera, C.M. (1984d) Tipos morfológicos y funcionales en plantas del matorral mediterráneo del sur de España. *Studia Oecologica* 5, 7–34.

Herrera, C.M. (1987) Vertebrate-dispersed plants of the Iberian Peninsula: a study of fruit characteristics. *Ecological Monographs* 57, 305–331.

Herrera, C.M. (1988a) The fruiting ecology of *Osyris quadripartita*: individual variation and evolutionary potential. *Ecology* 69, 233–249.

Herrera, C.M. (1988b) Plant size, spacing patterns, and host-plant selection in *Osyris quadripartita*, a hemiparasitic dioecious shrub. *Journal of Ecology* 76, 995–1006.

Herrera, C.M. (1988c) Variaciones anuales en las poblaciones de pájaros frugívoros y su relación con la abundancia de frutos. *Ardeola* 35, 135–142.

Herrera, C.M. (1989) Frugivory and seed dispersal by carnivorous mammals, and associated fruit characteristics, in undisturbed Mediterranean habitats. *Oikos* 55, 250–262.

Herrera, C.M. (1992) Interspecific variation in fruit shape: allometry, phylogeny, and adaptation to dispersal agents. *Ecology* 73, 1832–1841.

Herrera, C.M. (1995) Plant–vertebrate seed dispersal systems in the Mediterranean: ecological, evolutionary, and historical determinants. *Annual Review of Ecology and Systematics* 26, 705–727.

Herrera, C.M. (1998) Long-term dynamics of Mediterranean frugivorous birds and fleshy fruits: a 12-year study. *Ecological Monographs* 68, 511–538.

Herrera, C.M. and Jordano, P. (1981) *Prunus mahaleb* and birds: the high efficiency seed dispersal system of a temperate fruiting tree. *Ecological Monographs* 51, 203–221.

Herrera, C.M., Jordano, P., López Soria, L. and Amat, J.A. (1994) Recruitment of a mast-fruiting, bird-dispersed tree: bridging frugivore activity and seedling establishment. *Ecological Monographs* 64, 315–344.

Hilty, S.L. (1977) Food supply in a tropical frugivorous bird community. Unpublished PhD thesis, University of Arizona.

Hilty, S.L. (1980) Flowering and fruiting periodicity in a premontane rain forest in Pacific Colombia. *Biotropica* 12, 292–306.

Hladik, A. (1978) Phenology of leaf production in rainforest of Gabon: distribution and composition of food for folivores. In: Montgomery, G.G. (ed.) *The Ecology of Arboreal Folivores*. Smithsonian Institution Press, Washington, DC, pp. 51–71.

Hladik, C.M. (1981) Diet and the evolution of feeding strategies among forest primates. In: Harding, R.S.O. and Teleki, G. (eds) *Omnivorous Primates*. Columbia University Press, New York, pp. 215–254.

Hladik, C.M. and Hladik, A. (1967) Observations sur le rôle des primates dans la dissémination des végétaux de la forêt gabonaise. *Biologia Gabonica* 3, 43–58.

Hladik, C.M., Hladik, A., Bousset, J., Valdebouze, P., Viroben, G. and Delrot-Laval, J. (1971) Le régime alimentaire des primates de l'île de Barro-Colorado (Panama). *Folia Primatologica* 16, 85–122.

Hoffmann, A.J. and Armesto, J.J. (1995) Modes of seed dispersal in the Mediterranean regions in Chile, California, and Australia. In: Arroyo, M.T.K., Zedler, P.H. and Fox, M.D. (eds) *Ecology and Biogeography of Mediterranean Ecosystems in Chile, California and Australia*. Springer-Verlag, New York, USA, pp. 289–310.

Holl, K.D. (1998) Do bird perching structures elevate seed rain and seedling establishment in abandoned tropical pasture? *Restoration Ecology* 6, 253–261.

Hopkins, M.S. and Graham, A.W. (1989) Community phenological patterns of a lowland tropical rainforest in north-eastern Australia. *Australian Journal of Ecology* 14, 399–413.

Hoppes, W.G. (1987) Pre- and post-foraging movements of frugivorous birds in an eastern deciduous forest woodland, USA. *Oikos* 49, 281–290.

Hoppes, W.G. (1988) Seedfall pattern of several species of bird-dispersed plants in an Illinois woodland. *Ecology* 69, 320–329.

Houssard, C., Escarré, J. and Romane, F. (1980) Development of species diversity in some Mediterranean plant communities. *Vegetatio* 43, 59–72.

Howe, H.F. (1980) Monkey dispersal and waste of a neotropical fruit. *Ecology* 61, 944–959.

Howe, H.F. (1981) Dispersal of neotropical nutmeg (*Virola sebifera*) by birds. *Auk* 98, 88–98.

Howe, H.F. (1983) Annual variation in a neotropical seed-dispersal system. In: Sutton, S.L., Whitmore, T.C. and Chadwick, A.C. (eds) *Tropical Rainforest: Ecology and Management*. Blackwell Scientific Publications, London, pp. 211–227.

Howe, H.F. (1984) Constraints on the evolution of mutualisms. *American Naturalist* 123, 764–777.

Howe, H.F. (1986) Seed dispersal by fruit-eating birds and mammals. In: Murray, D.R. (ed.) *Seed Dispersal*. Academic Press, Sydney, Australia, pp. 123–190.

Howe, H.F. (1989) Scatter- and clump-dispersal and seedling demography: hypothesis and implications. *Oecologia* 79, 417–426.

Howe, H.F. (1993) Specialized and generalized dispersal systems: where does 'the paradigm' stand? In: Fleming, T.H. and Estrada, A. (eds) *Frugivory and Seed Dispersal: Ecological and Evolutionary Aspects*. Kluwer Academic Publishers, Dordrecht, The Netherlands, pp. 3–13.

Howe, H.F. and Estabrook, G.F. (1977) On intraspecific competition for avian dispersers in tropical trees. *American Naturalist* 111, 817–832.

Howe, H.F. and Richter, W.M. (1982) Effects of seed size on seedling size in *Virola surinamensis*: a within and between tree analysis. *Oecologia* 53, 347–351.

Howe, H.F. and Smallwood, J. (1982) Ecology of seed dispersal. *Annual Review of Ecology and Systematics* 13, 201–228.

Howe, H.F. and Vande Kerckhove, G.A. (1981) Removal of wild nutmeg (*Virola surinamensis*) crops by birds. *Ecology* 62, 1093–1106.

Howe, H.F., Schupp, E.W. and Westley, L.C. (1985) Early consequences of seed dispersal for a neotropical tree (*Virola surinamensis*). *Ecology* 66, 781–791.

Hudson, D.A., Levin, R.J. and Smith, D.H. (1971) Absorption from the alimentary tract. In: Bell, D.J. and Freeman, B.M. (eds) *Physiology and Biochemistry of the Domestic Fowl*, Vol. I. Academic Press, London, pp. 51–71.

Hughes, L., Westoby, M. and Johnson, A.D. (1993) Nutrient costs of vertebrate-dispersed and ant-dispersed fruits. *Functional Ecology* 7, 54–62.

Hume, I.D. (1989) Optimal digestive strategies in mammalian herbivores. *Physiological Zoology* 62, 1145–1163.

Hylander, W.L. (1975) Incisor size and diet in anthropoids with special reference to Cercopithecidae. *Science* 189, 1095–1098.

Idani, G. (1986) Seed dispersal by pygmy chimpanzees (*Pan paniscus*): a preliminary report. *Primates* 27, 441–447.

Innis, G.J. (1989) Feeding ecology of fruit pigeons in subtropical rainforests of south-eastern Queensland. *Australian Wildlife Research* 16, 365–394.

Izhaki, I. and Safriel, U.N. (1989) Why are there so few exclusively frugivorous birds? Experiments on fruit digestibility. *Oikos* 54, 23–32.

Izhaki, I., Walton, P.B. and Safriel, U.N. (1991) Seed shadows generated by frugivorous birds in an eastern mediterranean scrub. *Journal of Ecology* 79, 575–590.

Janson, C.H. (1983) Adaptation of fruit morphology to dispersal agents in a neotropical forest. *Science* 219, 187–189.

Janson, C.H., Stiles, E.W. and White, D.W. (1986) Selection on plant fruiting traits by brown capuchin monkeys: a multivariate approach. In: Estrada, A. and Fleming, T.H. (eds) *Frugivores and Seed Dispersal.* Junk, Dordrecht, pp. 83–92.

Janzen, D.H. (1967) Synchronization of sexual reproduction of trees within the dry season in Central America. *Evolution* 21, 620–637.

Janzen, D.H. (1981a) Digestive seed predation by a Costa Rican Baird's tapir. *Biotropica* (*Suppl.*) 13, 59–63.

Janzen, D.H. (1981b) Guanacaste tree seed-swallowing by Costa Rican range horses. *Ecology* 62, 587–592.

Janzen, D.H. (1981c) *Enterolobium cyclocarpum* seed passage rate and survival in horses, Costa Rican Pleistocene seed dispersal agents. *Ecology* 62, 593–601.

Janzen, D.H. (1982) Differential seed survival and passage rates in cows and horses, surrogate pleistocene dispersal agents. *Oikos* 38, 150–156.

Janzen, D.H. (1983) Dispersal of seeds by vertebrate guts. In: Futuyma, D.J. and Slatkin, M. (eds) *Coevoution.* Sinauer Associates, Sunderland, Massachusetts, pp. 232–262.

Janzen, D.H., Miller, G.A., Hackforth-Jones, J., Pond, C.M., Hooper, K. and Janos, D.P. (1976) Two Costa Rican bat-generated seed shadows of *Andira inermis* (Leguminosae). *Ecology* 57, 1068–1075.

Johnson, A.S. and Landers, J.L. (1978) Fruit production in slash pine plantations in Georgia. *Journal of Wildlife Management* 42, 606–613.

Johnson, R.A., Willson, M.F., Thompson, J.N. and Bertin, R.I. (1985) Nutritional values of wild fruits and consumption by migrant frugivorous birds. *Ecology* 66, 819–827.

Jordano, P. (1981) Alimentación y relaciones tróficas entre los paseriformes en paso otoñal por una localidad de Andalucía central. *Doñana Acta Vertebrata* 8, 103–124.

Jordano, P. (1982) Migrant birds are the main seed dispersers of blackberries in southern Spain. *Oikos* 38, 183–193.

Jordano, P. (1983) Fig-seed predation and dispersal by birds. *Biotropica* 15, 38–41.

Jordano, P. (1984) Relaciones entre plantas y aves frugívoras en el matorral mediterráneo del área de Doñana. PhD thesis, Universidad de Sevilla, Sevilla.

Jordano, P. (1985) El ciclo anual de los paseriformes frugívoros en el matorral mediterráneo del sur de España: importancia de su invernada y variaciones interanuales. *Ardeola* 32, 69–94.

Jordano, P. (1987a) Avian fruit removal: effects of fruit variation, crop size, and insect damage. *Ecology* 68, 1711–1723.

Jordano, P. (1987b) Frugivory, external morphology and digestive system in Mediterranean sylviid warblers *Sylvia* spp. *Ibis* 129, 175–189.

Jordano, P. (1987c) Notas sobre la dieta no-insectívora de algunos Muscicapidae. *Ardeola* 34, 89–98.

Jordano, P. (1988) Diet, fruit choice and variation in body condition of frugivorous warblers in Mediterranean scrubland. *Ardea* 76, 193–209.

Jordano, P. (1989) Pre-dispersal biology of *Pistacia lentiscus* (Anacardiaceae): cumulative effects on seed removal by birds. *Oikos* 55, 375–386.

Jordano, P. (1995a) Angiosperm fleshy fruits and seed dispersers: a comparative analysis of adaptation and constraints in plant–animal interactions. *American Naturalist* 145, 163–191.

Jordano, P. (1995b) Frugivore-mediated selection on fruit and seed size: birds and St Lucie's cherry, *Prunus mahaleb. Ecology* 76, 2627–2639.

Jordano, P. and Herrera, C.M. (1995) Shuffling the offspring: uncoupling and spatial discordance of multiple stages in vertebrate seed dispersal. *Ecoscience* 2, 230–237.

Jordano, P. and Schupp, E.W. (2000) Determinants of seed disperser effectiveness: the quantity component and patterns of seed rain for *Prunus mahaleb. Ecological Monographs* 70.

Julliot, C. (1997) Impact of seed dispersal of red howler monkeys *Alouatta seniculus* on the seedling population in the understorey of tropical rain forest. *Journal of Ecology* 85, 431–440.

Kantak, G.E. (1979) Observations on some fruit-eating birds in Mexico. *Auk* 96, 183–186.

Kaplin, B.A. and Moermond, T.C. (1998) Variation in seed handling by two species of forest monkeys in Rwanda. *American Journal of Primatology* 45, 83–101.

Karasov, W.H. (1996) Digestive plasticity in avian energetics and feeding ecology. In: Carey, C. (ed.) *Avian Energetics and Nutritional Ecology*. Chapman and Hall, New York, pp. 61–84.

Karasov, W.H. and Levey, D.J. (1990) Digestive system trade-offs and adaptations of frugivorous passerine birds. *Physiological Zoology* 63, 1248–1270.

Karr, J.R. and James, F.C. (1975) Eco-morphological configurations and convergent evolution in species and communities. In: Cody, M.L. and Diamond, J.M. (eds) *Ecology and Evolution of Communities*. Belknap Press, Cambridge, Massachusetts, pp. 258–291.

Katusic-Malmborg, P. and Willson, M.F. (1988) Foraging ecology of avian frugivores and some consequences for seed dispersal in an Illinois woodlot. *Condor* 90, 173–186.

Kenward, R.E. and Sibly, R.M. (1977) A woodpigeon (*Columba palumbus*) feeding preference explained by a digestive bottle-neck. *Journal of Applied Ecology* 14, 815–826.

Kinnaird, M.F. (1998) Evidence for effective seed dispersal by the Sulawesi red-knobbed hornbill, *Aceros cassidix*. *Biotropica* 30, 50–55.

Ko, I.W.P., Corlett, R.T. and Xu, R.J. (1998) Sugar composition of wild fruits in Hong Kong, China. *Journal of Tropical Ecology* 14, 381–387.

Kochmer, J.P. and Handel, S.N. (1986) Constraints and competition in the evolution of flowering phenology. *Ecological Monographs* 56, 303–325.

Koelmeyer, K.O. (1959) The periodicity of leaf change and flowering in the principal forest communities of Ceylon. *Ceylon Forester* 4, 157–189.

Kollmann, J. and Poschlod, P. (1997) Population processes at the grassland–scrub interface. *Phytocoenologia* 27, 235–256.

Korine, C., Arad, Z. and Arieli, A. (1996) Nitrogen and energy balance of the fruit bat *Rousettus aegyptiacus* on natural fruit diets. *Physiological Zoology* 69, 618–634.

Kunz, T.H. and Diaz, C.A. (1995) Folivory in fruit-eating bats, with new evidence from *Artibeus jamaicensis* (Chiroptera: Phyllostomidae). *Biotropica* 27, 106–120.

Ladley, J.J. and Kelly, D. (1996) Dispersal, germination and survival of New Zealand mistletoes (Loranthaceae): dependence on birds. *New Zealand Journal of Ecology* 20, 69–79.

Lambert, F. (1989a) Fig-eating by birds in a Malaysian lowland rain forest. *Journal of Tropical Ecology* 5, 401–412.

Lambert, F. (1989b) Pigeons as seed predators and dispersers of figs in a Malaysian lowland forest. *Ibis* 131, 521–527.

Langer, P. (1986) Large mammalian herbivores in tropical forests with either hindgut- or forestomach-fermentation. *Zeitschrift für Saugetierkunde* 51, 173–187.

Lee, W.G., Grubb, P.J. and Wilson, J.B. (1991) Patterns of resource allocation in fleshy fruits of nine European tall-shrub species. *Oikos* 61, 307–315.

Leigh, E.G., Jr (1975) Structure and climate in tropical rain forest. *Annual Review of Ecology and Systematics* 6, 67–86.

Leighton, M. and Leighton, D.R. (1984) Vertebrate responses to fruiting seasonality within a Bornean rainforest. In: Sutton, S.L., Whitmore, T.C. and Chadwick, A.C. (eds) *Tropical Rainforests: Ecology and Management*. Blackwell Scientific Publications, Oxford, pp. 181–209.

Levey, D.J. (1986) Methods of seed processing by birds and seed deposition patterns. In: Estrada, A. and Fleming, T.H. (eds) *Frugivores and Seed Dispersal*. Junk, Dordrecht, pp. 147–158.

Levey, D.J. (1987) Seed size and fruit-handling techniques of avian frugivores. *American Naturalist* 129, 471–485.

Levey, D.J. (1988a) Spatial and temporal variation in Costa Rican fruit and fruit-eating bird abundance. *Ecological Monographs* 58, 251–269.

Levey, D.J. (1988b) Tropical wet forest treefall gaps and distributions of understory birds and plants. *Ecology* 69, 1076–1089.

Levey, D.J. (1990) Habitat-dependent fruiting behaviour of an understorey tree, *Miconia centrodesma*, and tropical treefall gaps as keystone habitats for frugivores in Costa Rica. *Journal of Tropical Ecology* 6, 409–420.

Levey, D.J. and Duke, G.E. (1992) How do frugivores process fruit: gastrointestinal transit and glucose absorption in cedar waxwings (*Bombycilla cedrorum*). *Auk* 109, 722–730.

Levey, D.J. and Grajal, A. (1991) Evolutionary implications of fruit-processing limitations in cedar waxwings. *American Naturalist* 138, 171–189.

Levey, D.J. and Stiles, F.G. (1992) Evolutionary precursors of long-distance migration: resource availability and movement patterns in neotropical landbirds. *American Naturalist* 140, 447–476.

Levey, D.J., Moermond, T.C. and Denslow, J.S. (1984) Fruit choice in neotropical birds: the effect of distance between fruits on preference patterns. *Ecology* 65, 844–850.

Lieberman, D. (1982) Seasonality and phenology in a dry tropical forest in Ghana. *Journal of Ecology* 70, 791–806.

Lieberman, D., Lieberman, M. and Martin, C. (1987) Notes on seeds in elephant dung from Bia National Park, Ghana. *Biotropica* 19, 365–369.

Lieberman, M. and Lieberman, D. (1980) The origin of gardening as an extension of infra-human seed dispersal. *Biotropica* 12, 316.

Loiselle, B.A. (1990) Seeds in droppings of tropical fruit-eating birds: importance of considering seed composition. *Oecologia* 82, 494–500.

Loiselle, B.A. and Blake, J.G. (1991) Temporal variation in birds and fruits along an elevational gradient in Costa Rica. *Ecology* 72, 180–193.

Losos, J.B. and Greene, H.W. (1988) Ecological and evolutionary implications of diet in monitor lizards. *Biological Journal of the Linnean Society* 35, 379–407.

McDiarmid, R.W., Ricklefs, R.E. and Foster, M.S. (1977) Dispersal of *Stemmadennia donnell-smithii* (Apocyanaceae) by birds. *Biotropica* 9, 9–25.

McDonnell, M.J. and Stiles, E.W. (1983) The structural complexity of old field vegetation and the recruitment of bird-dispersed plant species. *Oecologia* 56, 109–116.

McKey, D. (1975) The ecology of coevolved seed dispersal systems. In: Gilbert, L.E. and Raven, P.H. (eds) *Coevolution of Animals and Plants*. University of Texas Press, Austin, pp. 159–191.

McKey, D.B., Gartlan, J.S., Waterman, P.G. and Choo, G.M. (1981) Food selection by black colobus monkeys (*Colobus satanas*) in relation to plant chemistry. *Biological Journal of the Linnean Society* 16, 115–146.

McWilliams, S.R. and Karasov, W.H. (1998) Test of a digestion optimization model: effects of costs of feeding on digestive parameters. *Physiological Zoology* 71, 168–178.

Mack, A.L. (1990) Is frugivory limited by secondary compounds in fruits? *Oikos* 57, 135–138.

Mack, A.L. (1993) The sizes of vertebrate-dispersed fruits: a neotropical–paleotropical comparison. *American Naturalist* 142, 840–856.

Magnan, A. (1912) Essai de morphologie stomacal en fonction du régime alimentaire chez les oiseaux. *Annales des Sciences Naturelles, Zoologie, 9e Série* 15, 1–41.

Manasse, R.S. and Howe, H.F. (1983) Competition for dispersal agents among tropical trees: influences of neighbors. *Oecologia* 59, 185–190.

Marks, P.L. (1974) The role of pin cherry (*Prunus pennsylvanica* L.) in the maintenance of stability in northern hardwood ecosystems. *Ecological Monographs* 44, 73–88.

Marks, P.L. and Harcombe, P.A. (1981) Forest vegetation of the Big Thicket, Southeast Texas. *Ecological Monographs* 51, 287–305.

Marshall, A.G. (1983) Bats, flowers and fruit: evolutionary relationships in the Old World. *Biological Journal of the Linnean Society* 20, 115–135.

Marshall, A.G. and McWilliam, A.N. (1982) Ecological observations on Epomorphorinae fruit-bats (Megachiroptera) in West African savannah woodland. *Journal of Zoology, London* 198, 53–67.

Martin, T.E. (1985) Selection of second-growth woodlands by frugivorous migrating birds in Panama: an effect of fruit size and plant density? *Journal of Tropical Ecology* 1, 157–170.

Martin, T.E. and Karr, J.R. (1986) Temporal dynamics of neotropical birds with special reference to frugivores in second-growth woods. *Wilson Bulletin* 98, 38–60.

Martínez del Rio, C. and Karasov, W.H. (1990) Digestion strategies in nectar- and fruit-eating birds and the sugar composition of plant rewards. *American Naturalist* 136, 618–637.

Martínez del Rio, C. and Restrepo, C. (1993) Ecological and behavioral consequences of digestion in frugivorous animals. In: Fleming, T.H. and Estrada, A. (eds) *Frugivory and Seed Dispersal: Ecological and Evolutionary Aspects*. Kluwer Academic Publisher, Dordrecht, The Netherlands, pp. 205–216.

Mazer, S.J. and Wheelwright, N.T. (1993) Fruit size and shape: allometry at different taxonomic levels in bird-dispersed plants. *Evolutionary Ecology* 7, 556–575.

Medway, L. (1972) Phenology of a tropical rain forest in Malaya. *Biological Journal of the Linnean Society* 4, 117–146.

Merz, G. (1981) Recherches sur la biologie de nutrition et les habitats préférés de l'éléphant de forêt, *Loxodonta africana cyclotis* Matschie. *Mammalia* 45, 299–312.

Milewski, A.V. (1982) The occurrence of seeds and fruits taken by ants versus birds in Mediterranean Australia and Southern Africa, in relation to the availability of soil potassium. *Journal of Biogeography* 9, 505–516.

Milewski, A.V. and Bond, W.J. (1982) Convergence in myrmecochory in Mediterranean Australia and South Africa. In: Buckley, R.C. (ed.) *Ant–Plant Interactions in Australia*. Junk, Dordrecht, pp. 89–98.

Milton, K.L. (1981) Food choice and digestive strategies of two sympatric primate species. *American Naturalist* 117, 496–505.

Moermond, T.C. and Denslow, J.S. (1985) Neotropical avian frugivores: patterns of behavior, morphology, and nutrition, with consequences for fruit selection. In: Buckley, P.A., Foster, M.S.,

Morton, E.S., Ridgely, R.S. and Buckley, F.G. (eds) *Neotropical Ornithology*. Ornithological Monographs No. 36, American Ornithologist Union, Washington, pp. 865–897.

Moermond, T.C., Denslow, J.S., Levey, D.J. and Santana, E. (1986) The influence of morphology on fruit choice in neotropical birds. In: Estrada, A. and Fleming, T.H. (eds) *Frugivores and Seed Dispersal*. Junk, Dordrecht, pp. 137–146.

Mooney, H.A., Kummerov, J., Johnson, A.W., Parsons, D.J., Keeley, S.A., Hoffmann, A., Hays, R.I., Gilberto, J. and Chu, C. (1977) The producers – their resources and adaptive responses. In: Mooney, H.A. (ed.) *Convergent Evolution in Chile and California*. Dowden, Hutchinson and Ross, Stroudsburg, Pennsylvania, pp. 85–143.

Morel, G. and Morel, M.Y. (1972) Recherches écologiques sur une savane sahélienne du Ferlo septentrional, Sénégal: l'avifaune et son cycle annuel. *La Terre et la Vie* 26, 410–439.

Morris, J.G. and Rogers, Q.R. (1983) Nutritionally related metabolic adaptations of carnivores and ruminants. In: Margaris, N.S., Arianoutsou-Faraggitaki, M. and Reiter, R.J. (eds) *Plant, Animal and Microbial Adaptations to Terrestrial Environment*. Plenum, New York, pp. 165–180.

Morrison, D.W. (1980) Efficiency of food utilization by fruit bats. *Oecologia* 45, 270–273.

Morton, E.S. (1973) On the evolutionary advantages and disadvantages of fruit eating in tropical birds. *American Naturalist* 107, 8–22.

Murray, K.G. (1988) Avian seed dispersal of three neotropical gap-dependent plants. *Ecological Monographs* 58, 271–298.

Nagy, K.A. and Milton, K. (1979) Aspects of dietary quality, nutrient assimilation and water balance in wild howler monkeys (*Alouatta palliata*). *Oecologia* 39, 249–258.

Noma, N. and Yumoto, T. (1997) Fruiting phenology of animal-dispersed plants in response to winter migration of frugivores in a warm temperate forest on Yakushima Island, Japan. *Ecological Research* 12, 119–129.

Oates, J.F. (1978) Water-plant and soil consumption by guereza monkeys (*Colobus guereza*): a relationship with minerals and toxins in the diet? *Biotropica* 10, 241–253.

Oates, J.F., Waterman, P.G. and Choo, G.M. (1980) Food selection by the South Indian leaf-monkey, *Presbytis johnii*, in relation to food chemistry. *Oecologia* 45, 45–56.

O'Dowd, D.J. and Gill, A.M. (1986) Seed dispersal syndromes in Australian *Acacia*. In: Murray, D.R. (ed.) *Seed Dispersal*. Academic Press, Sydney, Australia, pp. 87–121.

Opler, P.A., Frankie, G.W. and Baker, H.G. (1980) Comparative phenological studies of treelet and shrub species in tropical wet and dry forests in the lowlands of Costa Rica. *Journal of Ecology* 68, 167–188.

Pannell, C.M. and Koziol, M.J. (1987) Ecological and phytochemical diversity of arillate seeds in *Aglaia* (Meliaceae): a study of vertebrate dispersal in tropical trees. *Philosophical Transactions of the Royal Society, London, Series B* 316, 303–333.

Parrish, J.D. (1997) Patterns of frugivory and energetic condition in Nearctic landbirds during autumn migration. *Condor* 99, 681–697.

Piper, J.K. (1986a) Effects of habitat and size of fruit display on removal of *Smilacina stellata* (Liliaceae) fruits. *Canadian Journal of Botany* 64, 1050–1054.

Piper, J.K. (1986b) Seasonality of fruit characters and seed removal by birds. *Oikos* 46, 303–310.

Poupon, H. and Bille, J.C. (1974) Recherches écologiques sur une savane sahélienne du Ferlo septentrional, Sénégal: influence de la sécheresse de l'année 1972–1973 sur la strate ligneuse. *La Terre et la Vie* 28, 49–75.

Pratt, T.K. (1983) Diet of the dwarf cassowary *Casuarius bennetti picticollis* at Wau, Papua New Guinea. *Emu* 82, 283–285.

Pratt, T.K. (1984) Examples of tropical frugivores defending fruit-bearing plants. *Condor* 86, 123–129.

Pratt, T.K. and Stiles, E.W. (1983) How long fruit-eating birds stay in the plants where they feed: implications for seed dispersal. *American Naturalist* 122, 797–805.

Pratt, T.K. and Stiles, E.W. (1985) The influence of fruit size and structure on composition of frugivore assemblages in New Guinea. *Biotropica* 17, 314–321.

Primack, R.B. (1987) Relationships among flowers, fruits, and seeds. *Annual Review of Ecology and Systematics* 18, 409–430.

Pulliainen, E., Helle, P. and Tunkkari, P. (1981) Adaptive radiation of the digestive system, heart and wings of *Turdus pilaris*, *Bombycilla garrulus*, *Sturnus vulgaris*, *Pyrrhula pyrrhula*, *Pinicola enucleator* and *Loxia pyttyopsittacus*. *Ornis Fennica* 58, 21–28.

Putz, F.E. (1979) Aseasonality in malaysian tree phenology. *Malaysian Forester* 42, 1–24.

Reid, N. (1989) Dispersal of mistletoes by honeyeaters and flowerpeckers: components of seed dispersal quality. *Ecology* 70, 137–145.

Reid, N. (1990) Mutualistic interdependence between mistletoes (*Amyema quandang*), and spiny-cheeked honeyeaters and mistletoebirds in an arid woodland. *Australian Journal of Ecology* 15, 175–190.

Restrepo, C. and Gómez, N. (1998) Responses of understory birds to anthropogenic edges in a neotropical montane forest. *Ecological Applications* 8, 170–183.

Rey, P.J. and Gutiérrez, J.E. (1996) Pecking of olives by frugivorous birds: a shift in feeding behaviour to overcome gape limitation. *Journal of Avian Biology* 27, 327–333.

Rey, P.J., Gutiérrez, J.E., Alcántara, J. and Valera, F. (1997) Fruit size in wild olives: implications for avian seed dispersal. *Functional Ecology* 11, 611–618.

Robbins, C.T. (1983) *Wildlife Feeding and Nutrition*, Academic Press, New York.

Rogers, M.E., Maisels, F., Williamson, E.A., Fernández, M. and Tutin, C.E.G. (1990) Gorilla diet in the Lopé Reserve, Gabon: a nutritional analysis. *Oecologia* 84, 326–339.

Sallabanks, R. (1992) Fruit fate, frugivory, and fruit characteristics: a study of the hawthorn, *Crataegus monogyna* (Rosaceae). *Oecologia* 91, 296–304.

Santana, E. and Milligan, B.G. (1984) Behavior of toucanets, bellbirds, and quetzals feeding on lauraceous fruits. *Biotropica* 16, 152–154.

Schlesinger, W.H. (1978) Community structure, dynamics and nutrient cycling in the Okefenokee Cypress swamp forest. *Ecological Monographs* 48, 43–65.

Schupp, E.W. (1988) Seed and early seedling predation in the forest understory and in treefall gaps. *Oikos* 51, 71–78.

Schupp, E.W. (1993) Quantity, quality, and the effectiveness of seed dispersal by animals. In: Fleming, T.H. and Estrada, A. (eds) *Frugivory and Seed Dispersal: Ecological and Evolutionary Aspects*. Kluwer Academic Publishers, Dordrecht, The Netherlands, pp. 15–29.

Schupp, E.W. and Fuentes, M. (1995) Spatial patterns of seed dispersal and the unification of plant population ecology. *Ecoscience* 2, 267–275.

Sherburne, J.A. (1972) Effects of seasonal changes in the abundance and chemistry of the fleshy fruits of northern woody shrubs on patterns of exploitation by frugivorous birds. Unpublished PhD thesis, Cornell University.

Short, J. (1981) Diet and feeding behaviour of the forest elephant. *Mammalia* 45, 177–185.

Sibly, R.M. (1981) Strategies of digestion and defecation. In: Townsend, C.R. and Calow, P. (eds) *Physiological Ecology: an Evolutionary Approach to Resource Use*. Blackwell Scientific Publications, London, pp. 109–138.

Simons, D. and Bairlein, F. (1990) The significance of seasonal frugivory in migratory garden warblers *Sylvia borin*. *Journal für Ornithologie* 131, 381–401.

Smith, A.J. (1975) Invasion and ecesis of bird-disseminated woody plants in a temperate forest sere. *Ecology* 56, 19–34.

Snow, B.K. (1979) The oilbirds of Los Tayos. *Wilson Bulletin* 91, 457–461.

Snow, B.K. and Snow, D.W. (1984) Long-term defence of fruit by mistle thrushes *Turdus viscivorus*. *Ibis* 126, 39–49.

Snow, B.K. and Snow, D.W. (1988) *Birds and Berries*. T. & A.D. Poyser, Calton, UK.

Snow, D.W. (1962a) A field study of the black and white manakin, *Manacus manacus*, in Trinidad. *Zoologica* 47, 65–109.

Snow, D.W. (1962b) A field study of the golden-headed manakin, *Pipra erythrocephala*, in Trinidad, W.I. *Zoologica* 47, 183–198.

Snow, D.W. (1962c) The natural histroy of the oilbird, *Steatornis caripensis*, in Trinidad, W.I. Part 2. Population, breeding ecology and food. *Zoologica* 47, 199–221.

Snow, D.W. (1971) Evolutionary aspects of fruit-eating by birds. *Ibis* 113, 194–202.

Snow, D.W. (1981) Tropical frugivorous birds and their food plants: a world survey. *Biotropica* 13, 1–14.

Sorensen, A.E. (1981) Interactions between birds and fruit in a temperate woodland. *Oecologia* 50, 242–249.

Sorensen, A.E. (1983) Taste aversion and frugivore preference. *Oecologia* 56, 117–120.

Sorensen, A.E. (1984) Nutrition, energy and passage time: experiments with fruit preference in European blackbirds (*Turdus merula*). *Journal of Animal Ecology* 53, 545–557.

Sourd, C. and Gauthier-Hion, A. (1986) Fruit selection by a forest guenon. *Journal of Animal Ecology* 55, 235–244.

Stiles, E.W. (1980) Patterns of fruit presentation and seed dispersal in bird-disseminated woody plants in the eastern deciduous forest. *American Naturalist* 116, 670–688.

Stiles, E.W. and White, D.W. (1986) Seed deposition patterns: influence of season, nutrients, and vegetation structure. In: Estrada, A. and Fleming, T.H. (eds) *Frugivores and Seed Dispersal*. Junk, Dordrecht, pp. 45–54.

Stocker, G.C. and Irvine, A.K. (1983) Seed dispersal by cassowaries (*Casuarius casuarius*) in North Queensland's rainforests. *Biotropica* 15, 170–176.

Stransky, J.J. and Halls, L.K. (1980) Fruiting of woody plants affected by site preparation and prior land use. *Journal of Wildlife Management* 44, 258–263.

Studier, E.H., Szuch, E.J., Thompkins, T.M. and Cope, V.M. (1988) Nutritional budgets in free flying birds: cedar waxwings (*Bombycilla cedrorum*) feeding on Washington hawthorn fruit (*Crataegus phaenopyrum*). *Comparative Biochemistry and Physiology* 89A, 471–474.

Sukumar, R. (1990) Ecology of the Asian elephant in southern India. II. Feeding habits and crop raiding patterns. *Journal of Tropical Ecology* 6, 33–53.

Sun, C. and Moermond, T.C. (1997) Foraging ecology of three sympatric turacos in a montane forest in Rwanda. *Auk* 114, 396–404.

Sun, C., Ives, A.R., Kraeuter, H.J. and Moermond, T.C. (1997) Effectiveness of three turacos as seed dispersers in a tropical montane forest. *Oecologia* 112, 94–103.

Swain, T. (1979) Tannins and lignins. In: Rosenthal, G.A. and Janzen, D.H. (eds) *Herbivores: Their Interaction with Secondary Plant Metabolites.* Academic Press, New York, pp. 657–682.

Tanner, E.V.J. (1982) Species diversity and reproductive mechanisms in Jamaican trees. *Biological Journal of the Linnean Society* 18, 263–278.

Terborgh, J. (1983) *Five New World Primates. A Study in Comparative Ecology,* Princeton University Press, Princeton, New Jersey.

Tester, M., Paton, D., Reid, N. and Lange, R.T. (1987) Seed dispersal by birds and densities of shrubs under trees in arid south Australia. *Transactions of the Royal Society of South Australia* 111, 1–5.

Théry, M. and Larpin, D. (1993) Seed dispersal and vegetation dynamics at a cock-of-the-rock's lek in the tropical forest of French-Guiana. *Journal of Tropical Ecology* 9, 109–116.

Thomas, D.W. (1984) Fruit intake and energy budgets of frugivorous bats. *Physiological Zoology* 57, 457–467.

Thompson, J.N. and Willson, M.F. (1978) Disturbance and the dispersal of fleshy fruits. *Science* 200, 1161–1163.

Traveset, A. (1994) Influence of type of avian frugivory on the fitness of *Pistacia terebinthus. Evolutionary Ecology* 8, 618–627.

Turcek, F.J. (1961) *Okologische Beziehungen der Vögel und Gehölze.* Slowakische Akademie der Wiesenschaften, Bratislava.

Tutin, C.E.G., Williamson, E.A., Rogers, M.E. and Fernandez, M. (1991) A case study of a plant–animal relationship: *Cola lizae* and lowland gorillas in the Lopé Reserve, Gabon. *Journal of Tropical Ecology* 7, 181–199.

Tutin, C.E.G., Parnell, R.J. and White, F. (1996) Protecting seeds from primates: examples from *Diospyros* spp. in the Lopé Reserve, Gabon. *Journal of Tropical Ecology* 12, 371–384.

van Schaik, C.P., Terborgh, J.W. and Wright, S.J. (1993) The phenology of tropical forests: adaptive significance and consequences for primary consumers. *Annual Review of Ecology and Systematics* 24, 353–377.

Viljoen, S. (1983) Feeding habits and comparative feeding rates of three Southern African arboreal squirrels. *South African Journal of Zoology* 18, 378–387.

Voronov, H.R. and Voronov, P.H. (1978) [Morphometric study of the digestive system of the Waxwing (*Bombycilla garrulus* L.) (Aves, Bombycillidae)]. *Vestny k Zoology* 5, 28–31.

Walsberg, G.E. (1975) Digestive adaptations of *Phainopepla nitens* associated with the eating of mistletoe berries. *Condor* 77, 169–174.

Waterman, P.G., Mbi, C.N., McKey, D.B. and Gartlan, J.S. (1980) African rainforest vegetation and rumen microbes: phenolic compounds and nutrients as correlates of digestibility. *Oecologia* 47, 22–33.

Wenny, D.G. and Levey, D.J. (1998) Directed seed dispersal by bellbirds in a tropical cloud forest. *Proceedings of the National Academy of Sciences USA* 95, 6204–6207.

Wetmore, A. (1914) The development of stomach in the euphonias. *Auk* 31, 458–461.

Wheelwright, N.T. (1983) Fruits and the ecology of resplendent quetzals. *Auk* 100, 286–301.

Wheelwright, N.T. (1985) Fruit size, gape width, and the diets of fruit-eating birds. *Ecology* 66, 808–818.

Wheelwright, N.T. (1986) The diet of American robins: an analysis of US Biological Survey records. *Auk* 103, 710–725.

Wheelwright, N.T. (1988) Seasonal changes in food preferences of American robins in captivity. *Auk* 105, 374–378.

Wheelwright, N.T., Haber, W.A., Murray, K.G. and Guindon, C. (1984) Tropical fruit-eating birds and their food plants: a survey of a Costa Rican lower montane forest. *Biotropica* 16, 173–192.

White, D.W. and Stiles, E.W. (1990) Co-occurrences of foods in stomachs and feces of fruit-eating birds. *Condor* 92, 291–303.

White, D.W. and Stiles, E.W. (1991) Fruit harvesting by American robins: influence of fruit size. *Wilson Bulletin* 103, 690–692.

White, S.C. (1974) Ecological aspects of growth and nutrition in tropical fruit-eating birds. Unpublished PhD thesis, University of Pennsylvania.

Whitney, K.D. and Smith, T.B. (1998) Habitat use and resource tracking by African *Ceratogymna* hornbills: implications for seed dispersal and forest conservation. *Animal Conservation* 1, 107–117.

Williamson, E.A., Tutin, C.E.G., Rogers, M.E. and Fernandez, M. (1990) Composition of the diet of
    lowland gorillas at Lopé in Gabon. *American Journal of Primatology* 21, 265–277.
Willson, M.F. (1986) Avian frugivory and seed dispersal in eastern North America. In: Johnston, R.F.
    (ed.) *Current Ornithology*, Vol. 3. Plenum, New York, pp. 223–279.
Willson, M.F. (1988) Spatial heterogeneity of post-dispersal survivorship of Queensland rainforest
    seeds. *Australian Journal of Ecology* 13, 137–145.
Willson, M.F., Porter, E.A. and Condit, R.S. (1982) Avian frugivore activity in relation to forest light
    gaps. *Caribbean Journal of Science* 18, 1–4.
Willson, M.F., Irvine, A.K. and Walsh, N.G. (1989) Vertebrate dispersal syndromes in some
    Australian and New Zealand plant communities, with geographic comparisons. *Biotropica* 21,
    133–147.
Witmer, M.C. (1996) Annual diet of cedar waxwings based on US Biological Survey records
    (1885–1950) compared to diet of American robins: contrasts in dietary patterns and natural his-
    tory. *Auk* 113, 414–430.
Witmer, M.C. (1998a) Ecological and evolutionary implications of energy and protein requeriments
    of avian frugivores eating sugary diets. *Physiological Zoology* 71, 599–610.
Witmer, M.C. (1998b) Do seeds hinder digestive processing of fruit pulp? Implications for plant/fru-
    givore mutualisms. *Auk* 115, 319–326.
Witmer, M.C. and van Soest, P.J. (1998) Contrasting digestive strategies of fruit-eating birds.
    Functional Ecology 12, 728–741.
Wood, C.A. (1924) The Polynesian fruit pigeon, *Globicera pacifica*, its food and digestive apparatus.
    Auk 41, 433–438.
Worthington, A.H. (1982) Population sizes and breeding rhythms of two species of manakins in rela-
    tion to food supply. In: Leigh, E.G., Rand, A.S. and Windsor, D.M. (eds) *The Ecology of a
    Tropical Forest: Seasonal Rhythms and Long-term Changes*. Smithsonian Institution Press,
    Washington, DC, pp. 213–225.
Worthington, A.H. (1989) Adaptations for avian frugivory: assimilation efficiency and gut transit time
    of *Manacus vitellinus* and *Pipra mentalis*. *Oecologia* 80, 381–389.
Wrangham, R.W., Chapman, C.A. and Chapman, L.J. (1994) Seed dispersal by forest chimpanzees in
    Uganda. *Journal of Tropical Ecology* 10, 355–368.
Wrangham, R.W., Conklin-Brittain, N.L. and Hunt, K.D. (1998) Dietary responses of chimpanzees and
    cercopithecines to seasonal variation in fruit abundance. I. Antifeedants. *International Journal
    of Primatology* 19, 949–970.
Zhang, S.Y. and Wang, L.X. (1995) Comparison of three fruit census methods in French Guiana.
    Journal of Tropical Ecology 11, 281–294.

Appendix

Summary statistics (sample size, mean and SE of the mean for each family and variable) of fruit characteristics and pulp constituents of vertebrate-dispersed plants, by families.

Family	Fruit diameter (mm)	Pulp dry mass (g)	Seed dry mass (g)	Relative yield	$kcal\ g^{-1}$ dry mass	$kcal\ fruit^{-1}$	% Water	Lipids	Protein	Carbo-hydrates	Ash
Anacardiaceae (*n* = 12)											
	5	6	5	6	10	5	9	10	10	10	7
	7.6	0.047	0.117	21.25	5.410	0.122	57.12	0.240	0.054	0.638	0.033
	2.3	0.029	0.093	6.90	0.473	0.051	7.69	0.080	0.005	0.090	0.007
Annonaceae (*n* = 11)											
	3	5	3	4	5	3	7	8	9	7	5
	15.1	0.374	0.405	16.28	3.043	1.458	71.67	0.114	0.042	0.636	0.022
	1.8	0.156	0.233	4.28	0.629	1.181	6.53	0.039	0.009	0.093	0.008
Apocynaceae (*n* = 10)											
	2	3	3	3	8	2	7	9	9	9	6
	6.1	0.313	0.147	15.80	4.734	2.026	79.09	0.143	0.047	0.762	0.032
	2.4	0.290	0.099	6.05	0.412	1.904	3.47	0.069	0.014	0.094	0.011
Caprifoliaceae (*n* = 26)											
	16	17	16	17	21	14	25	17	21	17	15
	6.6	0.088	0.127	15.97	4.175	0.426	71.60	0.057	0.060	0.756	0.060
	0.4	0.057	0.104	1.80	0.086	0.284	3.27	0.016	0.010	0.049	0.007
Ericaceae (*n* = 10)											
	8	8	8	8	6	4	10	6	6	6	6
	9.9	0.199	0.026	17.25	4.200	1.275	78.61	0.047	0.034	0.899	0.024
	1.4	0.129	0.007	2.70	0.029	1.091	2.85	0.006	0.002	0.012	0.006
Lauraceae (*n* = 46)											
	36	39	26	39	27	21	41	39	40	28	4
	15.6	0.510	0.680	14.32	4.337	1.956	68.03	0.271	0.061	0.274	0.032
	0.9	0.089	0.134	0.93	0.360	0.396	2.05	0.021	0.007	0.044	0.004
Liliaceae (*n* = 13)											
	11	13	12	13	8	8	13	8	8	8	10
	9.3	0.055	0.091	14.18	4.056	0.243	69.06	0.030	0.046	0.782	0.061
	0.6	0.008	0.022	1.94	0.078	0.049	2.88	0.008	0.006	0.067	0.008
Melastomataceae (*n* = 7)											
	2	3	3	3	6	2	7	4	6	6	3
	4.9	0.035	0.009	22.03	3.407	0.202	75.11	0.044	0.035	0.738	0.057
	0.4	0.027	0.006	8.30	0.386	0.176	4.85	0.016	0.009	0.080	0.012
Meliaceae (*n* = 19)											
	4	7	4	7	15	4	9	17	18	15	8
	12.4	0.237	0.120	20.96	5.627	1.232	53.88	0.305	0.075	0.588	0.032
	2.7	0.052	0.015	4.19	0.346	0.283	7.16	0.059	0.016	0.071	0.008
Moraceae (*n* = 39)											
	14	8	7	7	20	6	18	19	25	18	12
	13.4	0.588	0.286	10.77	3.462	2.997	79.67	0.044	0.055	0.653	0.071
	2.0	0.254	0.177	1.19	0.238	1.378	1.50	0.008	0.007	0.057	0.008
Myrsinaceae (*n* = 4)											
	3	4	3	4	3	3	4	4	4	3	2
	8.9	0.029	0.030	11.98	3.376	0.126	82.45	0.062	0.041	0.629	0.066
	2.4	0.009	0.013	1.39	0.942	0.052	2.94	0.021	0.019	0.165	0.013
Myrtaceae (*n* = 18)											
	8	8	4	8	11	3	14	14	16	12	9
	15.5	0.730	0.477	10.86	3.265	0.805	82.29	0.022	0.040	0.722	0.037
	3.1	0.433	0.313	1.85	0.347	0.374	2.02	0.004	0.003	0.077	0.005

Family	Fruit diameter (mm)	Pulp dry mass (g)	Seed dry mass (g)	Relative yield	kcal g⁻¹ dry mass	kcal fruit⁻¹	% Water	Lipids	Protein	Carbo-hydrates	Ash
Oleaceae (n = 9)											
	7	6	5	6	8	5	6	8	9	7	7
	7.4	0.123	0.072	15.62	4.254	0.207	62.98	0.079	0.046	0.796	0.029
	0.7	0.084	0.038	1.13	0.334	0.094	4.79	0.049	0.005	0.060	0.005
Palmae (n = 17)											
	6	7	6	7	13	3	11	14	14	13	6
	13.7	0.582	1.436	12.34	4.356	5.396	54.30	0.181	0.061	0.592	0.079
	1.2	0.412	1.015	4.35	0.361	3.999	9.00	0.048	0.012	0.069	0.021
Piperaceae (n = 11)											
	1	2	1	2	10	1	11	11	11	10	1
	5.1	0.118	0.170	13.55	2.468	0.964	83.27	0.057	0.074	0.389	0.125
	–	0.103	–	5.75	0.285	–	2.24	0.014	0.007	0.044	–
Rhamnaceae (n = 13)											
	7	7	7	7	10	6	11	10	11	10	7
	8.2	0.110	0.090	16.20	3.785	0.494	66.50	0.014	0.053	0.839	0.051
	1.2	0.084	0.044	3.12	0.120	0.389	3.44	0.004	0.011	0.031	0.013
Rosaceae (n = 47)											
	37	34	31	34	36	26	40	31	38	30	26
	12.3	0.390	0.120	21.83	3.928	1.757	66.78	0.023	0.044	0.787	0.044
	0.9	0.116	0.025	1.75	0.109	0.594	2.08	0.002	0.004	0.045	0.004
Rubiaceae (n = 23)											
	8	15	9	15	10	5	19	10	16	10	7
	7.8	0.019	0.013	11.31	3.875	0.035	81.99	0.047	0.045	0.728	0.043
	1.6	0.006	0.005	2.22	0.171	0.007	2.82	0.016	0.011	0.052	0.010
Rutaceae (n = 6)											
	3	3	2	3	4	2	3	5	4	4	4
	16.5	0.503	0.862	4.29	4.285	2.178	72.50	0.104	0.100	0.650	0.066
	6.8	0.276	0.826	4.47	0.371	1.931	8.88	0.030	0.007	0.043	0.011
Sapotaceae (n = 10)											
	2	4	3	4	7	2	6	9	9	8	7
	16.2	0.477	0.145	21.13	3.761	1.327	74.08	0.073	0.063	0.742	0.045
	4.4	0.228	0.065	3.64	0.309	0.902	3.59	0.016	0.013	0.066	0.011
Smilacaceae (n = 4)											
	3	4	2	4	4	4	4	4	4	4	2
	7.4	0.036	0.051	12.55	4.215	0.153	77.45	0.011	0.050	0.488	0.069
	0.1	0.005	0.012	0.62	0.214	0.026	3.67	0.004	0.006	0.215	0.019
Solanaceae (n = 25)											
	13	13	10	13	19	8	24	21	22	19	7
	11.2	0.099	0.085	10.54	3.019	0.522	81.33	0.044	0.093	0.487	0.056
	0.8	0.021	0.021	1.04	0.345	0.123	1.43	0.022	0.009	0.063	0.012
Tiliaceae (n = 6)											
	0	0	0	0	6	0	0	6	6	6	6
	–	–	–	–	2.945	–	–	0.010	0.064	0.650	0.039
	–	–	–	–	0.249	–	–	0.003	0.012	0.054	0.006
Ulmaceae (n = 5)											
	3	3	2	3	5	3	3	4	5	4	3
	8.9	0.118	0.068	33.67	5.044	0.494	44.87	0.241	0.084	0.380	0.082
	0.5	0.063	0.066	5.02	0.628	0.243	11.20	0.136	0.027	0.183	0.017
Viscaceae (n = 9)											
	6	7	4	6	6	3	7	6	7	5	3
	5.6	0.041	0.010	15.55	4.847	0.161	74.13	0.163	0.084	0.671	0.040
	0.5	0.021	0.003	3.08	0.430	0.049	5.12	0.075	0.023	0.081	0.003

Family	Fruit diameter (mm)	Pulp dry mass (g)	Seed dry mass (g)	Relative yield	kcal g^{-1} dry mass	kcal fruit^{-1}	% Water	Lipids	Protein	Carbo-hydrates	Ash
Vitaceae (*n* = 8)											
	6	5	2	5	5	3	7	5	5	4	2
	9.2	0.071	0.050	13.72	4.528	0.279	81.86	0.138	0.041	0.509	0.016
	0.5	0.022	0.000	2.20	0.286	0.127	3.16	0.060	0.017	0.227	0.010

Only families with more than four species sampled have been included. For each family, numbers above the mean of each variable indicate the number of species with data available for that variable. Figures for pulp constituents are proportions relative to pulp dry mass.

Data from Snow (1962c); Sherburne (1972); White (1974); Crome (1975); McDiarmid *et al.* (1977); Nagy and Milton (1979); Snow (1979); Frost (1980); Morrison (1980); Howe (1981); Howe and Vande Kerckhove (1981); Beehler (1983); Foster and McDiarmid (1983); Jordano (1983); Viljoen (1983); Estrada *et al.* (1984); Wheelwright *et al.* (1984); Johnson *et al.* (1985); Moermond and Denslow (1985); Dinerstein (1986); Piper (1986b); Sourd and Gauthier-Hion (1986); Debussche *et al.* (1987); Herrera (1987); Pannell and Koziol (1987); Atramentowicz (1988); Dowsett-Lemaire (1988); Abrahamson and Abrahamson (1989); Izhaki and Safriel (1989); Worthington (1989); F.H.J. Crome, personal communication; C.M. Herrera and P. Jordano, unpublished data.

n, Number of species sampled per family.

Chapter 7

Seed Predators and Plant Population Dynamics

Michael J. Crawley

Department of Biology, Imperial College of Science, Technology and Medicine,
Ascot, Berkshire, UK

Introduction

The enormous seed production of most plants, coupled with the general paucity of seedlings and saplings, is vivid testimony to the intensity of seed mortality. The degree to which this mortality results from seed predation (the consumption and killing of seeds by granivorous animals) is the subject of the present review (for earlier references, see Crawley, 1992). To people who are unfamiliar with Darwinian thinking, it appears obvious that seed mortality is so high, because 'plants need only to leave one surviving offspring in a lifetime'. In fact, of course, every individual plant is struggling to ensure that its own offspring make up as big a fraction as possible of the plants in the next generation, and for each individual plant there is a huge evolutionary gain to be achieved by leaving more surviving seedlings, i.e. more copies of its genes, in the next generation. The mass mortality of seeds is part of natural selection in action, and the group selectionist argument that plants only need to replace themselves is simply wrong. Since the numbers of seeds produced are so large, it only takes a small percentage change in seed mortality to make a massive difference to the number of seedlings that survive.

Estimates of the magnitude of seed predation are common in agricultural folklore. For example, the traditional Northumbrian recipe for sowing cereal seeds runs like this:

> One for the rook,
> One for the crow,
> One for the pigeon,
> And one to grow.

This affords an estimate of 75% seed predation, which agrees remarkably closely with results from carefully controlled experiments.

It is useful to deal with predispersal and postdispersal seed predators separately. Perhaps the most important reason is that the cast of characters is so different. Most of the species involved in predispersal seed predation are small, sedentary, specialist feeders belonging to the insect orders Diptera, Lepidoptera, Coleoptera and Hymenoptera. In contrast, the postdispersal seed predators tend to be larger, more mobile, generalist herbivores, such as rodents and granivorous birds. Some insects, such as ants, are, of course, important postdispersal seed predators, especially in deserts and nutrient-poor communities (LeCorff, 1996; Espadaler and Gomez, 1997; Predavec, 1997). Other invertebrate herbivores, such as slugs and snails, are important as both pre- and postdispersal predators (Hulme, 1996a).

From the seed's perspective it is also clear that pre- and postdispersal predation are different, if only because the costs of predispersal seed defence can be drawn from the resources of the parent plant, whereas the costs of postdispersal defence must all be borne by the independent seed. This has major implications for the chemistry and resource costs of pre- and postdispersal defence (Janzen, 1969).

Finally, the spatial distributions of the seeds and the implications for predator foraging are quite distinct (Augspurger and Kitajima, 1992). Predispersal seed predators exploit a spatially and temporally aggregated resource and can use searching cues based on a conspicuous parent plant. Postdispersal seed predators usually have no such cues and must search for inconspicuous items scattered or buried, often at low density, against a cryptic background.

Seed predation differs from other kinds of herbivory in several ways (Janzen, 1969; Crawley, 1983). Even the most fervent advocates of the unimportance of competition among herbivores, such as Hairston (1989), have to admit that granivores are different. Indeed, some of the most convincing demonstrations of interspecific competition for resources have come from studies on granivorous desert rodents and ants. Notable among them is the classic series of experiments carried out by Jim Brown and his colleagues (Brown and Heske, 1990). Among the reasons why granivory differs from other kinds of herbivory are: (i) food quality is relatively high, e.g. protein nitrogen levels are higher in seeds than in many other plant tissues; (ii) the food is parcelled into discrete packets, which are often too small to allow the full development of an insect herbivore in a single seed; (iii) the seeds are available on the plant for only a brief period; and (iv) the production of seeds in any one year is typically less predictable than the production of other plant resources, such as new leaves.

Patterns of seed production

Seed production varies with the weather, plant density, the size structure of the plant population, the proportion of plants of flowering size, pollination rates, the level of defoliation and, in masting or alternate-bearing species, with the recent history of seed production. As perceived by seed-feeding animals, therefore, the annual seed crop is a highly variable and unpredictable resource.

Most of the classic long-term studies of annual plant demography have paid little or no attention to herbivory of any kind, let alone to the impact of seed predation (Rees et al., 1996). Little is known about the degree to which seed predators influence the population dynamics of annual plants over the long term, but a number of studies point to important short-term effects (Harmon and Stamp, 1992; Samson et al., 1992; Robinson et al., 1995; Eriksson and Eriksson, 1997). These studies all involve generalist, postdispersal granivores, and it remains to be demonstrated that specialist predispersal seed feeders have important effects on the population dynamics of annual plants. For many annual plant populations, the impact of seed predation is buffered by recruitment from the bank of dormant seeds in the soil (Price and Joyner, 1997) or by the immigration of wind-borne propagules from elsewhere (Aguiar and Sala, 1997). Thus, quite large changes in the seed predation rate may have no measurable impact on plant recruitment, at least in the short term.

Herbaceous perennials often show wide variation in the proportion of individuals producing seeds in any year (Eriksson, 1989) and in the proportion of ramets that flower within a given individual (Maron and Simms, 1997). The size structure of the plant population is important, because fluctuations in the proportion of plants in the reproductive size classes can be substantial (Eriksson and Froborg, 1996; Hurtt and Pacala, 1996; Rees and Paynter, 1997; Tilman, 1997). Thus, the size of the seed crop per plant can vary by several orders of magnitude from one year to the next (Kelly,

1994). In some extreme cases where seed set is pollinator-limited, the production of any seed at all may be the exception, as in the lady's slipper orchids studied by Gill (1989), where in only 1 year out of 10 did more than 5% of the plants produce seed and in 4 years out of 10 there was complete reproductive failure, despite the fact that hand-pollinated individuals set good seed crops every year. Such extreme examples of pollinator limitation are uncommon, and the consensus is that pollination limitation of seed production is the exception rather than the rule (Bertness and Shumway, 1992; Ehrlen and Eriksson, 1995; Brody and Mitchell, 1997; Shibata *et al.*, 1998).

We have rather few long-term records on the fecundity of individual trees, and records are skimpy even for forest seed orchards. Data on commercial fruit production in orchards are of limited value in understanding tree fecundity, because it is not clear how changes in the numbers of seeds are related to changes in the yield of fruit. The few data that are available confirm the view, obtained from shorter-lived plants, that seed production is highly variable from year to year and from individual plant to plant. Thus, while it is clear that there are good years and bad years for seed production, individual trees are consistently more fecund than others, suggesting both genetic and microenvironmental causes (Armstrong and Westoby, 1993; Gautierhion *et al.*, 1993; Crawley and Long, 1995; Hammond, 1995; Chapman and Chapman, 1996; Hulme, 1996b; Maze and Bond, 1996; Wolff, 1996; Cunningham, 1997; Hoshizaki *et al.*, 1997; Clark *et al.*, 1998; Russell and Schupp, 1998; Shibata *et al.*, 1998).

The variation in seed crop is exacerbated in trees like ash, *Fraxinus excelsior* (Tapper, 1996), and beech, *Fagus sylvatica* (Nilsson and Wastljung, 1987), by the propensity of the plants to produce large quantities of empty fruits. Thus, what look superficially to be good seed years turn out on close inspection of the fruits to be years of low seed production. The cause of this failure to fill the seeds may involve low pollination success, but whether or not the

behaviour is adaptive as an anti-predator strategy, e.g. through effects on increasing the handling time, is not yet known.

Predispersal seed predation

A great many studies have documented the impact of specialist seed feeders on plant fecundity (see Table 6.1 in Crawley, 1992). The references were selected to illustrate the range of impacts attributable to different kinds of seed predators in different habitats, and to demonstrate the substantial variation in seed predation rates in the same system from year to year and place to place.

Differential seed predation may reverse the relative production of viable seed in closely related, coexisting plant species, so that the species with the higher potential seed production produces fewer seeds because it suffers a disproportionately high rate of seed predation. For example, of the two *Astragalus* spp. studied by Green and Palmbald (1975), in the more fecund *A. cibarius* (1400 seeds per plant) suffered total seed predation of 93% over 3 years – 74% from bruchids and 19% from chalcid wasps and hemipterans – whereas in the less fecund *A. utahensis* (1100 seeds) suffered an average 60% bruchid predation. This meant that an average individual of *A. cibarius* dispersed only 19 intact seeds compared with 327 for *A. utahensis*. This is just the kind of predator-induced switch in demographic performance that can reduce the likelihood of competitive exclusion and lead to enhanced species richness.

An important problem with many of these studies is the lack of adequate spatial replication and their extremely short-term nature. This means that they provide little more than snapshots of what is clearly a highly dynamic process (Edwards and Crawley, 1999a). In particular, it has only rarely been established that there is a positive relationship between seed production and the recruitment of juvenile plants, which would suggest that predispersal seed predation might have an important role in plant population dynamics (Crawley and Long, 1995; Crawley, 1997). Even less often has the per-seed probability of recruitment been estimated for the same plant species in a variety of habitats over a representative sample of years (Louda and Potvin, 1995).

The nature of the soil in which the seeds are buried is also important, and seed recovery rates by rodents have been shown to depend on the texture of the soil and the particle-size distribution relative to the size of the seeds in question (Vanderwall, 1998).

Postdispersal seed predation

Once the seeds have been scattered to the winds, they fall prey to a different suite of seed predators. Generalist herbivores, such as small mammals, become extremely important at this stage, but larger insects, such as ants, lygaeid bugs and carabid beetles, are significant postdispersal seed predators in many habitats (Augspurger and Kitajima, 1992; Harmon and Stamp, 1992; Samson et al., 1992; Gautierhion et al., 1993; Hammond, 1995; Cardina et al., 1996; Chapman and Chapman, 1996; Bai and Romo, 1997; Forget, 1997; Hulme, 1997; Predavec, 1997; Wurm, 1998). As with predispersal seed predators, the literature on the impact of these seed predators is voluminous. The protocols of some of the studies are rather suspect, because they involve commercially available seeds that are unfamiliar to the resident seed predators. These might be expected to suffer either untypically high rates of predation if the seeds were especially attractive or untypically low rates if their unfamiliarity led to their being avoided. This difficulty aside, it is clear that postdispersal seed predation rates can be very high and that they are extremely variable in space and time (see Table 6.2 in Crawley, 1992). Notice, in particular, how often 100% postdispersal seed losses are recorded.

Seed burial is a major determinant of the predation rate, and loss rates are substantially lower for buried seed than for seed exposed on the surface (Crawley and Long, 1995). For buried seed, the probability of predation increases rapidly with seed size. Once buried, small seeds are relatively secure from small mammal predation, but rodents will dig up larger seeds from considerable depths (Hulme, 1996a).

Seed-bank predation

There is a great scarcity of data on mortality attributable to predators of dormant seeds in the soil. Protecting buried seeds in cages with a variety of mesh sizes should enable data on subterranean seed predation rates to be gathered without much difficulty, but, so far as I know, such experiments have not been carried out.

Given the tremendous longevity of many species' seeds, especially the seeds of ruderal plants (Rees, 1994), we would predict that the protracted exposure to the risk of predation would select for seeds that were relatively unpalatable to generalist seed predators and too small for specialist, arthropod seed feeders to complete their larval development within a single seed. The data appear to conform to these predictions (Leck and Simpson, 1995), but much more work is needed on this component of seed predation (Edwards and Crawley, 1999b).

Seed dormancy provides an extremely powerful buffer against the ravages of seed predation, and the existence of a large bank of dormant seeds may mean that wide fluctuations in predispersal seed predation have little or no impact on plant dynamics (Crawley, 1990). Thus, the old gardening proverb:

> One year's seeding
> Is seven years' weeding

illustrates the degree to which heavy mortality inflicted by seed predators in one year can be compensated by recruitment from the seed bank in the following years. The failure of many apparently successful establishments of biological weed-control agents is often attributed to the existence of a seed bank, so that even seed-feeding insects that destroy over 90% of the annual seed crop may have no long-term impact in reducing the rate of plant recruitment. For example, gorse (*Ulex europaeus*) in New Zealand loses most of its seeds to the introduced weevil *Apion ulicis*, but it remains a serious pest of pastureland because its seeds can survive for a quarter of a century or more in the seed bank (Miller, 1970).

Mast fruiting

Masting is a traditional forester's term that has taken on a rather precise ecological meaning (Crawley and Long, 1995). The core of the idea is simply that synchronous plant reproduction over wide geographical areas leads to satiation of both the specialist and generalist seed predators, with the added bonus that specialist seed predators are kept scarce during the intermast period. The rationale for the evolution of the masting habit is based on five putative selective advantages (Smith *et al.*, 1990):

1. Populations of seed predators are reduced during years of small seed crops and hence unable to exploit a high percentage of the seeds during mast years.
2. Larger mast crops cause seeds to be dispersed over greater distances.
3. The use of weather cues for the timing of mast crops may also provide optimal conditions for reproductive growth.
4. Weather cues may predict optimum future conditions for seed germination and seedling growth.
5. Concentrations of pollen production in mast years increase the probability of pollination for wind-pollinated species.

Taking these conditions together leads to the prediction that masting species should be long-lived, wind-pollinated plants with large, edible seeds that would otherwise suffer high rates of seed predation. While many masting species fit comfortably into this classification, there is a sufficiently large number of exceptions to caution against complacency. It is not at all obvious, for example, why insect-pollinated plants with small seeds and fleshy, animal-

dispersed fruits should be mast producers (Norconk *et al.*, 1998). Also, while predator satiation usually works reasonably well in most mast-fruiting species (Nilsson and Wastljung, 1987; Augspurger and Kitajima, 1992; Crawley and Long, 1995; Tapper, 1996; Wolff, 1996; Hulme, 1997; Kelly and Sullivan, 1997; Shibata *et al.*, 1998), this is not always the case, as for example in cycads (Ballardie and Whelan, 1986).

Alternate bearing is distinct from mast fruiting. Not only is the period shorter, but the interval between seed crops is relatively predictable. This has important evolutionary implications for the life-history evolution of the insects that specialize on the seeds of alternate-bearing and masting plants, e.g. their investment in protracted dormancy. Also, weather cues are substantially more important in masting than in alternate-bearing species (Crawley and Long, 1995; Tapper, 1996). It seems useful, therefore, to treat these two categories of fruiting behaviour separately, defining them on the basis of the duration of the intercrop interval, as follows: (i) alternate bearing – 2 (sometimes 3) years; and (ii) masting – 3 or more years but with the interval unpredictable.

There is certainly plenty of evidence for the effectiveness of predator satiation in heavy seed crops. Both pre- and postdispersal seed predators tend to inflict substantially lower percentage mortality during the mast year than in either the preceding or subsequent years. What is less clear, however, is the relationship between plant recruitment and masting. Rather few studies have shown that successful recruits are significantly more likely to come from a mast crop. It is plausible that they should do so, but largely unproved. See Crawley and Long (1995) for a case, in which recruitment peaked following peak years of acorn production in oak, *Quercus robur*.

Variability in seed mortality

A recurrent theme in the literature on seed mortality is the variability in the risk of death from predation to which each seed is exposed. The probability of death varies with plant density, seed crop size, within-season phenology, pollination rate, spatial location, weather conditions, predator density and the availability of alternative foods for generalist seed predators. This whole suite of probabilities varies from year to year with changes in the weather, and the same change in weather may produce a different change in seed death rate in different habitats. The mortality suffered by dormant seeds in the soil is likely also to vary in response to events that occurred during ripening on the parent plant (e.g. phenotypic variation in seed size and dormancy state), as well as to events occurring after addition to the pool of seeds in the soil (e.g. patterns of wetting and drying, degree of chill experienced and the extent of seed-coat scarification) (Edwards and Crawley, 1999b).

Numerous studies have documented the extent of phenotypic variation in seed predation rates suffered by different individual plants in the same year (Thompson

and Baster, 1992; Ehrlen, 1995; Ollerton and Lack, 1996; Manson *et al.*, 1998; Russell and Schupp, 1998). Rather few studies have investigated whether there is a genetic basis to these differences, and so we have no clear picture of the relative importance of microhabitat heterogeneity and genetic polymorphism in generating the observed patterns of heterogeneity in seed predation. Nevertheless, this variation in the risk of death can be a potent force in promoting coexistence and enhancing plant diversity, because it allows the inferior competitors a refuge from competitive exclusion (Chesson and Warner, 1985; Pacala and Crawley, 1992).

The impact of seed predation on plant population dynamics

The impact of seed predation on plant recruitment is summarized in Fig. 7.1. When seed densities are low, as they are when plant population density is low, or

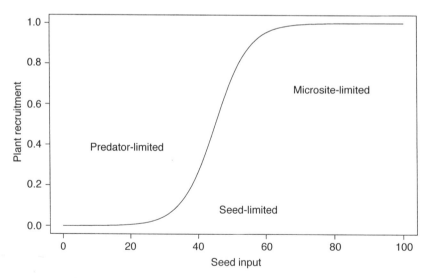

Fig. 7.1. Recruitment limitation in plants. Recruits are defined as juvenile plants that are sufficiently well established for them to have a reasonable chance of surviving to reproductive maturity (as distinct from seedlings, most or all of which die within a few months of germination). There are three distinct circumstances. At a low rate of seed input, seed predators may consume all of the seed input, with the result that recruitment is prevented. This is the region of predator-limited recruitment. At intermediate seed densities, recruitment will be proportional to seed input. This region of seed-limited recruitment is where predispersal, seed-feeding herbivores might be expected to influence plant dynamics. At high rates of seed input, recruitment is limited by microsite availability, and substantial reductions in seed production by predispersal seed predators might have no measurable impact on plant dynamics.

when a site is a long way from patchily distributed parent plants, seed predation will
inevitably reduce plant recruitment. This is
because, at low seed densities, there is no
opportunity for compensatory reductions
in other mortality factors; i.e. compensation is only possible when recruitment is
subject to density-dependent control.

When seed densities are high (Fig.
7.1), as in monocultures or close to parent
plants in species-rich habitats, there may
be intense competition for access to suitable recruitment microsites. Under these
circumstances, large reductions in seed
density resulting from seed predation may
lead to no measurable reduction in the
number of plants that recruit to the juvenile population. Thus, when recruitment is
microsite-limited, seed predation is
unlikely to have any impact on mature
plant density.

Whether or not reduced seedling
recruitment leads to reduced densities of
adult plants depends upon the spacing of
recruitment microsites and on the composite interactions of the adult plants. This
topic is not considered further here, but see
Harper (1977) and Crawley (1990) for discussion.

The first issue, therefore, is to establish
whether plant recruitment is seed-limited.
The traditional view is that annual plants
are seed-limited, because they need to
reproduce each year, and that herbaceous
perennials are not seed-limited, because
seedlings are so rarely found in undisturbed perennial vegetation. Both these
views are oversimplified. For many annual
plants, the evidence suggests that recruitment is not seed-limited, either because the
seed bank is sufficiently large to make good
any temporary shortfall in seed input
caused by seed predation (Peart, 1989;
Eriksson and Ehrlen, 1992; Armstrong and
Westoby, 1993; Bai and Romo, 1997;
Eriksson and Eriksson, 1997; Price and
Joyner, 1997; Tilman, 1997) or because of
microsite limitation (see above). Again, a
more detailed study of the literature on
seedling recruitment in perennial communities shows that seedlings are present in
about 40% of cases (Turnbull *et al.*, 2000).

Of course, the mere presence of seedlings
does not mean that recruitment is seed-limited, but it does indicate that seedling
recruitment is possible.

The simple experiment of sowing extra
seeds and measuring whether there is any
extra plant recruitment has been carried
out all too infrequently. In one recent
study, it was found that, of 20 species of
native grassland plants sown into undisturbed mesic grassland at a rate of 1000
extra seeds m^{-2}, only two species (*Lotus
corniculatus* and *Rumex acetosella*)
showed enhanced recruitment. More
recently, we have discovered that
Arrhenatherum elatius, *Heracleum sphondylium* and *Centaurea nigra* are seed-limited in Nash's Field (Edwards and Crawley,
1999a), but that 60 other species of herbs
and grasses are not seed-limited (M.J.
Crawley, unpublished). In former times,
prior to major human impact, it may well
have been that bigger, more abundant herbivores caused higher rates of disturbance
than are currently observed in grassland
communities, and that seed limitation was
consequently more frequent than it is
today.

The balance of this meagre evidence
suggests that seed-limited recruitment is
the exception rather than the rule, and that
the expectation, therefore, is that seed predation will tend to have rather little impact
on plant recruitment (Impson *et al.*, 1999).
This prediction is borne out by the weed
biocontrol literature, which shows that
seed-feeding agents have a relatively low
rate of success (McFadyen, 1998).

Convincing experimental demonstrations for the role of seed predators on plant
dynamics have tended to come from arid
and semi-arid communities, where the
open vegetation means that microsite limitation is less likely to be important (Louda
and Potvin, 1995). Most of these studies
were relatively short term, and it is possible that the observed dynamics were
merely transient, rather than equilibrium
responses to changes in seed predation
rate. Long-term exclusion experiments in
the south-western deserts of the USA leave
no doubt, however, about the potential

importance of seed-feeding animals in plant community structure (Brown and Heske, 1990).

It is important to contrast the evolutionary impact and the demographic impact of seed predation. If there is genetic variation for traits affecting the rate of seed predation and there is differential seed predation from one plant individual to another, the potential exists for the evolution of seed-defending mechanisms. This evolution can occur whether or not the seed predators have any impact on plant population dynamics.

There is abundant evidence of differential seed predation between individual plants (see above), but the genetic basis of these differences has rarely been established, and it is not possible to disentangle the effects of spatial variations in seed quality and predator density from any genetic component. A review of the impact of predation on seed evolution is beyond the remit of the present chapter, but a useful summary is to be found in Estrada and Fleming (1986).

Compensation

Plants often produce many more flowers than they could ever turn into ripened fruits packed with proteins and carbohydrates. This allows substantial scope for compensating for predispersal seed predation through the differential abortion of damaged fruits prior to seed fill (Crawley, 1997). This kind of compensation requires that fruit production is not pollinator-limited, and compensation for predispersal seed predation is not likely to be important in habitats or in years when pollinators are scarce (Nilsson and Wastljung, 1987; Gill, 1989; Brody and Mitchell, 1997).

There have been rather few carefully controlled, long-term manipulative experiments on compensation for predispersal seed predation (see review in Crawley, 1997). There is tremendous scope for further experimentation using well-replicated pairs of plants, one of which is kept free of seed predators by cage exclosure or with

systemic insecticides and the other is infested by natural densities of seed predators.

What of the potential for postdispersal compensation? For the plant population as a whole, there may be considerable scope for compensation through predator satiation (see Fig. 7.1). But, in terms of individual plant fitness, the options are dramatically reduced as soon as the seed leaves the vicinity of the parent plant. In order to benefit from competitor release, the seedling would need to obtain an advantage from growing among a lower density of conspecific seedlings of the same parentage. At a distance from the parent plant, any given genotype of seed is almost certain to be at the left-hand end of the x axis in Fig. 7.1, where reductions in density lead to reduced probability of recruitment. Thus, what is good for the species (e.g. predator satiation) is not likely to be good for the particular genotypes that are eaten.

Mutualist seed predators

The natural-history literature is rich in examples of animals that increase plant fitness by dispersing their seeds. When dispersal is achieved by seed predators, the plant pays the price of a certain degree of seed mortality for the evident benefits of removal from the parent plant and the escape from competitors and natural enemies that this normally entails. A parallel may be drawn with the mutualism that involves pollinators, as in the classic examples of fig wasps, yucca moths and the dipteran pollinators of globe flowers, where ovules are sacrificed to obtain cross-pollination. See Crawley (1997) for references.

In the case of seed-dispersing predators, we know something about the distances that seeds are moved (jays are recorded as taking the acorns of Q. robur over 20 km and mice can move acorns into Danish heathlands at a rate of about 35 m year^{-1}), but rather little about the mortality that is inflicted. The best information comes from seed caching species, and periodic sampling of rodent caches provides accurate data on seed survival (Vanderwall, 1994). In the case of scatter-hoarding vertebrates, such as jays, this kind of work is more difficult. Thus, while it is reasonable to suppose that seed-dispersing seed predators do indeed increase the fitness of the plants from which they feed, we have no convincing data to demonstrate it.

Case histories

Small mammals in grassland

There is no doubt that small mammals are important postdispersal seed predators (see Table 6.2 in Crawley, 1992). The question is whether or not they have an important impact on plant community structure and dynamics, or whether they simply kill seeds that would have died in any case for other reasons. Some long-term studies have provided unequivocal evidence of the impact of granivorous rodents on desert plant communities (Brown and Heske, 1990), but their role in mesic grasslands is less well known.

Hulme (1996a) carried out a variety of seed addition and rodent exclusion studies in two grasslands in Silwood Park, Berkshire. He investigated seed and seedling predation by *Apodemus sylvaticus* and *Microtus agrestis* on a wide range of native and exotic plant species, whose seeds varied in size between 0.11 mg (*Agrostis capillaris*) and 2.53 mg (*Festuca pratensis*). Small mammals removed an average of 43% of the seeds from the dry grassland site and 37% from the meadow. Native plant species 'familiar' seeds were taken at a significantly higher rate than alien seeds in the dry grassland (49 versus 35%), but both were taken equally in the meadow habitat. Between plant species, predation rates varied from 80% for *Festuca ovina* (not present in either community) and *Dactylis glomerata* (present in both communities) to 8% for *L. corniculatus* and *Plantago lanceolata* (both present in both communities). As in other studies of small-mammal predation, there was

tremendous spatial variation in predation rates.

Experimental seed burial reduced predation rates in both sites and at all times of year; in the dry grassland burial reduced seed losses by over 98% and in the meadow by between 40 and 90%. The larger the buried seed, the more likely it was to be eaten by small mammals. *Festuca pratensis, Lolium perenne* and *Trifolium pratense* suffered the highest rates of post-burial predation and all had relatively heavy seeds – approximately 2 mg.

Small-mammal exclosure increased seedling numbers, but increased seedling densities mapped through to increased adult plant densities in only two cases (*Arrhenatherum elatius* and *Centaurea nigra*) (Edwards and Crawley, 1999c). In most cases, seedling densities were typically low, so there appears to be little scope for compensatory reductions in other mortality factors, and it is likely that small-mammal predation does, indeed, reduce the density, at least of large-seeded plant species, in these Silwood grasslands. It is interesting that none of the dominant species in these sites suffered significant seedling density reductions under small-mammal predation (e.g. *Holcus lanatus, Festuca rubra* and *P. lanceolata*), whereas the rare and absent species suffered heavy predation on both seeds and seedlings (Edwards and Crawley, 1999c).

Acorns of *Quercus robur*

The seed production and seed mortality of pedunculate oak, *Quercus robur*, has been studied at Silwood Park for 20 years. The pattern of acorn production is alternate-bearing, with high and low crops alternating in strict sequence, with only a single phase shift during this time, i.e. there were two successive low acorn crops in 1983 and 1984 (Crawley and Long, 1995). A severe air frost in late April can completely defoliate early-flushing trees and kill their entire crop of female flowers for the year. The regrowth shoots that are produced in May and June do not bear any flowers of either sex. In a given year, it is unlikely that the total seed crop would be destroyed by spring frosts, because of the wide phenological span of the oak. Some of the latest-leafing trees will always tend to escape unscathed, because frosts do not damage flowers that are still protected within the bud.

The flowers that survive frost defoliation may be either wind- or insect-pollinated. Hand pollination, using pollen from a variety of male parent trees and adding the pollen on different occasions and to different flowers of different ages, had no effect on final seed production, so there was no evidence of pollen limitation. Between pollination in May and acorn fill in August, a great many of the female flowers are aborted. Whole peduncles may wilt and fall off, and many of the flowers on each peduncle may fail to develop. Thus out of the maximum four peduncles per shoot and six flowers per peduncle (a potential of 24 acorns per shoot), perhaps only one acorn is filled. The modal number of acorns per shoot is zero, even in years of high acorn production. The rate of peduncle abortion varies from tree to tree in a consistent pattern across years, and certain individual trees always abort a higher proportion of their female flowers than other trees in the same microhabitat (Crawley and Long, 1995). The cycle of alternate bearing is therefore a cycle of differential peduncle abortion, rather than a 2-year cycle in flower production.

The young acorns are prey to two important predispersal seed predators: a native weevil, *Curculio glandium*, and an alien cynipid gall-wasp, *Andricus quercus-calicis*. These two insects exhibit strongly asymmetric interspecific competition, because the gall-wasp attacks the female flower early in the summer before the acorn has begun to fill. The 'knopper' gall, which develops inside the female flower, competes for resources with the developing acorn and usually outgrows it, so that, by the time the female *Curculio* are laying their eggs in late July, the galled acorns are completely destroyed. This pre-emptive exploitation competition means that, in years when acorns are scarce, the weevils

are faced with greatly reduced resource availability as a result of gall formation. The numbers of both insects appear to be resource-limited, so, in years when acorns are abundant, both weevils and gall-formers inflict much lower rates of acorn loss. Despite this, their population densities increase because they are taking a smaller proportion of a vastly greater total crop. In the next year of the alternate-bearing cycle, there are high insect numbers and a low acorn crop, leading to high percentage attack by the gall-former and intense competition between the weevils for access to sound acorns. And so the cycle goes on.

Insects are not the only predispersal seed predators of oak. A guild of vertebrates – again, curiously, one native and one alien species – takes the ripe acorns from the tree in late summer; the jay, *Garrulus garrulus glandarius*, and the grey squirrel, *Sciurus carolinensis*. The jay is interesting because it is believed to have a positive impact on oak fitness, through the long-distance dispersal of acorns (see above). The impact of grey squirrels is strongly negative. Apart from ring barking and killing mature trees, the squirrels kill large quantities of acorns. Their acorn storage behaviour is of no benefit to the oak, because they nip the embryo out of the seed before caching in order to prevent germination. It is notable that the squirrels do not nip out the embryo of red oak acorns, such as *Quercus borealis;* this is a spring-germinating species, whose acorns are readily stored in caches through the winter. Estimates of acorn loss from the canopy for the combined guild of vertebrate acorn feeders range between less than 10% in high seed crops and more than 70% in low years. As with insect attack rates, these predispersal losses differ enormously from tree to tree.

Once the surviving acorns have fallen to the ground, they are fed upon by many of the large vertebrate herbivores (rabbits, pigs, deer, etc.). In low acorn years, these animals (especially rabbits in Silwood Park) can inflict almost 100% mortality on the seeds. Acorns placed in the field in experimental piles of different sizes and at different distances from cover are almost always removed during the first night. For example, out of ten piles of 1000 acorns, only two piles had any acorns left after 24 h and, in both these piles, the few acorns left behind were all weevily (M.J. Crawley, unpublished). In high acorn years, the vertebrate herbivores appear to be satiated, and large numbers of sound acorns survive through the winter to produce their first seedling shoots in the following spring.

In summary, the total rate of acorn predation varies in a roughly 2-year cycle, with close to 100% seed loss in the low years and about 50% seed predation in the years of high acorn abundance. Recruitment is proportional to acorn production in some habitats, but not in all (Crawley and Long, 1995). In many cases, there is no seedling recruitment, except after the very largest peaks of acorn production, e.g. after the massive crop of autumn of 1995 (M.J. Crawley, unpublished). It must be stressed, however, that these averages hide large, and potentially important, differences in average predation rates from tree to tree and from location to location.

Seed predators of ragwort, *Senecio jacobaea*

Ragwort, *Senecio jacobaea*, is toxic to most vertebrate herbivores and is a serious weed in semi-arid grazing land in both northern and southern hemispheres (Cameron, 1935; Dempster, 1971). The plant becomes progressively more competitive as the grasses are eaten back and gaps in the sward are opened up in which seedling recruitment can occur. The two major predispersal seed predators are the cinnabar moth, *Tyria jacobaeae*, and the seed-head fly, *Pegohylemyia seneciella*, both of which have been established as potential weed biocontrol agents. As with the gall-wasp and the weevil on oaks, their interaction is highly asymmetric, because cinnabar moth feeds on the growing ragwort shoots and the colonially feeding larvae can strip a plant of its flower-heads (capitula) before the seed-head fly has completed its development. The density of the seed-head

fly is significantly increased in areas that are kept free of cinnabar moth experimentally, but there is no reciprocal effect of the seed-head fly on the cinnabar moth (Crawley and Pattrasudhi, 1988).

Flowering in ragwort varies greatly from year to year. This is due partly to the weather, partly to cinnabar moth defoliation and partly to the recent history of flowering. Autumn and spring rains promote flowering in the following summer by ensuring that rosette growth is rapid, so that a high proportion of rosettes are above the threshold size for flowering. Cinnabar moths affect flowering directly by eating flowers and flower-buds and indirectly by reducing rosette size, although their impact on rosettes is great only in years when they have stripped the entire crop of flowers and buds. The history of flowering is important, because plants that have produced a large seed crop die back to the base. Many of these plants die, and those that survive have to grow back from the small basal shoots. Since it is unusual for rosettes to grow large enough to flower in their first year, a mass flowering is almost always followed by a more or less protracted run of years in which flowering is sparse. At first, this is due to mass post-reproductive dieback of the larger reproductive plants. Later on, cinnabar moth defoliation can lead to 100% seed losses for several successive years (Dempster, 1971; Crawley and Gillman, 1989).

Cinnabar moth feeding tends to reduce ragwort death rates, because defoliation is less of a drain on resources than seed set. For a plant of a given size, there is a clear negative relationship between the probability of surviving the winter and the size of the seed crop produced the previous summer (Gillman and Crawley, 1990). In mass-flowering years, such as 1985 and 1990, the cinnabar moth population is satiated and there is mass seed production, wide dispersal and, presumably, replenishment of the seed bank. These years, however, are not inevitably associated with peaks of ragwort seedling recruitment, and it appears that in many habitats ragwort recruitment is not seed-limited (Crawley and Gillman, 1989).

Rather little is known about postdispersal seed predation of ragwort. Crawley and Nachapong (1985) sowed ragwort seed into ragwort-infested grasslands without observing any increase in seedling recruitment, so it appears that recruitment was not seed-limited. However, recruitment is more likely to be seed-limited in open, sandy habitats, such as coastal dunes or breckland.

Discussion

A cynic might argue that the only pattern to emerge from this review is that seed predation rates tend to lie somewhere between 0 and 100%. It is true that the search for simple patterns to describe the causes and consequences of variation in seed predation rates has been relatively unsuccessful. Neither seed density (Connell, 1971) nor distance from parent plants (Janzen, 1970) has proved to have the degree of explanatory power once hoped for (Augspurger and Kitajima, 1992; Donaldson, 1993; Gautierhion et al., 1993; Hammond, 1995; Hurtt and Pacala, 1996; Cintra, 1997; Cunningham, 1997; Pacala and Rees, 1998). On the other hand, it has become plain that variation in the rate of seed predation is the norm. Thus, we find pronounced variation from individual to individual, with some plants exhibiting consistently high rates of seed predation and others appearing to be more or less immune. There is pronounced spatial variation in predation rates, although the amount of this variation that can be explained by differences in seed density or by isolation from seed parents may be rather low in most cases. Year-to-year differences in seed production may run to several orders of magnitude. This, in turn, leads to predator satiation in the high seed years, and is the mechanism that underlies the typical pattern of negative density dependence found in time-series data on seed predation. This variation in the risk of predation can have important effects in promoting species richness by reducing the probability of competitive exclusion

(Chesson and Warner, 1985; Pacala and Crawley, 1992; and see below).

Even taking full account of the statistical shortcomings of the data, there does appear to be an interesting difference between the patterns of predispersal and postdispersal predation. While the mean rates of reported seed losses are not markedly different (the mean is a little higher in studies of postdispersal predation, at 50%, than in predispersal studies, at 45%), the shapes of the distributions of loss rates are characteristically different. For postdispersal predation, the distribution is distinctly U-shaped, with strong classes at both the 0 and 100% ends of the spectrum. The predispersal predation data show no such tendency, and their strongest classes are in the middle range, with relatively few reports of 100 or 0% losses. These figures have not changed significantly since the earlier review (Crawley, 1992).

This is consistent with what we know about the natural history of seed predation and the degree of variation in loss rates from year to year and from plant to plant. The predispersal predators tend to be invertebrate herbivores with narrow host ranges, whose numbers are rather closely coupled to seed production. In a number of cases, their populations can be reasonably well predicted from knowledge of seed production and percentage seed destruction in the previous year. Because of these time-lagged numerical responses to changes in seed density, their impact on seed production tends to fluctuate in an inverse density-dependent manner, with high proportions of small seed crops destroyed, and low proportions of large crops. Many of these specialist predispersal predators also suffer relatively high rates of attack by insect parasitoids, which may explain why the range of fluctuation in seed attack rates is relatively small compared with the more generalist, postdispersal predators. This might also help to explain why alien predispersal seed predators, having been freed from their natural enemies, tend to inflict a wider range of seed mortalities than their native counterparts, e.g. the knopper gall insect on oaks in England (Hails and Crawley, 1991).

The consequences of these patterns for plant population dynamics are potentially great. It is now becoming clear, from an expanding body of theoretical work, that spatial and temporal variation in death rates can have important consequences for plant coexistence (Chesson and Warner, 1985). Variation in seed predation rates of the sort reported here is more than sufficient to promote species diversity in model communities. All that is required is that there is a sufficiently large number of places within a spatially heterogeneous environment where the rates of seed predation suffered by the inferior competitor happen, by chance alone, to be low at the same time as those suffered by the superior competitor happen to be high. This stochastic variation in predation rate can prevent the inherent competitive superiority of the dominant species from being expressed to the full, and hence can permit coexistence (Chesson and Warner, 1985; Pacala and Crawley, 1992). Notice that this mechanism does not require the existence of any clear-cut deterministic processes affecting variation in seed mortality, e.g. density dependence (Connell, 1971) or frequency dependence (Janzen, 1970), although, of course, frequency dependence in seed predation would be a powerful additional force in promoting plant species coexistence. All that is required is that differences exist in the probabilities of death in different places and that these probabilities have a sufficiently low covariance for a refuge from competition to be created (Pacala 1997). Given this view of the world, the variation in predation rates becomes the focus of interest. Instead of being a nuisance that tends to blur the estimation of crisp, deterministic parameter values, the variation becomes an important mechanism of population dynamics in its own right.

References

Aguiar, M.R. and Sala, O.E. (1997) Seed distribution constrains the dynamics of the Patagonian steppe. *Ecology* 78, 93–100.

Armstrong, D.P. and Westoby, M. (1993) Seedlings from large seeds tolerate defoliation better: a test using phylogenetically independent contrasts. *Ecology* 74, 1092–1100.

Augspurger, C.K. and Kitajima, K. (1992) Experimental studies of seedling recruitment from contrasting seed distributions. *Ecology* 73, 1270–1284.

Bai, Y.G. and Romo, J.T. (1997) Seed production, seed rain, and the seedbank of fringed sagebrush. *Journal of Range Management* 50, 151–155.

Ballardie, R.T. and Whelan, R.J. (1986) Masting, seed dispersal and seed predation in the cycad *Macrozamia communis. Oecologia* 70, 100–105.

Bertness, M.D. and Shumway, S.W. (1992) Consumer driven pollen limitation of seed production in marsh grasses. *American Journal of Botany* 79, 288–293.

Brody, A.K. and Mitchell, R.J. (1997) Effects of experimental manipulation of inflorescence size on pollination and pre-dispersal seed predation in the hummingbird- pollinated plant *Ipomopsis aggregata. Oecologia* 110, 86–93.

Brown, J.H. and Heske, E.J. (1990) Control of a desert–grassland transition by a keystone rodent guild. *Science* 250, 1705–1707.

Cameron, E. (1935) A study of the natural control of ragwort (*Senecio jacobaea* L.). *Journal of Ecology* 23, 265–322.

Cardina, J., Norquay, H.M., Stinner, B.R. and McCartney, D.A. (1996) Postdispersal predation of velvetleaf (*Abutilon theophrasti*) seeds. *Weed Science* 44, 534–539.

Chapman, C.A. and Chapman, L.J. (1996) Frugivory and the fate of dispersed and non-dispersed seeds of six African tree species. *Journal of Tropical Ecology* 12, 491–504.

Chesson, P.L. and Warner, R.R. (1985) Coexistence mediated by recruitment fluctuations: a field guide to the storage effect. *American Naturalist* 125, 769–787.

Cintra, R. (1997) A test of the Janzen–Connell model with two common tree species in Amazonian forest. *Journal of Tropical Ecology* 13, 641–658.

Clark, J.S., Macklin, E. and Wood, L. (1998) Stages and spatial scales of recruitment limitation in southern Appalachian forests. *Ecological Monographs* 68, 213–235.

Connell, J.H. (1971) On the role of natural enemies in preventing competitive exclusion in some marine animals and in rain forest trees. In: den Boer, P.J. and Gradwell, G. (eds) *Dynamics of Populations*. Pudoc, Wageningen, pp. 298–312.

Crawley, M.J. (1983) *Herbivory: the Dynamics of Animal–Plant Interactions*. Blackwell Scientific Publications, Oxford.

Crawley, M.J. (1990) The population dynamics of plants. *Philosophical Transactions of the Royal Society of London Series B – Biological Sciences* 330, 125–140.

Crawley, M.J. (1992) Seed predators and plant population dynamics. In: Fenner, M. (ed.) *Seeds: The Ecology of Regeneration in Plant Communities*. CAB International, Wallingford, UK, pp. 157–191.

Crawley, M.J. (1997) Plant–herbivore dynamics. In: Crawley, M.J. (ed.) *Plant Ecology*. Blackwell Scientific Publications, Oxford, pp. 401–474.

Crawley, M.J. and Gillman, M.P. (1989) Population dynamics of cinnabar moth and ragwort in grassland. *Journal of Animal Ecology* 58, 1035–1050.

Crawley, M.J. and Long, C.R. (1995) Alternate bearing, predator satiation and seedling recruitment in *Quercus robur* L. *Journal of Ecology* 83, 683–696.

Crawley, M.J. and Nachapong, M. (1985) The establishment of seedlings from primary and regrowth seeds of ragwort (*Senecio jacobaea*). *Journal of Ecology* 73, 255–261.

Crawley, M.J. and Pattrasudhi, R. (1988) Interspecific competition between insect herbivores – asymmetric competition between cinnabar moth and the ragwort seed-head fly. *Ecological Entomology* 13, 243–249.

Cunningham, S.A. (1997) Predator control of seed production by a rain forest understory palm. *Oikos* 79, 282–290.

Dempster, J.P. (1971) The population ecology of the cinnabar moth, *Tyria jacobaeae* L. (Lepidoptra: Arctiidae). *Oecologia* 7, 26–67.

Donaldson, J.S. (1993) Mast-seeding in the cycad genus *Encephalartos* – a test of the predator satiation hypothesis. *Oecologia* 94, 262–271.

Edwards, G.R. and Crawley, M.J. (1999a) Effects of disturbance and rabbit grazing on seedling recruitment of six mesic grassland species. *Seed Science Research* 9, 145–156.

Edwards, G.R. and Crawley, M.J. (1999b) Herbivores, seed banks and seedling recruitment in mesic grassland. *Journal of Ecology* 87, 423–435.

Edwards, G.R. and Crawley, M.J. (1999c) Rodent seed predation and seedling recruitment in mesic grassland. *Oecologia* 118, 288–296.

Ehrlen, J. (1995) Demography of the perennial herb *Lathyrus vernus*. 2. Herbivory and population dynamics. *Journal of Ecology* 83, 297–308.

Ehrlen, J. and Eriksson, O. (1995) Pollen limitation and population-growth in a herbaceous perennial legume. *Ecology* 76, 652–656.

Eriksson, A. and Eriksson, O. (1997) Seedling recruitment in semi-natural pastures: the effects of disturbance, seed size, phenology and seed bank. *Nordic Journal of Botany* 17, 469–482.

Eriksson, O. (1989) Seedling dynamics and life histories in clonal plants. *Oikos* 55, 231–238.

Eriksson, O. and Ehrlen, J. (1992) Seed and microsite limitation of recruitment in plant populations. *Oecologia* 91, 360–364.

Eriksson, O. and Froborg, H. (1996) 'Windows of opportunity' for recruitment in long-lived clonal plants: experimental studies of seedling establishment in *Vaccinium* shrubs. *Canadian Journal of Botany* 74, 1369–1374.

Espadaler, X. and Gomez, C. (1997) Soil surface searching and transport of *Euphorbia characias* seeds by ants. *Acta Oecologica – International Journal of Ecology* 18, 39–46.

Estrada, A. and Fleming, T.H. (1986) *Frugivores and Seed Dispersal*. Junk, Dordrecht.

Forget, P.M. (1997) Effect of microhabitat on seed fate and seedling performance in two rodent-dispersed tree species in rain forest in French Guiana. *Journal of Ecology* 85, 693–703.

Gautierhion, A., Gautier, J.P. and Maisels, F. (1993) Seed dispersal versus seed predation – an intersite comparison of two related African monkeys. *Vegetatio* 108, 237–244.

Gill, D.E. (1989) Fruiting failure, pollinator inefficiency, and speciation in orchids. In: Otte, D. and Endler, J.A. (eds) *Speciation and its Consequences*. Sinauer, Sunderland, Massachusetts, pp. 458–481.

Gillman, M.P. and Crawley, M.J. (1990) The cost of sexual reproduction in ragwort (*Senecio jacobaea* L). *Functional Ecology* 4, 585–589.

Green, T.W. and Palmbald, I.G. (1975) Effects of insect seed predators on *Astragalus cibarius* and *Astragalus utahensis* (Leguminosae). *Ecology* 56, 1435–1440.

Hails, R.S. and Crawley, M.J. (1991) The population-dynamics of an alien insect – *Andricus quercuscalicis* (Hymenoptera, Cynipidae). *Journal of Animal Ecology* 60, 545–562.

Hairston, N.G. (1989) *Ecological Experiments: Purpose, Design and Execution*. Cambridge University Press, Cambridge.

Hammond, D.S. (1995) Postdispersal seed and seedling mortality of tropical dry forest trees after shifting agriculture, Chiapas, Mexico. *Journal of Tropical Ecology* 11, 295–313.

Harmon, G.D. and Stamp, N.E. (1992) Effects of postdispersal seed predation on spatial inequality and size variability in an annual plant, *Erodium cicutarium* (Geraniaceae). *American Journal of Botany* 79, 300–305.

Harper, J.L. (1977) *The Population Biology of Plants*. Academic Press, London.

Hoshizaki, K., Suzuki, W. and Sasaki, S. (1997) Impacts of secondary seed dispersal and herbivory on seedling survival in *Aesculus turbinata*. *Journal of Vegetation Science* 8, 735–742.

Hulme, P.E. (1996a) Herbivory, plant regeneration, and species coexistence. *Journal of Ecology* 84, 609–615.

Hulme, P.E. (1996b) Natural regeneration of yew (*Taxus baccata* L): microsite, seed or herbivore limitation? *Journal of Ecology* 84, 853–861.

Hulme, P.E. (1997) Post-dispersal seed predation and the establishment of vertebrate dispersed plants in Mediterranean scrublands. *Oecologia* 111, 91–98.

Hurtt, G.C. and Pacala, S.W. (1996) The consequences of recruitment limitation: reconciling chance, history and competitive differences between plants. *Journal of Theoretical Biology* 176, 1–12.

Impson, F.A.C., Moran, V.C. and Hoffmann, J.H. (1999) A review of the effectiveness of seed-feeding bruchid beetles in the biological control of mesquite *Prosopis* species (Fabaceae) in South Africa. *African Entomology* 8, 81–88.

Janzen, D.H. (1969) Seed eaters versus seed size, number, toxicity and dispersal. *Evolution* 23, 1–27.

Janzen, D.H. (1970) Herbivores and the number of tree species in tropical forests. *American Naturalist* 104, 501–528.

Kelly, D. (1994) The evolutionary ecology of mast seeding. *Trends in Ecology and Evolution* 9, 465–470.

Kelly, D. and Sullivan, J.J. (1997) Quantifying the benefits of mast seeding on predator satiation and wind pollination in *Chionochloa pallens* (Poaceae). *Oikos* 78, 143–150.

Leck, M.A. and Simpson, R.L. (1995) Ten-year seed bank and vegetation dynamics of a tidal freshwater marsh. *American Journal of Botany* 82, 1547–1557.

LeCorff, J. (1996) Establishment of chasmogamous and cleistogamous seedlings of an ant-dispersed understory herb, *Calathea micans* (Marantaceae). *American Journal of Botany* 83, 155–161.

Louda, S.M. and Potvin, M.A. (1995) Effect of inflorescence-feeding insects on the demography and lifetime fitness of a native plant. *Ecology* 76, 229–245.

McFadyen, R.E.C. (1998) Biological control of weeds. *Annual Review of Entomology* 43, 369–393.

Manson, R.H., Ostfeld, R.S. and Canham, C.D. (1998) The effects of tree seed and seedling density on predation rates by rodents in old fields. *Ecoscience* 5, 183–190.

Maron, J.L. and Simms, E.L. (1997) Effect of seed predation on seed bank size and seedling recruitment of bush lupine (*Lupinus arboreus*). *Oecologia* 111, 76–83.

Maze, K.E. and Bond, W.J. (1996) Are *Protea* populations seed limited? Implications for wildflower harvesting in Cape fynbos. *Australian Journal of Ecology* 21, 96–105.

Miller, D. (1970) *Biological Control of Weeds in New Zealand 1927–48*. New Zealand Department of Scientific and Industrial Research Information Series, Wellington.

Nilsson, S.G. and Wastljung, U. (1987) Seed predation and cross pollination in mast seeding beech (*Fagus sylvatica*) patches. *Ecology* 68, 260–265.

Norconk, M.A., Grafton, B.W. and Conklin-Brittain, N.L. (1998) Seed dispersal by neotropical seed predators. *American Journal of Primatology* 45, 103–126.

Ollerton, J. and Lack, A. (1996) Partial predispersal seed predation in *Lotus corniculatus* L (Fabaceae). *Seed Science Research* 6, 65–69.

Pacala, S.W. (1997) Dynamics of plant communities. In: Crawley, M.J. (ed.) *Plant Ecology*. Blackwell Science, Oxford, pp. 532–555.

Pacala, S.W. and Crawley, M.J. (1992) Herbivores and plant diversity. *American Naturalist* 140, 243–260.

Pacala, S.W. and Rees, M. (1998) Models suggesting field experiments to test two hypotheses explaining successional diversity. *American Naturalist* 152, 729–737.

Peart, D.R. (1989) Species interactions in a successional grassland, 1. Seed rain and seedling recruitment. *Journal of Ecology* 77, 236–251.

Predavec, M. (1997) Seed removal by rodents, ants and birds in the Simpson Desert, central Australia. *Journal of Arid Environments* 36, 327–332.

Price, M.V. and Joyner, J.W. (1997) What resources are available to desert granivores: seed rain or soil seed bank? *Ecology* 78, 764–773.

Rees, M. (1994) Delayed germination of seeds: a look at the effects of adult longevity, the timing of reproduction, and population age/stage structure. *American Naturalist* 144, 43–64.

Rees, M. and Paynter, Q. (1997) Biological control of Scotch broom: modelling the determinants of abundance and the potential impact of introduced insect herbivores. *Journal of Applied Ecology* 34, 1203–1221.

Rees, M., Grubb, P.J. and Kelly, D. (1996) Quantifying the impact of competition and spatial hetero-geneity on the structure and dynamics of a four-species guild of winter annuals. *American Naturalist* 147, 1–32.

Robinson, G.R., Quinn, J.F. and Stanton, M.L. (1995) Invasibility of experimental habitat islands in a California winter annual grassland. *Ecology* 76, 786–794.

Russell, S.K. and Schupp, E.W. (1998) Effects of microhabitat patchiness on patterns of seed disper-sal and seed predation of *Cercocarpus ledifolius* (Rosaceae). *Oikos* 81, 434–443.

Samson, D.A., Philippi, T.E. and Davidson, D.W. (1992) Granivory and competition as determinants of annual plant diversity in the Chihuahuan Desert. *Oikos* 65, 61–80.

Shibata, M., Tanaka, H. and Nakashizuka, T. (1998) Causes and consequences of mast seed produc-tion of four co-occurring *Carpinus* species in Japan. *Ecology* 79, 54–64.

Smith, C.C., Hamrick, J.L. and Kramer, C.L. (1990) The advantage of mast years for wind pollination. *American Naturalist* 136, 154–166.

Tapper, P.G. (1996) Long-term patterns of mast fruiting in *Fraxinus excelsior*. *Ecology* 77, 2567–2572.

Thompson, K. and Baster, K. (1992) Establishment from seed of selected Umbelliferae in unmanaged grassland. *Functional Ecology* 6, 346–352.

Tilman, D. (1997) Community invasibility, recruitment limitation, and grassland biodiversity. *Ecology* 78, 81–92.

Turnbull, L.A., Crawley, M.J. and Rees, M. (2000) Are plant populations seed-limited? A review of seed sowing experiments. *Oikos* 88, 225–238.

Vanderwall, S.B. (1994) Seed fate pathways of antelope bitterbrush – dispersal by seed-caching yel-low pine chipmunks. *Ecology* 75, 1911–1926.

Vanderwall, S.B. (1998) Foraging success of granivorous rodents: effects of variation in seed and soil water on olfaction. *Ecology* 79, 233–241.

Wolff, J.O. (1996) Population fluctuations of mast-eating rodents are correlated with production of acorns. *Journal of Mammalogy* 77, 850–856.

Wurm, P.A.S. (1998) A surplus of seeds: high rates of post-dispersal seed predation in a flooded grassland in monsoonal Australia. *Australian Journal of Ecology* 23, 385–392.

Chapter 8
Dormancy, Viability and Longevity

Alistair J. Murdoch and Richard H. Ellis
Department of Agriculture, University of Reading, Reading, Berkshire, UK

Introduction

The regeneration of plant communities from seed depends on seeds being in the right physiological state to germinate in the right place at the right time. In some species, this requirement is satisfied by a regeneration strategy in which seeds germinate as soon as they are shed. In others, seeds may survive for long periods in the soil seed bank, with intermittent germination of a part of the population.

There are two basic physiological prerequisites for seeds to survive in soil: germination must be avoided by dormancy or quiescence, while viability must be maintained. Moreover, for such seeds to contribute to regeneration, this dormancy or quiescence must be relieved and germination promoted, perhaps within a limited period, when there is a good chance of successful seedling establishment. In explaining how different regeneration strategies can result from varying physiology, our approach in considering dormancy and viability and their consequences for seed longevity is to examine the responses of seed populations to environmental factors. The remarkable longevity of some individual seeds – the record breakers in the survival game – is legendary. In this chapter, we review some of these accounts, assessing these long-lived individuals in the con-

text of the overall seed population. In so doing, we try to avoid 'the accumulation of ever more individualistic special cases', in which the boundaries of ignorance are 'defined by a cloud of unique spots of knowledge, rather than by broad sweeping lines of generalization' (Harper, 1988).

Dormancy

Seed dormancy is caused by a block to the process of germination within the imbibed seed. Dormancy is most easily observed, measured and defined negatively as the failure of a viable seed to germinate, given moisture, air and a suitable constant temperature for radicle emergence and seedling growth (Amen, 1968). If a seed remains ungerminated because these minimum requirements for germination are lacking, it is better described as quiescent, especially if metabolism is arrested or retarded (Bewley and Black, 1994).

Recent attempts to frame a positive definition of dormancy include 'a seed characteristic, the degree of which defines what conditions should be met to make the seed germinate' (Vleeshouwers, 1997). Measurement of a hypothetical phytochrome receptor protein was proposed as a possible future method of directly measuring dormancy. At the time of writing, however, dormancy must be measured in terms of (non-)germination. The simplest measure is $(V - G)$, where V is the seed lot viability and G is the proportion of seeds germinating, usually at a specified constant temperature in darkness with water and air. This measurement implies incorrectly that dormancy is a quantal response: a seed either germinates or does not and, by inference, it might be thought it is either dormant or it is not. In reality, it is only the expression of dormancy that is recorded as a quantal response. The intrinsic nature of dormancy in the individual seed is probably more like 'a hill that seeds approach from one side (i.e. as the requirements to break dormancy are met) and then progress down the other, gathering momentum (increasing germination rate)' (Bradford, 1995, 1996). We do not yet know how to measure the depth of dormancy in individual seeds, but its variation is reflected in the seed-to-seed variation in the expression of dormancy in a seed population. This variation is itself the basis of other possible measures of seed dormancy, which include the mean dormancy period or the rate of loss of dormancy during after-ripening at constant temperature and moisture content (Sharif-Zadeh and Murdoch, 2000), or the median base water potential of the seed lot (Bradford, 1995; Christensen et al., 1996).

The dormancy of seeds is often usefully classified by the origin of the dormancy (e.g. see Table 1.5 in Simpson, 1990) or by the putative mechanism (Baskin and Baskin, 1998). When considering the ecology of regeneration, a distinction of particular value is that between primary and secondary dormancy. Primary (or innate) dormancy develops during seed maturation on the mother plant, while secondary dormancy is induced after shedding.

Primary dormancy

Primary dormancy may simply occur because the embryo is immature and, in some cases, largely undifferentiated, such that further development is needed prior to germination. Examples include seeds with linear embryos, as in the *Ranunculaceae* and *Umbelliferae* (Atwater, 1980; Ellis *et al.*, 1985; Baskin and Baskin, 1998). Irrespective of embryo maturity, seeds may be dormant in the sense that germination is blocked physiologically (Baskin and Baskin, 1998). The mechanism(s) of this physiological block have not been elucidated, though they may be in the seed-coat and/or the embryo itself. Several hypotheses have been proposed to explain the mechanisms of dormancy and the modes of action of dormancy-relieving treatments (e.g. Bewley and Black, 1994; Cohn, 1996; Foley, 1996; Bewley, 1997).

The functions of primary dormancy appear to be twofold. First, along with the inhibition of germination of developing seeds by the mother plant, primary dormancy helps to prevent precocious germination on the mother plant (Bewley and Downie, 1996). For example, *Arabidopsis* mutants deficient in abscisic acid (ABA) production or in responsiveness to ABA during seed development yield seeds that lack dormancy and germinate precociously (Karssen *et al.*, 1983), as do ABA-deficient mutants in other species (Bewley, 1997). Secondly, in many species, primary dormancy persists after maturation and shedding, resulting in the temporal dispersal of seeds by preventing the immediate and approximately synchronous germination of seeds.

Heritability of dormancy is complex, because parts of the seeds differ genetically. For example, a diploid embryo may receive nutrients from a triploid endosperm (with two maternal and one paternal sets of chromosomes) and is surrounded by maternal tissues (testa and fruit structures). The genetics of dormancy in the *Gramineae* has been reviewed by Simpson (1990).

Primary dormancy not only varies with genotype but also with maturation environment, a topic discussed in this volume by Gutterman (Chapter 3) and also reviewed by Fenner (1991, 1992). For example, heavier, thinner-coated, non-dormant seeds of *Chenopodium album* were produced in short days compared with long days (Karssen, 1970). Warmer temperatures during maturation usually reduce dormancy (Peters, 1982a, b), whereas the response to water stress varies with species. In *Avena fatua* and *Sorghum halepense*, dormancy is lower if seeds mature under water stress (Peters, 1982a, b; Benech Arnold *et al.*, 1992), implying that seeds produced in warm, dry summers are thus likely to have less dormancy than those produced in cool, moist ones. In contrast, in barley (*Hordeum vulgare*) and *Cenchrus ciliaris* (a perennial grass that occurs in arid and semi-arid environments), dormancy was greater for seeds produced under water stress, a characteristic that may be of adaptive value in plants growing in arid ecosystems (Aspinall, 1965; Sharif-Zadeh and Murdoch, 2000). With respect to interspecific differences, seeds that mature within green tissues tend to be more light-sensitive than those where chlorophyll declines at an early stage during maturation (Cresswell and Grime, 1981).

Polymorphism is shown when developmentally or morphologically different seeds differ in primary dormancy. These seeds may be produced on the same or on different plants of a given species. The classic example of *Xanthium pensylvanicum* achieves dispersal in time. Two seeds are dispersed together in each capsule, the upper seed being much more dormant than the lower one, and at least 12 months normally separate the germination of the two seeds (Esashi and Leopold, 1968; Harper, 1977). In *Chenopodium album*, between 0.2 and 5.0% of seeds, collected from five localities in England and Wales, were large, brown, thin-coated and non-dormant, in comparison with the majority, which were smaller, black, thick-coated and dormant (Table 8.1). The black seeds had smooth or reticulate testae, the deeper dormancy of the former being partly relieved by chilling (Table 8.1; Williams and Harper, 1965). Two types of seeds are produced by ragwort (*Senecio jacobaea*) (McEvoy, 1984). Those from around the edge of the inflorescence (ray achenes) differ from those at the centre (disc achenes). Ray achenes are heavy and smooth, shed their long hairs and have greater dormancy. They remain on the mother plant for longer, are dispersed close to it and may therefore replace the mother plant at the same site. In contrast, the disc achenes are lighter and retain a group of long hairs, which assists in wind dispersal to new sites (McEvoy, 1984). Similarly, a comparison among diverse species of *Argyranthemum* showed that ray achenes tended to show more dormancy than disc achenes, although differences were only significant in 11 of the 21 species studied and only substantial in four species (Francisco-Ortega *et al.*, 1994). Further examples of polymorphic seeds are discussed by Fenner (1991) and Baskin and Baskin (1998).

After-ripening

Primary dormancy declines both prior to shedding and subsequently. When this loss of dormancy occurs in 'air-dry' seeds, it is termed after-ripening. The progressive loss of dormancy during after-ripening is a

Table 8.1. Germination polymorphism in *Chenopodium album*. Seeds were collected from various sites in England and Wales and separated according to testa colour and texture. Testa thickness and thousand-seed weights were measured and seeds were germinated at 20°C with or without prechilling of imbibed seeds for at least 3 weeks at 0°C. (From data in Williams and Harper, 1965.)

Testa colour and texture	Testa thickness (µm)	Thousand-seed weight (g)	Germination (%)	
			Unchilled	Chilled
Brown	16	1.57	> 90	> 90
Black, reticulate	60	1.33	62	64
Black, smooth	60	1.13	32	61

function of environment as well as time. Thus the rate of after-ripening increases with temperature in a predictable manner. Circumstantial evidence suggests that many seeds behave like those members of the *Gramineae* in which the logarithm of the mean dormancy period is a negative linear function of temperature, the Q_{10} for the relation being typically in the range 2.5–3.8 (Roberts, 1965, 1988). Thus, the longer and warmer the storage environment, the greater the loss of dormancy. For example, in one seed lot of *Oryza glaberrima* at 11.2% moisture content, half the seeds lost dormancy after 65 days at 30°C, while at 40°C only about 20 days were required (Ellis *et al.*, 1983), and so the Q_{10} is close to 3. A similar approach has been published for loss of primary dormancy in barley (Favier and Woods, 1993). There is also evidence that seed moisture content can influence the rate of loss in dormancy during after-ripening, at least over certain moisture-content ranges (Quail and Carter, 1969; Tokumasu *et al.*, 1975; Ellis *et al.*, 1983; Sharif-Zadeh, 1999). In cereals, at very low moisture contents (< 8%) the rate of dry after-ripening is minimal, while it is greatest between 11 and 15% moisture content (Roberts, 1962, 1988; Ellis *et al.*, 1983). In *Cenchrus ciliaris*, evidence for an optimum moisture level for after-ripening was clear, because seeds in equilibrium with 40% equilibrium relative humidity (e.r.h.) at 20°C after-ripened more slowly than those at 50% e.r.h., whereas those at 70% e.r.h. remained viable but largely failed to after-ripen (Sharif-Zadeh, 1999).

Not surprisingly, procedures used during seed collection and postharvest seed drying, together with subsequent storage prior to investigations of seed dormancy, will influence dormancy. It follows that different methods of preparing seeds for burial experiments may result in differing degrees of dormancy at burial and, since dormancy is the major factor influencing survival in soil, thus influence the subsequent rate of depletion of the viable seed population. Accordingly, it is possible that reports of the longevity of seeds in soil from experiments that treated collected seeds in environments in which considerable after-ripening might have occurred could underestimate the natural longevity under such circumstances.

Secondary dormancy

Dormancy may be induced in both dormant and non-dormant seeds after shedding. When moist (though not necessarily fully imbibed) seeds are kept in sufficient environmental stress to prevent germination or with insufficient promotion to relieve dormancy, secondary dormancy may be induced in both dormant and non-dormant seeds. The main causes of this secondary dormancy in buried seeds were originally thought to be low oxygen and/or high carbon dioxide (CO_2) levels (Harper, 1957; Roberts, 1972c), and the efficacy of insufficient air and anaerobiosis to induce dormancy is well known. For example, in one line of *Avena fatua*, subsequent germination of caryopses (dehulled florets) decreased from 90 to 17% after allowing the caryopses to imbibe for 3 h in an anoxic atmosphere. The secondary dormancy was relieved by dry storage, nitrite, nitrate, ethanol, azide and gibberellin (GA_3) (Symons *et al.*, 1986).

Soil atmosphere in cultivated soils may deviate little from ambient air (Murdoch and Roberts, 1996). Nevertheless, the occurrence of anaerobic microsites within a largely aerobic mass has been deduced from evidence for anaerobic microbial denitrification of well-aerated pasture soils (Burford and Millington, 1968; Burford and Stefanson, 1973). Very high CO_2 concentrations (> 40%) may occur in soil with stagnating water columns (Enoch and Dasberg, 1971). Hence, anaerobic atmospheres may contribute to the induction of secondary dormancy of seeds in soil.

Secondary dormancy may also be induced in prolonged moist aerobic treatments, in which germination does not occur (Totterdell and Roberts, 1979; Jones *et al.*, 1997; Kebreab and Murdoch, 1999a). During such treatments (e.g. prechilling or

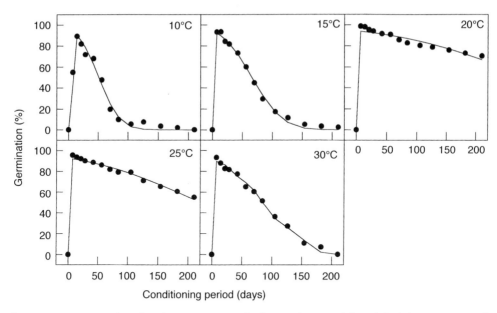

Fig. 8.1. Germination of *Orobanche aegyptiaca* seeds after conditioning (fully imbibed) for various periods at 10, 15, 20, 25 and 30°C. Seeds were surface-sterilized with sodium hypochlorite solution before conditioning. Germination tests lasted 10 days at 20°C with 3 p.p.m. of GR24. GR24 is an analogue of strigol – an exudate of cotton roots, which stimulates germination of seeds of some parasitic plants. Lines were fitted according to a multiplicative probability model. (After Kebreab and Murdoch, 1999a.)

at higher temperatures, warm stratification), loss of primary dormancy is initially evidenced by progressive sensitization to a subsequent germination-promoting stimulus. However, prolonging the period of moist storage results in a decrease in the response to subsequent stimulation, due to secondary dormancy (Fig. 8.1). Temperature has a marked effect on the induction of secondary dormancy. While an increase of temperature increases the rate of after-ripening of air-dry seeds, it also affects the rate of induction of secondary dormancy in imbibed seeds. In imbibed *Rumex crispus* seeds, for example, secondary dormancy was evident after about 21, 7–14, 7 and 3 days at 1.5, 10, 15 and 20°C, respectively (Totterdell and Roberts, 1979; compare Totterdell and Roberts, 1981). Interestingly, for *Orobanche* spp. from a warm temperate environment, rates of induction were fastest at 3°C and decreased to a minimum above about 20°C (Fig. 8.1; Kebreab and Murdoch, 1999a).

In aerobic conditions, germination may also be inhibited by exposing seeds to

prolonged white light, especially at high radiant flux densities, or to far-red light (Bartley and Frankland, 1982; Fig. 8.2). In contrast, induction of secondary dormancy may be delayed, reduced or prevented by intermittent low-intensity laboratory light and nitrate in *Capsella bursa-pastoris* (Fig. 8.3) and by nitrate in *Sisymbrium officinale* (Hilhorst, 1990b).

A final and somewhat surprising circumstance in which dormancy is reimposed has been shown in Sitka spruce (*Picea sitchensis*). Conditional dormancy (defined as failure to germinate at 10°C) was reimposed when prechilled seeds were then dried to 6% moisture content and stored at 10°C (Jones *et al.*, 1998). The effect was reversible and repeatable and was most rapid at lower (4–10%) compared with higher moisture contents (15–20%). This result is contrary to the conventional expectations of dry after-ripening already described. It is also too early to suggest the ecological consequences of this response or how widely it applies.

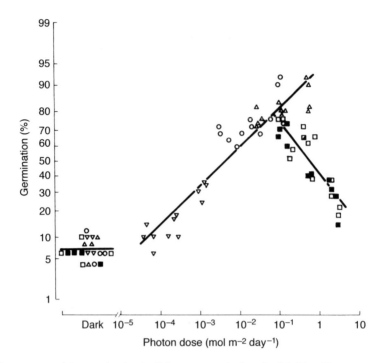

Fig. 8.2. The response of the germination (radicle emergence) of seeds of *Echinochloa turnerana* to white fluorescent light in 7-day tests at an alternating temperature of 20/30°C (16 h/8 h). The results from testing in the dark are shown on the left. The seeds were exposed to different photon flux densities for 24 h day^{-1} ($\square$,$\blacksquare$), 8 h day^{-1} ($\triangle$), 1 h day^{-1} ($\bigcirc$) or 1 min day^{-1} ($\triangledown$) on filter-paper moistened with water (open symbols) or a 0.2% potassium nitrate solution (solid symbols). The lines shown are cumulative log-normal dose–response curves fitted by probit analysis. The positive response, in which increase in photon dose promoted germination – presumably as a result of the low-energy reaction (see Pons, Chapter 10, this volume) – shows reciprocity where day lengths ″ 8 h day^{-1}. The negative response, in which increase in photon dose inhibited germination – presumably as a result of the high-irradiance reaction (see Pons, Chapter 10, this volume) – was only detected in continuous light in this seed lot. The two slopes show the variation in the sensitivity of individual seeds to the minimum photon dose required to promote and inhibit germination, respectively. The standard deviation for promotion was $10^{1.52}$ mol m^{-2} day^{-1}, while that for inhibition was $10^{1.16}$ mol m^{-2} day^{-1}. (From Ellis *et al.*, 1986.)

Are the mechanisms of primary and secondary dormancies the same? Dry after-ripening and the same chemicals may relieve both primary and secondary dormancy (Symons *et al.*, 1986), implying some similarity. It is, however, interesting that, while short periods of imbibition are associated with relief of primary dormancy, longer periods lead to secondary dormancy. As shown below, empirical mathematical models imply that the dual processes of loss and induction of dormancy may be concurrent in individual seeds and so must have a different physiology. In *Orobanche*, the loss of primary dormancy could be due to increased

sensitivity to the exogenous stimulant GR24 (Kebreab and Murdoch, 1999a), while induction of secondary dormancy may be associated with accumulation of an inhibitor, since in this case the seeds must be dried to overcome secondary dormancy.

Seed-to-seed variation in dormancy

Unless 0 or 100% of seeds germinate, the results of germination tests divide seeds into two groups – those which germinate and those which do not. This separation may imply polymorphism but, if this term is to retain its usefulness, it should only be

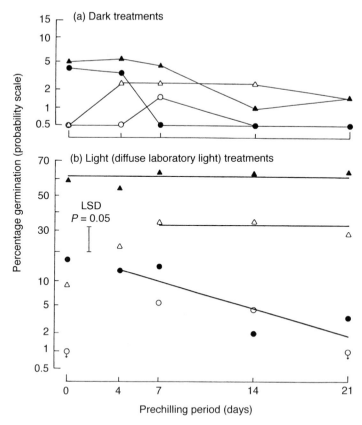

Fig. 8.3. Mean germination of seeds of *Capsella bursa-pastoris* after different periods of prechilling. Subsequent germination tests lasted 4 weeks and were carried out at alternating (10/25°C, 16 h/8 h: solid symbols) or at constant (25°C: open symbols) temperatures and in water (○,●) or 0.01 M potassium nitrate solution (△,▲). Seeds were (a) kept in darkness or (b) intermittently exposed to diffuse laboratory light. In (b), lines were fitted by probit analysis with a common line for induction of secondary dormancy in water. The least significant difference value is not valid for comparisons with treatments designated (↓). Unpublished data from A.J. Murdoch and E.H. Roberts.

applied where there is evidence of a discontinuous distribution of dormancy periods or in the degree of dormancy (Roberts, 1972c). There is often, however, a continuous seed-to-seed variation in the dormancy of individual seeds, even in apparently uniform seed lots. For example, the chilling requirements for the loss of primary dormancy of individual *Rumex* seeds varied from about 1 day to 3 weeks at 1.5°C. After-ripening periods varied from 0 to about 130 days in rice seeds (cultivar 'Masalaci') when stored at 13.5% moisture content and 27°C (Roberts, 1963). In Fig. 8.2, the seed-to-seed variation in the response of *Echinochloa turnerana* to white light ranges over four orders of magni-

tude from those requiring a photon dose of 10^{-5} mol m^{-2} day^{-1} to those requiring 10^{-1} mol m^{-2} day^{-1} (Ellis *et al.*, 1986). The variation in percentage germination as a function of the dose of dormancy-breaking treatments usually approximates to the cumulative normal frequency distribution. As with any normal distribution, the germination : dose curve can therefore be described by its mean and variance. Two crucial points are that such curves may therefore be linearized if percentage germination values are transformed to normal equivalent deviates (ned) or probits (Figs 8.2 and 8.3; Hewlett and Plackett, 1979), and that the slope of the probit germination : dose line therefore has the units ned per unit dose. The reciprocal of the slope quantifies the standard deviation of the dormancy response to the applied treatment in the seed population. This analysis has been found appropriate or proposed for germination responses of dormant seeds to light in the very-low-fluence response (Hartmann et al., 1998), as well as the low-energy reaction (Fig. 8.2; Borthwick et al., 1954; Duke, 1978; Frankland and Taylorson, 1983), dry after-ripening period (Roberts, 1961, 1965; Probert et al., 1985; Favier and Woods, 1993), imbibition pretreatments, such as chilling or conditioning (Fig. 8.1; Totterdell and Roberts, 1979; Kebreab and Murdoch, 1999a), ethylene before and after burial of seeds in soil (Schonbeck and Egley, 1980; Egley, 1989b), alternating temperatures (Probert et al., 1985, 1987; Murdoch et al., 1989; Kuo, 1994), GA_3 and nitrate (Hilhorst and Karssen, 1988; Hilhorst et al., 1996).

Similarly, variation in percentage germination as a function of the dose of dormancy-inducing and germination-inhibiting treatments may approximate to the negative cumulative normal distribution, for example, with aerobioconditioning or chilling of imbibed seeds (Figs 8.1 and 8.3; Totterdell and Roberts, 1979; Kebreab and Murdoch, 1999a) and the high-irradiance reaction, in which high photon flux densities of light of any wavelength may inhibit germination (Fig. 8.2; Bartley and Frankland, 1982). An exception with respect to induction of dormancy was an exponential decrease in light sensitivity of Sisymbrium officinale as a function of the duration of dark imbibition at 15°C prior to exposure to red light (Hilhorst, 1990a). Nevertheless, whichever model is used, it is clear that considerable seed-to-seed variation still exists.

Seed-to-seed variation within seed lots is therefore generally normally distributed. Sometimes, however, responses are nonlinear, because more than one process may be occurring simultaneously (Figs 8.1–8.3). The loss of primary dormancy and induction of secondary dormancy may still be normally distributed in the seed population, but, because they occur concurrently,

the germination after any period is given by a multiplicative probability model. The final germination is in fact the product of the probabilities that a seed: (i) has lost primary dormancy; (ii) has not entered secondary dormancy; and (iii) is still viable. For example, the seeds of the parasitic weed Orobanche aegyptiaca show optimum preconditioning periods of about 3–10 days, depending on temperature, after which secondary dormancy is induced (Kebreab and Murdoch, 1999a; Fig. 8.1). The rate of induction of secondary dormancy appears to be minimized at 20°C, but the more rapid decline in subsequent germination at 25 and 30°C (Fig. 8.1) is due to a third concurrent process of loss of viability (Kebreab and Murdoch, 1999a).

Key hypotheses of the non-linear models are: (i) seed-to-seed variation with respect to each component process is normally distributed with respect to the 'dose' variable (e.g. time); and (ii) the processes operate concurrently and independently within each individual seed. Further examples are discussed by Murdoch et al. (2000).

Annual cycles in dormancy

Annual cycles, in which physiologically based dormancy is relieved and induced during the course of a year, occur in buried seeds of some annual plants in both temperate and tropical soil environments (Courtney, 1968; Baskin and Baskin, 1985, 1998; Bouwmeester, 1990; Murdoch, 1998). Germination of seeds occurs at times of low or no dormancy. Non-dormant seeds simply await the availability of moisture and suitable temperatures for germination (Roberts and Potter, 1980; Karssen, 1982; Bouwmeester, 1990; Vleeshouwers, 1997). Exposure to light (Taylorson, 1970, 1972), nitrate (Murdoch and Roberts, 1996), fluctuations of temperature (Benech Arnold and Sanchez, 1995) or combinations of these treatments (Roberts et al., 1987; Murdoch, 1998) may be needed to relieve residual primary and secondary dormancy at times of low dormancy. The combination of the annual dormancy cycle with the seasonal variation in temperature and moisture gives a net germinability at any given time. Sauer and Struick (1964) and Wesson and Wareing (1967, 1969a, b) produced evidence which suggested that buried seeds often acquired a positive requirement for light, which prevented germination until seeds were brought to the surface. While agreeing with the overriding importance of light, Vincent and Roberts (1977) contended that the reduced temperature fluctuations experienced by buried seeds were also likely to be important, especially because the light sensitivity of many annual weed seeds was only exhibited when they were also exposed to alternating temperatures (Roberts *et al.*, 1987). These responses help to explain the higher seedling emergence that is associated with cultivation treatments particularly at certain times of the year (Courtney, 1968; Vincent and Roberts, 1977; Roberts and Ricketts, 1979; Roberts and Potter, 1980), as well as the periodicity of seedling emergence that characterizes many species (Chepil, 1946; Roberts, 1964). An annual germinability cycle is thus superimposed.

Ecologically, it is useful to identify factors in the environment that may promote germination *in situ* at any given time. An understanding of these stimuli, and of how they act in seed populations with different levels of dormancy, is crucial if seedling emergence from persistent soil seed banks is to be predicted (Murdoch, 1998). In the case of *Chenopodium album*, responses to light were, not surprisingly, rare in seeds retrieved from the soil surface, but their behaviour was quite interesting. Only two classes of seeds existed among the surface seeds for much of the time (Fig. 8.4): a non-dormant minority, which would presumably germinate *in situ* given moisture; and an ungerminable majority, whose dormancy could seldom be broken by light or nitrate.

In buried seeds, various intermediate classes were found and proportions would germinate on retrieval, given light or nitrate (Fig. 8.4). Typically, the responses to both of these factors were additive, so

that positive and negative interactions were unusual (Fig. 8.4). The germinability of retrieved seeds tended to increase with depth, and high proportions during the spring and summer reflect times of low dormancy. With moisture, some seeds would be expected to germinate *in situ* at 0–25 mm without nitrate or light. At their original depths of burial, however, a lower amplitude of temperature fluctuation might limit germination of deeply buried seeds.

The seeds that would not germinate in any treatment fell into two categories. First, there were those whose residual dormancy could not be broken in the temperature regime employed. Additional tests carried out at an approximately optimal 30/10°C (16/8 h) alternation showed that this proportion was often large (data not shown in Fig. 8.4). Secondly, there were apparently viable (firm) seeds that were ungerminable in any dormancy-relieving treatment. The proportion of these decreased with depth.

Quiescence

A viable seed may therefore avoid germination by being dormant. An alternative strategy is that of quiescence. Quiescence may be enforced by the environment, as, for example, in air-dry seeds or in imbibed seeds below the base temperature for the rate of germination of non-dormant seed (Labouriau, 1970; Roberts, 1988; Kebreab and Murdoch, 1999c). Imbibed seeds above the ceiling temperature for the rate of germination (Roberts, 1988) are also quiescent, but viability is unlikely to be maintained (Ellis *et al.*, 1987).

In the soil, environmentally enforced quiescence – 'the inability to germinate due to an environmental restraint – shortage of water, low temperature, poor aeration' (Harper, 1957, 1977) – has previously been called enforced dormancy. These restraints enforce quiescence, rather than dormancy, since one or more of the three minimum requirements for germination of non-dormant seeds is lacking.

The seed-coats of some species may be impermeable to water, and such hard seeds

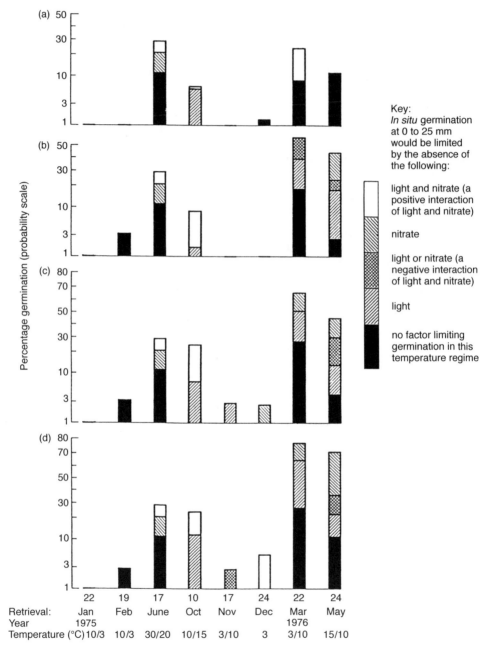

Fig. 8.4. Environmental factors limiting germination of *Chenopodium album* seeds retrieved at night from (a) the soil surface or depths of (b) 25, (c) 75 and (d) 230 mm. Seeds were germinated in light or darkness and in water or 0.01M potassium nitrate in a 16/8 h temperature regime approximating the soil temperature at 0–25 mm depth 7–10 days prior to retrieval. (Soil temperature was recorded every 20 min.) Significant effects and interactions are shown (from analysis of variance). Thus differences between depths are only shown if the effect of depth or interactions with it were significant. Germination in winter was often low and analysis was restricted to subsets of data and, where applicable, responses are shown. Otherwise, overall mean values are given. Dark treatments were not tested on 10 October 1975. (From Murdoch, 1998.)

are intrinsically quiescent. Hard-seeded-ness (called physical dormancy by Baskin and Baskin, 1998) ensures the persistence of the seed banks of both wild (Egley and Chandler, 1983) and cultivated species (Lewis, 1973; Saunders, 1981; Egley, 1989a; Mott *et al.*, 1989; Standifer *et al.*, 1989; Russi *et al.*, 1992a). Although best known in the *Leguminosae*, hard impermeable seeds are found in species of several other families including the *Anacardiaceae* (e.g. *Rhus* spp.), *Bixaceae* (e.g. *Bixa orellana*), *Cannaceae* (e.g. *Canna* spp.), *Convolvulaceae* (e.g. *Ipomoea* spp.), *Ebenaceae* (e.g. *Diospyros virginiana*), *Geraniaceae* (e.g. *Pelargonium* spp.), *Liliaceae* (e.g. *Asparagus densiflorus*), *Malvaceae* (e.g. *Gossypium* spp.), *Myrtaceae* (e.g. *Psidium* spp.), *Rhamnaceae* (e.g. *Ceanothus*), *Sapindaceae* (e.g. *Cardiospermum halicacabum*), *Solanaceae* (e.g. *Datura* spp.) and *Zingiberaceae* (e.g. *Elattaria cardamomum*) (Ballard, 1973; Atwater, 1980; Ellis *et al.*, 1985). The genetic basis of impermeability has been demonstrated in several species, but environmental conditions during seed development, such as humidity, temperature, day length and mineral nutrition, modify its expression (Egley, 1989a).

Impermeability develops during maturation drying on the mother plant and it may increase after shedding in dry environments. For example, Standifer *et al.* (1989) have demonstrated that hard-seededness in okra (*Abelmoschus esculentus*) increased as seeds were dried progressively from 11 to 3% moisture content. When seeds of the legumes *Trifolium pratense*, *T. repens* and *Lupinus arboreus* were dried below about 14% moisture content, intact and undamaged testae became impermeable to water (Hyde, 1954). Further drying occurs through the hilum, which acts as a one-way valve, permitting water vapour to escape at low ambient relative humidities, but closing at high relative humidity. Seed moisture will therefore equilibrate with the lowest relative humidity experienced. At seed moisture contents of about 10–12%, sudden increases in humidity cause the hilar fissure to close. However, a gradual increase in humidity leaves the fissure

open (Hyde, 1954), and imbibition may occur. Hence, this hard-seededness is reversible. Drying to even lower moisture contents (e.g. 8–9% in *Lupinus cosentini*) renders the seed absolutely and irreversibly impermeable until the hard seed-coat is broken (Quinlivan, 1971). This pattern appears to be common to most, if not all, papilionoid legumes. A simpler mechanical closure of the hilum has been described in the *Caesalpinoideae* and some of the *Mimosoideae* (Werker, 1980/81).

Hard seeds can easily be made permeable in the laboratory by mechanical abrasion. Natural softening is often protracted, as shown later in this chapter by the longevities of hard seeds in soil. Observations of buried seeds of six papilionoid legumes (three medics and three clovers) in Syria showed that: (i) different proportions of softened seeds were found in each species each year; (ii) some species were more susceptible to softening (e.g. *Trifolium stellatum* compared with *Medicago orbicularis*); (iii) the seed-coat was usually broken at the lens (or strophiole); and (iv) burial of seeds reduced the number of softened seeds present at the end of summer (Russi *et al.*, 1992a, b). The wide diurnal temperature fluctuations (15°/60°C) of the dry summers of some Mediterranean climates are thought to soften legume seeds (Quinlivan, 1971; Saunders, 1981; Egley, 1989a; Russi *et al.*, 1992b). Taylor (1981) proposed that high soil temperatures weaken the cells of the lens and that temperature fluctuations cause a gradual expansion and contraction of cells, which eventually disrupts the testa at the weak points – e.g. the lens – thus relieving impermeability. Egley (1989a) also proposed that temperature fluctuations cause softening in the *Malvaceae* by expansion and contraction of cells in the chalazal region.

The preservation of seed viability in relation to environment

Survival of seeds in the soil does not depend merely on avoiding germination by dormancy or quiescence: viability must be preserved. Loss of seed viability is the final stage in seed deterioration. Prior to death, ageing results in a decline in many aspects of potential performance, such as the rate of germination (Ellis and Roberts, 1981). The mechanism of seed ageing is not known, but several theories have been proposed; for example, see the reviews by Roberts (1972b) and Priestley (1986).

Comparisons of the considerable differences in seed longevity among seed-producing species (Ewart, 1908; Harrington, 1972) are not only hampered by differences in environmental conditions but also by differences in seed storage physiology.

Classification of seed storage behaviour

Three types of seed storage behaviour are recognized, of which Roberts (1973) defined two: orthodox and recalcitrant.

Of 6919 species representing 251 plant families listed in their compendium of seed storage behaviour, Hong *et al.* (1996) classified 88.6% as having or probably having orthodox seed storage behaviour. Such seeds can be dried without damage to low moisture contents and, over a wide range of conditions, their longevity increases with decrease in seed storage moisture content and temperature in a quantifiable and predictable way (Roberts, 1973; Roberts and Ellis, 1982). Orthodox seed longevity is improved by desiccation to around −350 MPa (Ellis *et al.*, 1989). Seeds of the majority of crops and many non-cultivated species show orthodox storage characteristics.

Recalcitrant seeds, on the other hand, do not obey these rules, since they do not survive desiccation (Roberts, 1973). Such seeds are killed if the water potential drops below about −1.5 to −5.0 MPa (Probert and Longley, 1989; Roberts and Ellis, 1989; Pritchard, 1991), i.e. roughly equivalent to the permanent wilting point of many growing tissues (Kramer, 1983). Hong *et al.* (1996) classified 7.4% of 6919 species as probably having recalcitrant seed storage behaviour. Recalcitrant seeds do not occur naturally in arid habitats, deserts or savan-

nah. Most seeds in such environments show orthodox seed storage behaviour (Hong *et al.*, 1996). The corollary is not true, since species native to moist environments may show orthodox, intermediate (see below) or recalcitrant seed storage behaviour. Recalcitrant seeds are produced in a number of large-seeded woody perennials, including some tropical plantation crops, e.g. cocoa (*Theobroma cacao*) and rubber (*Hevea brasiliensis*); many tropical fruits, e.g. avocado (*Persea americana*) and mango (*Mangifera indica*); a number of timber species, e.g. from the temperate latitudes, oak (*Quercus* spp.) and chestnut (*Castanea* spp.), and from the tropics, some of the *Dipterocarpaceae* and *Araucariaceae*. Such seeds are sometimes categorized as being of temperate or tropical origin, as the former can be stored for several years at near-freezing temperatures, whereas the latter are damaged by exposure to cool temperatures of 10–15°C or less (Bonner, 1990; Hong and Ellis, 1996).

An intermediate category of seed storage behaviour is also recognized, in which dry seeds (−90 to −250 MPa) are injured by low temperatures and also by further desiccation (Ellis *et al.*, 1990a). Clear experimental evidence for this category of seed storage behaviour has been shown in several species, including arabica coffee (*Coffea arabica*), papaya (*Carica papaya*) and oil-palm (*Elaeis guineensis*) (Ellis *et al.*, 1990a, 1991a, b), and intermediate species probably comprised 1.9% (134) of 6919 species listed by Hong *et al.* (1996).

Survival of orthodox seeds in dry controlled environments

Survival curves of homogeneous populations (i.e. single genotypes harvested at the same time from the same place) of orthodox seeds in constant dry storage conditions are negative cumulative normal distributions. The slope of the survival curve depends on seed-to-seed variation in lifespan and, after transforming the viability percentages to normal equivalent deviates or probits, the slope of the survival curve is the reciprocal of the standard deviation of seed lifespans, σ, so that the survival curve is described by the following equation:

$$v = K_i - p/\sigma \qquad (8.1)$$

where v is probit percentage viability after p days in storage and K_i is a constant specific to the seed lot (the estimated viability in probits at the start of storage). The standard deviation (σ) is a measure of seed longevity, since it is the period in days during which percentage viability is reduced by one unit on the probit scale (e.g. from 97.7 to 84.1%, or from 84.1 to 50.0%).

It has been suggested that Equation 8.1 should be modified by adding an extra 'control mortality' parameter (e.g. Mead and Gray, 1999). This parameter is needed in toxicological studies to quantify target organisms that would have died in the absence of an applied chemical. In seed viability studies, control mortality is one of the factors being quantified, and so it is inappropriate to include the additional parameter (unless the seed lot includes empty seeds).

Storage at lower moisture content and/or temperature increases the mean viability period and also the variation in lifespans of individual seeds. Clearly, the greater the longevity, the shallower the slope of the seed survival curve. It is, therefore, possible to quantify the response of seed longevity to moisture content and temperature in terms of σ, as shown in the following equation,

$$\sigma = 10^{K_E - C_W \log_{10} m - C_H t - C_Q t^2} \qquad (8.2)$$

where m is seed moisture content (percentage fresh weight), t is temperature (°C) and K_E, C_W, C_H and C_Q are species viability constants (Ellis and Roberts, 1980a).

The viability equation

In order to quantify the relation between seed survival and storage period, temperature and moisture content, the evaluation of σ in Equation 8.2 can be substituted in Equation 8.1, providing what is commonly

known as the viability equation:

$$v = K_i - p/10^{K_E - C_W \log_{10} m - C_H t - C_Q t^2} \qquad (8.3)$$

(Ellis and Roberts, 1980a). Equation 8.3 was developed from original equations developed for a range of species over a limited range of environments by E.H. Roberts in 1960 (see Roberts (1972a) for a historical review). The improved equation was obtained by analysing data for the survival of seeds of several lots of barley (*Hordeum vulgare*), which had been stored in a wider range of storage environments, namely moisture contents between 5 and 25% and temperatures between 3 and 90°C (Ellis and Roberts, 1980b). The equation has since been successfully fitted to similarly comprehensive data sets for seeds of a further nine species (Ellis *et al.*, 1982; Tompsett, 1986; Kraak and Vos, 1987; Dickie *et al.*, 1990; Zewdie and Ellis, 1991b). Interestingly, Equation 8.3 also applies to other propagules in anhydrous biology – for example, to fungal spores (Hong *et al.*, 1997) and pollen (Hong *et al.*, 1999).

Predictions of seed longevity based on the viability equation imply that dry seeds at cool temperatures will survive for very long periods, especially in cereals and legumes. For example, it has been estimated that it would take 440 years for barley seed viability to decline from 80 to 75% if seeds were kept at −18°C with 6% moisture content (Ellis and Roberts, 1980a). Such estimates can, of course, be provided only by considerable extrapolation of the models beyond the data from which they have been derived. Nevertheless, it has been shown that the viability equation and the viability constants for barley are consistent with data provided by Aufhammer and Simon (1957) for the viability of barley seeds stored for 123 years in the foundation stone of the Nuremburg City Theatre (Ellis and Roberts, 1980a). In other words, there is evidence that the viability equation can be applied to considerable periods of storage and that the survival of air-dry seeds can be very substantial.

Extrapolation is, however, inadvisable to very low temperatures. In the viability equation, the expression $-C_H t - C_Q t^2$ quantifies the effect of temperature on seed longevity (Ellis and Roberts, 1980a). This expression implies that longevity increases as the seed storage temperature is decreased. However, a diminishing law of returns effect applies, due to the empirical inclusion of the quadratic temperature term, $C_Q t^2$. Moreover, at typical values of the temperature coefficients, C_H and C_Q, a reversal in the relation between longevity and temperature is expected, such that, below about −40°C, the viability equation implies that a further reduction in temperature will reduce longevity. At present, there is no evidence for such a reversal and it is suggested that the equations should not be applied at temperatures cooler than −20°C (Dickie *et al.*, 1990).

Interspecific differences in the response of seed longevity to temperature

The hypothesis that the relative effect of temperature on longevity is the same in different species was tested by Dickie *et al.* (1990), who used the viability equation to compare seed survival data for eight species, namely barley, chickpea (*Cicer arietinum*), cowpea (*Vigna unguiculata*), soybean (*Glycine max*), elm (*Ulmus carpinifolia*), mahogany (*Swietenia humilis*), terb (*Terminalia brassii*) and lettuce (*Lactuca sativa*) over a wide range of storage environments (temperatures from −13 to 90°C, seed moisture contents from 1.8 to 25% fresh weight). The estimates of the temperature coefficients (C_H and C_Q) of the viability equation did not differ significantly between these species and were equal to 0.0329 and 0.000478, respectively. The relative effect of temperature on longevity thus appears to be the same in these eight diverse species, which represent four of the ten superorders of flowering plants (Dickie *et al.*, 1990).

Interspecific differences in the response of seed longevity to moisture content

The viability equation describes a negative logarithmic relation between seed longevity and seed moisture content. In Fig. 8.5, the negative linear relation between $\log_{10}(p_{50})$ and $\log_{10} m$ is observed for lettuce seeds at moisture contents between 2.6 and about 15%. The slope of this line quantifies the effect of moisture content on seed longevity and, in Equations 8.2 and 8.3, this is the coefficient C_W. In contrast to the effect of temperature on seed longevity, the effect of moisture content on longevity (i.e. the coefficient C_W) differs significantly between species (Ellis *et al.*, 1982, 1989, 1990b; Tompsett, 1986). If, however, the responses of seed longevity to changes in seed equilibrium relative humidity or seed water potential are compared, there are no significant differences between the responses of quite diverse species (Ellis *et al.*, 1989, 1990b; Roberts and Ellis, 1989). Results for 20 contrasting species indicate that longevity is doubled for each 8.4–8.7% reduction in seed equilibrium relative humidity between about 90 and 10% (Ellis *et al.*, 1989, 1990b). Most of the interspecific variation in C_W appears to result from differences in seed composition; the higher the seed oil content, the lower the value of C_W.

There are limits to the negative logarithmic relation between seed longevity and moisture detected in air-dry storage (Fig. 8.5). The curve showing the effect of moisture on longevity can be divided into three zones. It is the central of these three zones over which the viability equation applies. This zone is relevant to the survival of hard (impermeable) seeds in the soil and to other seeds in dry environments.

The survival of very dry seeds

Maximum seed longevity appears to be achieved at the low moisture content limit for the application of the viability equation. This limit varies among species (Ellis *et al.*, 1988, 1989, 1990b) and the extreme estimates provided to date are 2.0% moisture content in groundnut (*Arachis hypogaea*) (Ellis *et al.*, 1990b) and 6.2% moisture content in pea (*Pisum sativum*) (Ellis *et al.*, 1989). These differences appear to be associated with differences in

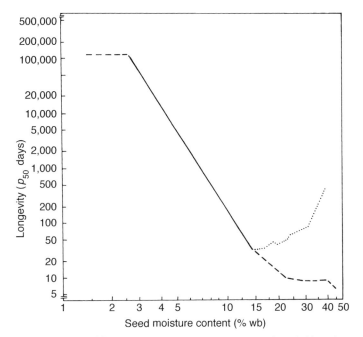

Fig. 8.5. Effect of moisture on seed longevity at a constant temperature. The solid line shows the relation between lettuce seed longevity (p_{50}, the time in days for viability to fall to 50%) and seed moisture content (%, wet weight basis (wb)) in hermetic storage at 25°C provided by the viability equation (Equation 8.3) and the values of the viability constants reported by Kraak and Vos (1987) for lettuce with the value of the seed lot constant K_i set to 2. Both axes have logarithmic scales. Consequently the solid line shows the negative logarithmic relation between seed longevity and moisture. The relations at moisture contents > 15% are redrawn from data of Ibrahim *et al.* (1983) for lettuce seeds in aerobic (···) and anaerobic storage (---) at 25°C. Longevity in aerobic storage at moisture contents greater than 35–40% is considerable, but, since little loss in viability has been detected over 2–3 years, estimates of p_{50} cannot be provided. The relation shown at moisture contents < 2.6% assumes that the value of the low moisture content limit to negative logarithmic relations between longevity and moisture and the relation between moisture and longevity at lower moisture contents determined at 65°C by Ellis *et al.* (1989) applies at 25°C.

seed composition, as the equilibrium relative humidities of seeds of 21 different species at the low moisture limit appear to coincide at about 10% relative humidity at 20°C, or roughly −350 MPa (Ellis *et al.*, 1989, 1990b; Roberts and Ellis, 1989). Below the lower limit, there is a zone where further reduction in moisture content generally has little effect on longevity (Ellis *et al.*, 1988, 1989, 1990b); that is, longevity remains at about its maximum for a given temperature in many species (see, for example, Fig. 8.5).

The survival of moist seeds

There is also an upper moisture content limit to the application of the viability equation. In aerobic storage, increase of lettuce seed moisture content above the 15% upper limit increases longevity. The longevity of fully imbibed seeds in air is considerable, provided they do not germinate and are not damaged by fungi (Fig. 8.5; Ibrahim *et al.*, 1983). In anaerobic storage, lettuce seed longevity declines with increase in moisture content from 15 to about 20%: at this value, longevity reaches a minimum and further increase in moisture has little effect on longevity (Fig. 8.5; Ibrahim and Roberts, 1983; Ibrahim *et al.*, 1983).

The upper moisture content limit to the application of the viability equation also appears to vary among species. Although the data are limited, the upper limit is equivalent to equilibrium relative humidities of 85–93%, or −10 to −20 MPa (Roberts and Ellis, 1989; Zewdie and Ellis, 1991a).

Since detectable respiration occurs in seeds at high water potentials (e.g. Vertucci and Leopold, 1984), do they rapidly exhaust their dry-matter reserves? Apparently not; although respiration rates are high as the seeds imbibe, after a few days at high water potentials, seed respiration rates stabilize at a much lower level, such that seed reserves are not rapidly depleted (Barton, 1945; Ibrahim *et al.*, 1983).

In natural environments, seeds may only experience intermittent hydration. However, Villiers and Edgecumbe (1975) showed that, at least over a period of 5 months for seeds stored at 30°C, an intermittent hydration cycle of 2 days every 2 weeks during dry storage was as beneficial as continuously moist storage. Hence, it is not necessary for seeds to be kept continuously moist to obtain these benefits.

These effects of moisture on the longevity of very dry, dry and hydrated seeds may be associated with the different types of water present in the seeds (Roberts and Ellis, 1989). In very dry seeds, most of the water is strongly bound and has little chemical potential, and so its removal has little effect on longevity. In the second zone, most of the water is weakly bound, and it is the variation in this water fraction that results in the negative logarithmic relation between seed longevity and moisture. Finally, at high moisture contents, free water becomes available, significant respiration can occur and damage can be repaired.

It is clear that seeds may avoid germination by dormancy or quiescence and, where germination is avoided, some seeds may survive in controlled laboratory conditions for substantial periods. In the much more variable soil environment, it is, therefore, implied that the longevity of seeds in the soil is likely to vary considerably between species and seed lots, but that some seeds may be very long-lived. Is this so?

The longevity of seeds in soil

The longevity of seeds in soil is epitomized by the gardeners' adage 'One year's seeding: seven years' weeding'. In 1760, Eliot wrote in an essay on field husbandry in New England, 'The seeds of weeds are numerous and hardy, they will lye many years in the ground' (Rasmussen, 1975). De Candolle (1832) observed that 'certain portions of soil, which, by reason of terracing work, were exposed to the air after several centuries, covered themselves the first year with a multitude of individuals belonging to certain species, sometimes uncommon in the vicinity'. He concluded that the seeds of these species had remained viable for considerable periods in the soil and there is much anecdotal evidence like this in the literature (Turner, 1933). However, long-lived seeds are not all weeds: in most temperate and tropical habitats some species form persistent seed banks (Thompson and Grime, 1979; Garwood, 1989; Thompson *et al.*, 1997).

Circumstantial evidence for longevity

Qualitative circumstantial evidence for the longevity of seeds in soil is considerable. For example at Rothamsted Experimental Station, Hertfordshire, UK, a field called Laboratory House Meadow was established as an ungrazed hay meadow in 1859 and, after 58 years without ploughing, 'a large enough number of arable weed seedlings [i.e. 30 seedlings equivalent to 320 seeds m^{-2}] appeared from the soil samples to lead one to assume that a considerable proportion must have been buried in the soil since the time of grassing down'. In contrast, four arable weed seedlings (40 m^{-2}) appeared in soil samples from Park Grass, an old pasture that had not been ploughed for at least 300 years and which had not been grazed by livestock for 40 years (Brenchley, 1918). In a similar type of study, buried viable seeds were examined in successional field and forest stands in the Harvard Forest, Massachusetts. The sites ranged from a field that had been

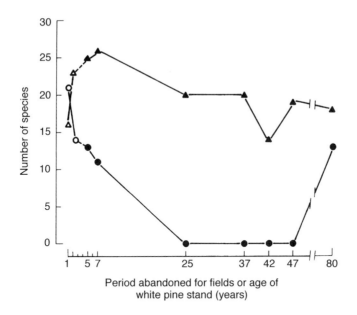

Fig. 8.6. Numbers of species in the ground-cover vegetation (○,●) and in the buried viable seed bank (△,▲) of fields abandoned for different periods (open symbols) and white pine – *Pinus strobus* – stands of different ages (closed symbols) at the Harvard Forest, Massachusetts. To avoid contamination when soil sampling, all leaf litter and humus were removed down to the mineral soil, which was then sampled to a depth of 11.4 cm. (Plotted from data in Livingston and Allessio, 1968.)

abandoned for 1 year to a mature and open 80-year-old pine stand with woody seedlings and ground-cover species. There were no ground-cover plants in the 25–47-year-old plantations (Fig. 8.6), but an average of 189 seedlings (2325 m^{-2}) from about 20 species appeared in soil samples taken from these sites (Fig. 8.6); and many of these were characteristic field species (e.g. *Panicum capillare* and *Potentilla* spp.), rather than woodland plants. The authors proposed that:

> 'the occurrence of viable seed in soils of forest stands of increasing age, together with the complete absence of the same species as ground cover plants is certainly indicative [that the seeds had] remained viable during long burial in the soils'

(Livingston and Allessio, 1968)

These studies are not unique. Where fields have been replaced by forests, disturbances such as fire, cultivation, tree felling or severe thinning are usually followed almost immediately by the appearance of earlier successional species, whose reappearance sometimes relates to viable seeds in the soil (see reviews by Priestley, 1986; Garwood, 1989; Pickett and McDonnell, 1989; Rice, 1989). The converse, however, does not occur. Late successional species do not re-establish from a persistent seed bank (Partridge, 1989).

Further circumstantial evidence for the longevity of seeds in soil is the presence of viable seeds in archaeologically dated soil samples. One of the best known, though least satisfactory, is Porsild *et al.*'s (1967) claim that seeds of *Lupinus arcticus* have remained viable for about 10,000 years in permafrost in the Yukon Territory, Canada. The seeds were retrieved during mining operations from collared lemming burrows buried at depths of 3–6 m in frozen silt. Porsild *et al.* dated the seeds by analogy with other radiocarbon-dated burrows in central Alaska and by inference from the biogeography of the animals, which are no longer found in that region. The claimed longevity has, however, been disputed, since neither the seeds nor the associated animal remains from the Yukon site were dated physically (McGraw and Vavrek, 1989). Odum (1965) recorded one seed of *Chenopodium album* and three of *Spergula arvensis* germinating in a soil sample dated AD 200 from a depth of 120–146 cm at Vestervig in Denmark. Many such examples are available; and some show seeds recovered from considerable depths, even though no seeds were detected at shallower depths and the species was not growing on the site. For example, a soil sample from a previously undisturbed 11th-century grave 1.5 m under the floor of the tower of a medieval church in the village of Uggelose in north Sjaelland in Denmark contained at least one viable seed of *Verbascum thapsiforme*. Odum (1974) concluded that 'the deep-lying and well-protected soil' certifies that this 'viable seed is at least 850 years old'.

These authors have argued that seed-bank species not present in the vegetation at a site must have originated from previous vegetation at that site. Opportunities for short-distance dispersal are, however, easily ignored. Seed-bank species not present in the vegetation 'on site' may occur nearby. For example, in a study of four New Zealand vegetation types (grassland, bracken, scrub and forest), an average of 35% of the seed bank species were growing 'on site', while 60% and 72% were found within 5 m and 10 m, respectively (Partridge, 1989). The simplest explanation for the presence in the seed bank of about half the species not growing 'on site' is that they were dispersed to the site and so there is no conclusive evidence for their longevity.

In the absence of physical dating (e.g. by radiocarbon), it is easy to criticize the circumstantial evidence for the longevity of seeds in soil, even though the sampling procedures may be flawless. Seeds may be carried into grassland, forests or, indeed, any site by birds (Piper, 1986), on people's feet or on agricultural equipment. Seeds may be worked into the ground by animals, including earthworms, whose burrows may extend 210 to 240 cm below the surface (Harper, 1977). And the mere dating of

associated objects and structures is inadequate evidence of longevity. More satisfactory evidence for viable ancient seeds is provided by radiocarbon dating of seeds and ideally of dead parts of the seed, such as the testa (Priestley and Posthumus, 1982).

Radiocarbon dating of two viable seeds of East Indian lotus (*Nelumbo nucifera*) from an ancient lake-bed deposit at Pulantien in southern Manchuria suggested probable ages of 466 and 705 years at the time of germination (Priestley and Posthumus, 1982, and Shen-Miller *et al.*, 1983, respectively). *N. nucifera* seeds are dry fruits, in which the true seed is surrounded by a highly impervious pericarp, which must be scarified prior to germination. Hence, it is the longevity of dry, hard fruits and not of imbibed seeds in soil that is demonstrated by these results.

Classical experiments on seed longevity

Experimental evidence for the longevity of seeds in soil has come from the classical burial trials initiated by W.J. Beal in 1879 at East Lansing, Michigan, and by J.W.T. Duvel in 1902 at Arlington, Virginia. Beal buried 20 bottles, each containing 50 seeds from 20 species. The seeds had been mixed with moist sand and the bottles (473 ml capacity) were buried at a depth of 46 cm in sandy soil with the mouth slanting downwards. After 100 years, 21 viable seeds of *Verbascum blattaria*, and one each of *V. thapsus* and *Malva rotundifolia* were recovered (Kivilaan and Bandurski, 1981), while *Oenothera biennis* and *Rumex crispus* had survived for 80 years (Darlington and Steinbauer, 1961). Duvel's experiment was much larger and the burial conditions more natural. Batches of 100 or 200 seeds of 107 species were mixed with sterilized soil and buried in porous flower pots covered with inverted porous saucers at three depths (20, 56 and 107 cm). Thirty-six species survived for 39 years (Fig. 8.7), with over 80% survival of *Datura stramonium*, *Solanum nigrum* and *Phytolacca americana*. There was negligible survival

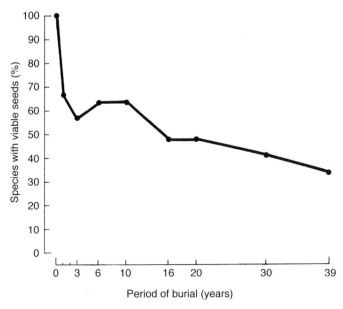

Fig. 8.7. Percentage of species with surviving seeds after various periods in Duvel's buried-seed experiment. Seeds of 107 species were buried in soil. See text for further details of the experiment. (Plotted from data in Toole and Brown, 1946.)

of 22 crop species even for 1 year; but some seeds of tobacco (*Nicotiana tabacum*), celery (*Apium graveolens*), red clover (*Trifolium pratense*) and Kentucky blue-grass (*Poa pratensis*) persisted to the end of the experiment (Toole and Brown, 1946). The short longevity of most crop species in soil seed banks, other than some forage/pasture grasses and legumes, has been confirmed in several other studies (Madsen, 1962; Rampton and Ching, 1970; Lewis, 1973). Results of similar long-term experiments have been published after 26 years' burial in soil in Denmark (Madsen, 1962) and after 20 years' burial in three soil types in England (Lewis, 1973). These experiments were all in cool temperate environments. More recently, 50-year studies have started in the warm, humid, southern USA (Egley and Chandler, 1983) and in subarctic Alaska (Conn, 1990).

Extensive information on the survival of seeds of weeds and ruderals in cultivated soils after shorter periods (5 or more years) is also available – for example, from studies in Canada, England, France and the USA (Chepil, 1946; Roberts, 1964, 1979,

1986; Roberts and Chancellor, 1979; Roberts and Neilson, 1980, 1981; Roberts and Boddrell, 1983a, b; 1984a, b; Barralis *et al.*, 1988; Egley and Williams, 1990).

Quantifying the longevity of seeds in soil

The above studies confirm, albeit qualitatively, that many species form persistent seed banks in the soil in a wide range of environments. Quantifying their longevity is useful, in order to put what may be the performance of extreme individuals into perspective.

In the absence of further seed influx, depletion of persistent seed banks does not correspond to the negative cumulative normal distribution that characterizes the survival curves of seeds in dry and moist laboratory storage. Instead, persistent seed banks tend to decrease exponentially on a year-to-year basis over periods of at least 5 years. In the UK, this negative exponential model was applicable to the depletion of natural populations of mixed species and for the principal species separately

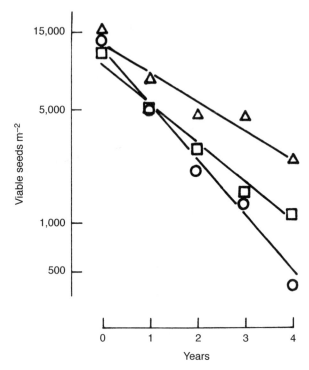

Fig. 8.8. Depletion of naturally occurring populations of viable seeds in the top 23 cm of soils dug seven times a year (○); dug twice a year (□); and undisturbed (△). Weeds were prevented from seeding during the 4 years of the experiment by spraying the plots with paraquat/diquat and by hand-weeding. Surrounding areas were also kept clean by cultivation and spraying to minimize the seed influx. (Adapted from Roberts and Feast, 1973b.)

(Roberts and Dawkins, 1967; Roberts and Feast, 1973b; Fig. 8.8). It has also been applied satisfactorily to some, but not all, species when seeds have been planted in England (Roberts and Feast, 1972, 1973a), Japan (Watanabe and Hirokawa, 1975a) and France (Barralis *et al.*, 1988) and in Mississippi and Alaska in the USA (Egley and Chandler, 1983; Conn, 1990). According to the model, the viable seed population density, *S*, after *t* years is estimated as follows:

$$S = S_0 \, e^{-gt} \qquad (8.4)$$

where S_0 is the initial seed population density and *g* is the annual rate of depletion in a given environment (Roberts and Dawkins, 1967; Roberts, 1972c).

Egley and Chandler (1983) found that the negative exponential model satisfactorily accounted for the depletion of buried seeds of only ten out of 20 persistent weed

species. The seven most persistent species (*Abutilon theophrasti*, *Anoda cristata*, *Cassia obtusifolia*, *Ipomoea lacunosa*, *I. turbinata*, *Sesbania exaltata* and *Sorghum halepense*) showed some deviation from this model, and the first six of these exhibited hard-seededness, which may confuse the trends, since the physiological basis of longevity differs between dry, hard seed and imbibed, soft seed (see above; Fig. 8.5). Moreover, only in *C. obtusifolia* did depletion exceed 90% over 5.5 years and we would suggest that deviations from linearity are difficult to deduce unless there is depletion by one or two orders of magnitude (i.e. from 100 to about 10 or 1%). Depletion of *C. obtusifolia* is, however, difficult to explain by any model, since viability was recorded as increasing between 1.5 and 3.5 years in the soil, perhaps because of loss of dormancy during this period, combined with inadequate dormancy-breaking treatments in the seed germination tests intended to estimate viability. Variability also explains the poor fit for *Stellaria media* and *Desmodium tortuosum*. Other workers have reported exponential depletion of naturally and artificially buried seeds of *S. media* (Roberts and Dawkins, 1967; Roberts and Feast, 1973a; Conn, 1990). The tenth species that showed non-linear depletion in Egley and Chandler's results was *Echinochloa crus-galli*. A fairly non-dormant seed lot was used, which showed higher than average depletion in the first 6 months of burial, while subsequent depletion seems to be approximately linear in the published results. Watanabe and Hirokawa (1975a) found exponential depletion of *E. crus-galli* in Japan. Higher or lower than average depletion immediately after burial also explains some non-linearity in the depletion of 17 weed species in France (Barralis *et al.*, 1988). Barralis *et al.* buried seeds by ploughing after hand-sowing 17 weed species. Depletion was monitored for 5 years in cultivated soils, which were either uncropped or cropped with spring barley or winter wheat monocultures. Over 5 years, there was a tendency for the rate of depletion to decrease slightly with increasing period of burial, suggesting a greater aptitude for survival in old seeds. This conclusion was, however, confused with greater than average depletion in the first year in some species, and the survival curves were based on recovered intact seeds. Where survival curves were also presented for recovered viable seeds, the slight non-linearity seems to be reduced.

A rectangular hyperbola has been proposed as an alternative empirical model for seed survival in soil, and has sometimes given better fits to experimental data than the negative exponential model (Burnside *et al.*, 1981; Conn, 1990). In the rectangular hyperbolic model, the viable seed population density, S, after t years is estimated as follows:

$$S = a / (t + b) \qquad (8.5)$$

where a and b are constants without direct biological significance (Burnside *et al.*,

1981). The model is logical in that the seed population is inversely proportional to time. The model implies that the rate of depletion decreases with increasing periods of burial, and this may explain why it sometimes appears to give a better fit than the negative exponential model, since different rates of depletion are sometimes observed between the first and subsequent years of burial. There is no parameter that estimates the rate of decline – the constant a is the change in the population density per unit change in $1 / (t + b)$. The initial seed density (S_0) is the ratio $a : b$. Thus, for a given initial seed population, the parameters a and b could both be either positive or negative and either large or small. Hence, the parameters cannot be used to compare the depletion of different seed lots or species and provide little information of biological significance (Conn, 1990).

The usefulness of the simple negative exponential model is not, therefore, in question. But three cautions are needed. First, unless seed populations decline by one or two orders of magnitude over the period of an experiment, it may be difficult to test for conformity to the negative exponential model. Secondly, seed losses immediately after burial associated with the germination of relatively non-dormant seeds may need to be distinguished from the subsequent losses. For example, when modelling the economics of controlling *Avena fatua* in winter wheat and spring barley crops, 50 or 60% depletion was assumed for the first year of burial, compared with 65 or 75% per annum thereafter, according to a negative exponential model of depletion in ploughed or shallow-tine-cultivated land, respectively (Cousens *et al.*, 1986). Finally and most importantly, these rates of decay only apply on a year-to-year basis for a given environment. Contrary to some reports in the literature (for example, Sarukhan, 1974; Yadav and Tripathi, 1982), exponential decay does not apply within a given year, since depletion usually corresponds to the timing of germination and emergence, which is well known to be seasonal (Watanabe and Hirokawa, 1975b; Wilson, 1978, 1981, 1985).

The rate of depletion (g) varies with

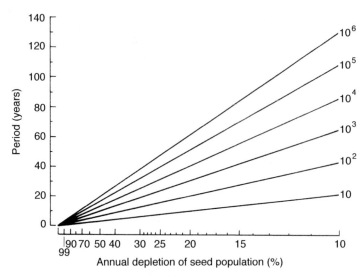

Fig. 8.9. The time required to deplete soil seed banks by factors of ten to 1 million (10 to 10^6 shown against each line) as a function of the annual rate of depletion. The negative exponential model described in the text is assumed, so that the rate of depletion remains the same from year to year in a given environment. The depletion factor is the ratio of the initial to the final seed population (i.e. S_0/S in Equation 8.4). Linearity is achieved by plotting the predicted depletion period against the reciprocal of the rate of depletion (i.e. $1/g$ in Equation 8.4).

species and increases with increasing frequency of cultivation (Roberts and Dawkins, 1967; Roberts and Feast, 1973b; Fig. 8.8). The exponential decay means that the probability of depletion remains constant from year to year and that persistent seed banks have a constant half-life. Moreover, if the survival curves are extrapolated, the time taken for depletion by given amounts at different depletion rates can be predicted (Fig. 8.9). Hence, although elimination is impossible according to the exponential decay model, a population that decreased from 1000 seeds m^{-2} to 100 ha^{-1} – that is, by a factor of 10^5 – could be considered eliminated. Such a decrease would only be predicted to occur within a reasonable time-scale (say, 7 years) if the rate of depletion exceeded 80% per annum (Fig. 8.9). At the rates of depletion observed by Roberts and Feast (1973b) for undisturbed soil and for soils cultivated two and seven times a year (34, 42 and 56% per annum: Fig. 8.8), it would take 28, 22 and 14 years, respectively, to deplete the populations by a factor of 10^5 (Fig. 8.9).

Seeds may be lost from the seed bank by various means. Loss of viability is only one of several means of depletion and is likely to be greater for seeds at or near the soil surface or in semi-arid or arid regions. For example, the predicted annual loss of viability for a seed lot of *Orobanche aegyptiaca* with high initial viability ($K_i = 2.89$) was as high as 38% in Eritrea (Table 8.2; Kebreab and Murdoch, 1999b). Although predation and loss of viability can be important, especially for larger seeds at the soil surface, most studies do, however, suggest that germination is the primary cause of depletion of buried seeds (Schafer and Chilcote, 1970; Taylorson, 1970, 1972; Murdoch and Roberts, 1982; Zorner *et al.*, 1984). Hence it is not surprising that survival curves for seeds in the soil differ from the normal distributions found in the laboratory, where loss of viability is the sole means of depletion. In most investigations, large numbers of seeds disappear without emerging (Chepil, 1946; Roberts, 1964, 1979, 1986; Roberts and Chancellor, 1979; Roberts and Neilson, 1980, 1981; Roberts and Boddrell, 1983a, b, 1984a, b) Fatal germination (i.e. germination of the seed at

Table 8.2. Prediction of loss of viability of *Orobanche aegyptiaca* seeds under rain-fed farm conditions in the highlands of Eritrea in a full year. Meteorological data from FAO (1994). (From Kebreab and Murdoch, 1999b.)

Season	Mean temperature (°C)	Relative humidity (%)	Period (months)	Predicted loss of viability (%)[a]
Early dry season	20	53	5	3
Late dry season	26	40	5	2.5
Rainy season	25	Imbibed	2	33
Total (over 1 year)				38.5

[a] Assumes $K_i = 2.89$ (Equations 8.1 and 8.3).

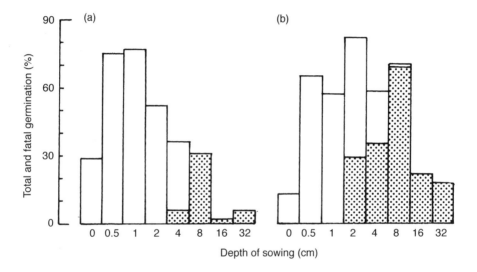

Fig. 8.10. Emergence (open areas of histograms) and fatal germination (shaded areas of histograms) of seedlings from seeds sown at various depths. The experiment was carried out in the field on red sandy loam at Nairobi, Kenya, using two common local ruderals, (a) *Bidens pilosa* and (b) *Achyranthes aspera*. Seedlings were counted 2 weeks after sowing. Germination decreased with depth below 1 or 2 cm, but almost all seeds germinating at depths of 8, 16 and 32 cm failed to emerge. (From Fenner, 1985.)

a depth from which it fails to emerge) is the usual explanation and this is supported by experimental evidence (Murdoch, 1983; Fenner, 1985; Fig. 8.10). However, in most long-term studies of the soil seed bank, loss of viability due to disease and ageing is indistinguishable from fatal germination, because dead seeds and non-emergent seedlings both decay in moist soil. For the same reason, disease and ageing cannot be separated.

Contrasting strategies for plant regeneration by seed

The longevity of seeds in soil is probably the most useful way to classify the seed banks of individual species, since the transience or persistence of seeds in soil is part of the regeneration strategy adopted by a species. However, it is possible for the same longevity of seeds in soil seed banks to be achieved by different strategies. This review has attempted to identify the key aspects of seed dormancy, viability and longevity that influence the duration of the seed–seedling phase of temporal dispersal. Individual dormant seeds of orthodox species may survive for very long periods in soil, but this is only one of four contrasting strategies for plant regeneration by seed.

Non-dormant recalcitrant seeds

Primary dormancy appears to be totally lacking in those recalcitrant seeds from the tropics that cannot be chilled below 10–15°C without damage (King and Roberts, 1979). Berjak et al. (1989) have proposed the hypothesis that the germination of such recalcitrant seeds (which are imbibed when shed) is initiated at or around shedding, when the inhibitory effect of the maternal environment may be relieved. Thus, in extreme examples of this strategy, there may be virtually no temporal dispersal, with seed development, germination and establishment representing a continuum.

One of the most likely adaptive functions of primary dormancy – dispersal of germination and hence regeneration over time – may nevertheless be achieved by such species despite this strategy. For example, the large non-dormant recalcitrant seeds of the primary species of some tropical forests germinate rapidly but produce persistent seedling banks. Some of these seedlings may eventually benefit from a gap in the vegetation caused, for example, by a fallen tree (Whitmore, 1988; Garwood, 1989).

Dormant recalcitrant seeds

Some recalcitrant seeds do show dormancy, particularly those from temperate latitudes. For example, seed dormancy in sycamore (Acer pseudoplatanus) can be considerable when the fruits are shed (Hong and Ellis, 1990), and this is largely testa-imposed dormancy (Pinfield et al., 1987, 1990). Nevertheless, the seeds' per-

sistence is not great, as dormancy is generally lost by early spring as a result of prechilling.

Non-dormant orthodox seeds

Some of the greatest longevities of seeds in the soil are achieved by seeds in which there may be no primary dormancy, but the seeds are innately quiescent because hard-seededness prevents imbibition. In a sense, this strategy is closest to conventional crop seed storage. Hard-seededness enables the seeds to maintain their moisture content at values approaching the lower moisture content limit to the negative logarithmic relation between seed longevity and moisture, and so their longevity is considerable. Another example of this strategy is represented by non-dormant orthodox seeds in dry regions (e.g. the Sahel or in the Mediterranean summers), where persistence is associated with enforced quiescence. However, most species in this category germinate rapidly after shedding, provided moisture is available, and are therefore transient in the soil seed bank.

Dormant orthodox seeds

In this strategy, the seeds may be imbibed for some or all of the time. The maintenance of viability depends upon the regular repair of deteriorative changes during periods spent at high water potentials. Longevity depends upon the primary dormancy present at shedding either not being lost or being reinforced by secondary dormancy. In this strategy, variation in the degree of dormancy among the individual seeds within each influx provides both a regular proportion of (surviving) seeds that germinate each year and the potential for considerable persistence in extreme individuals. Long-lived orthodox seeds (whether their longevity depends on dormancy or quiescence) can have a crucial role in regeneration of plant communities if land-use changes have led to loss of those species from the vegetation.

Each of these strategies is based on differences in seed physiology and each is successful, since at least some seeds are in the right physiological state to germinate and emerge in the right place at the right time. Others have quantified the relative success of different strategies (e.g. Russi *et al.*, 1992a). But, by using the simple models described in this chapter, it is possible to quantify and compare the responses of seed populations with similar physiology to environmental factors. In this way, the behaviour of extreme individuals can be placed in perspective. In addition, these models may ultimately be used to predict the responses of buried seed populations in different environments.

References

Amen, R.D. (1968) A model of seed dormancy. *Botanical Review* 34, 1–31.

Aspinall, D.J. (1965) Effects of soil moisture stress on the growth of barley. III. Germination of grain from plants subject to water stress. *Journal of the Institute of Brewing* 72, 174–176.

Atwater, B.R. (1980) Germination, dormancy and morphology of the seeds of herbaceous ornamental plants. *Seed Science and Technology* 8, 523–573.

Aufhammer, G. and Simon, U. (1957) Die Samen landwirtschaftlicher Kulturpflanzen im Grundstein des chamaligen Nürnberger Stadttheaters und ihre Keimfähigkeit. *Zeitschrift für Acker- und Pflanzenbau* 103, 454–472.

Ballard, L.A.T. (1973) Physical barriers to germination. *Seed Science and Technology* 1, 285–303.

Barralis, G., Chadoeuf, R. and Lonchamp, J.P. (1988) Longévité des semences de mauvaises herbes annuelles dans un sol cultivé. *Weed Research* 28, 407–418.

Bartley, M.R. and Frankland, B. (1982) Analysis of the dual role of phytochrome in the photoinhibition of seed germination. *Nature (London)* 300, 750–752.

Barton, L.V. (1945) Respiration and germination studies of seeds in moist storage. *Annals of the New York Academy of Science* 46, 185–200.

Baskin, C.C. and Baskin, J.M. (1998) *Seeds: Ecology, Biogeography and Evolution of Dormancy and Germination.* Academic Press, San Diego, California, 666 pp.

Baskin, J.M. and Baskin, C.C. (1985) The annual dormancy cycle in buried weed seeds: a continuum. *Bioscience* 35, 492–498.

Benech Arnold, R.L. and Sanchez, R.A. (1995) Modelling weed seed germination. In: Kigel, J. and Galili, G. (eds) *Seed Development and Germination.* Marcel Dekker, New York, pp. 545–566.

Benech Arnold, R.L., Fenner, M. and Edwards, P.J. (1992) Changes in dormancy level in *Sorghum halepense* seeds induced by water stress during seed development. *Functional Ecology* 6, 596–605.

Berjak, P., Farrant, J.M. and Pammenter, N.W. (1989) The basis of recalcitrant seed behaviour: cell biology of the homoiohydrous seed condition. In: Taylorson, R.B. (ed.) *Recent Advances in the Development and Germination of Seeds.* Plenum Press, New York, pp. 89–108.

Bewley, J.D. (1997) Seed germination and dormancy. *The Plant Cell* 9, 1055–1066.

Bewley, J.D. and Black, M. (1994) *Seeds: Physiology of Development and Germination.* Plenum Press, New York, 445 pp.

Bewley, J.D. and Downie, B. (1996) Is failure of seeds to germinate during development a dormancy-related phenomenon? In: Lang, G.A. (ed.) *Plant Dormancy: Physiology, Biochemistry and Molecular Biology.* CABI, Wallingford, UK, pp. 17–27.

Bonner, F.T. (1990) Storage of seeds: potential and limitations for germplasm conservation. *Forest Ecology Management* 35, 35–43.

Borthwick, H.A., Hendricks, S.B., Toole, E.H. and Toole, V.K. (1954) Action of light on lettuce-seed germination. *Botanical Gazeteer* 115, 205–225.

Bouwmeester, H.J. (1990) The effect of environmental conditions on the seasonal dormancy pattern and germination of weed seeds. PhD thesis, Agricultural University in Wageningen, The Netherlands.

Bradford, K.J. (1995) Water relations in seed germination. In: Kigel, J. and Galili, G. (eds) *Seed Development and Germination.* Marcel Dekker, New York, pp. 351–396.

Bradford, K.J. (1996) Population-based models describing seed dormancy behaviour: implications for experimental design and interpretation. In: Lang, G.A. (ed.) *Plant Dormancy: Physiology, Biochemistry and Molecular Biology.* CABI, Wallingford, UK, pp. 313–339.

Brenchley, W.E. (1918) Buried weed seeds. *Journal of Agricultural Science* 9, 1–31.

Burford, J.R. and Millington, R.J. (1968) Nitrous oxide in the atmosphere of a red-brown earth. In: *Transactions of the 9th International Congress of Soil Science, Adelaide,* Vol. 2, pp. 505–511.

Burford, J.R. and Stefanson, R.C. (1973) Measurement of gaseous losses of nitrogen from soils. *Soil Biology and Biochemistry* 5, 133–141.

Burnside, O.V., Fenster, C.R., Evetts, L.L. and Mumm, R.F. (1981) Germination of exhumed weed seed in Nebraska. *Weed Science* 29, 577–586.

Chepil, W.S. (1946) Germination of weed seeds. 1. Longevity, periodicity of germination and vitality of seeds in cultivated soil. *Scientific Agriculture* 26, 307–346.

Christensen, M., Meyer, S.E. and Allen, P.S. (1996) A hydrothermal time model of seed after-ripening in *Bromus tectorum* L. *Seed Science Research* 6, 155–163.

Cohn, M.A. (1996) Chemical mechanisms of breaking seed dormancy. *Seed Science Research* 6, 95–99.

Conn, J.S. (1990) Seed viability and dormancy of 17 weed species after burial for 4.7 years in Alaska. *Weed Science* 38, 134–138.

Courtney, A.D. (1968) Seed dormancy and field emergence in *Polygonum aviculare*. *Journal of Applied Ecology* 5, 675–684.

Cousens, R., Doyle, C.J., Wilson, B.J. and Cussans, G.W. (1986) Modelling the economics of controlling *Avena fatua* in winter wheat. *Pesticide Science* 17, 1–12.

Cresswell, E.G. and Grime, J.P. (1981) Induction of a light requirement during seed development and its ecological consequences. *Nature* 291, 583–585.

Darlington, H.T. and Steinbauer, G.P. (1961) The eighty-year period for Dr. Beal's seed viability experiment. *American Journal of Botany* 48, 321–325.

De Candolle, A.P. (1832) *Physiologie Végétale* ii, 620. (Cited by Turner, 1933.)

Dickie, J.B., Ellis, R.H., Kraak, H.L., Ryder, K. and Tompsett, P.B. (1990) Temperature and seed storage longevity. *Annals of Botany* 55, 147–151.

Duke, S.O. (1978) Significance of fluence-response data in phytochrome-initiated seed germination. *Photochemistry and Photobiology* 28, 383–388.

Egley, G.H. (1989a) Water-impermeable seed coverings as barriers to germination. In: Taylorson, R.B. (ed.) *Recent Advances in the Development and Germination of Seeds*. Plenum Press, New York, pp. 207–223.

Egley, G.H. (1989b) Some effects of nitrate-treated soil upon the sensitivity of buried redroot pigweed (*Amaranthus retroflexus* L.) seeds to ethylene, temperature, light and carbon dioxide. *Plant, Cell and Environment* 12, 581–588.

Egley, G.H. and Chandler, J.M. (1983) Longevity of weed seeds after 5.5 years in the Stoneville 50-year buried-seed study. *Weed Science* 32, 264–270.

Egley, G.H. and Williams, R.D. (1990) Decline of weed seeds and seedling emergence over five years as affected by soil disturbances. *Weed Science* 38, 504–510.

Eliot, J. (1760) *Essays upon field husbandry in New England, as it is or may be ordered.* Boston, Mass. U.S.A. (Reprinted in Rasmussen, 1975).

Ellis, R.H. and Roberts, E.H. (1980a) Improved equations for the prediction of seed longevity. *Annals of Botany* 45, 13–30.

Ellis, R.H. and Roberts, E.H. (1980b) The influence of temperature and moisture on seed viability period in barley (*Hordeum distichum* L.). *Annals of Botany* 45, 31–37.

Ellis, R.H. and Roberts, E.H. (1981) The quantification of ageing and survival in orthodox seeds. *Seed Science and Technology* 9, 373–409.

Ellis, R.H., Osei-Bonsu, K. and Roberts, E.H. (1982) The influence of genotype, temperature and moisture on seed longevity in chickpea, cowpea and soya bean. *Annals of Botany* 50, 69–82.

Ellis, R.H., Hong, T.D. and Roberts, E.H. (1983) Procedures for the safe removal of dormancy from rice seed. *Seed Science and Technology* 11, 72–112.

Ellis, R.H., Hong, T.D. and Roberts, E.H. (1985) *Handbook of Seed Technology for Genebanks*. Vol. II. *Compendium of Specific Germination Information and Test Recommendations*. International Board for Plant Genetic Resources, Rome, pp. 211–667.

Ellis, R.H., Hong, T.D. and Roberts, E.H. (1986) Quantal response of seed germination in *Brachiaria humidicola*, *Echinochloa turnerana*, *Eragrostis tef* and *Panicum maximum* to photon dose for the low energy reaction and the high irradiance reaction. *Journal of Experimental Botany* 37, 742–753.

Ellis, R.H., Simon, G. and Covell, S. (1987) The influence of temperature on seed germination rate in grain legumes. III. A comparison of five faba bean genotypes using a new screening method. *Journal of Experimental Botany* 38, 1033–1043.

Ellis, R.H., Hong, T.D. and Roberts, E.H. (1988) A low-moisture-content limit to logarithmic relations between seed moisture content and longevity. *Annals of Botany* 61, 405–408.

Ellis, R.H., Hong, T.D. and Roberts, E.H. (1989) A comparison of the low-moisture-content limit to the logarithmic relation between seed moisture and longevity in twelve species. *Annals of Botany* 63, 601–611.

Ellis, R.H., Hong, T.D. and Roberts, E.H. (1990a) An intermediate category of seed storage behaviour? I. Coffee. *Journal of Experimental Botany* 41, 1167–1174.

Ellis, R.H., Hong, T.D., Roberts, E.H. and Tao, K.-L. (1990b) Low-moisture-content limits to relations between seed longevity and moisture. *Annals of Botany* 65, 493–504.

Ellis, R.H., Hong, T.D. and Roberts, E.H. (1991a) Effect of storage temperature and moisture on the germination of papaya seeds. *Seed Science Research* 1, 69–72.

Ellis, R.H., Hong, T.D., Roberts, E.H. and Soetisna, U. (1991b) Seed storage behaviour in *Elaeis guineensis*. *Seed Science Research* 1, 99–104.

Enoch, H. and Dasberg, S. (1971) The occurrence of high CO_2 concentrations in soil air. *Geoderma* 6, 17–21.

Esashi, Y. and Leopold, C. (1968) Physical forces in dormancy and germination of *Xanthium* seeds. *Plant Physiology* 43, 871–876.

Ewart, A.J. (1908) On the longevity of seeds. *Proceedings of the Royal Society of Victoria* 21, 1–210.

FAO (1994) *Eritrea – Agricultural Sector Review and Project Identification. Report for the Government of Eritrea*. Food and Agriculture Organization of the United Nations, Rome.

Favier, J.F. and Woods, J.L. (1993) The quantification of dormancy loss in barley (*Hordeum vulgare* L.). *Seed Science and Technology* 21, 653–674.

Fenner, M. (1985) *Seed Ecology*. Chapman and Hall, London, 151 pp.

Fenner, M. (1991) The effects of the parent environment on seed germinability. *Seed Science Research* 1, 75–84.

Fenner, M. (1992) Environmental influences on seed size and composition. *Horticulture Reviews* 13, 183–213.

Foley, M. (1996) Carbohydrate metabolism as a physiological regulator of seed dormancy. In: Lang, G.A. (ed.) *Plant Dormancy: Physiology, Biochemistry and Molecular Biology*. CABI, Wallingford, UK, pp. 245–256.

Francisco-Ortega, J., Ellis, R.H., González-Feria, E. and Santos-Guerra, A. (1994) Overcoming seed dormancy in *ex situ* plant germplasm conservation programmes: an example in the endemic *Argyranthemum* (Asteraceae: Anthemideae) species from the Canary Islands. *Biodiversity and Conservation* 3, 341–353.

Frankland, B. and Taylorson, R.B. (1983) Light control of seed germination. In: Shropshire, W., Jr and Mohr, H. (eds) *Encyclopaedia of Plant Physiology*, Vol. 16. Springer-Verlag, Heidelberg, pp. 428–456.

Garwood, N.C. (1989) Tropical soil seed banks: a review. In: Leck, M.A., Parker, V.T. and Simpson, R.L. (eds) *Ecology of Soil Seed Banks*. Academic Press, San Diego, California, pp. 149–209.

Harper, J.L. (1957) The ecological significance of dormancy and its importance in weed control. *International Congress of Plant Protection* 4, 415–420.

Harper, J.L. (1977) *Population Biology of Plants*. Academic Press, London.

Harper, J.L. (1988) An apophasis of plant population biology. In: Davy, A.J., Hutchings, M.J. and Watkinson, A.R. (eds) *Plant Population Ecology*. Blackwell Scientific Publications, Oxford, pp. 435–452.

Harrington, J.F. (1972). Seed storage and longevity. In: Kozlowski, T.T. (ed.) *Seed Biology*, Vol. III. Academic Press, New York, pp. 145–245.

Hartmann, K.M., Mollwo, A. and Tebbe, A. (1998) Photo control of germination by moon- and starlight. *Zeitschrift fur Pflanzenkrankheit und PflanzenSchutz* 16, 119–127.

Hewlett, P.S. and Plackett, R.L. (1979) *An Introduction to the Interpretation of Quantal Responses in Biology*. Edward Arnold, London.

Hilhorst, H.W.M. (1990a) Dose–response analysis of factors involved in germination and secondary dormancy of seeds of *Sisymbrium officinale*. I. Phytochrome. *Plant Physiology* 94, 1090–1095.

Hilhorst, H.W.M. (1990b) Dose–response analysis of factors involved in germination and secondary dormancy of seeds of *Sisymbrium officinale*. II. Nitrate. *Plant Physiology* 94, 1096–1102.

Hilhorst, H.W.M. and Karssen, C.M. (1988) Dual effect of light on the gibberellin- and nitrate-stimulated seed germination of *Sisymbrium officinale* and *Arabidopsis thaliana*. *Plant Physiology* 86, 591–597.

Hilhorst, H.W.M., Derkx, M.P.M. and Karssen, C.E. (1996) An integrating model for seed dormancy cycling: characterisation of reversible dormancy. In: Lang, G.A. (ed.) *Plant Dormancy: Physiology, Biochemistry and Molecular Biology*. CABI, Wallingford, UK, pp. 341–360.

Hong, T.D. and Ellis, R.H. (1990) A comparison of maturation drying, germination, and desiccation tolerance between developing seeds of *Acer pseudoplatanus* L. and *Acer platanoides* L. *New Phytologist* 116, 589–596.

Hong, T.D. and Ellis, R.H. (1996) *A Protocol to Determine Seed Storage Behaviour*. International Plant Genetic Resources Institute, Rome, 64 pp.

Hong, T.D., Linington, S. and Ellis, R.H. (1996) *Seed Storage Behaviour: a Compendium*. Handbooks for Genebanks, No. 4, International Plant Genetic Resources Institute, Rome, 656 pp.

Hong, T.D., Ellis, R.H. and Moore, D. (1997) Development of a model to predict the effect of temperature and moisture on fungal spore longevity. *Annals of Botany* 79, 121–128.

Hong, T.D., Ellis, R.H., Buitink, J., Walters, C., Hoekstra, F.A. and Crane, J. (1999) A model of the effect of temperature and moisture on pollen longevity in air-dry storage environments. *Annals of Botany* 83, 167–173.

Hyde, E.O.C. (1954) The function of the hilum in some Papilionaceae in relation to the ripening of the seed and the permeability of the testa. *Annals of Botany* 18, 241–256.

Ibrahim, A.E. and Roberts, E.H. (1983) Viability of lettuce seeds. I. Survival in hermetic storage. *Journal of Experimental Botany* 34, 620–630.

Ibrahim, A.E., Roberts, E.H. and Murdoch, A.J. (1983) Viability of lettuce seeds. II. Survival and oxygen uptake in osmotically controlled storage. *Journal of Experimental Botany* 34, 631–640.

Jones, S.K., Ellis, R.H. and Gosling, P.G. (1997) Loss and induction of conditional dormancy in seeds of Sitka spruce maintained moist at different temperatures. *Seed Science Research* 7, 351–358.

Jones, S.K., Gosling, P.G. and Ellis, R.H. (1998) Reimposition of conditional dormancy during air-dry storage of prechilled Sitka spruce seeds. *Seed Science Research* 8, 113–122.

Karssen, C.M. (1970) The light promoted germination of the seeds of *Chenopodium album* L. III. Effect of the photoperiod during growth and development of the plants on the dormancy of the produced seeds. *Acta Botanica Neerlandica* 19, 81–94.

Karssen, C.M. (1982) Seasonal patterns of dormancy in weed seeds. In: Khan, A.A. (ed.) *The Physiology and Biochemistry of Seed Development, Dormancy and Germination*. Elsevier/North Holland Biomedical Press, Amsterdam, pp. 243–271.

Karssen, C.M., Brinkhorst-Van der Swan, D.L.C., Breekland, A.E. and Koorneef, M. (1983) Induction of dormancy during seed development by endogenous abscisic acid: studies on abscisic acid deficient genotypes of *Arabidopsis thaliana* (L.) Heynh. *Planta* 157, 158.

Kebreab, E. and Murdoch, A.J. (1999a) A quantitative model for loss of primary dormancy and induction of secondary dormancy in imbibed seeds of *Orobanche* spp. *Journal of Experimental Botany* 50, 211–219.

Kebreab, E. and Murdoch, A.J. (1999b) Effect of moisture and temperature on the longevity of *Orobanche* seeds. *Weed Research* 39, 199–211.

Kebreab, E. and Murdoch, A.J. (1999c) Modelling the effects of water stress and temperature on germination rate of *Orobanche aegyptiaca* seeds. *Journal of Experimental Botany* 50, 655–664.

King, M.W. and Roberts, E.H. (1979) *The Storage of Recalcitrant Seeds: Achievements and Possible Approaches*. International Board for Plant Genetic Resources, Rome.

Kivilaan, A. and Bandurski, R.S. (1981) The one hundred-year period for Dr Beal's seed viability experiment. *American Journal of Botany* 68, 1290–1292.

Kraak, H.L. and Vos, J. (1987) Seed viability constants for lettuce. *Annals of Botany* 59, 353–359.

Kramer, P.J. (1983) *Water Relations of Plants*. Academic Press, New York.

Kuo, W.H.J. (1994) Seed germination of *Cyrtococcum patens* under alternating temperature regimes. *Seed Science and Technology* 22, 43–50.

Labouriau, L.G. (1970) On the physiology of seed germination in *Vicia graminea* Sm. *Anais da Academia Brasileira de Ciências* 42, 235–262

Lewis, J. (1973) Longevity of crop and weed seeds: survival after 20 years in soil. *Weed Research* 13, 179–191.

Livingston, R.B. and Allessio, M.L. (1968) Buried viable seed in successional field and forest stands, Harvard Forest, Massachusetts. *Bulletin Torrey Botanical Club* 95, 58–69.

McEvoy, P.B. (1984) Dormancy and dispersal in dimorphic achenes of tansy ragwort, *Senecio jacobaea* L. (Compositae). *Oecologia (Berlin)* 61, 160–168.

McGraw, J.B. and Vavrek, M.C. (1989) The role of buried viable seeds in Arctic and alpine plant communities. In: Leck, M.A., Parker, V.T. and Simpson, R.L. (eds) *Ecology of Soil Seed Banks*. Academic Press, San Diego, California, pp. 91–105.

Madsen, S.B. (1962) Germination of buried and dry stored seeds, III, 1934–60. *Proceedings of the International Seed Testing Association* 27, 920–928.

Mead, A. and Gray, D. (1999). Prediction of seed longevity: a modification of the shape of the Ellis and Roberts seed survival curves. *Seed Science Research* 9, 63–73.

Mott, J.J., Winter, W.H. and McLean, R.W. (1989) Management options for increasing the productivity of tropical savannah pastures. IV. Population biology of introduced *Stylosanthes* spp. *Australian Journal of Agricultural Research* 40, 1227–1240.

Murdoch, A.J. (1983) Environmental control of germination and emergence in *Avena fatua*. *Aspects of Applied Biology* 4, 63–69.

Murdoch, A.J. (1998) Dormancy cycles of weed seeds in soil. *Aspects of Applied Biology* 51, 119–126.

Murdoch, A.J. and Roberts, E.H. (1982) Biological and financial criteria of long-term control strategies for annual weeds. In: *Proceedings of the 1982 British Crop Protection Conference – Weeds*. British Crop Protection Council, Farnham, pp. 741–748.

Murdoch, A.J. and Roberts, E.H. (1996) Dormancy cycle of *Avena fatua* seeds in soil. In: *Proceedings of the Second International Weed Control Congress Copenhagen, Denmark*, Vol. 1. Department of Weed Control and Pesticide Ecology, Flakkebjerg, Denmark, pp. 147–152.

Murdoch, A.J., Roberts, E.H. and Goedert, C.O. (1989) A model for germination responses to alternating temperatures. *Annals of Botany* 63, 97–111.

Murdoch, A.J., Sonko, L. and Kebreab, E. (2000) Population responses to temperature for loss and induction of seed dormancy and consequences for predictive empirical modelling. In: Viémont, J.-D. and Crabbé, J. (eds) *Dormancy in Plants*. CAB International, Wallingford, UK, pp. 57–68.

Odum, S. (1965) Germination of ancient seeds: floristical observations and experiments with archaeologically dated soil samples. *Dansk Botanisk Arkiv* 24, 1–70.

Odum, S. (1974) Seeds in ruderal soils, their longevity and contribution to the flora of disturbed ground in Denmark. In: *Proceedings of the 12th British Weed Control Conference*. British Crop Protection Council, Farnham, pp. 1131–1144.

Partridge, T.R. (1989) Soil seed banks of secondary vegetation on the Port Hills and Banks Peninsula, Canterbury, New Zealand, and their role in succession. *New Zealand Journal of Botany* 27, 421–436.

Peters, N.C.B. (1982a) Production and dormancy of wild oat (*Avena fatua*) seed from plants grown under soil waterstress. *Annals of Applied Biology* 100, 189–196.

Peters, N.C.B. (1982b) The dormancy of wild oat seed (*Avena fatua* L.) from plants grown under various temperature and soil moisture conditions. *Weed Research* 22, 205–212.

Pickett, S.T.A. and McDonnell, M.J. (1989) Seed bank dynamics in temperate deciduous forest. In: Leck, M.A., Parker, V.T. and Simpson, R.L. (eds) *Ecology of Soil Seed Banks*. Academic Press, San Diego, California, pp. 123–147.

Pinfield, N.J., Stutchbury, P.A. and Bazaid, S.M. (1987) Seed dormancy in *Acer*: is there a common mechanism for all *Acer* species and what part is played in it by abscisic acid? *Physiologia Plantarum* 71, 365–371.

Pinfield, N.J., Stutchbury, P.A., Bazaid, S.A. and Gwarazimba, V.E.E. (1990) Abscisic acid and the regulation of embryo dormancy in the genus *Acer*. *Tree Physiology* 6, 79–85.

Piper, J.K. (1986) Germination and growth of bird-dispersed plants: effects of seed size and light on seedling vigor and biomass allocation. *American Journal of Botany* 73, 959–965.

Porsild, A.E., Harington, C.R. and Mulligan, G.A. (1967) *Lupinus arcticus* Wats. grown from seeds of Pleistocene age. *Science (Washington, DC)* 158, 113–114.

Priestley, D.A. (1986) *Seed Aging: Implications for Seed Storage and Persistence in the Soil*. Cornell University Press, Ithaca.

Priestley, D.A. and Posthumus, M.A. (1982) Extreme longevity of lotus seeds from Pulantien. *Nature (London)* 299, 148–149.

Pritchard, H.W. (1991) Water potential and embryonic axis viability in recalcitrant seeds of *Quercus rubra*. *Annals of Botany* 67, 43–49.

Probert, R.J. and Longley, P.L. (1989) Recalcitrant seed storage physiology in three aquatic grasses (*Zizania palustris, Spartina anglica* and *Porteresia coarctata*). *Annals of Botany* 63, 53–63.

Probert, R.J., Smith, R.D. and Birch, P. (1985) Germination responses to light and alternating temperatures in European populations of *Dactylis glomerata* L. IV. The effects of storage. *New Phytologist* 101, 521–529.

Probert, R.J., Gajjar, K.H. and Haslam, I.K. (1987) The interactive effects of phytochrome, nitrate and thiourea on the germination response to alternating temperatures in seeds of *Ranunculus sceleratus* L.: a quantal approach. *Journal of Experimental Botany* 38, 1012–1025.

Quail, P.H. and Carter, O.G. (1969) Dormancy in seeds of *Avena ludoviciana* and *A. fatua*. *Australian Journal of Agricultural Research* 20, 1–11.

Quinlivan, B.J. (1971) Seed coat impermeability in legumes. *Journal of the Australian Institute of Agricultural Science* 37, 283–295.

Rampton, H.H. and Ching, T.M. (1970) Persistence of crop seeds in soil. *Agronomy Journal* 62, 272–277.

Rasmussen, W.D. (1975) The tilling of land, 1760. In: Rasmussen, W.D. (ed.) *Agriculture in the United States: A Documentary History*, Vol. 1. Random House, New York, pp. 188–203.

Rice, K.J. (1989) Impacts of seed banks on grassland community structure and population dynamics. In: Leck, M.A., Parker, V.T. and Simpson, R.L. (eds) *Ecology of Soil Seed Banks*. Academic Press, San Diego, California, pp. 211–230.

Roberts, E.H. (1961) Dormancy in rice seed. I. The distribution of dormancy periods. *Journal of Experimental Botany* 12, 319–329.

Roberts, E.H. (1962) Dormancy in rice seed. III. The influence of temperature, moisture, and gaseous environment. *Journal of Experimental Botany* 13, 75–94.

Roberts, E.H. (1963) An investigation of inter-varietal differences in dormancy and viability of rice seed. *Annals of Botany* 27, 365–369.

Roberts, E.H. (1965) Dormancy in rice seed. IV. Varietal responses to storage and germination temperatures. *Journal of Experimental Botany* 16, 341–349.

Roberts, E.H. (1972a) Storage environment and the control of viability. In: Roberts, E.H. (ed.) *Viability of Seeds*. Chapman and Hall, London, pp. 14–58.

Roberts, E.H. (1972b) Cytological, genetical, and metabolic changes associated with loss of viability. In: Roberts, E.H. (ed.) *Viability of Seeds*. Chapman and Hall, London, pp. 253–306.

Roberts, E.H. (1972c) Dormancy: a factor affecting seed survival in the soil. In: Roberts, E.H. (ed.) *Viability of Seeds*. Chapman and Hall, London, pp. 321–359.

Roberts, E.H. (1973) Predicting the storage life of seeds. *Seed Science and Technology* 1, 499–514.

Roberts, E.H. (1988) Temperature and seed germination. In: Long, S.P. and Woodward, F.E. (eds) *Plants and Temperature*. Symposia of the Society of Experimental Biology 42. Company of Biologists, Cambridge, UK, pp. 109–132.

Roberts, E.H. and Ellis, R.H. (1982) Physiological, ultrastructural and metabolic aspects of seed viability. In: Khan, A.A. (ed.) *The Physiology and Biochemistry of Seed Development, Dormancy and Germination*. Elsevier/North Holland Biomedical Press, Amsterdam, pp. 465–485.

Roberts, E.H. and Ellis, R.H. (1989) Water and seed survival. *Annals of Botany* 63, 39–52.

Roberts, E.H., Murdoch, A.J. and Ellis, R.H. (1987) The interaction of environmental factors on seed dormancy. In: *Proceedings of the 1987 British Crop Protection Conference – Weeds*. British Crop Protection Council, Farnham, pp. 687–694.

Roberts, H.A. (1964) Emergence and longevity in cultivated soils of seeds of some annual weeds. *Weed Research* 4, 296–307.

Roberts, H.A. (1979) Periodicity of seedling emergence and seed survival in some Umbelliferae. *Journal of Applied Ecology* 16, 195–201.

Roberts, H.A. (1986) Seed persistence in soil and seasonal emergence in plant species from different habitats. *Journal of Applied Ecology* 23, 639–656.

Roberts, H.A. and Boddrell, J.E. (1983a) Seed survival and periodicity of seedling emergence in ten species of annual weeds. *Annals of Applied Biology* 102, 523–532.

Roberts, H.A. and Boddrell, J.E. (1983b) Seed survival and periodicity of seedling emergence in eight species of Cruciferae. *Annals of Applied Biology* 103, 301–304.

Roberts, H.A. and Boddrell, J.E. (1984a) Seed survival and seasonal emergence of seedlings of some ruderal plants. *Journal of Applied Ecology* 21, 617–628.

Roberts, H.A. and Boddrell, J.E. (1984b) Seed survival and periodicity of seedling emergence in four weedy species of *Papaver*. *Weed Research* 24, 195–200

Roberts, H.A. and Chancellor, R.J. (1979) Periodicity of emergence and achene survival in some species of *Carduus, Cirsium* and *Onopordum*. *Journal of Applied Ecology* 16, 641–647.

Roberts, H.A. and Dawkins, P.A. (1967) Effect of cultivation on the numbers of viable weed seeds in soil. *Weed Research* 7, 290–301.

Roberts, H.A. and Feast, P.M. (1972) Fate of seeds of some annual weeds in different depths of cultivated and undisturbed soil. *Weed Research* 12, 316–324.

Roberts, H.A. and Feast, P.M. (1973a) Emergence and longevity of seeds of annual weeds in cultivated and undisturbed soil. *Journal of Applied Ecology* 10, 133–143.

Roberts, H.A. and Feast, P.M. (1973b) Changes in the numbers of viable weed seeds in soil under different regimes. *Weed Research* 13, 298–303.

Roberts, H.A. and Neilson, J.E. (1980) Seed survival and periodicity of seedling emergence in some species of *Atriplex, Chenopodium, Polygonum* and *Rumex*. *Annals of Applied Biology* 94, 111–120.

Roberts, H.A. and Neilson, J.E. (1981) Seed survival and periodicity of seedling emergence in twelve weedy species of Compositae. *Annals of Applied Biology* 97, 325–334.

Roberts, H.A. and Potter, M.E. (1980) Emergence patterns of weed seedlings in relation to cultivation and rainfall. *Weed Research* 20, 377–386.

Roberts, H.A. and Ricketts, M.E. (1979) Quantitative relationships between the weed flora after cultivation and the seed population in the soil. *Weed Research* 19, 269–275.

Russi, L., Cocks, P.S. and Roberts, E.H. (1992a) Hard-seededness and seed bank dynamics of six pasture legumes. *Seed Science Research* 2, 231–241.

Russi, L., Cocks, P.S. and Roberts, E.H. (1992b) Coat thickness and hard-seededness in some *Medicago* and *Trifolium* species. *Seed Science Research* 2, 243–249.

Sarukhan, J. (1974) Studies on plant demography: *Ranunculus repens* L., *R. bulbosus* L. and *R. acris* L. II. Reproductive strategies and seed population dynamics. *Journal of Ecology* 62, 151–177.

Sauer, J. and Struik, G. (1964) A possible ecological relation between soil disturbance, light-flash and seed germination. *Ecology* 45, 884–886.

Saunders, D.A. (1981) Investigations into seed coat impermeability in annual *Medicago* species. Unpublished PhD thesis, University of Reading, UK.

Schafer, D.E. and Chilcote, D.O. (1970) Factors influencing persistence and depletion of buried seed populations. II. The effects of soil temperature and moisture. *Crop Science* 10, 342–345.

Schonbeck, M.W. and Egley, G.H. (1980) Redroot pigweed (*Amaranthus retroflexus*) seed germination responses to afterripening, temperature, ethylene, and some other environmental factors. *Weed Science* 28, 543–548.

Sharif-Zadeh, F. (1999) Mechanism and modelling of seed dormancy and germination in *Cenchrus ciliaris* as affected by different conditions of maturation and storage. Unpublished PhD thesis, University of Reading, UK.

Sharif-Zadeh, F. and Murdoch, A.J. (2000) The effects of different maturation conditions on seed dormancy and germination of *Cenchrus ciliaris*. *Seed Science Research* (in press).

Shen-Miller, J., Schopf, J.W. and Berger, R. (1983) Germination of a *ca.* 700 year-old lotus seed from China: evidence of exceptional longevity of seed viability. *American Journal of Botany* 70, 78.

Simpson, G.M. (1990) *Seed Dormancy in Grasses*. Cambridge University Press.

Standifer, L.C., Wilson, P.W. and Drummond, A. (1989) The effects of seed moisture content on hard-seededness and germination in four cultivars of okra (*Abelmoschus esculentus* (L.) Moench). *Plant Varieties and Seeds* 2, 149–154.

Symons, S.J., Naylor, J.M., Simpson, G.M. and Adkins, S.W. (1986) Secondary dormancy in *Avena fatua*: induction and characteristics in genetically pure dormant lines. *Physiologia Plantarum* 68, 27–33.

Taylor, G.B. (1981) Effect of constant temperature treatments followed by fluctuating temperatures on the softening of hard seeds of *Trifolium subterraneum* L. *Australian Journal of Plant Physiology* 35, 201–210.

Taylorson, R.B. (1970) Changes in dormancy and viability of weed seeds in soils. *Weed Science* 18, 265–269.

Taylorson, R.B. (1972) Phytochrome controlled changes in dormancy and germination of buried weed seeds. *Weed Science* 20, 417–422.

Thompson, K. and Grime, J.P. (1979). Seasonal variation in the seed banks of herbaceous species in ten contrasting habitats. *Journal of Ecology* 67, 893–921.

Thompson, K., Bakker, J.P. and Bekker, R.M. (1997) *The Soil Seed Banks of North West Europe: Methodology, Density and Longevity*. Cambridge University Press, Cambridge.

Tokumasu, S., Kata, M. and Yano, F. (1975) [The dormancy of seed as affected by different humidities during storage in *Brassica*.] *Japanese Journal of Breeding* 25, 197–202 (in Japanese).

Tompsett, P.B. (1986) The effect of temperature and moisture content on the seed of *Ulmus carpinifolia* and *Terminalia brassii*. *Annals of Botany* 57, 875–883.

Toole, E.H. and Brown, E. (1946) Final results of the Duvel buried seed experiment. *Journal of Agricultural Research* 72, 201–210.

Totterdell, S. and Roberts, E.H. (1979) Effects of low temperatures on the loss of innate dormancy and the development of induced dormancy in seeds of *Rumex obtusifolius* L. and *Rumex crispus* L. *Plant, Cell and Environment* 2, 131–137.

Totterdell, S. and Roberts, E.H. (1981) Ontogenetic variation in response to temperature change in the control of seed dormancy of *Rumex obtusifolius* L. and *Rumex crispus* L. *Plant, Cell and Environment* 4, 75–80.

Turner, J.H. (1933) The viability of seeds. *Bulletin of Miscellaneous Information, Royal Botanic Gardens, Kew*, 34, 257–269.

Vertucci, C.W. and Leopold, A.C. (1984) Bound water in soybean seed and its relation to respiration and imbibitional damage. *Plant Physiology* 75, 114–117.

Villiers, T.A. and Edgecumbe, D.J. (1975) On the cause of seed deterioration in dry storage. *Seed Science and Technology* 3, 761–774.

Vincent, E.M. and Roberts, E.H. (1977) The interaction of light, nitrate and alternating temperature in promoting the germination of dormant seeds of common weed species. *Seed Science and Technology* 5, 659–670.

Vleeshouwers, L.M. (1997) Modelling weed emergence patterns. PhD thesis, Wageningen Agricultural University, The Netherlands.

Watanabe, Y. and Hirokawa, F. (1975a) [Ecological studies on the germination and emergence of annual weeds. 3. Changes in emergence and viable seeds in cultivated and uncultivated soil.] *Weed Research, Japan* 19, 14–19 (in Japanese).

Watanabe, Y. and Hirokawa, F. (1975b) Ecological studies on the germination and emergence of annual weeds. 4. Seasonal changes in dormancy status of viable seeds in cultivated and uncultivated soil. *Weed Research, Japan* 19, 20–24 (In Japanese).

Werker, E. (1980/81) Seed dormancy as explained by the anatomy of embryo envelopes. *Israel Journal of Botany* 29, 22–44.

Wesson, G. and Wareing, P.F. (1967) Light requirements of buried seeds. *Nature* 213, 600–601.

Wesson, G. and Wareing, P.F. (1969a) The role of light in the germination of naturally occurring populations of buried weed seeds. *Journal of Experimental Botany* 20, 402–413.

Wesson, G. and Wareing, P.F. (1969b) The induction of light sensitivity in weed seeds by burial. *Journal of Experimental Botany* 20, 414–425.

Whitmore, T.C. (1988) The influence of tree population dynamics on forest species composition. In: Davy, A.J., Hutchings, M.J. and Watkinson, A.R. (eds) *Plant Population Ecology*. Blackwell Scientific Publications, Oxford, pp. 271–291.

Williams, J.T. and Harper, J.L. (1965) Seed polymorphism and germination. I. The influence of nitrates and low temperatures on the germination of *Chenopodium album*. *Weed Research* 5, 141–150.

Wilson, B.J. (1978) The long-term decline of a population of *Avena fatua* L. with different cultivations associated with spring barley cropping. *Weed Research* 18, 25–31.

Wilson, B.J. (1981) The influence of reduced cultivations and direct drilling on the long-term decline of a population of *Avena fatua* L. in spring barley. *Weed Research* 21, 23–28.

Wilson, B.J. (1985) Effect of seed age and cultivation on seedling emergence and seed decline of *Avena fatua* L. in winter barley. *Weed Research* 25, 213–219.

Yadav, A.S. and Tripathi, R.S. (1982) A study on seed population dynamics of three weedy species of *Eupatorium*. *Weed Research* 22, 69–76.

Zewdie, M. and Ellis, R.H. (1991a) The upper-moisture-content limit to negative relations between seed longevity and moisture in niger and tef. *Seed Science and Technology* 19, 295–305.

Zewdie, M. and Ellis, R.H. (1991b) Response of tef and niger seed longevity to storage temperature and moisture. *Seed Science and Technology* 19, 319–329.

Zorner, P.S., Zimdahl, R.L. and Schweizer, E.E. (1984) Sources of viable seed loss in buried dormant and non-dormant populations of wild oat (*Avena fatua* L.) seed in Colorado. *Weed Research* 24, 143–150.

Chapter 9
The Functional Ecology of Soil Seed Banks

Ken Thompson

Unit of Comparative Plant Ecology, Department of Animal and Plant Sciences,
University of Sheffield, Sheffield, UK

What is persistence?

Since this chapter is concerned with seed persistence in the soil, it might be helpful to begin with what is implied by the term and, more importantly, what is not implied. When a seed arrives at the soil surface, it may suffer one of several fates. It may germinate immediately – for recalcitrant seeds, rapid germination is more or less obligatory. Alternatively, it may persist in the soil or on the soil surface for a shorter or longer period. It is useful to make a distinction between, on the one hand, short delays, whose primary purpose is to synchronize germination with a season suitable for establishment and growth, and, on the other hand, longer delays, measured in years rather than months. This distinction is important, if for no other reason than that it is easy to see why seeds avoid germinating in or immediately before hard winters or long dry periods. Why seeds should persist for years is harder to explain.

Another reason for observing the distinction is that a relatively short seasonal delay is often, though not always, associated with some form of dormancy, while long-term persistence in the soil is much less often dependent on dormancy. The situation described by Washitani and Masuda (1990) in a temperate Japanese grassland is

typical. In this tall grassland, accumulation of biomass during the summer, followed by winter burning, virtually precludes autumn germination; germination is almost entirely confined to spring and early summer. All species dispersing their seeds in spring or summer, plus some of those dispersed in autumn, avoid autumn germination by producing dormant seeds. Dormancy is broken by a period of low temperatures (chilling). However, some autumn dispersers shed seeds capable of immediate germination (Fig. 9.1); in these species, autumn germination is prevented by a relatively high temperature requirement for germination. Note that this information about dormancy, in itself, tells us nothing about which species have persistent seed banks.

In temperate, boreal and arctic climates, the adaptive significance of seed dormancy is most often avoidance of low temperatures during seedling establishment. In warmer climates, escaping a dry season is more often the cause. For a detailed account of the diversity of seed dormancy mechanisms in both types of environment, see Baskin and Baskin (1998). Those species in which persistence is confined to moving the timing of germination from one side of an unfavourable season to the other are often said to have a transient seed bank (Thompson and Grime, 1979). This chapter is not concerned,

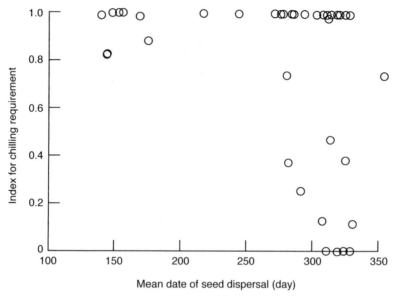

Fig. 9.1. Relationship between an index of chilling requirement and the mean date of seed dispersal, counted from 1 January. The larger the index, the greater the requirement for chilling. (From Washitani and Masuda, 1990. Reproduced by permission of Blackwell Science.)

except in passing, with such short-term between-season persistence.

A digression: persistence, dormancy and storage

Is there any simple relationship between persistence and dormancy? The short answer to this question is no. As Baskin and Baskin (1998) convincingly demonstrate, species with long-persistent seeds may undergo repeated seasonal cycles of dormancy (e.g. *Lamium purpureum*), be initially dormant but soon become and remain non-dormant (e.g. *Rumex crispus*) or be non-dormant from the outset (e.g. *Digitalis purpurea*). Equally, seeds that remain in the soil for only a short time may be profoundly dormant when shed (e.g. *Heracleum sphondylium*). Dormancy is therefore neither a necessary nor a sufficient condition for persistence in the soil. Indeed, most types of dormancy (including the most frequent, physiological dormancy) play little direct part in seed persistence in the soil. Many species persist in the soil for years or decades in a non-dormant state.

Even in those species that experience cycles of dormancy, persistence is due not to dormancy, but to the absence of appropriate germination cues during that part of the cycle when the seeds are non-dormant. The only dormancy type that plays a large part in seed persistence is hard-seeded or 'physical' (Baskin and Baskin, 1998) dormancy. The relatively insignificant role of dormancy in seed persistence has been previously described on many occasions. Baskin and Baskin (1989) noted that 'a primary reason seeds do not germinate in the soil while non-dormant is that most have a light requirement for germination'.

There is no harm in using 'seed dormancy' simply to indicate that germination is spread across years (Rees, 1994), although to do so reduces 'dormant' to meaning nothing more than 'ungerminated', but care must then be taken not to cite data from sources (e.g. Baskin and Baskin, 1988) where the term is being used quite differently. For a good discussion of what plant ecophysiologists understand (and do not understand) by seed dormancy, see Vleeshouwers *et al.* (1995). Briefly, Vleeshouwers *et al.* argue (correctly) that

(i) dormancy should not be identified with the absence of germination; (ii) dormancy is a characteristic of the seed (and not of the environment) and defines the conditions necessary for germination; and (iii) so far only temperature has been shown to alter the degree of dormancy in seeds.

Among those unfamiliar with the subject, and occasionally among those who should know better, there is sometimes confusion about the distinction between seed storage in the laboratory and seed persistence in the soil. Over a wide range of temperatures and moisture contents, longevity of orthodox seeds in dry storage can be increased by lowering either or both (see Murdoch and Ellis, Chapter 8, this volume). Such conditions dramatically slow the rate of accumulation of the genetic and membrane damage that reduce viability and eventually lead to death. Above a certain moisture content, however, longevity increases with increasing moisture content, provided oxygen is available and germination and attack by pathogens are prevented. In such imbibed seeds, damage occurs but is rapidly repaired by the metabolically active seed (Villiers, 1974). Nor does a seed have to be continuously imbibed to enjoy these benefits (Villiers and Edgecumbe, 1975). In humid climates, where seeds in the soil are, at least intermittently, fully imbibed, persistence presumably depends on such active repair mechanisms (Priestley, 1986). It is therefore perfectly possible for imbibed seeds to survive much longer than dry seeds. For example, seeds of five tropical pioneer species (all known to survive for several years in soil) remained viable after 7 years of imbibed storage, while dry seeds were dead after 3 years (Vàzquez-Yanes and Orozco-Segovia, 1990).

Generally speaking, therefore, longevity in dry storage is a poor predictor of persistence in the soil. In only two groups of species does this generalization break down. Recalcitrant seeds are short-lived and hard-seeded species are long-lived under both conditions. Recalcitrant seeds normally germinate rapidly and die if dried below a critical moisture content,

while the embryos of hard seeds are maintained in a dry state both in the laboratory and in the soil (see Murdoch and Ellis, Chapter 8, this volume). Hard seeds are responsible for some of the most extreme examples of longevity recorded both in dry storage (Youngman, 1951) and in the soil (Shen-Miller *et al.*, 1995).

It is worth noting that the existence of essentially two separate mechanisms of persistence causes difficulties for attempts to predict seed longevity in the soil. In temperate regions, where the great majority of persistent seeds are imbibed, seed persistence is strongly linked to seed size (Thompson and Grime, 1979; Leck, 1989; Tsuyuzaki, 1991) and it is possible to use seed size and shape as predictors of persistence (Thompson *et al.*, 1993; Bekker *et al.*, 1998b; Funes *et al.*, 1999). In more arid parts of the world, most persistent seeds are of the hard-seeded type, and here the relationship between size and persistence breaks down completely (Leishman and Westoby, 1998). How seeds in arid climates break the rules obeyed by plants of temperate regions, which are founded on well-understood interactions between seed size, burial and predation, is one of the more urgent questions in seed-bank ecology.

Why persist?

Why should seeds persist in the soil? There are at least two good reasons why they should not. First, seeds that fail to germinate at the first opportunity may die before they are able to do so. Secondly, because of the nature of population growth, seeds that germinate and reproduce early leave more descendants than those that germinate later (Rees, 1994). Seed persistence therefore incurs potentially large costs, so what might offset these costs? First, there is variation in establishment success and reproductive output – in other words, 'good' and 'bad' years. A plant population with no seed persistence will maximize its arithmetic rate of increase in good years, but just one bad year in which either establishment or reproduction fails completely will lead to its extinction. Equally, a plant with 100% persistence (and therefore no germination) would also have zero fitness. We therefore expect most species to show some intermediate behaviour, with a fraction of seed production germinating immediately and the remainder persisting in the soil. In the simplest case of an annual plant with no density dependence, the fraction that germinates immediately should be approximately equal to the probability of a good year (Cohen, 1966).

Kalisz and McPeek (1993) applied Cohen's (1966) model to data from a natural population of the winter annual *Collinsia verna* (*Scrophulariaceae*). Data from this population showed that *Collinsia* has good (geometric population growth rate) ($\lambda = 1.80$) and bad ($\lambda = 0.40$) years, although Kalisz and McPeek were careful not to claim that these represent the extreme values possible in this population. Models of the demography of *Collinsia* varied the frequency of good and bad years and the autocorrelation between conditions in consecutive years, with and without a persistent seed bank. The results confirmed theoretical expectations. A persistent seed bank does have a cost in terms of reduced population growth rate, but this cost is small and manifest only when the population is growing rapidly (good years frequent, autocorrelation not too positive). As the frequency of good years declines and positive autocorrelation increases (i.e.

potentially longer runs of consecutive bad years), populations with a seed bank grow faster than those without. However, the most important consequence of a seed bank is to increase the time to extinction (Table 9.1), and this effect is most pronounced when the environment is most unpredictable (proportion of good years = 0.5, autocorrelation = 0). Thus, as expected, a seed bank buffers plant populations against environmental variability.

Essentially, the fraction of seeds persisting in the soil reflects the relative risks of a seed remaining in the soil (and dying before germination) or germinating under conditions that do not permit reproduction. So far, we have considered that this latter risk is an absolute, i.e. a year is either 'good' or 'bad'. In reality, this risk could be much reduced if seeds had access to information about the likely conditions before germination. Conditions may vary temporally from year to year and spatially within a year, and may reflect both abiotic variables (chiefly climate) and the presence and activities of other plants and animals. Seeds have evolved a wide diversity of mechanisms that enable them to detect all these variables. In deserts, for example, seeds of some species in the soil are able to detect the quantity of rainfall by germinating only after water-soluble germination inhibitors have been leached from the seed-coat (Went, 1949). Other species retain ripe seeds within fruits on the dead

Table 9.1. The differences between the average time to extinction (years) for simulated populations of *Collinsia verna* with and without a seed bank (time to extinction with a seed bank − time to extinction without a seed bank) for all combinations of frequencies of good/bad years and autocorrelations used in the simulation. *** means no populations with seed banks went extinct but some populations without seed banks did go extinct for that set of conditions. NE means no population with or without seed banks went extinct. It is unclear why the time to extinction is greater for populations without seed banks in two cases. (From Kalisz and McPeek 1993.)

Autocorrelation	Frequency of good years				
	0.9	0.7	0.5	0.3	0.1
0.9	5.64	9.16	2.91	0.48	1.84
0.6	***	−7.58	9.95	3.76	0.66
0.3	NE	−8.6	17.63	1.5	0.6
0	NE	***	24.83	0.22	1.1
−0.3	–	NE	16.16	1.62	–
−0.6	–	–	13.7	–	–
−0.9	–	–	4.8	–	–

plant, releasing them only after substantial rainfall (Gutterman, 1993). In fire-prone habitats, seeds may be released from the plant only after fire (Lamont *et al.*, 1991) or seeds in the soil may respond to heat (Gimingham, 1972) or chemicals released by burning plant material (Thanos and Rundel, 1995; Keeley and Fotheringham, 1998). In more mesic habitats, the chief hazard for seeds is germinating in the presence of established perennials. Here seeds have evolved responses to a number of cues that indicate gaps in the cover of perennial vegetation, including intensity, duration and wavelength of light (Gorski *et al.*, 1977; Cresswell and Grime, 1981; Vàzquez-Yanes and Orozco-Segovia, 1990; Milberg and Andersson, 1997), temperature alternations (Thompson and Grime, 1983) and soil nitrate concentration (Pons, 1989b). A response to interactions between several gap-detection stimuli may give seeds the ability to determine both the timing and location of germination with some precision (Hilton, 1984; Milberg, 1997). Successful targeting of germination on optimum conditions for establishment makes the evolution of persistence much more likely. Note, however, that a plant's perception of environmental hazards is strongly influenced by other traits, particularly seed size. If large seeds are better able to tolerate drought or competition (Westoby *et al.*, 1996), large-seeded species will experience less selection for the evolution of seed persistence. Effectively, large seed size and persistence are doing the same job, and are thus expected to show negative covariation (Venable and Brown, 1988). I consider the interaction of seed persistence with other traits in a later section.

A further process that may contribute to the evolution of seed persistence is competition between seedlings, particularly between siblings. If most seeds are dispersed only a short distance, parent plants may experience strong selection pressure to delay germination of a fraction of their seed output by manipulating the structure or chemistry of the testa or pericarp (both maternal tissues). The theoretical case for this hypothesis is strong (Ellner, 1986; Nilsson *et al.*, 1994), but at the moment there is very little evidence (Zammit and Zedler, 1990; Cheplick, 1992; Hyatt and Evans, 1998).

Persistence, seed size and dispersal

Seed persistence did not evolve and does not operate in isolation. Rather, it forms part of a cohesive and presumably well-adapted phenotype. Since seed persistence, dispersal and large seed size all reduce the impact of environmental variation, the evolution of any one of them can only be properly understood in the context of the other two. Modelling the joint evolution of all three traits demonstrates that, as expected, increasing any one of the seed traits normally has the effect of reducing the other two (Venable and Brown, 1988). Where this does not happen, it is usually because one of the traits is being pushed very far from its optimum value.

Intuitively, dispersal is a better response to environmental variability than persistence or large seed size. This is because increased seed size has an inherent cost of reduced fecundity, while persistence inevitably involves delayed reproduction. Dispersal, on the other hand, has no inherent cost – it is theoretically possible, though unlikely, for every seed to land in a site suitable for establishment. More detailed models (McPeek and Kalisz, 1998) also support the idea that, because dispersal reduces variation in fitness more than persistence (by averaging across all habitat patches), but has no inherent within-year fitness cost, dispersal is the favoured strategy under most circumstances. Persistent seed banks are favoured when habitable patches are few, dispersal rate is low and the probability of widespread catastrophic failure of reproduction is high. Venable and Lawlor (1980) came to broadly similar conclusions.

There is general agreement that both seed size and dispersal should be negatively correlated with persistence, and there is some evidence in support of this hypothesis. Rees (1993) analysed data for 171 British herbaceous species. Using data from the classic experiments of H.A. Roberts (Roberts, 1986, and numerous other papers by the same author), Rees used the proportion of seeds that emerge in the first year after burial under standard conditions as a (negative) measure of seed persistence. Seed weights and dispersal ability were obtained from Grime *et al.* (1988). Dispersal capacity was treated as a discrete variable; all species with obvious morphological adaptations for dispersal (other than by ants) or that were thought to be dispersed by water were assumed to have efficient dispersal. The results were clear: persistent seeds are both smaller and less likely to be effectively dispersed. A more recent analysis, however, casts doubt on some of these findings (D.J. Hodkinson, unpublished). Hodkinson used seed persistence data from the database of Thompson *et al.* (1997) and estimated wind-dispersal capacity by measuring the terminal velocity of seeds or fruits in still air (Askew *et al.*, 1997). This analysis confirmed the negative relationship between seed size and persistence, but failed to find any relationship between persistence and dispersal. A number of factors may contribute to the contrasting results of the two analyses. First, treating dispersal capacity as a discrete variable captures little of the complexity of dispersal in the real world. Secondly, the 'efficient dispersal' category of Rees (1993) included all forms of dispersal adaptations, while Hodkinson *et al.* considered only wind dispersal. This difference may have important consequences. Structures that reduce terminal velocity (usually some form of parachute in herbs) are often relatively easily shed and therefore may not impede burial once the seed lands on the ground. Hooks and barbs, in contrast, are usually tougher and probably do interfere significantly with burial. Furthermore, dispersal by animals (externally or internally) is compatible with larger seeds than wind dispersal (Willson *et al.*, 1990) and, as we have seen, seed size is itself correlated with persistence. It may, therefore, be unreasonable to expect a universal trade-off between dispersal and persistence; the strength (or even the existence) of such a trade-off may depend upon other variables in quite a complex manner.

The above discussion also highlights the issue of adaptations and constraints. Relationships between seed size, persistence and dispersal have so far been discussed entirely in terms of adaptive trade-offs, but they may also be strongly linked by simple biophysical constraints that set strict limits to what is physically possible. Large seeds, for example, may be much less readily buried than small seeds (Peart, 1984). The combination of large size and absence of the protection afforded by burial may greatly increase predation levels (Thompson, 1987; Hulme, 1993). In other words, some simple, direct consequences of increased seed size may select against persistence, independently of the adaptive argument outlined above. In this case, of course, constraint and adaptation point in the same direction, and it is perhaps not surprising that the negative correlation between seed size and persistence seems to be both strong and universal (e.g. Leck, 1989; Dalling *et al.*, 1998).

Suppose, however, that constraint and adaptation acted in opposing directions. For the reasons outlined above, small seeds are more likely to become buried and less likely to suffer predation than large ones. Weight constraints also mean that small seeds are likely to be better dispersed than large ones, particularly by wind. Small seeds can also be produced in larger numbers, and greater seed production is itself an important contributor to effective dispersal, increasing the probability that the tail of the seed distribution contains at least a few seeds. So, all things being equal, small seeds should be both better dispersed and more persistent (and some small-seeded species, e.g. *Typha latifolia*, do manage to excel at both). If constraints predict that seed persistence and wind dispersal should be positively correlated, while adaptive arguments predict a negative correlation, maybe it is not surprising that the outcome is sometimes no relationship at all.

Other traits: life history and chemistry

If reproduction fails completely, an annual plant with no seed bank becomes extinct, but a perennial does not. In other words, perenniality itself significantly reduces the effect of environmental uncertainty. We would therefore expect adult longevity to trade off against seed persistence in much the same way as seed size and dispersal. This intuitive expectation is supported by explicit models; under most realistic conditions, increased adult longevity always selects against seed persistence (Rees, 1994). Note also that there is a strong tendency for large, long-lived plants to have large seeds (Salisbury, 1942). Here, then, constraints and adaptations reinforce each other, and the evidence strongly supports the negative relationship between adult and seed longevity (Rees, 1993; Thompson *et al.*, 1998; Fig. 9.2).

The counter-argument has been expressed (Waller, 1988) that, since seedling establishment is unlikely in communities of long-lived perennials, this should select for increased persistence (and perhaps also

dispersal). Waller had in mind the situation where seedling establishment was prevented by the spread of long-lived, clonal individuals, but the same argument should apply to any closed community of perennial plants. This argument fails to appreciate that it is not unfavourability that selects for increased persistence, but unpredictability. In a constant environment, immediate germination is always the evolutionarily stable strategy (Rees, 1994). Long-lived perennials may accumulate persistent seed banks (e.g. *Calluna vulgaris*, *T. latifolia*), but these are associated with infrequent but major disturbances – in this case, fire (Gimingham, 1972) and flooding (Van Der Valk and Davis, 1976), respectively. In environments that remain undisturbed for very long periods, persistent seed banks are rarely found. For example, no mature forest tree species in north-western Europe has persistent seeds (Thompson *et al.*, 1997), for the simple reason that the lifespan of the dominant trees exceeds that of even the most long-lived seeds. Murray (1988) came to a similar conclusion from a detailed study of three gap-dependent tropical shrubs. All three were effectively bird-dis-

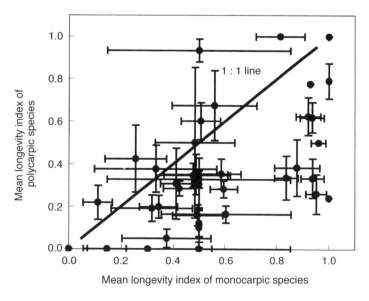

Fig. 9.2. Mean seed longevity indices for confamilial monocarpic and polycarpic species. Seed longevity data from the seed-bank database of Thompson *et al.* (1997). Bars are standard errors. (From Thompson *et al.*, 1998. Reproduced by permission of Blackwell Science.)

persed and had persistent seed banks. Simulations showed that fitness was greatly increased by dispersal and additionally by seed persistence, but that persistence alone had little effect. All three species had an absolute requirement for canopy gaps for establishment, and the mean interval between consecutive gaps in the same place was just too long for seed persistence alone to make a significant contribution to the dynamics of any of the three species. Similarly, in attempting to answer the question 'Where were old field plants in north-eastern USA before European colonization?', Marks (1983) concluded that very few evolved as specialist colonizers of forest openings. The slender list of candidates (*Prunus pensylvanica*, *Phytolacca americana*, *Rubus strigosus*) all have persistent seeds and effective dispersal.

Paradoxically, some shade-tolerant plants of undisturbed woodland appear to accumulate persistent seed banks (Piroznikow, 1983; Metcalfe and Grubb, 1995; Jankowska-Blaszczuk and Grubb, 1997). Here recruitment appears to be linked to unpredictable gaps in the litter rather than in the canopy, caused in Europe by large animals, such as wild boar or bison.

Independently of the above arguments, it has been suggested that seed chemical composition might influence both dispersal and persistence (Lokesha *et al.*, 1992). Fats yield more energy per unit weight than proteins or carbohydrates, so storage of lipids would allow seed weight to be reduced without altering energy content, thus favouring the evolution of wind dispersal. Lokesha *et al.* also proposed that oily seeds might have inherently lower longevity, owing to lipid peroxidation, and that this would select for lower persistence in soil. Unfortunately, this latter argument illustrates the problems that can arise from confusion of dry storage and persistence in soil. The evidence cited by Lokesha and others (e.g. Ponquett *et al.*, 1992; see also references in Lokesha *et al.*, 1992) relates specifically to dry storage. There is no experimental evidence that lipid peroxidation is a cause of reduced longevity in

imbibed oily seeds. Lokesha *et al.* were unable to examine the relationship between seed persistence and fat content, owing to a lack of data on seed persistence. Moreover, their analysis of the relationship between chemical composition and wind dispersal failed to take account of phylogeny and was therefore seriously flawed. The data set they analysed contained numerous species of *Leguminosae* (largely low in fat and poorly wind-dispersed) and *Compositae* (high in fat and often wind-dispersed), and it is not possible to treat these species as independent data points. A new analysis, using phylogenetically independent contrasts, suggests that there is no relationship between seed chemical composition and dispersal or persistence (D.J. Hodkinson, unpublished). In fact, there is a suggestion that persistent seeds may actually be higher in fat than short-lived seeds. Perhaps this is not too surprising; given the consistent tendency for persistent seeds to be small, one might expect selection to favour the storage of carbon in its most compact form.

Death, decay and defence

Pons (1989a; 1991a) buried seeds of seven heathland species and 13 chalk grassland species and monitored their fate by exhuming them over subsequent months and years. His data show that the short-lived seeds in both groups (e.g. *Deschampsia flexuosa*, *Carlina vulgaris*) had suffered the same fate – germination *in situ*. In contrast, seeds of those species that were destined to form a persistent seed bank (e.g. *Calluna vulgaris*, *Origanum vulgare*) showed little or no *in situ* germination. Recent work has therefore done much to reinforce the opinion of Cook (1980), based on earlier work (e.g. Schafer and Chilcote, 1970), that 'most mortality is due to the breakdown of dormancy mechanisms and subsequent germination while buried in the soil'. Cook was certainly correct that failure of transient seeds to persist arises largely from an inability to prevent germination, even when deeply buried.

Nevertheless, Cook's conclusion requires some qualification. First, while *in situ* germination is clearly a major source of mortality in short-lived seeds, it is probably much less important for long-lived seeds. Secondly, Cook was using dormancy in its loose and potentially confusing sense of 'ungerminated'. I have already mentioned that long-term seed persistence in the soil owes little to dormancy and is much more a consequence of exacting germination requirements and, crucially, an absolute light requirement. Thirdly, even if we assume that Cook intended 'breakdown of dormancy mechanisms' to mean 'relaxation of cues required to stimulate germination', there is little evidence to support this statement. Indeed, the opposite seems to be true, since seeds frequently acquire a light requirement after burial (Wesson and Wareing, 1969; Pons, 1991b; Milberg and Andersson, 1997). Nor is there much evidence that a pre-existing light requirement gradually decays during burial. Seeds of *D. purpurea* suffered no mortality or germination during 2 years in the soil and remained absolutely dependent on light for germination throughout that time (van Baalen, 1982). Rees and Long (1993) found that, in a large data set of 145 species, the probability of recruitment from buried seeds in the majority of species was either constant or declined through time. In only a small minority of species did probability of recruitment increase over time.

Persistence clearly depends, as a necessary first step, on avoiding germination under inappropriate circumstances. Nevertheless, even among seeds that manage to avoid immediate germination, interspecific variation in longevity spans about two orders of magnitude, from a few years to a few centuries (Thompson *et al.*, 1997). What accounts for this variation, and what are the main hazards faced by buried seeds? Predation and attack by pathogens seem the most likely candidates, and indeed a striking feature of many of the most persistent species (e.g. several members of the *Solanaceae*) is the possession of substantial chemical defences. While such defences may repel both predators and pathogens, small size and burial itself are probably adequate defence against most predators. Seed-eating insects (chiefly ants and beetles) are mainly surface feeders and small mammals do not excavate very small seeds (Thompson, 1987). It therefore seems likely that the main threat to small buried seeds is attack by microbial pathogens. The only attempt to systematically investigate the relationship between persistence and chemical defence is that of Hendry *et al.* (1994). Hendry measured *ortho*-dihydroxyphenol concentrations in seeds of 81 species and found that levels were significantly higher in species with persistent seeds (Fig. 9.3). *Ortho*-dihydroxyphenols are among the commonest plant phenols, including tannins, and are widely reported to have bactericidal and fungicidal activity and to be toxic to a variety of herbivores (references in Hendry *et al.*, 1994). Nevertheless, plants utilize many other secondary defence compounds, and some of the most persistent species in Hendry's data set (e.g. *Urtica* spp.) are very low in phenols. Other studies have implicated tannins in seeds as defences against microbial attack (e.g. Kremer *et al.*, 1984; Fellows and Roeth, 1992), although in some cases the chemicals involved in defence remain unknown (Siemens *et al.*, 1992). In short, our present understanding of the defensive role of secondary compounds in seeds under field conditions is no better (and perhaps worse) than that for plants as a whole (Harborne, 1997; Hartley and Jones, 1997).

Although the main hazards faced by buried seeds are biotic, it would be a mistake to completely ignore the effect of the abiotic environment, although this is very incompletely understood. The subjective impression is easily gained that different soil types support different seed-bank densities (e.g. Staaf *et al.*, 1987), but, of course, differences in pH and ion concentrations are confounded by floristic differences, and the subject has not been studied experimentally. Differences in soil water content might be expected to have quite strong impacts on seed survival, since waterlogged soils are often anoxic and seeds of

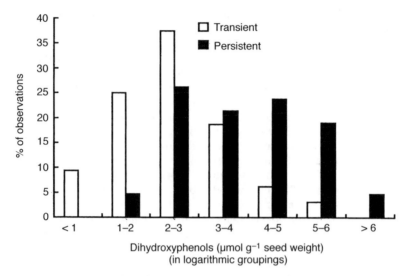

Fig. 9.3. Frequency of *ortho*-dihydroxyphenol concentrations in persistent and transient seed banks. Difference significant at $P < 0.001$ (Mann–Whitney). Percentages calculated on numbers of observations of persistent ($n = 41$) and transient ($n = 36$) species, respectively, and not on total observations. (From Hendry *et al.* 1994. Reproduced by permission of Blackwell Science.)

intolerant terrestrial species are quite quickly killed by anoxia in the laboratory (Ibrahim and Roberts, 1983). There is also much field evidence that survival of seeds of plants from terrestrial habitats is dramatically reduced by burial in wet soil; Skoglund and Verwijst (1989) found that the half-life of *Betula pubescens* seeds varied from < 1 year in a wet meadow to > 13 years in a dry forest soil. Nevertheless, dense seed banks of wetland species do accumulate in waterlogged soils (Van Der Valk and Davis, 1976), and presumably such species have metabolic adaptations that permit them to survive and even germinate in the almost complete absence of oxygen (Kennedy *et al.*, 1980). Seeds of wetland species often have germination preferences for water depths that reflect those occupied by the mature plants in the field (Moore and Keddy, 1988). In the first study of its kind, Bekker *et al.* (1998a) found that, in turfs containing a natural seed bank and kept under two moisture conditions, seeds of most wetland plants survived better in waterlogged conditions. In contrast to the effects of the water-table, which certainly deserve further study, there is little evidence that soil fertility

affects seed longevity. Despite the well-known stimulation of germination by nitrate ions, Bekker *et al.* (1998c) found no effect of added N, P or K on the survival of seeds of 14 fen meadow species at five sites in Britain and The Netherlands.

Equally little is known about the potential effects of climate change on seed banks. Experimental results so far are mixed. Hogenbirk and Wein (1992) subjected Canadian wetland seed banks to temperature treatments designed to simulate global warming and concluded that a drier, warmer climate would favour increased emergence of introduced weedy species. However, Akinola *et al.* (1998) found no effect of experimental soil warming or cooling on the seed banks of several species. In attempting to correlate seed persistence with climate in the field, it is often impossible to separate direct effects on the seed bank from indirect effects acting on adult plants. For example, Pakeman *et al.* (1999) found strong correlations between maximum *Calluna vulgaris* seed bank density and climate in Britain, but were able to conclude only that postdispersal processes, rather than seed production, were probably responsible.

Seed persistence in plant communities

It is common knowledge that plant communities and the soil seed banks beneath them frequently differ, in terms of both species present and the relative proportions of those species. Such differences have two quite different sources. First, even in plant communities with a long history of stable species composition, many species present in the vegetation may be absent from the seed bank. It is rare, however, to find many seeds of species absent from the vegetation beneath such communities, and the few that are found are usually effectively dispersed 'vagrants'. Secondly, if the composition of the community has changed over time, seeds from previous stages may persist in the soil long after the species have vanished from the vegetation. Such successional changes, which may lead to very large differences between seed-bank and above-ground flora, have long fascinated plant ecologists. Some of the most frequently cited seed-bank papers document these successional changes (e.g. Oosting and Humphreys, 1940; Livingston and Allessio, 1968; Brown and Oosterhuis, 1981; Koniak and Everett, 1982). These studies generally support the paradigm of declining seed numbers and diversity and decreasing similarity between seed bank and vegetation as succession proceeds.

Recently, however, some studies have challenged this paradigm. Milberg (1995), in a study of Swedish grassland plots after 18 years of succession to birch/aspen woodland, found no change in species richness, density or composition of the seed bank over this time. One possible cause of these negative results could be the relatively short time involved. Another possibility is that our expectations of successional patterns come largely from work on (often American) old fields, which begin with dense accumulations of highly persistent seeds of weedy annuals. Milberg's grassland succession contained hardly any annuals at any stage. Many grassland species do not accumulate per-

sistent seed banks, and it is perhaps not surprising that most of the species that were lost from Milberg's plots left no trace in the seed bank. Some other studies also suggest that the real world is often messier than simple successional patterns would predict. Beatty (1991) studied the vegetation, seed bank and seed rain in a mosaic of old fields, plantations and young and old forest in New York. The data were used to test a number of hypotheses that derive from the successional hypothesis – specifically, that seed banks are dissimilar to the existing vegetation and to the seed rain, and that seed banks of different communities are similar (on account of sharing a common successional past). In fact, none of these hypotheses was supported, and the simplest explanation for the data was that the seed bank of any community was largely derived from its own recent seed rain and from that of adjacent communities.

Nevertheless, where unambiguous successional sequences can be identified, studies of the seed bank can reveal much about past community dynamics. Two recent examples illustrate this point. Grandin and Rydin (1998) examined the depth distribution of the seed banks of a number of islands created by the lowering of Lake Hjälmaren in Sweden between 1882 and 1886. The seed bank at all depths was not very similar to the vegetation on any date, but the surface seed bank was more similar to recent vegetation than to older vegetation (Fig. 9.4). In contrast, seeds from the deepest soil showed a decreasing similarity with island vegetation over time and had no species in common with the modern vegetation. The seed bank of the upper soil horizons was dominated by mid–late successional perennials with heavy, short-lived seeds, while the lower horizons consisted of pioneer annuals with light, persistent seeds. Twelve species found in the seed bank, all known from other evidence to possess persistent seed banks, had not been recorded in the vegetation since 1927. These data are consistent with the interpretation that deeply buried seeds are older than shallow seeds, and that the seed

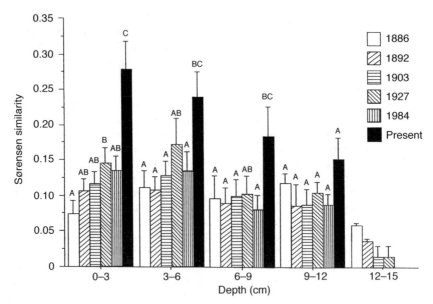

Fig. 9.4. Mean Sørensen similarity (+ 1 SE) between the seed bank of different soil horizons and species composition of the whole islands in Lake Hjälmaren, Sweden, in 1886–1985 and the vegetation of the sampling area in 1993. Note that the vegetation is analysed on different scales in 1993 compared with 1886–1985. Within each soil horizon, different letters above bars indicate significant differences (Fisher's protected least significant difference; $P < 0.05$). The deepest soil horizon was not treated statistically, owing to the low number of samples. (From Grandin and Rydin 1998. Reproduced by permission of Blackwell Science.)

bank can form an ecological memory over at least a century, if not longer.

Bekker *et al.* (1999) examined the seed bank of a dune slack on the island of Terschelling in The Netherlands. The present vegetation and seed bank (0–5 and 5–10 cm) were investigated at three sites that had developed for 5, 9 and 39 years after sod-cutting experiments and one site where succession had continued uninterrupted for over 80 years since a storm in 1915. The vegetation had developed from a pioneer *Samolus valerandi–Littorella uniflora* community, through several intermediate stages, to a climax community dominated by *Calamagrostis epigejos*. Figure 9.5 shows the trends through time of some of the commoner species. Most of the species in Fig. 9.5 have persistent seed banks, including pioneers, such as *Samolus*, which have been absent from the vegetation for many years and yet remain in the seed bank. Several species clearly show waves of abundance, which can be traced from the vegetation, through the

upper soil layer into the lower soil layer. Accordingly, although the number of species in the vegetation peaks early in succession, the number of species in the seed bank, particularly in the lower soil layer, continues to increase throughout the succession.

From both a theoretical and practical viewpoint, a crucial question in plant community ecology is the contribution made by seed banks to regeneration of the aboveground vegetation, following the creation of smaller or larger areas of bare ground. In annual communities, seed bank and vegetation are often almost identical, with the former regularly giving rise to the latter after disturbance by drought or cultivation. Thus, in arable fields, annual pastures (Levassor *et al.*, 1990), tidal wetlands (Leck and Simpson, 1987) and disturbed lake shores (Keddy and Reznicek, 1982), both transient and persistent seed banks play a major role in regeneration of mature vegetation. At the opposite extreme, few climax woodland species have persistent seed

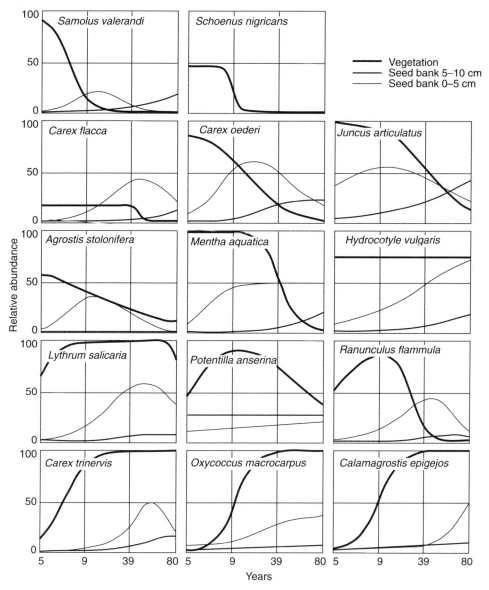

Fig. 9.5. Logistic curves showing the performance of 14 species in the vegetation and upper and lower seed-bank layers at different times after sod cutting in a Dutch dune slack. Seed abundance was standardized to the species maximum found in one quadrat. Only curves with a significant trend are shown ($P < 0.05$), except for *Hydrocotyle vulgaris* (vegetation) and *Potentilla anserina* (seed bank 5–10 cm). * Owing to the very large maximum number of seeds of *Samolus* found in the upper layer (1097), abundance in the lower layer was standardized to 34, the maximum found in this layer. (From Bekker *et al.*, 1999. Reproduced by permission of Opulus Press.)

banks, and such seed banks play little or no part in regeneration of the mature vegetation after disturbance (Pickett and McDonnell, 1989). Where seed banks are important in regeneration, the species involved are usually derived from an earlier successional stage – for example, *Prunus pensylvanica* in eastern North America (Marks, 1974) and *Rubus fruticosus* in eastern England (Brown and Oosterhuis, 1981).

Often the nature of the disturbance determines the relative importance of the seed bank and other sources of colonists. After clear-cutting and burning of conifer forest in Idaho, shrubs recolonized from the seed bank and from sprouts (Morgan and Neuenschwander, 1988). The contribution of *Ceanothus sanguineus*, a shade-intolerant species that was abundant in the seed bank, was increased by the severity of the burn, presumably because the seeds are scarified by high temperatures.

Note that the apparent presence of a seed bank does not guarantee that it plays much part in regeneration. The tropical pioneer tree *Cecropia obtusifolia* is widely reported to possess a persistent seed bank and rapidly colonizes gaps created by fallen trees. Alvarez-Buylla and Martínez-Ramos (1990) found that, although about half such colonists came from the seed bank and about half from freshly dispersed seeds, most seeds in the soil were less than 1 year old. The contribution of persistent seeds to regeneration of *Cecropia* is thus very small. Dalling *et al.* (1998) found much the same in a community of pioneer trees colonizing tree-fall gaps on Barro Colorado Island, Panama. The abundance of 24 pioneers in

the seed bank was strongly negatively related to seed size, but this relationship was less strong for young seedlings and non-significant for older seedlings and adults (Table 9.2). Rank abundance of seedlings, however, was closely related to abundance of adults. These changing relationships arise because some large-seeded species, although very rare in the seed bank, became common as seedlings, while the abundance of small-seeded species, such as *Miconia* and *Cecropia*, fell dramatically between the seed and seedling stage. In other words, seed-bank composition is a poor guide to the identity of both gap regeneration and mature plants.

Between the extremes of annual-dominated communities and woodlands lie a range of principally perennial communities, consisting of herbaceous and dwarf shrub species, here referred to loosely as 'grasslands'. Such communities commonly experience moderate but frequent disturbance from large and small herbivores and burrowing animals, plus occasional more serious disturbances, caused by drought, flooding or fire. This diversity makes it hard to generalize about the role of persistent seed banks in such communities. For

Table 9.2. Spearman's rank correlation tests among rank abundances of seeds in the soil seed bank, seedling abundances and adult abundances, for a group of pioneer species in 53 recently formed gaps in a 50 ha plot on Barro Colorado Island, Panama. Sample sizes are smaller for comparisons with the soil seed bank, since not all seedling taxa germinated from the soil. (From Dalling *et al.*, 1998.)

	n	r_s	P
Seed size vs.:			
Seed bank density	17	−0.74	< 0.01
Seedlings < 10 cm	24	−0.58	< 0.01
Seedlings > 10 cm	24	−0.32	NS
Adults	24	−0.19	NS
Seed bank density vs.:			
Seedlings < 10 cm	17	0.54	< 0.05
Seedlings > 10 cm	17	0.26	NS
Adults	17	0.08	NS
Seedlings < 10 cm vs.:			
Seedlings > 10 cm	24	0.86	< 0.01
Adults	24	0.83	< 0.01
Seedlings > 10 cm vs.:			
Adults	24	0.90	< 0.01

n, Number of species, r_s, correlation coefficient, P, level of significance.

example, in upland Britain, large areas of pasture are dominated by two grass species, *Festuca ovina* and *Agrostis capillaris*, the latter with long-persistent seeds and the former without. In the lowlands, there is a similar dichotomy, with *Lolium perenne* and *Dactylis glomerata* (lacking seed banks) rubbing shoulders with *Holcus lanatus* and *Poa pratensis* (both with seed banks) (Thompson *et al.*, 1997). Among the canopy dominants of Japanese grasslands, *Zoysia japonica* accumulates a dense seed bank, while *Miscanthus sinensis* and *Arundinella hirta* do not (Hayashi and Numata, 1975). In Queensland, the balance between coexisting grasses with and without persistent seed banks seems to depend on the frequency of fire. Fires both kill the surface-lying seeds of grasses without seed banks and stimulate the germination of the buried seeds of those species with seed banks (Peart, 1984). Similarly, in American prairie, *Sporobolus cryptandrus* (one of the few dominant grasses to accumulate a persistent seed bank) seems to be favoured by fire (Lippert and Hopkins, 1950).

Not surprisingly, experimental investigations of recruitment in grasslands have produced contradictory results. In a few cases, an attempt has been made to determine the origin of colonists by replacing small plots of soil with sterile, seed-free soil. In a wet semi-natural grassland in Sweden, only one species (*Taraxacum officinale*) colonized such sterile plots (Milberg, 1993). Although many seedlings emerged from the seed bank, the majority of taxa in the seed bank did not appear as seedlings. A similar experiment in New York (Marks and Mohler, 1985) confirmed the negligible importance of the seed rain in colonizing small gaps in grassland. Here the seed bank and regrowth from vegetative fragments contributed about equally to recruitment in 1 m² experimental plots. In a sandy grassland in Sweden, the seed rain was again found to be relatively unimportant (Zimmergren, 1980). Here the seed bank was the major source of colonists, but the species involved were mainly annuals derived from former arable cultivation. The two dominants (*D. glomerata* and *Coryne-*

phorus canescens, both lacking a persistent seed bank) did not recruit from the seed bank. In an acid grassland in eastern England, Pakeman *et al.* (1998) found that the seed bank accounted for about 40% of recolonization of 0.5 m × 0.5 m disturbed plots. Wind and other non-endozoochory accounted for about the same, while seeds in the droppings of rabbits (the main herbivore present) accounted for the rest. Interestingly, although different species used predominantly different dispersal pathways, no species was entirely dependent on a single dispersal method.

A recurrent theme of such experiments is the overwhelming importance of vegetative regeneration in recolonizing small (≤ 1 m²) gaps in grassland. Hartman (1988) found that bare patches in a New England salt-marsh were swiftly colonized by vegetative expansion of *Spartina alterniflora*, while *Cynodon dactylon* and *Rumex acetosella* rapidly invaded gaps in a North Carolina grassland (Fowler, 1981). Milberg (1993) found that, only 3 months after removal of the upper 1–2 cm of soil in a wet grassland, vegetative regrowth covered 80–95% of the ground, while seedlings covered only 1%. Marks and Mohler (1985) took elaborate care to prevent vegetative invasion from the edges of their plots, but they comment that this proved almost impossible.

In one of the most detailed studies available, Hillier (1990) detected complex interactions between the timing, size and location of gaps. She cut gaps, varying from 5 to 40 cm in diameter, in autumn 1978 and spring 1979 on the north (moist)- and south (dry)-facing slopes of a limestone ridge in Derbyshire, England. Vegetative regrowth was important in all gaps, but more so in small gaps and in the denser vegetation of the north-facing slope. Gaps also promoted germination more on the north-facing slope, since germination was frequent in the sparser vegetation of the south-facing slope, even in the absence of gaps. Autumn gaps favoured seedlings of grasses, chiefly with transient seeds, while spring gaps favoured forbs, chiefly with persistent seed banks (Table 9.3), but the

Table 9.3. A comparison of some of the more common species recorded in autumn- and spring-cut gaps at two derelict calcareous grassland sites in Millers Dale, north Derbyshire, UK (from Hillier, 1990).

	Total number of seedlings recorded		
Species	Autumn gaps	Significance of difference between seasons	Spring gaps
North-facing slope			
Grasses			
Agrostis capillaris	29	*	50
Anthoxanthum odoratum	2	***	18
Danthonia decumbens	0	***	20
Holcus lanatus	36	NS	52
Forbs			
Campanula rotundifolia	3	***	51
Carex flacca	1	***	83
Centaurea nigra	10	***	54
Origanum vulgare	95	***	238
Plantago lanceolata	124	***	349
South-facing slope			
Grasses			
Helictotrichon pratense	43	**	18
Briza media	59	***	4
Festuca arundinacea	26	***	3
Festuca ovina	371	***	76
Koeleria macrantha	146	***	31
Forbs			
Arabidopsis thaliana	53	***	0
Campanula rotundifolia	35	NS	29
Gentianella amarella	241	***	1
Pilosella officinarum	48	***	10
Linum catharticum	1	***	401
Lotus corniculatus	2	***	48
Pimpinella saxifraga	2	***	66
Sonchus asper	20	*	9
Dwarf shrubs			
Helianthemum nummularium	9	*	22
Thymus polytrichus	43	***	14

exact timing of gap creation was critical; gap formation relatively late in spring or autumn could itself destroy whole cohorts of seedlings. The results clearly showed the value of monitoring gaps for at least 2 years after creation; numbers of seedlings of different species appearing soon after gap creation were no guide to numbers of survivors (and hence relative contributions of transient and persistent seeds) 2 years later. In another detailed study, Edwards and Crawley (1999) examined the effects of seed sowing, disturbance and rabbit grazing on recruitment from seed in six grassland species, three with and three without persistent seed banks. Their study confirmed the importance of disturbance in determining the role of seed banks in regeneration. In disturbed soil, seedling densities of all six species increased enormously with increasing density of sown seeds, suggesting that recruitment of disturbed sites would normally be dominated by the current seed rain (Edwards and Crawley prevented the current year's seed rain in their study). However, in the absence of sown seeds, disturbance increased recruitment only in the species with persistent seed banks. In undisturbed turf, recruitment was much lower and, in two species with persistent seed banks, the seed bank alone was enough to saturate the opportunities for establishment. These results suggest that the seed bank might make a major contribution to regeneration in the absence of disturbance or if disturbance occurred at a time (e.g. early summer) when there was little seed rain.

Gap colonization in tundra shows the same lack of consistent pattern. Large areas disturbed by vehicles in Alaska were colonized about equally from the seed bank and seed rain (Ebersole, 1989). All but one species found to be abundant in the seed bank were also frequent colonists, while many species known to possess short-lived seeds (including several *Salix* spp.) colonized by seed dispersal. As in grasslands, the nature of the disturbance may determine the source of colonists. Disturbed organic soils were colonized by *Carex bigelowii* and *Eriophorum vaginatum* (the seed-bank dominants), but exposed mineral soil (which lacks a seed bank) was colonized more slowly by a number of grasses from the current seed rain (Gartner *et al.*, 1983).

It is hard to escape the conclusion that climate, nutrients and herbivory are the principal determinants of the composition of grasslands and related communities. Within the limits imposed by these major biotic and abiotic controls, however, there is clearly scope for variation in regeneration niche (*sensu* Grubb, 1977) to promote increased diversity (Keddy and Reznicek, 1982; Gartner *et al.*, 1983; Peart, 1984; Hillier, 1990; Lavorel *et al.*, 1994). Most studies have found that vegetative regrowth, seed rain and persistent seed banks are all involved, to varying degrees, in colonization of gaps in grassland. Although the persistent seed bank is frequently the least important numerical source of colonists, its contribution can be dramatically altered by quite subtle changes in severity of disturbance and in the location and timing of gap creation (Thompson *et al.*, 1996). The role of gaps in regeneration is further explored by Bullock in Chapter 16 in this volume.

References

Akinola, M.O., Thompson, K. and Hillier, S.H. (1998) Development of soil seed banks beneath synthesised meadow communities after seven years of climate and management manipulations. *Seed Science Research* 8, 493–500.

Alvarez-Buylla, E.R. and Martínez-Ramos, M. (1990) Seed bank versus seed rain in the regeneration of a tropical pioneer tree. *Oecologia*, 84, 314–325.

Askew, A.P., Corker, D., Hodkinson, D.J. and Thompson, K. (1997) A new apparatus to measure the rate of fall of seeds. *Functional Ecology* 11, 121–125.

Baskin, C.C. and Baskin, J.M. (1988) Germination ecophysiology of herbaceous plant species in a temperate region. *American Journal of Botany* 75, 286–305.

Baskin, C.C. and Baskin, J.M. (1989) Physiology of dormancy and germination in relation to seed bank ecology. In: Leck, M.A., Parker, V.T. and Simpson, R.L. (eds) *Ecology of Soil Seed Banks*. Academic Press, San Diego, pp. 53–66.

Baskin, C.C. and Baskin, J.M. (1998) *Seeds: Ecology, Biogeography and Evolution of Dormancy and Germination*. Academic Press, San Diego, 666 pp.

Beatty, S.W. (1991) Colonization dynamics in a mosaic landscape: the buried seed pool. *Journal of Biogeography* 18, 553–563.

Bekker, R.M., Knevel, I.C., Tallowin, J.B.R., Troost, E.M.L. and Bakker, J.P. (1998a) Soil nutrient input effects on seed longevity: a burial experiment with fen meadow species. *Functional Ecology* 12, 673–682.

Bekker, R.M., Bakker, J.P., Grandin, U., Kalamees, R., Milberg, P., Poschlod, P., Thompson, K. and Willems, J.H. (1998b) Seed size, shape and vertical distribution in the soil: indicators of seed longevity. *Functional Ecology* 12, 834–842.

Bekker, R.M., Oomes, M.J.M. and Bakker, J.P. (1998c) The impact of groundwater level on soil seed bank survival. *Seed Science Research* 8, 399–404.

Bekker, R.M., Lammerts, E.J., Schutter, A. and Grootjans, A.P. (1999) Vegetation development in dune slacks: the role of persistent seed banks. *Journal of Vegetation Science* 10, 745–754.

Brown, A.H.F. and Oosterhuis, L. (1981) The role of buried seeds in coppicewoods. *Biological Conservation* 21, 19–38.

Cheplick, G.P. (1992) Sibling competition in plants. *Journal of Ecology* 80, 567–575.

Cohen, D. (1966) Optimizing reproduction in a randomly varying environment. *Journal of Theoretical Biology* 12, 119–129.

Cook, R. (1980) The biology of seeds in the soil. In: Solbrig, O.T. (ed.) *Demography and Evolution in Plant Populations.* Blackwell Scientific Publications, New York, pp. 107–129.

Cresswell, E. and Grime, J.P. (1981) Induction of a light requirement during seed development and its ecological consequences. *Nature* 291, 583–585.

Dalling, J.W., Hubbell, S.P. and Silvera, K. (1998) Seed dispersal, seedling establishment and gap partitioning among tropical pioneer trees. *Journal of Ecology* 86, 674–689.

Ebersole, J.J. (1989) Role of the seed bank in providing colonizers on a tundra disturbance in Alaska. *Canadian Journal of Botany* 67, 466–471.

Edwards, G.R. and Crawley, M.J. (1999) Effects of disturbance and rabbit grazing on seedling recruitment of six mesic grassland species. *Seed Science Research* 9, 145–156.

Ellner, S. (1986) Germination dimorphisms and parent–offspring conflict in seed germination. *Journal of Theoretical Biology* 123, 173–186

Fellows, G.M. and Roeth, F.W. (1992) Factors influencing shattercane (*Sorghum bicolor*) seed survival. *Weed Science* 40, 434–440.

Fowler, N. (1981) Competition and coexistence in a North Carolina grassland. II. The effects of the experimental removal of species. *Journal of Ecology* 69, 843–854.

Funes, G., Basconcelo, S., Diaz, S. and Cabido, M. (1999) Seed size and shape are good predictors of seed persistence in soil in temperate mountain grasslands of Argentina. *Seed Science Research* 9, 341–345.

Gartner, B.L., Chapin, S.F. and Shaver, G.R. (1983) Demographic patterns of seedling establishment and growth of native graminoids in an Alaskan tundra disturbance. *Journal of Applied Ecology* 20, 965–980.

Gimingham, C.H. (1972) *Ecology of Heathlands.* Chapman and Hall, London.

Gorski, T., Gorska, K. and Nowicki, J. (1977) Germination of seeds of various species under leaf canopy. *Flora* 166, 249–259.

Grandin, U. and Rydin, H. (1998) Attributes of the seed bank after a century of primary succession on islands in Lake Hjälmaren, Sweden. *Journal of Ecology* 86, 293–303.

Grime, J.P., Hodgson, J.G. and Hunt, R. (1988) *Comparative Plant Ecology: a Functional Approach to Common British Plants.* Unwin Hyman, London, 742 pp.

Grubb, P.J. (1977) The maintenance of species-richness in plant communities: the importance of the regeneration niche. *Biological Reviews* 52, 107–145.

Gutterman, Y. (1993) *Seed Germination in Desert Plants.* Springer-Verlag, Berlin.

Harborne, J.B. (1997) Plant secondary metabolism. In: Crawley, M. (ed.) *Plant Ecology*, 2nd edn. Blackwell Science, Oxford, pp. 132–155.

Hartley, S.E. and Jones, C.G. (1997) Plant chemistry and herbivory, or why the world is green. In: Crawley, M. (ed.) *Plant Ecology*, 2nd edn. Blackwell Science, Oxford, pp. 284–324.

Hartman, J.M. (1988) Recolonization of small disturbance patches in a New England salt marsh. *American Journal of Botany* 75, 1625–1631.

Hayashi, I. and Numata, M. (1975) Viable buried seed populations in grasslands in Japan. In: Numata, M. (ed.) *JIBP Synthesis*, Vol. 13. *Ecological Studies in Japanese Grasslands with Special Reference to the IBP Area Productivity of Terrestrial Ecosystems.* University of Tokyo Press, Tokyo, pp. 58–69.

Hendry, G.A.F., Thompson, K., Moss, C.J., Edwards, E. and Thorpe, P.C. (1994) Seed persistence: a correlation between seed longevity in the soil and *ortho*-dihydroxyphenol concentration. *Functional Ecology* 8, 658–664.

Hillier, S.H. (1990) Gaps, seed banks and plant species diversity in calcareous grasslands. In: Hillier, S.H., Walton, D.W.H. and Wells, D.A. (eds) *Calcareous Grasslands – Ecology and Management.* Bluntisham Books, Huntingdon, pp. 57–66.

Hilton, J.R. (1984) The influence of light and potassium nitrate on the dormancy and germination of *Avena fatua* L. (wild oat) seed and its ecological significance. *New Phytologist* 96, 31–34.

Hogenbirk, J.C. and Wein, R.W. (1992) Temperature effects on seedling emergence from boreal wetland soils: implications for climate change. *Aquatic Botany* 42, 361–373.

Hulme, P.E. (1993) Post-dispersal seed predation by small mammals. *Symposium of the Zoological Society of London* 65, 269–287.

Hyatt, L.A. and Evans, A.S. (1998) Is decreased germination fraction associated with risk of sibling competition? *Oikos* 83, 29–35.

Ibrahim, A.E. and Roberts, E.H. (1983) Viability of lettuce seeds. I. Survival in hermetic storage. *Journal of Experimental Botany* 34, 620–630.

Jankowska-Blaszczuk, M. and Grubb, P.J. (1997) Soil seed banks in primary and secondary deciduous forest in Bialowieza, Poland. *Seed Science Research* 7, 281–292.

Kalisz, S. and McPeek, M.A. (1993) Extinction dynamics, population growth and seed banks. An example using an age-structured annual. *Oecologia* 95, 314–320.

Keddy, P.A. and Reznicek, A.A. (1982) The role of seed banks in the persistence of Ontario's coastal plain flora. *American Journal of Botany* 69, 13–22.

Keeley, J.E. and Fotheringham, C.J. (1998) Mechanism of smoke-induced seed germination in a post-fire chaparral annual. *Journal of Ecology* 86, 27–36.

Kennedy, R.A., Barrett, S.C.H., Delmar, V.Z. and Rumpho, M.E. (1980) Germination and seedling growth under anaerobic conditions in *Echinochloa crus-galli* (barnyard grass). *Plant, Cell and Environment* 3, 243–248.

Koniak, S. and Everett, R.L. (1982) Seed reserves in soils of successional stages of pinyon woodlands. *American Midland Naturalist* 108, 295–303.

Kremer, R.J., Hughes, L.B., Jr and Aldrich, R.J. (1984) Examination of microorganisms and deterioration resistance mechanisms associated with velvetleaf (*Abutilon theophrasti*) seed. *Agronomy Journal* 76, 745–749.

Lamont, B.B., Le Maitre, D.C., Cowling, R.M. and Enright, N.J. (1991) Canopy seed storage in woody plants. *The Botanical Review* 57, 278–311.

Lavorel, S., Lepart, J., Debussche, M., Lebreton, J.D. and Beffy, J.L. (1994) Small scale disturbances and the maintenance of species diversity in Mediterranean old fields. *Oikos* 70, 455–472.

Leck, M.A. (1989) Wetland seed banks. In: Leck, M.A., Parker, V.T. and Simpson, R.L. (eds) *Ecology of Soil Seed Banks*. Academic Press, San Diego, pp. 283–305.

Leck, M.A. and Simpson, R.L. (1987) Seed bank of a freshwater tidal wetland: turnover and relationship to vegetation change. *American Journal of Botany* 74, 360–370.

Leishman, M.R. and Westoby, M. (1998) Seed size and shape are not related to persistence in soil in Australia in the same way as in Britain. *Functional Ecology* 12, 480–485.

Levassor, C., Ortega, M. and Peco, B. (1990) Seed bank dynamics of Mediterranean pastures subjected to mechanical disturbance. *Journal of Vegetation Science* 1, 339–344.

Lippert, R.D. and Hopkins, H.H. (1950) Study of viable seeds in various habitats in mixed prairie. *Transactions of the Kansas Academy of Science* 53, 355–364.

Livingston, R.B. and Allessio, M.L. (1968) Buried viable seed in successional field and forest stands, Harvard Forest, Massachusetts. *Bulletin of the Torrey Botanical Club* 95, 58–69.

Lokesha, R., Hegde, S.G., Shaanker, R.U. and Ganeshaiah, K.N. (1992) Dispersal mode as a selective force in shaping the chemical composition of seeds. *American Naturalist* 140, 520–525.

McPeek, M.A. and Kalisz, S. (1998) The joint evolution of dispersal and dormancy in metapopulations. *Archive für Hydrobiologie* 52, 33–51.

Marks, P.L. (1974) The role of pin cherry (*Prunus pensylvanica* L.) in the maintenance of stability in northern hardwood ecosystems. *Ecological Monographs* 44, 73–88.

Marks, P.L. (1983) On the origin of the field plants of the northeastern United States. *American Naturalist* 122, 210–228.

Marks, P.L. and Mohler, C.L. (1985). Succession after elimination of buried seeds from a recently plowed field. *Bulletin of the Torrey Botanical Club* 112, 376–382.

Metcalfe, D.J. and Grubb, P.J. (1995) Seed mass and light requirement for regeneration in south east Asian rainforest. *Canadian Journal of Botany* 73, 817–826.

Milberg, P. (1993). Seed bank and seedlings emerging after soil disturbance in a wet semi-natural grassland in Sweden. *Annales Botanici Fennici* 30, 9–13.

Milberg, P. (1995) Soil seed bank after eighteen years of succession from grassland to forest. *Oikos* 72, 3–13.

Milberg, P. (1997) Weed seed germination after a short-term light exposure: germination rate, photon fluence response and interaction with nitrate. *Weed Research* 37, 157–164.

Milberg, P. and Andersson, L. (1997) Seasonal variation in dormancy and light sensitivity in buried seeds of eight annual weed species. *Canadian Journal of Botany* 75, 1998–2004.

Moore, D.R.J. and Keddy, P.A. (1988) Effects of a water-depth gradient on the germination of lakeshore plants. *Canadian Journal of Botany* 66, 548–552.

Morgan, P. and Neuenschwander, L.F. (1988) Seed-bank contributions to regeneration of shrub species after clear-cutting and burning. *Canadian Journal of Botany* 66, 169–172.

Murray, K.G. (1988). Avian seed dispersal of 3 neotropical gap-dependent plants. *Ecological Monographs* 58, 271–298.

Nilsson, P., Fagerstrom, T., Tuomi, J. and Astrom, M. (1994) Does seed dormancy benefit the mother plant by reducing sib competition? *Evolutionary Ecology* 8, 422–430.

Oosting, H.T. and Humphreys, M.E. (1940) Buried viable seeds in a successional series of old field and forest soils. *Bulletin of the Torrey Botanical Club* 67, 253–273.

Pakeman, R.J., Attwood, J.P. and Engelen, J. (1998) Sources of plants colonizing experimentally disturbed patches in an acidic grassland, in eastern England. *Journal of Ecology* 86, 1032–1041.

Pakeman, R.J., Cummins, R.P., Miller, G.R. and Roy, D.B. (1999) Potential climatic control of seed-bank density. *Seed Science Research* 9, 101–110.

Peart, M.H. (1984) The effects of morphology, orientation and position of grass diaspores on seedling survival. *Journal of Ecology* 72, 437–453.

Pickett, S.T.A. and McDonnell, M.J. (1989) Physiology of dormancy and germination in relation to seed bank ecology. In: Leck, M.A., Parker, V.T. and Simpson, R.L. (eds) *Ecology of Soil Seed Banks*. Academic Press, San Diego, pp. 123–147.

Piroznikow, E. (1983). Seed bank in the soil of stabilized ecosystem of a deciduous forest (*Tilio–Carpinetum*) in the Bialowieza National Park. *Ekologia Polska* 31, 145–172.

Ponquett, R.T., Smith, M.T. and Ross, G. (1992) Lipid autoxidation and seed ageing: putative relationships between seed longevity and lipid stability. *Seed Science Research* 2, 51–54.

Pons, T.L. (1989a) Dormancy, germination and mortality of seeds buried in heathland and inland sand dunes. *Acta Botanica Neerlandica* 38, 327–335.

Pons, T.L. (1989b) Breaking of seed dormancy by nitrate as a gap detection mechanism. *Annals of Botany* 63, 139–143.

Pons, T.L. (1991a) Dormancy, germination and mortality of seeds in a chalk-grassland flora. *Journal of Ecology* 79, 765–780.

Pons, T.L. (1991b) Induction of dark dormancy in seeds: its importance for the seed bank in the soil. *Functional Ecology* 5, 669–675.

Priestley, D.A. (1986) *Seed Aging: Implications for Seed Storage and Persistence in the Soil*. Cornell University Press, Ithaca, New York.

Rees, M. (1993) Trade-offs among dispersal strategies in the British flora. *Nature* 366, 150–152.

Rees, M. (1994) Delayed germination of seeds: a look at the effects of adult longevity, the timing of reproduction, and population age/stage structure. *American Naturalist* 144, 43–64.

Rees, M. and Long, M.J. (1993) The analysis and interpretation of seedling recruitment curves. *American Naturalist* 141, 233–262.

Roberts, H.A. (1986). Seed persistence in soil and seasonal emergence in plant species from different habitats. *Journal of Applied Ecology* 23, 638–656.

Salisbury, E.J. (1942) *The Reproductive Capacity of Plants*. G. Bell & Sons, London, 244 pp.

Schafer, D.E. and Chilcote, D.O. (1970) Factors influencing persistence and depletion in buried seed populations. II. The effects of soil temperature and moisture. *Crop Science* 10, 342–345.

Shen-Miller, J., Mudgett, M.B., Schopf, J.W., Clarke, S. and Berger, R. (1995) Exceptional seed longevity and robust growth: ancient sacred lotus from China. *American Journal of Botany* 82, 1367–1380.

Siemens, D.H., Johnson, C.D. and Ribardo, K.J. (1992) Alternative seed defense mechanisms in congeneric plants. *Ecology* 73, 2152–2166.

Skoglund, J. and Verwijst, T. (1989) Age structure of woody species populations in relation to seed rain, germination and establishment along the river Dalälven, Sweden. *Vegetatio* 82, 25–34.

Staaf, H., Jonsson, M. and Olsen, L.G. (1987) Buried germinative seeds in mature beech forests with different herbaceous vegetation and soil types. *Holarctic Ecology* 10, 268–277.

Thanos, C.A. and Rundel, P.W. (1995) Fire-followers in chaparral: nitrogenous compounds trigger seed germination. *Journal of Ecology* 83, 207–216.

Thompson, K. (1987) Seeds and seed banks. In: Rorison, I.H., Grime, J.P., Hunt, R., Hendry, G.A.F. and Lewis, D.H. (eds) *Frontiers of Comparative Plant Ecology*. New Phytologist 106 (Suppl.), pp. 23–34.

Thompson, K. and Grime, J.P. (1979) Seasonal variation in the seed banks of herbaceous species in ten contrasting habitats. *Journal of Ecology* 67, 893–921.

Thompson, K. and Grime, J.P. (1983) A comparative study of germination responses to diurnally-fluctuating temperatures. *Journal of Applied Ecology* 20, 141–156.

Thompson, K., Band, S.R. and Hodgson, J.G. (1993) Seed size and shape predict persistence in soil. *Functional Ecology* 7, 236–241.

Thompson, K., Hillier, S.H., Grime, J.P., Bossard, C.C. and Band, S.R. (1996) A functional analysis of a limestone grassland community. *Journal of Vegetation Science* 7, 371–380.

Thompson, K., Bakker, J.P. and Bekker, R.M. (1997) *The Soil Seed Banks of North West Europe: Methodology, Density and Longevity*. Cambridge University Press, Cambridge, 276 pp.

Thompson, K., Bakker, J.P., Bekker, R.M. and Hodgson, J.G. (1998) Ecological correlates of seed persistence in soil in the NW European flora. *Journal of Ecology* 86, 163–169.

Tsuyuzaki, S. (1991) Survival characteristics of buried seeds 10 years after the eruption of the Usu volcano in northern Japan. *Canadian Journal of Botany* 69, 2251–2256.

van Baalen, J. (1982) Germination ecology and seed population dynamics of *Digitalis purpurea*. *Oecologia* 53, 61–67.

Van Der Valk, A.G. and Davis, C.B. (1976) The seed banks of prairie glacial marshes. *Canadian Journal of Botany* 54, 1832–1838.

Vàzquez-Yanes, C. and Orozco-Segovia, A. (1990) Ecological significance of light controlled seed germination in two contrasting tropical habitats. *Oecologia* 83, 171–175.

Venable, D.L. and Brown, J.S. (1988) The selective interactions of dispersal, dormancy, and seed size as adaptations for reducing risk in variable environments. *American Naturalist* 131, 360–384.

Venable, D.L. and Lawlor, L. (1980) Delayed germination and dispersal in desert annuals: escape in space and time. *Oecologia* 46, 272–282.

Villiers, T.A. (1974) Seed aging: chromosome stability and extended viability of seeds stored fully imbibed. *Plant Physiology* 53, 875–878.

Villiers, T.A. and Edgecumbe, D.J. (1975) On the causes of seed deterioration in dry storage. *Seed Science and Technology* 3, 761–774.

Vleeshouwers, L.M., Bouwmeester, H.J. and Karssen, C.M. (1995) Redefining seed dormancy: an attempt to integrate physiology and ecology. *Journal of Ecology* 83, 1031–1037.

Waller, D.M. (1988) Plant morphology and reproduction. In: Lovett Doust, J. and Lovett Doust, L. (eds) *Plant Reproductive Ecology: Patterns and Strategies*. Oxford University Press, New York, pp. 203–227.

Washitani, I. and Masuda, M. (1990) A comparative study of the germination characteristics of seeds from a moist tall grassland community. *Functional Ecology* 4, 543–557.

Went, F.W. (1949) Ecology of desert plants. II. The effect of rain and temperature on germination and growth. *Ecology* 30, 1–13.

Wesson, G. and Wareing, P.F. (1969) The induction of light sensitivity in weed seeds by burial. *Journal of Experimental Botany* 20, 414–425.

Westoby, M., Leishman, M.R. and Lord, J.M. (1996) Comparative ecology of seed size and seed dispersal. *Philosophical Transactions of the Royal Society of London B, Biological Sciences* 351, 1309–1318.

Willson, M.F., Rice, B.L. and Westoby, M. (1990) Seed dispersal spectra: a comparison of temperate plant communities. *Journal of Vegetation Science* 1, 547–562.

Youngman, B.J. (1951) Germination of old seeds. *Kew Bulletin* 6, 423–426.

Zammit, C. and Zedler, P.H. (1990) Seed yield, seed size and germination behaviour in the annual *Pogogyne abramsii*. *Oecologia* 84, 24–28.

Zimmergren, D. (1980) The dynamics of seed banks in an area of sandy soil in southern Sweden. *Botaniska Notiser* 133, 633–641.

Chapter 10
Seed Responses to Light

Thijs L. Pons

Department of Plant Biology, Utrecht University, Utrecht, The Netherlands

Introduction

The light response of seeds can control the timing of germination in the field, a crucial factor in the survival of the resulting seedlings and growth and fitness in subsequent life stages. The ultimate effect of light on seeds depends on genotype and on environmental factors during ripening of the seeds, during dormancy and during germination itself. These environmental factors may be light or factors other than light. The picture is further complicated by the fact that the light climate itself has various aspects that have different effects on seeds, such as photon flux density (PFD), spectral composition and duration of exposure of the seeds. All the above-mentioned factors can interact in one way or another in their effect on seeds. Moreover, the factors are not constant in time in the field and are difficult to measure at the seed's position, thus further complicating the analysis of what is actually happening with a seed in a natural situation and the interpretation of a possible ecological significance of light responses.

Responses of seeds to light are not universal in the plant kingdom. However, light responses are very common among small-seeded species capable of emerging after some form of disturbance. In these seeds, light serves as a signal rather than as a resource. Only water, oxygen and a suitable temperature are prerequisites for growth of the essentially heterotrophic embryo. Following Bewley and Black (1982), seed responses to light are thus considered as signs of the control of light over dormancy, rather than as a direct control over germination.

In the first part of this chapter, the seed's light climate is described and the various light responses are dealt with in physiological terms, with particular attention to the effect that environmental factors other than light can have on light responses of seeds. Postulated mechanisms explaining these light responses are considered next. They can help to bring order to the enormous variety of phenomena observed. In the second part, an attempt is made to interpret the ecological significance of the various light responses in different (micro)habitats. The possible adaptive value of the response to the various aspects of the light climate is considered for seeds in the soil and for those on the soil surface, where they may or may not be shaded by leaves. A number of case-studies from different climatic regions are treated in the third part. Available data on light responses for selected species are reviewed, and their importance in the life cycle of the plant is discussed.

The light climate of the seed

The light climate of a seed depends on its position in the soil, under litter or under vegetation. There can be variation in spectral composition of the light, in PFD and in duration of exposure to light, if there is any light at all. The light conditions within a seed, where the pigment system that perceives light is located, furthermore depend on the absorption characteristics of the seed-coat and any other surrounding structures.

Light is strongly attenuated by soil. Hence, seeds in deeper soil layers are in perpetual darkness. Measurable quantities of light do not penetrate to greater depths than a few millimetres or centimetres (Tester and Morris, 1987). Sandy soils tend to have higher light transmission than loam, and humic substances are efficient absorbers too. Transmission spectra indicate that the shorter wavelengths are more strongly absorbed than longer ones in the spectral region of interest (Fig. 10.1). The steep gradient of PFD in the upper soil layers means that the actual light climate of the seed greatly depends on its exact position. As the surface is approached from below, light conditions may change from ineffective to stimulating and further to inhibiting for germination across only a few millimetres.

Exposures of seeds to light of short duration are commonly used in physiological studies. The only occasion when short exposure can occur under field conditions is when soil is disturbed. Seeds may be brought to the surface and become buried again immediately afterwards (Sauer and Struik, 1964; Scopel et al., 1994) or on a later occasion. Once seeds are at or close to the soil surface, day length determines further variation in duration of exposure.

The presence of a leaf canopy above a seed reduces PFD of all wavelengths relative to full daylight, but much more in the photosynthetically active part of the spectrum (400–700 nm) than in the near infra-red (700–1000 nm), due to strong absorption by chlorophyll (Fig. 10.2). Hence, canopy shade light is rich in far red (FR) and poor in red (R). A light climate is conveniently characterized by the R : FR ratio with regard to phytochrome functioning (Smith, 1982). This is defined as the ratio in PFD in 10 nm wide bands at 660 nm and 730 nm, the absorption maxima of the R- and FR-absorbing forms of phytochrome, respectively (P_r and P_{fr}). Other wavelengths are also absorbed by phytochrome; hence, the ultimate effect of broad-spectrum light on phytochrome depends also on the presence of other wavelengths. Unfiltered daylight has a typical R : FR of c. 1.2 and leaf canopies may reduce this value to 0.2, depending on the leaf area index (Federer and Tanner, 1966; Stoutjesdijk, 1972; Holmes and Smith, 1977; Pons, 1983; Fig. 10.2).

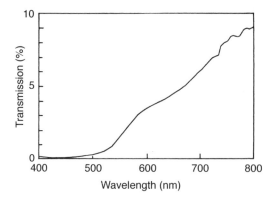

Fig. 10.1. Spectral transmission percentages in the 400–800 nm wavelength region under 3 mm of moist sand (after Bliss and Smith, 1985).

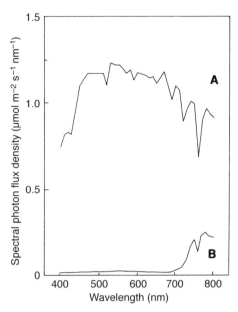

Fig. 10.2. The spectral photon distribution in the 400–800 nm wavelength region of (A) daylight and (B) leaf-canopy-filtered light. Measurements were made in overcast weather conditions under an ash canopy. The red : far-red photon ratio of daylight was 1.2, and under the canopy 0.18. (T.L. Pons, unpublished results.)

Phytochrome is located in the embryo. Hence, when interpreting light responses of seeds, the optical properties of the surrounding structures must be taken into account. Seed-coats can be intensely pigmented, which reduces the PFD and alters the spectral composition of the light in the seed (Widell and Vogelmann, 1988). The blue part of the spectrum can be particularly strongly absorbed.

Physiological aspects

Types of light responses

There is a wide variety of light responses in seeds. Two main effects of light can be distinguished:

1. Effects of short-duration exposures. Whether germination is stimulated or inhibited depends on wavelength, but is largely independent of PFD above a threshold value.

2. Effects of long-duration exposures. The stimulating or inhibiting effect depends on both spectral composition and PFD.

When light is given in short exposures (e.g. less than 1 h), R light of wavelengths around 660 nm tends to break dormancy and FR radiation of wavelengths around 730 nm tends to impose it (Flint and McAllister, 1937). The discovery of these responses contributed to the discovery of phytochrome (Borthwick *et al.*, 1954). The earlier experiments were largely carried out with lettuce seeds, but similar responses are now known from many other species also. The effects are reversible within a certain period of time (Fig. 10.3). Rather low photon doses are required for this response (*c.* 100 µmol m^{-2}); this response is known as the low-fluence response (LFR) (Fig. 10.4). After preconditioning at low or high temperature, light-sensitive seeds may even respond to extremely low photon doses (*c.* 0.1 µmol m^{-2}). This is known as the very-low-fluence response (VLFR) (Fig. 10.4; Blaauw-Jansen and Blaauw, 1975; Blaauw-Jansen, 1983; Cone and Kendrick, 1986).

Some seeds do not respond to short exposures of light, but require long exposure times for dormancy breaking (*Plantago major* in Fig. 10.3). In such cases, repeated short exposures are equally effective and the response is considered essentially an LFR. However, in most species, long exposure times tend to inhibit germination when PFD is high, particularly at low R : FR (Górski and Górska, 1979; Frankland, 1986; Fig. 10.4). The inhibiting effect of continuous illumination (or light interrupted by a night period) on germination is known as the high-irradiance response (HIR); wavelengths of 710–720 nm tend to be the most effective. Species that are characterized as negatively photoblastic (inhibited by light) show a strong HIR, as evident from inhibition of germination at low PFD, whereas short exposures, working through the LFR, may be stimulating in these seeds (Frankland, 1976).

The actual response of seeds to light varies greatly between species and seed lots, and further depends on pretreatment and conditions during the germination test.

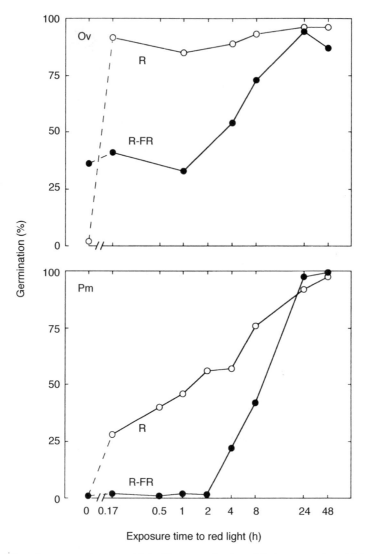

Fig. 10.3. Effect of exposure time to red light (R) on germination of *Origanum vulgare* (Ov) and *Plantago major* (Pm), and escape from phytochrome control, indicated by the effect of a subsequent exposure to far red (R-FR). Seeds were exposed to R for the period indicated. The exposure time for FR was 10 min. After the exposure(s), the seeds were tested for final germination percentage in darkness at 22°C (Ov) or 27°C (Pm). (After Pons, 1991a.)

The three responses as described above (VLFR, LFR, HIR) can sometimes be demonstrated in one species. For example, in lettuce seeds, the LFR can be easily demonstrated, but, when the seeds are pre-treated at high temperatures, the VLFR becomes effective (Blaauw-Jansen and Blaauw, 1975). When exposed for long periods to daylight, germination of lettuce

is negatively affected by high PFD (Fig. 10.4; Górski and Górska, 1979). Similar phenomena have been described *for P. major* and *Sinapis arvensis* (Frankland and Poo, 1980).

The different types of light responses as described above enable the seed to perceive different aspects of its light environment. The LFR and the VLFR are involved

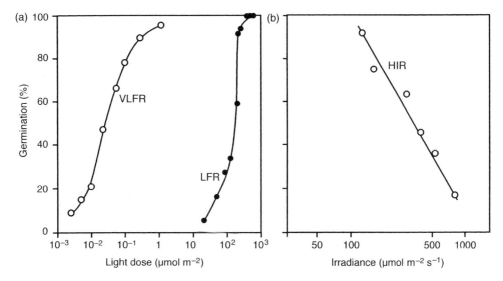

Fig. 10.4. The three main types of light responses of seeds demonstrated in one species (lettuce, *Lactuca sativa*). (a) Fluence response to red light of seeds pretreated at 37°C and to far red, showing the very low fluence response (VLFR) and the low fluence response (LFR), respectively (from Blaauw-Jansen and Blaauw, 1975). (b) Irradiance response to daylight, showing inhibition of germination by the high-irradiance response (HIR) (from Górski and Górska, 1979).

in the perception of the absence of light, which enforces dormancy as long as the situation persists (dark dormancy). The LFR and the HIR are involved in the perception of the spectral composition and intensity of the light to which the seeds are exposed.

Perception of light by phytochrome

The physiological mechanisms underlying the various seed responses to light will be summarized in brief. For more extensive information and references, the reader is referred to reviews by Bewley and Black (1982), Frankland and Taylorson (1983), Shinomura (1997) and Casal and Sánchez (1998). The LFR and the VLFR are conveniently explained by the perception of light by the phytochrome system and its subsequent transformations. The mechanisms responsible for the HIR are less well understood, but phytochrome is considered to play a key role there as well.

The perception of the light environment by phytochrome requires phototransformation of the pigment by the absorption

of either R or FR. When R is absorbed by P_r, P_{fr} is formed, via a number of intermediates. Absorption of FR by P_{fr} leads to the reverse transformation, but via different intermediates (Kendrick and Spruit, 1977). P_{fr} is considered as the form in which phytochrome exerts its action, such as dormancy breaking. Since the two forms of phytochrome have partly overlapping absorption spectra, irradiation with R and FR does not result in pure pools of P_{fr} and P_r, respectively. In daylight, R, FR and other wavelengths that are absorbed by phytochrome occur. Light establishes a certain photoequilibrium of phytochrome ($P_{fr}:P_{tot}$, the ratio of P_{fr} to total phytochrome), which depends on the spectral composition and on photon dose. A saturating dose of daylight results in *c.* 65% of phytochrome in P_{fr} form; in canopy-filtered light, this percentage can drop to 20% (Fig. 10.5). Pure R and FR result in *c.* 80 and 2% P_{fr}, respectively. A low photon dose may be too low to saturate the phototransformation of phytochrome, resulting in a different $P_{fr}:P_{tot}$ ratio than that expected from the spectral composition of the light. Transformation of P_{fr} to P_r may also occur in the absence of light, leading to

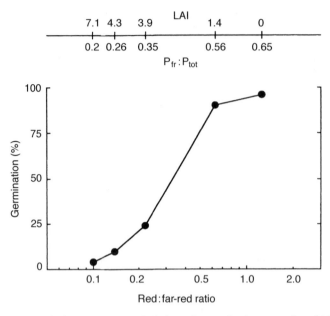

Fig. 10.5. Germination of *Plantago major* in daylight under stands of *Sinapis alba* of different densities, resulting in different red : far-red photon ratios (R : FR) of the transmitted light. Corresponding leaf area index (LAI) and phytochrome photoequilibria (P_{fr} : P_{tot} ratios) are indicated. (After Frankland and Poo, 1980.)

a pure P_r phytochrome pool after a prolonged period in darkness at a sufficiently high temperature.

Different types of phytochromes have been discovered. They have the chromophore in common, but have different apoproteins, which are encoded by different genes (Casal and Sánchez, 1998; Ballaré, 1999). The phytochromes have distinct characteristics associated with their functions, as evident from studies with mutants deficient in a particular phytochrome type (Koorneef and Kendrick, 1994; Smith, 1995). The stable phyB is present in seeds at imbibition. PhyA is more labile and is predominantly synthesized after imbibition. These two species of phytochrome have been studied extensively and their roles in seed responses to light have been characterized. More types of phytochrome exist (phyC–F), which are also involved in light responses, as evident from the fact that phytochrome responses still exist in a phyA–phyB-deficient double mutant (Poppe and Schäfer, 1997), but their roles in light responses in seeds are not so well defined.

PhyB is involved in the clasical LFR and is pre-existent in the light-sensitive seed. A certain threshold level of P_{fr} is required for the breaking of dormancy in this response type. The threshold may vary between species, seed lots and environmental conditions. Exposure of seeds to R or unmodified daylight typically results in a percentage P_{fr} that is well above the threshold value and thus breaks dormancy. Exposure to FR or leaf-canopy-shade light results in P_{fr} levels below the threshold in many seeds, which prevents breaking of dormancy.

Phytochrome may be pre-existent in the seeds as P_{fr}, which originates from before ripening of the seed. Phytochrome is roughly arrested at the photoequilibrium, as it is when the seed dries, which thus depends on the light environment of the mother plant and on chlorophyll content of the seed-coat and the fruit and other surrounding structures during ripening (Cresswell and Grime, 1981). The P_{fr} in the dry seed may result in germination in darkness upon imbibition. Repeated exposure to FR is sometimes required to prevent germination, since P_{fr} is formed in the imbibing seed from intermediates trapped in the drying seed (Kendrick and Spruit, 1977).

The presence of P_{fr} is required for a certain period of time – several hours or more – for the completion of its dormancy-breaking action, the so-called escape time. When P_{fr} is reversed in darkness to P_r before this time has passed, breaking of dormancy is prevented. After this 'escape time', P_{fr} is no longer necessary and germination cannot be prevented any more by a short exposure to FR (Fig. 10.3).

The reversion of P_{fr} to P_r has more consequences than the generation of a pure P_r pool of phytochrome in darkness. Seeds that have a high P_{fr} threshold and possibly a fast dark reversion may require long exposure times or repeated short exposures. P_{fr} disappears before it has completed its action or, in other words, within the escape time (*P. major* in Fig. 10.3). Another consequence is that, in continuous low PFD, phototransformation of P_r to P_{fr} may become quantitatively less important in comparison to dark reversion of P_{fr} to P_r. This results in a $P_{fr}:P_{tot}$ ratio lower than expected from phototransformations only (Frankland and Taylorson, 1983). A stimulating effect of increasing PFD on germination may be expected where this phenomenon plays a role.

PhyA is involved in the VLFR. It is synthesized in the P_r form after imbibition and is unstable once transformed to P_{fr}. The threshold level of P_{fr} required for dormancy breaking by phyA is much lower than by phyB. The very low photon doses that are effective in the VLFR are a manifestation of this very low requirement of phyA in P_{fr} form. Dormancy breaking by FR that establishes very low levels of P_{fr} may be interpreted as an indication of the involvement of the VLFR (*Origanum vulgare* in Fig. 10.3). The response of photosensitized lettuce seeds to moonlight or even starlight is another example of the high light sensitivity caused by phyA (Hartmann *et al.*, 1998). As a consequence of the high sensitivity to P_{fr}, seeds may respond to green light, which is often used for recording germination in darkness

(Baskin and Baskin, 1979; Grime *et al.*, 1981). This complicates the interpretation of the results of several studies where this green 'safe' light was used.

During exposure to light, phytochrome is constantly transformed from P_{fr} to P_r and back; the rate of cycling depends on PFD and spectral composition. Evidence indicates that this phytochrome cycling may somehow be responsible for the inhibiting effect on germination in the HIR (Frankland, 1986). The HIR is still effective after expiration of the escape time, which indicates that it interferes in a later step in the cascade of reactions leading to germination than the LFR. PhyA is involved in the HIR in seedlings, but the phytochrome species involved in the HIR in seeds has not been identified yet (Casal and Sánchez, 1998). Exposure of seeds to light in the field is generally of long duration. Hence, the HIR is potentially involved in most light responses under natural conditions.

The phytochrome system is a multifunctional pigment system, in the sense that it is involved in the perception of various aspects of the light climate (Smith and Whitelam, 1990). Hence, these aspects can show strong interaction in their effect on seeds; e.g. the effect of the spectral composition of light depends on PFD and period of exposure, etc. This complicates the interpretation of the ecological significance of a light response, if there is any ecological significance in that particular response at all.

Events further down the signal transduction pathway

In many seeds, germination is controlled by the balance between the growth potential of the embryo (particularly the radicle), on the one hand, and the constraint exerted by the surrounding structures, on the other (Bewley and Black, 1982). Light can influence both these factors, as evident from many studies done with different species. It is too early to arrive at a general scheme of phytochrome signal transduction, but the studies carried out so far have already increased our insight substantially (Casal and Sánchez, 1998).

The HIR causes a reduction in the growth potential of the radicle by an increase in the yield threshold (Schopfer and Plachy, 1993). The available evidence does not indicate a role for restriction of embryo growth by surrounding structures in this case.

P_{fr} increases the growth potential of the embryo by a decrease in osmotic potential and/or an increase in cell-wall extensibility. The endosperm surrounding the embryo is weakened by an increase in activity of enzymes that hydrolyse mannans in the cell walls. Not much is known about the signal transduction pathways at the molecular level. However, the available evidence suggest an important role for gibberellic acid (GA). Both synthesis of GA and sensitivity to it may be controlled by P_{fr} (Karssen et al., 1989; Derkx and Karssen, 1993a). GAs, synthesized in the embryo, can exert their influence there and migrate to the endosperm, where hydrolysing enzyme activity is stimulated (Sánchez and De Miguel, 1997). The radicle can now break through the weakened seed-coat. Decreases in abscisic acid (ABA) and increases in cytokinins (CKs) are also reported upon exposure of seeds to light (Thomas et al., 1997). ABA is known to decrease cell-wall extensibility and CKs stimulate cell division, protein synthesis and cell extension. However, primary and secondary events in the signal transduction must still be sorted out.

Interference of other environmental factors with light responses

The actual response of a seed to light greatly depends on other environmental conditions. Both the present and preceding conditions are relevant. These factors, such as temperature, water potential and chemicals, affect germination in their own right and are treated in other chapters of this volume. They are treated briefly here in relation to their interaction with light responses.

Temperature fluctuation can break dormancy in many seeds, and light can often substitute for this requirement, as in Nicotiana tabacum (Toole et al., 1955) and Rumex obtusifolius (Taylorson and Hendricks, 1972; Totterdel and Roberts, 1980). In other cases, such as Lycopus europaeus (Thompson, 1969) and Fimbristylis littoralis (Pons and Schröder, 1986), the joint action of temperature fluctuation and light is required. There are indications that temperature changes somehow interfere with P_{fr} action (Takaki et al., 1981; Probert and Smith, 1986).

Temperature itself has dual effects. It affects germination and the dormancy status of the seed. A seed may be light-requiring at one temperature but not at another one; e.g. lettuce seeds do not require light at low temperatures, but do at higher ones (Smith, 1975). Other species have a reverse response, such as Betula verrucosa (Black and Wareing, 1955). These effects can be ascribed to an effect of temperature on the threshold level of P_{fr} for dormancy breaking, on reversion of P_{fr} in darkness or on a combination of these. Temperature also has an effect on the response of seeds to the R:FR ratio (Fig. 10.6), supposedly also by an effect on the P_{fr} threshold (Pons, 1986; Senden et al., 1986; Van Tooren and Pons, 1988).

There are many examples that illustrate temperature effects on light responses through an influence on the dormancy status of seeds. Chilling or overwintering in soil can replace light in many species, or greatly modifies the light response. Germination in darkness increased after a chilling treatment in, for example, Lactuca sativa (VanderWoude and Toole, 1980), Amaranthus retroflexus (Taylorson and Hendricks, 1969), Cirsium palustre (Pons, 1983, 1984), Aster pilosus (Baskin and Baskin, 1985), Carlina vulgaris (Van Tooren and Pons, 1988) and Plantago lanceolata (Pons and Van der Toorn, 1988). A lower R:FR was needed to inhibit germination in P. major after chilling, indicating that a lower P_{fr} level was required for dormancy breaking (Pons, 1986; Fig. 10.6).

Seasonal changes in dormancy status

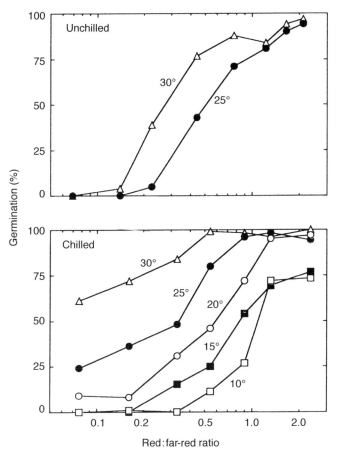

Fig. 10.6. Germination of *Plantago major* seeds in light of different red : far-red photon ratios and at different temperatures. Seeds were either used directly after dry storage (unchilled, above) or chilled for 8 weeks at 4°C in darkness (below). Unchilled seeds did not germinate below 25°C. (After Pons, 1986.)

of seeds are described for many species, which are caused by seasonal changes in temperature (see Probert, Chapter 11, this volume). This may be reflected in seasonal changes in the light requirement for dormancy breaking, as found for *Ambrosia artemisiifolia* (Baskin and Baskin, 1980), *R. obtusifolius* (Van Assche and Vanlerberghe, 1989), several species in a chalk grassland community (Pons, 1991b) and weed species (Milberg and Andersson, 1997). These species typically pass from light-requiring in autumn (in some cases, preceded by non-responsiveness to light) to non-light-requiring in spring, which is repeated in successive years. Derkx and Karssen (1993b) have studied the change in sensitivity to light during burial in more

detail in *Sisymbrium officinale*. Alleviation of primary dormancy was accompanied by the appearance of the VLFR. Indications were also found for the involvement of an increase in phytochrome receptors and GA synthesis.

Another factor that strongly interferes with the seed's light response is water potential. The threshold level of P_{fr} for dormancy breaking is increased at water potentials that do not totally inhibit germination (Scheibe and Lang, 1965; Karssen, 1970). This results in a requirement for higher R : FR ratios at low water potentials (Pons, 1986). A requirement for increased growth potential is supposedly involved here.

Nitrate, which is a naturally occurring

dormancy-breaking agent (see Hilhorst and Karssen, Chapter 12, this volume), can change the light response of seeds appreciably. As with temperature fluctuation, nitrate can replace a light requirement, e.g. in *P. lanceolata* (Steinbauer and Grigsby, 1957; Pons, 1989a) and *Sinapis arvensis* (Goudey *et al.*, 1987), or must act together with light to break dormancy, e.g. *Sisymbrium officinale* (Hilhorst *et al.*, 1986; Derkx and Karssen, 1993b) and *Ranunculus sceleratus* (Probert *et al.*, 1987). In *P. major*, nitrate lowered the R:FR required for breaking dormancy (Pons, 1986). There are indications, as with temperature fluctuation, that nitrate interferes with P_{fr} action.

These are examples of factors that interfere with the light response of seeds. The situation may become even more complex when more than two factors operate simultaneously. Chilling, nitrate, temperature fluctuation and light in various combinations showed additive effects or positive interactions in a number of weed species studied by Vincent and Roberts (1977). Simultaneous operation of several environmental factors together is more the rule than the exception under field conditions. On the one hand, the various interactions complicate ecological interpretations but, on the other hand, they can be viewed as tools that the seed can use for the fine-tuning of its germination to occur at the time and place best suited to the requirements of the subsequent developmental stages.

Light responses in distinct seed positions

Three main positions of seeds can be distinguished with respect to the soil–atmosphere interface and the presence of a leaf canopy overhead, where light responses play a role. Seeds can be deep in soil, where they are excluded from light; they can be at or close to the soil surface, where PFD varies greatly; and they may be exposed to different degrees of leaf-canopy-filtered light, where spectral composition is important. The position of a seed in the soil can change by burial or by movement

to the surface during disturbance. The degree of leaf-canopy shade can change seasonally or with disturbance. The light responses in the different seed positions with respect to the soil surface and a leaf canopy are schematically represented in Fig. 10.7.

A light requirement enforces dormancy in soil

The most obvious significance of a light requirement for dormancy breaking is avoidance of germination too deep in the soil for the seedlings to reach the surface on the available nutrient reserves. Only when a seed is brought to the surface is it exposed to light and its dormancy alleviated. The seed can germinate there when other factors are conducive. Considering the above argument, it is not surprising that most light-requiring seeds are small. In a survey by Grime *et al.* (1981), 69 species were tested in complete darkness. Many of the small-seeded species (< 0.1 mg) had a light requirement at least in some of their seeds. The majority of species with seed weights of 1 mg or more lacked a light-requirement (Table 10.1). Other studies also indicate the predominance of dark dormancy in small-seeded species (Kinzel, 1926; Toole, 1973).

Seeds largely rely on dark dormancy for avoidance of germination in the soil and, hence, for the formation of a persistent seed bank. However, the germination percentage of freshly harvested seeds in darkness is often not a very good predictor of avoidance of germination in the soil. Dark dormancy may be induced upon burial, as found in several arable weed species by Wesson and Wareing (1969), or a light requirement may have disappeared after overwintering in soil, as in *Leontodon hispidus* and several other grassland species (Pons, 1991b).

Germination in the soil may be caused by the presence of P_{fr} in the seed. This can disappear by dark reversion and thus dark dormancy can be induced. A prerequisite is that germination is prevented by

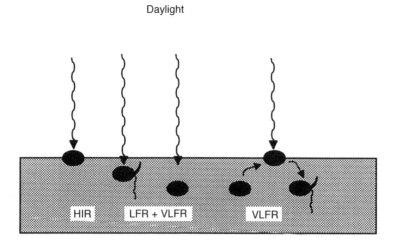

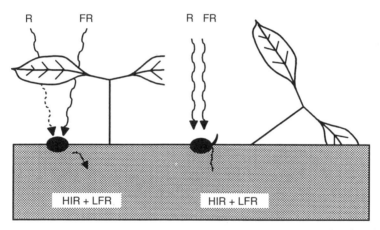

Fig. 10.7. Breaking of seed dormancy by light in seeds in distinct positions in soil and under a leaf canopy. The upper picture shows the involvement of the HIR, LFR and VLFR in seeds at and close under the soil surface, where a steep gradient in light intensity exists. At the right, the effect of a short exposure to light during soil disturbance in the range where the VLFR is active is shown. The lower picture shows the dormancy-enforcing effect of leaf-canopy-filtered light as a result of a shift in the red : far-red ratio. The HIR and the LFR are involved in this response. During leaf-canopy-shade-enforced dormancy, seeds can become buried in the soil, where dormancy is further enforced by absence of light.

Table 10.1. Distribution of 69 species over three seed mass classes and three germination categories. Dark germination is expressed as a percentage of germination in light. No green safe light was used. (After Grime *et al.*, 1981.)

Seed mass (mg)	Dark germination		
	< 20%	> 20%	NS
< 0.1	8	6	1
0.1–1.0	13	14	12
> 1.0	3	2	10

NS, germination in darkness not significantly different from germination in light.

conditions not conducive to germination. This may be either by forms of dormancy other than those mediated by phytochrome or by germination-inhibitory conditions, such as unsuitable temperature and low water potential (Fig. 10.8). Induction of dark dormancy at low water potentials has been demonstrated in *Rumex crispus* (Duke, 1978), *Chenopodium album* (Karssen, 1970), *C. bonus-henricus* (Kahn and Karssen, 1980), *P. major* and *Origanum vulgare* (Pons, 1991a).

Although seeds may have a light requirement when ripe, it is most probably more the rule than the exception that they have imbibed enough water for transformations in phytochrome to occur before they are excluded from light by burial. Hence, induction of dark dormancy may be essential for survival of the seeds in the soil when they are not dark-dormant (Wesson and Wareing, 1969), but also when seeds are light-requiring when ripe, but have been exposed to light before burial. Depth of burial of the seeds can influence the type of light response that is induced in the

seeds, as in *Datura ferox* (Botto *et al.*, 1998). The species in which dark dormancy can be induced are well known to form large seed reserves in the soil. The capacity for induction of dark dormancy thus appears to be of great importance for the formation of a persistent seed bank.

From the temperature dependence of dark reversion (Schäfer and Schmidt, 1974), it can be predicted that induction of dark dormancy is much slower at low than at high temperatures. It typically takes a few days at temperatures of 20°C or higher, but is very slow at 4°C or does not occur at all at low temperature (Kahn and Karssen, 1980; Pons, 1984, 1991a). As a consequence, seeds that are exposed to light in winter, e.g. with soil disturbance during tree felling, may still respond to the light stimulus in spring, when temperatures have become favourable for germination, as shown for *Cirsium palustre* in coppiced woodland (Pons, 1984).

Seasonal changes in light requirement do not necessarily lead to germination in the soil when the seeds are not fully dark-

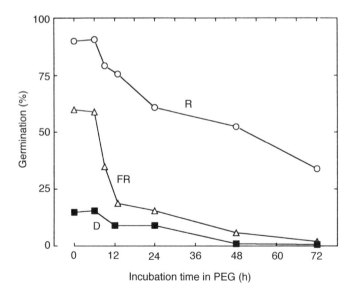

Fig. 10.8. Induction of dark dormancy in seeds of *Origanum vulgare* during incubation in polyethylene glycol (PEG-6000, 1.2 MPa). The seeds were irradiated with red (R) or far red (FR) for 10 min before incubation, or were not exposed to light at all (D). After various incubation periods, the seeds were rinsed with water and tested for germination. The experiment was carried out in darkness at 27°C. Controls that were irradiated with R or FR after the incubation indicated that the responsiveness to light did not change during incubation. (After Pons, 1991a.)

dormant. In the case of *Ambrosia artemisiifolia*, Baskin and Baskin (1980) have shown that the seeds can only germinate in darkness at temperatures that are higher than those prevailing in the soil in spring. Gradually rising soil temperatures later in spring induce a light requirement before they are sufficiently high to allow germination. However, an unknown proportion of the seeds, in which this mechanism apparently did not work, germinated in the first spring after their burial. Germination in the soil of a considerable proportion of the seeds was also found in the first spring after burial of *Scabiosa columbaria* (50%), *Carlina vulgaris* (50%) and *Leontodon hispidus* (95%) (Pons, 1991b). The remaining seeds showed a seasonal change in light requirement, while germination did not occur in the soil, presumably due to the same mechanism as described above for *A. artemisiifolia*.

A light requirement is also significant for gap detection. Seeds excluded from light in the soil can be brought to the surface, where they are exposed to light, during disturbance of the soil. Soil disturbance generally coincides with destruction of established vegetation and hence is a good predictor of reduced competition by established plants. Only short exposures are sufficient to break dormancy of many buried weed seeds (Sauer and Struik, 1964). Hence, seeds can germinate even when they are buried again by continued movement of the soil after being exposed at the surface (Pons, 1984). This may be essential when disturbance occurs infrequently – for example, only once in the lifetime of a seed, as in *C. palustre* growing in coppiced woodland (Pons and During, 1987).

The short (millisecond) exposure times during soil disturbance when seeds are brought to the surface and reburied again immediately during the same disturbance action result in light doses in the range where the VLFR is active. Scopel *et al.* (1991, 1994) showed that exposure to light was indeed essential during soil cultivation in an arable field for a large part of the seed population. Emergence of seedlings was stimulated more by cultivation during

daytime than during the night. Daytime cultivation with the soil surface protected from exposure to light resulted in less emergence than occurred on full exposure (Fig. 10.9).

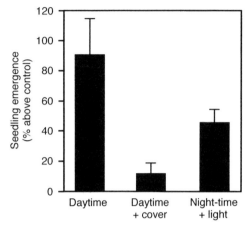

Fig. 10.9. The effect of manipulation of light conditions during soil cultivation on emergence of weed seedlings. Values are expressed as percentage above the control treatment, where cultivation was carried out during the night, which resulted in a seedling emergence of 366 m^{-2}. Soil cultivation during daytime and during the night with supplemental lighting stimulated germination of seeds brought to the surface, which were shortly exposed to light and buried again. Light doses are in the range where the VLFR is active. (After Scopel *et al.*, 1994.)

Exposure to light in soil

The steep gradient in PFD near the surface of soils has a profound effect on seeds. Studies indicate that dormancy of seeds can still be broken by the very low PFD penetrating the upper few millimetres of soil. Woolley and Stoller (1978) found that lettuce seeds germinated at a depth of 2 mm, to which *c.* 1% of daylight penetrated, but not at 6 mm. Van der Meyden and Van der Waals-Kooi (1979) studied the influence of soil cover on germination of *Senecio jacobaea* in two ways. Emergence of seedlings from seeds planted at different depths in dune sand was compared with germination in dishes covered with layers of sand of different thickness. The experiment indicated that germination was stimulated even by light penetrating at 16 mm; no data of PFD at that depth are given.

The dormancy of *Sinapis arvensis* and *Plantago major* can also be broken by the very low PFD in compost at 6 and 8 mm depth, respectively (Frankland and Poo, 1980). The authors interpret the decrease of germination with depth from the decrease in $P_{fr} : P_{tot}$ ratio as a result of the decrease in rate of PFD-dependent P_r to P_{fr} transformation, while the P_{fr} to P_r dark reversion remains constant. Equilibrium $P_{fr} : P_{tot}$ ratios were calculated to be 0.16 at a depth of 3 mm, while this would be 0.48 on the basis of the reduced R : FR alone. Daylight establishes a $P_{fr} : P_{tot}$ ratio of *c*. 0.6 (Fig. 10.5). These calculations show that reduced PFD can be more important than altered spectral composition of the light in soil for dormancy breaking, provided that the assumptions for the rate of dark reversion of P_{fr} are correct.

The influence of the light climate in soil on germination of a number of species was investigated by Bliss and Smith (1985). They observed widely different responses among the species to the steep light gradient in the upper soil layers (Fig. 10.10). *C. album* did not germinate under a cover of more than 2 mm sand; *Rumex obtusifolius,*

Cecropia obtusifolia and *P. major* also germinated under 6 mm, but high percentages of *Galium aparine* and particularly *Digitalis purpurea* seeds were stimulated to germinate by light penetrating 10 mm at an estimated PFD of only 0.026 µmol m^{-2} s^{-1}. The authors suggest that dormancy breaking at such a low PFD may involve the absence of dark reversion of P_{fr}, which would allow integration of light over longer periods. However, this would be in conflict with the significance of dark reversion for the formation of a seed bank, as argued above, and *D. purpurea* forms large seed banks (Van Baalen, 1982). A more likely explanation would seem to be that the VLFR is involved in the seeds that germinate at greater soil depths. This would add another ecologically significant role for this light-response type.

Germination on the soil surface

Inhibition of germination of seeds on the soil surface as compared with superficially buried seeds has been reported in several cases (Fenner, 1985). This inhibition could

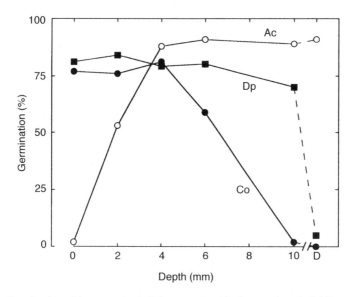

Fig. 10.10. Germination of three species in light transmitted by layers of sand of different thickness. The experiments were carried out at 20°C and a PFD of 280 µmol m^{-2} s^{-1} at the soil surface. Controls in darkness (D) were included. The species were *Amaranthus caudatus* (Ac), *Digitalis purpurea* (Dp) and *Cecropia obtusifolia* (Co). (After Bliss and Smith, 1985.)

be ascribed to reduced moisture availability, in the case of *S. jacobaea* (Van der Meyden and Van der Waals-Kooi, 1979) or to the inhibiting effect of oxygen, in the case of the rice-field weed *Scirpus juncoides* (Pons and Schröder, 1986). However, Bliss and Smith (1985) could ascribe the inhibition of germination at the soil surface of several species to the high PFD: *C. album* and *G. aparine* (> 70 µmol m^{-2} s^{-1}) and *Amaranthus caudatus* (> 3 µmol m^{-2} s^{-1}).

Inhibition of germination due to prolonged exposure to light has been described for several species. It is most obvious in seeds that are reported to be negatively photoblastic, e.g. *A. caudatus* (Kendrick and Frankland, 1969) and *Phacelia tanacetifolia* (Rollin and Maignan, 1967) and *Oldenlandia corymbosa* (Corbineau and Côme, 1982), in which prolonged exposure to rather low PFD already reduces germination below dark control values. Higher PFD is generally required for photoinhibition in otherwise positively photoblastic seeds, such as *Sinapis arvensis* (20 µmol m^{-2} s^{-1}: Frankland and Poo, 1980) and *Urtica urens* (200 µmol m^{-2} s^{-1}: Grime and Jarvis, 1975). The experiments of Ellis *et al.* (1986a, b, 1989) included species that were inhibited at low daily irradiances (< 10 mol m^{-2} day^{-1}) and species that were only inhibited at irradiance. Alternatively, there are also seeds that are stimulated at high as compared with low PFD. Several species showed this phenomenon in the survey of Grime *et al.* (1981). Pons (1989b) showed that freshly harvested seeds of *Calluna vulgaris* and *Erica tetralix*, but not previously dark-incubated seeds, showed a strong positive relation between PFD and germination up to 250 µmol m^{-2} s^{-1}.

The ecological significance of photoinhibition at the soil surface is not altogether clear. Some authors suggest that it may avoid germination on the soil surface on exposed sites, where conditions are not suitable for establishment, since the seedling may suffer desiccation. However, this is not necessarily the case. A dry soil surface itself can inhibit germination and

water relations will not be that different between a seedling germinated at the surface or at a few millimetres depth, unless penetration of the radicle is hampered in surface-germinating seeds. No studies are available where the survival value of photoinhibition at the soil surface is investigated. This matter awaits critical evaluation.

Spectral composition: a signal for the presence of established plants

After the discovery of the response of seeds to R and FR and the effect of leaves on the R : FR ratio of daylight (Fig. 10.2), the potential ecological significance of canopy-filtered light for seed dormancy and germination was recognized. Cumming (1963) carried out experiments with artificial light source and filter combinations to produce different R : FR ratios. Germination of the light-requiring *Chenopodium botrys* appeared to be inhibited at an R : FR ratio similar to those found under leaves.

Using leaves as filters, Taylorson and Borthwick (1969) and Van der Veen (1970) showed that germination was reduced below unfiltered-light control values in several light-requiring seeds (e.g. *C. album, R. obtusifolius, Plantago major, Ageratum conyzoides, Amaranthus retroflexus, Potentilla norvegis*). Germination of some species that germinated in darkness was even reduced below dark control values in the study of Van der Veen (1970) (e.g. *Ruellia tuberosa, Calotropa procera*). Other studies have further illustrated the potential significance of leaf-canopy-shade light for maintanance of seed dormancy. Stoutjesdijk (1972) found dormancy enforced by leaf-filtered light in five species from widely different habitats, but six were not influenced. King (1975) mentioned three short-lived grassland species that were inhibited by light filtered through leaves. Silvertown (1980) found dormancy enforced by leaf-filtered light in 17 out of 27 grassland species. Fenner (1980a) reported that 16 out of 18 African weed species showed this type of dormancy.

The most extensive survey on leaf-canopy-induced dormancy was done by Górski and co-workers (Górski, 1975; Górski et al., 1977, 1978). They tested 271 species under a rhubarb canopy and compared the germination with a similar test in diffuse white light of similar PFD. Leaf-canopy light reduced germination in 90% of the wild species that were tested, but in only 53% of cultivated plants (Table 10.2). Leaf-canopy-enforced dormancy was present in all positively photoblastic seeds, in 71% of the negatively photoblastic seeds and in 58% of the light-indifferent seeds.

The effect on seed germination of leaf-filtered daylight is generally similar to the effect of a mixture of artificial light sources giving the same R : FR ratio (Frankland and Letendre, 1978; Pons, 1983; Figs 10.5 and 10.6). Frankland and Poo (1980) showed clearly that the response of the seeds of *P. major* and *S. arvensis* to the R : FR operates through its effect on the photoequilibrium of phytochrome ($P_{fr} : P_{tot}$) (Fig. 10.5). However, germination was not so strongly reduced when the seeds were exposed to light with a low R : FR for only 1 h as compared with 20 h. The authors attributed the additional reduction in prolonged exposure to photoinhibition by the HIR.

The significance of the responses of seeds to leaf-canopy-filtered light interpreted on the information presented so far can be summarized as follows. With phytochrome as the photoreceptor, seeds have a device for the detection of shading by established plants and can respond by delayed germination. Only when gaps arise in the vegetation can the dormancy of a seed be broken by the high $P_{fr} : P_{tot}$ ratio established by unmodified daylight, allow-ing the seed to germinate. Gaps are considered to be important for the regeneration of many plants from seeds (see Bullock, Chapter 16, this volume), since shading and other competitive effects of established plants can greatly reduce the growth and survival of seedlings (Fenner, 1978; Van der Toorn and Pons, 1988). However, as Stoutjesdijk (1972) pointed out, this response is also found among species from temperate deciduous woodland, but the mechanism cannot be ecologically important in spring in that habitat, since the leaf canopy is absent when most seeds germinate. This indicates that a light response must be evaluated critically in its ecological context before conclusions can be drawn on its significance in a plant's life cycle.

Importance of light responses in the life cycle of selected plant species

As pointed out, the above-mentioned studies indicate potentials for the ecological significance of a particular light response only. A complicating factor for an ecological interpretation of results of seed germination experiments is that the photoreceptor has multiple roles in seeds. Hence, a particular response might just be a consequence of the role that phytochrome plays in another situation. Another complicating factor is the strong interaction of phytochrome-mediated processes with other environmental factors and the preconditioning of the seed. Most experiments have been carried out with freshly harvested or dry-stored seeds. Since the light response varies considerably with

Table 10.2. Distribution of 77 wild species and 75 cultivated species over three categories of leaf-canopy-enforced dormancy. The latter is expressed as germination in leaf-canopy-filtered light as a percentage of germination in diffuse unfiltered daylight. (After Górski et al., 1978.)

	Germination in leaf-canopy-filtered light		
	< 20%	> 20%	NS
Wild species	49	20	8
Cultivated species	7	33	35

NS, germination in canopy-filtered light not significantly different from germination in unfiltered light.

the seasons, the results may only apply to newly dispersed seeds, and may not be relevant for seeds in the seed bank, for instance. A third reason why conclusions may be confounded on the significance of a light response is that different mechanisms may have potentially similar functions. For instance, seeds have an array of detection mechanisms for disturbance events. Light responses represent only a few of those in the tool-box available to a seed. Only a critical evaluation can decide what the importance of a particular light response is. In this section, the available information on the light responses of a few selected plant species is critically evaluated with regard to the actual significance of phytochrome functioning for their life cycle.

P. major is a weedy, short-lived, perennial species occurring on open places in grassland and along paths and roadsides. Establishment from naturally dispersed seeds in a dense grass sward does not generally occur and, when the rather small seeds (0.2 mg) are experimentally sown in dense vegetation, successful establishment is again absent (Sagar and Harper, 1964; Blom, 1978; Van der Toorn and Pons, 1988). The species can form large seed banks (Van Altena and Minderhoud, 1972), with the seeds surviving long periods in soil (Toole and Brown, 1946; Kivilaan and Bandurski, 1981) They show a seasonal pattern of emergence (Froud-Williams *et al.*, 1984a), which is probably due to seasonal changes in dormancy (Pons and Van der Toorn, 1988). Several studies indicate an almost absolute light requirement for dormancy breaking (Steinbauer and Grigsby, 1957; Frankland and Poo, 1980; Grime *et al.*, 1981; Froud-Williams *et al.*, 1984b; Pons, 1986, 1991a; Fig. 10.3).

Far-red and leaf-canopy-filtered light inhibit germination of *P. major* (Van der Veen, 1970; Górski *et al.*, 1977; Frankland and Poo, 1980; Froud-Williams *et al.*, 1984b; Pons, 1986; Pons and Van der Toorn, 1988). Frankland and Poo (1980) exposed seeds to light under a *Sinapis alba* leaf canopy. Increasing plant density and thus increasing leaf area index resulted in a decreasing R:FR ratio of the light. This, in turn, caused decreasing $P_{fr}:P_{tot}$ ratios in the seeds, resulting in a decline in germination with increasing plant density, with complete inhibition at high densities (Fig. 10.5). Similar results were obtained with dry-stored seeds sown under grass by Pons and Van der Toorn (1988). However, several seeds sown in the field in autumn under grass germinated the following March and April, although germination in gaps was higher.

The experiments carried out with freshly harvested and/or cool- and dry-stored seeds can be considered as representative of summer conditions in the field. There are two important points of difference in spring as compared with summer: (i) leaf canopies are not so dense in spring as in summer, resulting in higher R:FR ratios; and (ii) lower R:FR ratios are required to enforce dormancy of chilled seeds than of freshly harvested seeds (Pons, 1986; Fig. 10.6). The two factors together permit germination in spring under grass of seeds that are at or near the soil surface. None of the seedlings that germinated under grass survived (Pons and Van der Toorn, 1988). Hence, the light climate in spring is not a reliable environmental signal for avoidance of ultimately fatal germination under established plants. The dormancy-enforcing effect of a low R:FR is likely to play a more important role in summer after seed fall. The newly ripened seeds have a high temperature and P_{fr} requirement, further increased by drought (Pons, 1986), which occurs more frequently in summer, and leaf canopies are denser than in spring. Germination will be retarded or avoided altogether and the small seeds may have a chance to slip into crevices or become buried in the soil in another way. The P_{fr} that may be present in the seeds will disappear (Pons, 1991a), rendering the seed obligately light-requiring. Only soil disturbance will bring seeds to the surface again, when some of them may germinate upon a short exposure, even when reburied. Others that do not respond to a short exposure (Frankland and Poo, 1980; Fig. 10.3) avoid possible fatal germination at depth and await another disturbance. Hence, the significance of leaf-canopy-induced dormancy may lie more in the prevention of germination shortly after dispersal of the seeds rather than during the time of normal germination in *P. major*.

The significance of induction of a light requirement by leaf-canopy shade before burial has also been suggested by Fenner (1980b) in the case of the weed *Bidens pilosa*. It was even more pronounced in the biennial *Cirsium palustre*, which emerges in ash coppice after felling (Pons, 1983, 1984). Although germination is inhibited at low R:FR, there is no leaf canopy in spring when germination takes place, which can serve as an environmental signal for reduced competition. Again, it was concluded in this case that the significance must be found in summer, after seed dispersal. Soil disturbance during felling is a principal trigger for germination, as concluded from the stimulation of emergence by artificial soil disturbance in a non-felled situation and its reduction by carefully avoiding disturbance during felling.

O. vulgare is a perennial growing on exposed rocky places, forest edges and chalk grasslands. Large numbers of small seeds (0.1 mg) are produced in the autumn. There is some germination in autumn, presumably of newly shed seeds, but most of the emergence of seedlings is concentrated in spring (Van Tooren and Pons, 1988). The species forms large persistent seed banks (Thompson and Grime, 1979; Willems, 1995).

Widely different percentages of germination in darkness have been reported for *O. vulgare*. Silvertown (1980) and Grime *et al.* (1981) reported high dark germination, but Van Tooren and Pons (1988) and Pons (1991a) found lower percentages, which decreased with decreasing temperature (Figs 10.3 and 10.6) and also depended on chilling of the seeds. Short exposures to R effectively broke dormancy in all seeds and even FR stimulated germination (Van Tooren and Pons, 1988; Pons, 1991a; Figs 10.3 and 10.6). Seeds buried experimentally in the soil in a chalk grassland did not germinate and quickly lost their germinability in darkness (Pons, 1991b).

Induction of dark dormancy was completed in a few days at high temperatures under conditions of osmotic stress (Fig. 10.6), but was slow at 4°C and in seeds that had been exposed to R (Pons, 1991a). There may also be slight seasonal changes in dormancy (Pons, 1991b). There was a strong inhibiting effect of a low R:FR ratio of the light at low temperatures in freshly harvested seeds, but this was virtually absent at high temperatures and in chilled seeds (Van Tooren and Pons, 1988).

From these studies, the picture arises that *Origanum* seeds may have a variable amount of P_{fr} in ripe seeds, which disappears in darkness at higher temperatures, when conditions are not conducive for germination, thus inducing dark dormancy. Furthermore, the P_{fr} threshold for dormancy breaking is low, particularly at high temperatures and after chilling. Germination after FR (Fig. 10.3) may indicate that the VLFR can be active.

The dormancy-enforcing effect of leaf-canopy-filtered light at low temperatures may successfully avoid germination of seeds dispersed in vegetation in autumn, but not in open places (Van Tooren and Pons, 1988). The small seeds may easily become buried in the soil, and dark dormancy is quickly induced in the seeds that were exposed to leaf-filtered light before burial, which will take somewhat longer in seeds previously exposed to unfiltered daylight. The low P_{fr} requirement of chilled seeds clearly points to the functioning of phytochrome as a light detector, rather than as a leaf-canopy-shade detector, in spring.

Similar results were obtained with the heather species *Calluna vulgaris* and *Erica tetralix* (Pons, 1989b, c). Freshly harvested seeds were inhibited by leaf-canopy-filtered light, due both to a low R:FR and to the low PFD. This may prevent their germination shortly after dispersal. However, germination was independent of light quality or intensity after a pretreatment resembling burial, but no germination occurred in darkness.

The above-mentioned studies have been carried out in temperate climates and in more or less deciduous vegetation types. The general picture that arises from them is that induction of dormancy by leaf-canopy-filtered light is most important in summer and/or autumn after dispersal of the seeds. The function of phytochrome shifts more towards a role as a light detector when dormancy is alleviated in winter and the leaf canopy has disappeared.

Members of the genus *Cecropia* are common in secondary forest in the tropical rainforest area of Central and South America. The rather short-lived trees produce large numbers of small, animal-dispersed seeds. They form persistent seed banks in forest soil (Holthuyzen and Boerboom, 1982). Experiments have shown that seeds can remain viable in the soil for periods of at least a year (Vázquez-Yanes and Smith, 1982), and presumably much longer. Seedlings cannot survive under closed forest canopy and establishment from seed occurs only when large gaps arise.

The germination of seeds of *Cecropia* and other species of secondary tropical forest succession is stimulated by exposure to normal daylight, but not by leaf-canopy-filtered light (Valio and Joly, 1979; Vázquez-Yanes, 1980; Vázquez-Yanes and Orozco-Segovia, 1994). Prolonged exposure to light is required, which can be replaced by intermittent short exposures (Vázquez-Yanes and Smith, 1982; Vázquez-Yanes and Orozco-Segovia, 1990). Low PFD is effective in breaking dormancy. The red part of the spectrum in dense forest stimulated germination when FR was excluded (Vázquez-Yanes, 1980), and light at 6 mm depth in sand stimulated germination too (Bliss and Smith, 1985). High irradiance does not seem to be inhibitory. The length of a daily exposure to high R:FR white light in the middle of a day consisting of FR only had to be over 6 h to break dormancy. The available data suggest a long escape time and a high P_{fr} threshold for dormancy breaking. These characteristics of the regulation of the light response by phytochrome can be interpreted as mechanisms for distinguishing large gaps, which are suitable for establishment, from short

sunflecks and small gaps, which are not. This reasoning applies only to newly dispersed seeds exposed to light at or close to the soil surface. It is not known what the position of most seeds is and thus what their light environment is, and whether changes in the light response occur after dispersal. Nevertheless, the picture arises of phytochrome functioning as a detector of leaf-canopy shade being more important in evergreen tropical forest than in temperate deciduous vegetation types.

In Mediterranean evergreen vegetation, phytochrome in seeds may also have significance as a leaf-canopy detector in the germination season. Germination of *Sarcopoterium spinosum* is stimulated after fire. Roy and Arianoutsou-Faraggitaki (1985) found that its germination is not stimulated by temperature shock or temperature fluctuation, but is inhibited by a low R:FR. It was concluded that a rise in R:FR is the main signal for timing of germination of seeds on the soil surface. However, in this study, only dry-stored seeds were used and the possibility remains that light responses change after dispersal in the field. Also, the possible role of chemical signals associated with fire (see Keeley and Fotheringham, Chapter 13, this volume) was not investigated in this study.

Concluding remarks with respect to the ecological significance of light responses

In this chapter, various light responses of seeds have been reviewed and attempts have been made to indicate their possible ecological significance. In some cases, the responses appear to have an adaptive function. In other cases, this is less clear, and further studies that are specifically designed to critically investigate the ecological significance of a particular light response are required.

There seems to be little doubt about the significance of phytochrome as a light detector for the maintenance of dark dormancy in soil. The LFR is the most important response type with respect to this role. Dark reversion of P_{fr} induces dark dormancy and the P_{fr} requirement for dormancy breaking maintains it. Also, phytochrome as a light detector with respect to soil disturbance events is straightforward. The LFR plays a role here as well. Probably more important is the VLFR, which enables the seed to detect the very short exposures during soil disturbance, and the extremely low PFDs just below the soil surface.

At the soil surface in unfiltered daylight, the HIR may avoid seedling desiccation by inhibiting germination. However, as already pointed out, its significance is doubtful. The HIR causes a further reduction of germination, in addition to the reduction caused by the LFR in canopy-filtered light. Hence, the inhibition of germi-

nation at high PFD caused by the HIR at the soil surface in full daylight may just be a consequence of a more ecologically relevant operation of the same system in leaf-canopy shade.

The role of enforcement of dormancy in leaf-canopy-filtered light, as a result of the combined action of LFR and HIR, varies with vegetation type. Evidence indicates that this light response plays an important role in gap detection in evergreen vegetation, such as tropical rainforest. In temperate regions, where the vegetation has a more deciduous character, the role of leaf-canopy-enforced dormancy appears to be more related to the prevention of germination shortly after dispersal, thus facilitating the incorporation of the seeds into the soil seed bank.

References

Ballaré, C.L. (1999) Keeping up with neighbours: phytochrome sensing and other signalling mechanisms. *Trends in Plant Science* 4, 97–102.

Baskin, J.M. and Baskin, C.C. (1979) Promotion of germination of *Stellaria media* seeds by light from a green safe lamp. *New Phytologist* 82, 381–383.

Baskin, J.M. and Baskin, C.C. (1980) Ecophysiology of secondary dormancy in seeds of *Ambrosia artemisiifolia*. *Ecology* 61, 475–480.

Baskin, J.M. and Baskin, C.C. (1985) The light requirement for germination of *Aster pilosus* seeds: temporal aspects and ecological consequences. *Journal of Ecology* 73, 765–773.

Bewley, J.D. and Black, M. (1982) *Physiology and Biochemistry of Seeds*, Vol. 2. Springer, Berlin.

Blaauw-Jansen, G. (1983) Thoughts on the possible role of phytochrome destruction in phytochrome-controlled responses. *Plant Cell and Environment* 6, 173–179.

Blaauw-Jansen, G. and Blaauw, O.H. (1975) A shift of the threshold to red irradiation in dormant lettuce seeds. *Acta Botanica Neerlandica* 24, 199–202.

Black, M. and Wareing, P.F. (1955) Growth studies in woody species. VII. Photoperiodic control of germination in *Betula pubescens*. *Physiologia Plantarum* 8, 300–316.

Bliss, D. and Smith, H. (1985) Penetration of light into soil and its role in the control of seed germination. *Plant Cell and Environment* 8, 475–483.

Blom, C.W.P.M. (1978) Germination, seedling emergence and establishment of some *Plantago* species under laboratory and field conditions. *Acta Botanica Neerlandica* 27, 257–271.

Borthwick, H.A., Hendricks, S.B., Toole, E.H. and Toole, V.K. (1954) Action of light on lettuce seed germination. *Botanical Gazette* 115, 205–225.

Botto, J.F., Sánchez, R.A. and Casal, J.J. (1998) Burial conditions affect light responses of *Datura ferox* seeds. *Seed Science Research* 8, 423–429.

Casal, J.J. and Sánchez, R.A. (1998) Phytochromes and seed germination. *Seed Science Research* 8, 317–329.

Cone, J.W. and Kendrick, R.E. (1986) Photocontrol of seed germination. In: Kendrick, R.E. and Kronenberg, G.H.M. (eds) *Photomorphogenesis in Plants*. Martinus Nijhoff, Dordrecht, pp. 443–465.

Corbineau, F. and Côme, D. (1982). Effect of intensity and duration of light at various temperatures on the germination *of Oldenlandia corymbosa* L. seeds. *Plant Physiology* 70, 1518–1520.

Cresswell, E.G. and Grime, J.P. (1981) Induction of light requirement during seed development and its ecological consequences. *Nature* 291, 583–585.

Cumming, B.G. (1963) The dependence of germination on photoperiod, light quality, and temperature in *Chenopodium* ssp. *Canadian Journal of Botany* 41, 1211–1233.

Derkx, M.P.M. and Karssen, C.M. (1993a) Effects of light and temperature on seed dormancy and gib-
berellin-stimulated germination in *Arabidopsis thaliana*: studies with gibberellin-deficient and -
insensitive mutants. *Physiologia Plantarum* 89, 360–368.

Derkx, M.P.M. and Karssen, C.M. (1993b) Changing sensitivity to light and nitrate but not to gibberellins regulates seasonal dormancy patterns in *Sisymbrium officinale* seeds. *Plant Cell and
Environment* 16, 469–479.

Duke, S.O. (1978) Interactions of seed water content with phytochrome-initiated germination of
Rumex crispus L. seeds. *Plant Cell Physiology* 19, 1043–1049.

Ellis, R.H., Hong, T.D. and Roberts, E.H. (1986a) Quantal response of seed germination in *Brachiaria
humidicola, Echinochloa turnerana, Eragrostis tef* and *Panicum maximum* to photon dose for
the low energy reaction and the high irradiance reaction. *Journal of Experimental Botany* 37,
742–753.

Ellis, R.H., Hong, T.D. and Roberts, E.H. (1986b) The response of seeds of *Bromus sterilis* L. and
Bromus mollis L. to white light of varying photon flux density and photoperiod. *New
Phytologist* 104, 485–496.

Ellis, R.H., Hong, T.D. and Roberts, E.H. (1989) Response of seed germination in three genera of
Compositae to white light of varying photon flux density and photoperiod. *Journal of
Experimental Botany* 40, 13–22.

Federer, C.A. and Tanner, C.B. (1966) Spectral distribution of light in the forest. *Ecology* 47, 555–560.

Fenner, M. (1978) A comparison of the abilities of colonizers and closed-turf species to establish
from seed in artificial swards. *Journal of Ecology* 66, 953–963.

Fenner, M. (1980a) Germination tests on thirty-two East African weed species. *Weed Research* 20,
135–138.

Fenner, M. (1980b) The induction of a light-requirement in *Bidens pilosa* seeds by leaf canopy shade.
New Phytologist 84, 103–106.

Fenner, M. (1985) *Seed Ecology*. Chapman and Hall, London.

Flint, L.H. and McAllister, E.D. (1937). Wavelengths of radiation in the visible spectrum promoting
the germination of light sensitive lettuce seeds. *Smithsonian Miscellaneous Collections* 96, 1–8.

Frankland, B. (1976). Phytochrome control of seed germination in relation to the light environment.
In: Smith, H. (ed.) *Light and Plant Development*. Butterworth, London, pp. 477–491.

Frankland, B. (1986) Perception of light quantity. In: Kendrick, R.E. and Kronenberg, G.H.M. (eds)
Photomorphogenesis in Plants. Martinus Nijhoff Publishers, Dordrecht, pp. 219–235.

Frankland, B. and Letendre, R.J. (1978) Phytochrome and effects of shading on the growth of woodland plants. *Photochemistry and Photobiology* 27, 223–230.

Frankland, B. and Poo, W.K. (1980). Phytochrome control of seed germination in relation to natural
shading. In: De Greef, J. (ed.) *Photoreceptors and Plant Development*. University Press,
Antwerp, pp. 357–366.

Frankland, B. and Taylorson, R. (1983) Light control of seed germination. In: Shropshire, W. and
Mohr, H. (eds) *Encyclopedia of Plant Physiology*, New Series, Vol. 16A. Springer, Berlin,
pp. 428–456.

Froud-Williams, R.J., Chancellor, R.J. and Drennan, D.S.H. (1984a) The effects of seed burial and soil
disturbance on emergence and survival of arable weeds in relation to minimal cultivation.
Journal of Applied Ecology 21, 629–641.

Froud-Williams, R.J., Drennan, D.S.H. and Chancellor, R.J. (1984b) The influence of burial and dry
storage upon cyclic changes in dormancy, germination and response to light in seeds of various
arable weeds. *New Phytologist* 96, 473–481.

Górski, T. (1975). Germination of seeds in the shadow of plants. *Physiologia Plantarum* 34, 342–346.

Górski, T. and Górska, K. (1979) Inhibitory effects of full daylight on the germination of *Lactuca
sativa* L. *Planta* 144, 121–124.

Górski, T., Górska, K. and Nowicki, J. (1977) Germination of seeds of various herbaceous species
under leaf canopy. *Flora* 166, 249–259.

Górski, T., Górska, K. and Rybicki, J. (1978) Studies on the germination of seeds under leaf canopy.
Flora 167, 289–299.

Goudey, J.S., Saini, H.S. and Spencer, M.S. (1987) Seed germination of wild mustard (*Sinapis arvensis*): factors required to break primary dormancy. *Canadian Journal of Botany* 65, 849–852.

Grime, J.P. and Jarvis, B.C. (1975) Shade avoidance and shade tolerance in flowering plants II. Effects
of light on the germination of contrasted ecology. In: Evans, G.C., Bainbridge, R. and Rackham,
O. (eds) *Light as an Ecological Factor*, Vol. II. Blackwell Scientific Publications, Oxford,
pp. 525–532.

Grime, J.P., Mason, G., Curtis, A.V., Rodman, J., Band, S.R., Mowforth, M.A.G., Neal, A.M. and Shaw,
S. (1981) A comparative study of germination characteristics in a local flora. *Journal of Ecology*
69, 1017–1059.

Hartmann, K.M., Mollwo, A. and Tebbe, A. (1998) Photocontrol of germination by moon- and starlight.
Zeitschrift für Pflanzenkrankheiten und Pflanzenschutz Sonderheft 16, 119–127.

Hilhorst, H.W.M., Smitt, A.I. and Karssen, C.M. (1986) Gibberellin-biosynthesis and -sensitivity mediated stimulation of seed germination of *Sisymbrium officinale* by red light and nitrate. *Physiologia Plantarum* 67, 285–290.

Holmes, M.G. and Smith, H. (1977) The function of phytochrome in the natural environment. I. The influence of vegetation canopies on the spectral energy distribution of natural daylight. *Photochemistry and Photobiology* 25, 539–545.

Holthuyzen, A.M.A. and Boerboom, J.H.A. (1982) The *Cecropia* seed bank in the Surinam lowland rainforest. *Biotropica* 14, 62–68.

Kahn, A.A. and Karssen, C.M. (1980) Induction of secondary dormancy in *Chenopodium bonus-henricus* L. seeds by osmotic and high temperature treatments and its prevention by light and growth regulators. *Plant Physiology* 66, 175–181.

Karssen, C.M. (1970) The light promoted germination of the seeds of *Chenopodium album* L. IV. Effects of red, far-red and white light on non-photoblastic seeds incubated in mannitol. *Acta Botanica Neerlandica* 19, 95–108.

Karssen, C.M., Zagórski, S., Kepczynski, J. and Groot, S.P.C. (1989) Key role for endogenous gibberellins in the control of seed germination. *Annals of Botany* 63, 71–80.

Kendrick, R.E. and Frankland, B. (1969) Photocontrol of germination in *Amaranthus caudatus*. *Planta* 85, 326–339.

Kendrick, R.E. and Spruit, C.J.P. (1977) Phototransformations of phytochrome. *Photochemistry and Photobiology* 26, 201–204.

King, T.J. (1975) Inhibition of seed germination under leaf canopies in *Arenaria serpyllifolia*, *Veronica arvensis* and *Cerastium holosteoides*. *New Phytologist* 75, 87–90.

Kinzel, W. (1926) *Frost und Licht als beeinflussende Kräfte der Samenkeimung*. Ulmer, Stuttgart.

Kivilaan, A. and Bandurski, R.S. (1981) The one hundred-year period for Dr Beal's seed viability experiment. *American Journal of Botany* 68, 1290–1292.

Koorneef, M. and Kendrick, R.E. (1994) Photomorphogenic mutants of higher plants. In: Kendrick, R.E. and Kronenberg, G.H.M. (eds) *Photomorphogenis in Plants*, 2nd edn. Kluwer, Dordrecht, pp. 601–630.

Milberg, P. and Andersson, L. (1997) Seasonal variation in dormancy and light sensitivity in buried seeds of eight annual weed species. *Canadian Journal of Botany* 75, 1998–2004.

Pons, T.L. (1983) Significance of inhibition of seed germination under the leaf canopy in ash coppice. *Plant Cell and Environment* 6, 385–392.

Pons, T.L. (1984) Possible significance of changes in the light requirement of *Cirsium palustre* seeds after dispersal in ash coppice. *Plant Cell and Environment* 7, 263–268.

Pons, T.L. (1986) Response of *Plantago major* seeds to the red/far-red ratio as influenced by other environmental factors. *Physiologia Plantarum* 68, 252–258.

Pons, T.L. (1989a) Breaking of seed dormancy by nitrate as a gap detection mechanism. *Annals of Botany* 63, 139–143.

Pons, T.L. (1989b) Dormancy and germination of *Calluna vulgaris* (L.) Hull and *Erica tetralix* L. seeds. *Acta Oecologica: Oecologia Plantarum* 10, 35–43.

Pons, T.L. (1989c) Dormancy, germination and mortality of seeds in heathland and inland sand dunes. *Acta Botanica Neerlandica* 38, 327–335.

Pons, T.L. (1991a) Induction of dark-dormancy in seeds: its importance for the seed bank in the soil. *Functional Ecology* 5, 669–675.

Pons, T.L. (1991b) Dormancy, germination and mortality of seeds in a chalk grassland flora. *Journal of Ecology* 79, 765–780

Pons, T.L. and During, H.J. (1987) Biennial behaviour of *Cirsium palustre* in ash coppice. *Holarctic Ecology* 10, 40–44.

Pons, T.L. and Schröder, H.F.J.M. (1986) Significance of temperature fluctuation and oxygen concentration of the rice field weeds *Fimbristylis littoralis* and *Scirpus juncoides*. *Oecologia* 68, 315–319.

Pons, T.L. and Van der Toorn, J. (1988) Establishment *of Plantago lanceolata* L. and *Plantago major* L. among grass. I. Significance of light for germination. *Oecologia* 75, 394–399.

Poppe, C. and Schäfer, E. (1997) Seed germination of *Arabidopsis thaliana* phyA/phyB double mutants is under phytochrome control. *Plant Physiology* 114, 1487–1492.

Probert, R.J. and Smith, R.D. (1986) The joint action of phytochrome and alternating temperatures in the control of seed germination in *Dactylis glomerata*. *Physiologia Plantarum* 67, 299–304.

Probert, R.J., Gajjar, K.H. and Haslam, I.K. (1987) The interactive effects of phytochrome, nitrate and thiourea on the germination response to alternating temperatures in seeds of *Ranunculus sceleratus* L.: a quantal approach. *Journal of Experimental Botany* 38, 1012–1025.

Rollin, P. and Maignan, G. (1967) Phytochrome and the photoinhibition of germination. *Nature* 214, 741–742.

Roy, J. and Arianoutsou-Faraggitaki, M. (1985) Light quality as the environmental trigger for the germination of the fire promoted species *Sarcopoterium spinosum* L. *Flora* 177, 345–349.

Sagar, G.R. and Harper, J.L. (1964) Biological flora of the British Isles. *Plantago major* L., *Plantago media* L., *Plantago lanceolata* L. *Journal of Ecology* 52, 189–221.

Sánchez, R.A. and De Miguel, L. (1997) Phytochrome promotion of mannan-degrading enzyme activities in the micropylar endosperm of *Datura ferox* seeds requires the presence of the embryo and gibberellin synthesis. *Seed Science Research* 7, 27–33.

Sauer, J. and Struik, G. (1964) A possible ecological relation between soil disturbance, light-flash, and seed germination. *Ecology* 45, 884–886.

Schäfer, E. and Schmidt, W. (1974) Temperature dependence of phytochrome dark reactions. *Planta* 116, 257–266.

Scheibe, J. and Lang, A. (1965) Lettuce seed germination: evidence for a reversible light-induced increase in growth potential and for phytochrome mediation of the low temperature effect. *Plant Physiology* 40, 485–492.

Schopfer, P. and Plachy, C. (1993) Photoinhibition of radish (*Raphanus sativus* L.) seed germination: control of growth potential by cell-wall yielding in the embryo. *Plant Cell and Environment* 16, 223–229.

Scopel, A.L., Ballaré, C.L. and Sánchez, R.A. (1991) Induction of extreme light sensitivity in buried weed seeds and its role in the perception of soil cultivations. *Plant Cell and Environment* 14, 501–508.

Scopel, A.L., Ballaré, C.L. and Radosevitch, S.R. (1994) Photostimulation of seed germination during soil tillage. *New Phytologist* 126, 145–152.

Senden, J.W., Schenkeveld, A.J. and Verkaar, H.J. (1986) The combined effect of temperature and red : far-red ratio on the germination of some short-lived chalk grassland species. *Acta Oecologica: Oecologia Plantarum* 7, 251–259.

Shinomura, T. (1997) Phytochrome regulation of seed germination. *Journal of Plant Research* 110, 151–161.

Silvertown, J. (1980) Leaf-canopy-induced seed dormancy in a grassland flora. *New Phytologist* 85, 109–118.

Smith, H. (1975) *Phytochrome and Photomorphogenesis*. McGraw-Hill, London.

Smith, H. (1982) Light quality, photoperception, and plant strategy. *Annual Review of Plant Physiology* 33, 481–518.

Smith, H. (1995) Physiological and ecological function within the phytochrome family. *Annual Review of Plant Physiology and Plant Molecular Biology* 46, 289–315.

Smith, H. and Whitelam, G.C. (1990) Phytochrome, a family of photoreceptors with multiple physiological roles. *Plant Cell and Environment* 13, 695–707.

Steinbauer, G.P. and Grigsby, B. (1957) Dormancy and germination characteristics of the seeds of four species of *Plantago*. *Proceedings of the Association of Official Seed Analysts of North America* 47, 158–164.

Stoutjesdijk, P. (1972) Spectral transmission curves of some types of leaf canopies with a note on seed germination. *Acta Botanica Neerlandica* 21, 185–191.

Takaki, M., Kendrick, R.E. and Dietrich, S.M.C. (1981) Interaction of light and temperature on the germination of *Rumex obtusifolius*. *Planta* 152, 209–214.

Taylorson, R.B. and Borthwick, H.A. (1969) Light filtration by foliar canopies: significance for light controlled weed seed germination. *Weed Science* 17, 48–51.

Taylorson, R.B. and Hendricks, S.B. (1969) Action of phytochrome during prechilling of *Amaranthus retroflexus* L. seeds. *Plant Physiology* 44, 821–825.

Taylorson, R.B. and Hendricks, S.B. (1972) Interaction of light and a temperature shift in seed germination. *Plant Physiology* 49, 127–130.

Tester, M. and Morris, C. (1987) The penetration of light through soil. *Plant Cell and Environment* 10, 281–286.

Thomas, T.H., Hare, P.D. and Van Staden, J. (1997) Phytochrome and cytokinin responses. *Plant Growth Regulation* 23, 105–122.

Thompson, K. and Grime, J.P. (1979) Seasonal variation in the seed banks of herbaceous species in contrasting habitats. *Journal of Ecology* 67, 893–921.

Thompson, P.A. (1969) Germination of *Lycopus europaeus* in response to fluctuating temperatures and light. *Journal of Experimental Botany* 20, 1–11.

Toole, E.H. and Brown, E. (1946) Final results of the Duvel buried seed experiment. *Journal of Agricultural Research* 72, 201–210.

Toole, E.H., Toole, V.K., Borthwick, H.A. and Hendricks, S.B. (1955) Interaction of temperature and light in germination of seeds. *Plant Physiology* 30, 473–478.

Toole, V.K. (1973) Effects of light, temperature and their interactions on the germination of seeds. *Seed Science and Technology* 1, 339–396.

Totterdel, S. and Roberts, E.H. (1980) Characteristics of alternating temperatures which stimulate loss of dormancy in seeds of *Rumex obtusifolius* L. and *Rumex crispus* L. *Plant Cell and Environment* 3, 3–12.

Valio, I.F.M. and Joly, C.A. (1979) Light sensitivity of the seeds on the distribution of *Cecropia glaziovi* Snethlage (Moraceae). *Zeitschrift für Pflanzenphysiologie* 91, 371–376.

Van Altena, S.C. and Minderhoud, J.W. (1972) Keimfähige Samen von Gräsern und Kräutern in der Narbenschicht der Niederländischen Weiden. *Zeitschrift für Acker- und Pflanzenbau* 136, 95–109.

Van Assche, J.A. and Vanlerberghe, K.A. (1989) The role of temperature on the dormancy cycle of seeds of *Rumex obtusifolius* L. *Functional Ecology* 3, 107–115.

Van Baalen, J. (1982) Germination ecology and seed population dynamics of *Digitalis purpurea*. *Oecologia* 53, 61–67.

Van der Meyden, E. and Van der Waals-Kooi, R.E. (1979) The population ecology of *Senecio jacobaea* in a sand dune system. I. Reproductive strategy and the biennial habit. *Journal of Ecology* 67, 131–153.

Van der Toorn, J. and Pons, T.L. (1988) Establishment *of Plantago lanceolata* L. and *Plantago major* L. among grass. II. Shade tolerance of seedlings and selection on time of germination. *Oecologia* 76, 341–347.

Van der Veen, R. (1970) The importance of the red-far red antagonism in photoblastic seeds. *Acta Botanica Neerlandica* 19, 809–812.

VanderWoude, W.J. and Toole, V.K. (1980) Studies of the mechanism of enhancement of phytochrome dependent lettuce seed germination by prechilling. *Plant Physiology* 66, 220–224.

Van Tooren, B.F. and Pons, T.L. (1988) Effects of temperature and light on the germination in chalk grassland species. *Functional Ecology* 2, 303–310.

Vázquez-Yanes, C. (1980) Light quality and seed germination in *Cecropia obtusifolia* and *Piper auritum* from a tropical rain forest in Mexico. *Fyton* 38, 33–35.

Vázquez-Yanes, C. and Orozco-Segovia, A. (1990) Ecological significance of light controlled seed germination in two contrasting tropical habitats. *Oecologia* 83, 171–175.

Vázquez-Yanes, C. and Orozco-Segovia, A. (1994) Signals for seeds to sense and respond to gaps. In: Caldwell, M.M. and Pearcy, R.W. (eds) *Exploitation of Environmental Heterogeneity by Plants. Ecophysiological Processes Above and Below-Ground.* Academic Press, New York, pp. 261–318.

Vázquez-Yanes, C. and Smith, H. (1982) Phytochrome control of seed germination in the tropical rainforest pioneer trees *Cecropia obtusifolia* and *Piper auritum* and its ecological significance. *New Phytologist* 92, 477–485.

Vincent, E.M. and Roberts, E.H. (1977) The interaction of light, nitrate and alternating temperature in promoting the germination of dormant seeds of common weed species. *Seed Science and Technology* 5, 659–670.

Wesson, G. and Wareing, P.F. (1969) The induction of light sensitivity in weed seeds by burial. *Journal of Experimental Botany* 20, 414–425.

Widell, K.O. and Vogelmann, T.C. (1988) Fibre optics studies of light gradients and spectral regime within *Lactuca sativa* achenes. *Physiologia Plantarum* 72, 706–712.

Willems, J.H. (1995) Soil seed bank, seedling recruitment and actual species composition in an old and isolated chalk grassland site. *Folia Geobotanica Phytotaxonomica, Praha* 30, 141–156.

Wooley, J.T. and Stoller, E.W. (1978) Light penetration and light-induced seed germination in soil. *Plant Physiology* 61, 597–600.

Chapter 11

The Role of Temperature in the Regulation of Seed Dormancy and Germination

Robin J. Probert

Seed Conservation Department, Royal Botanic Gardens, Kew, Wakehurst Place, Ardingly, West Sussex, UK

Introduction

Over the last three decades, a number of works have dealt with general aspects of the environmental control of germination and dormancy (e.g. Koller, 1972; Roberts, 1972; Heydecker, 1973; Bewley and Black, 1982, 1994; Mayer and Poljakoff-Mayber, 1989; Kigel and Galili, 1995). In addition there have been several contributions that have specifically addressed ecological aspects of germination behaviour (Harper, 1977; Angevine and Chabot, 1979; Grime, 1979; Thompson, 1981; Fenner, 1985, 1992) and recently Baskin and Baskin have assembled the most comprehensive review of seed germination and dormancy ever published, covering over 3500 wild plant species, (Baskin and Baskin, 1998). Other more focused reviews have also been published during this time. For example, Mott and Groves (1981) gave an account of germination strategies in Australian ecosystems; Egley and Duke (1985) concentrated on germination and dormancy in agricultural weeds; Gutterman (1993) described seed germination in desert plants; and Simpson (1990) has produced a comprehensive account of germination and dormancy within a single plant family – the grasses. While the importance of temperature is recognized in all of these publications, in the

last 30 years only a few reviews have dealt exclusively with the role of temperature (e.g. Hegarty, 1973; Thompson, 1973a, 1974; Simon, 1979; Roberts, 1988; Probert, 1992).

Angevine and Chabot (1979) pointed out that, when a seed germinates under natural conditions, the individual has, in a sense, 'bet its life' on the favourability of environmental conditions for seedling establishment. Consequently, selection favours environmental cueing mechanisms that decrease the probability of encountering unacceptable growth conditions following germination. Angevine and Chabot recognized a number of so-called germination syndromes, according to the physical and biotic stresses of the environment that influence seedling establishment. Seed responses to temperature play a pivotal role in several of these germination syndromes and it is therefore arguably the most important environmental variable responsible for the synchronization of germination with conditions suitable for seedling establishment.

Roberts (1988) recognized three separate physiological processes in seeds affected by temperature: first, temperature, together with moisture content, determines the rate of deterioration in all seeds; secondly, temperature affects the rate of dormancy loss in dry seeds and the pattern of dormancy change in moist seeds; and, thirdly, in non-dormant seeds temperature determines the rate of germination.

The role of temperature in seed deterioration is covered by Murdoch and Ellis in Chapter 8 in this volume. In view of the considerable importance of seed dormancy in the regeneration of wild species, attention will be mainly focused here on the effects of temperature on dormant seed populations. The effects of temperature on germination rate in non-dormant populations will also be briefly considered. It is clear that many physiological events associated with germination and dormancy can proceed in relatively dry seeds, and so, for convenience, this contribution is divided into two parts, according to seed moisture status. In the first section, the effects of temperature on air-dry seeds (that is to say, seeds with a moisture content less than approximately 20% of their fresh weight) will be considered. Accordingly, the second part will concentrate on the effects of temperature on imbibed seeds.

Experimental approaches to understanding seed germination behaviour are constantly being refined. One of the features of seed biology research during the last decade or so has been a much greater integration of laboratory and field-based studies than hitherto. Such approaches have improved the interpretation of data in an ecological context and thus our understanding of the role of temperature in the timing of germination in natural seed populations. In particular, I have attempted to highlight these recent advances.

The effects of temperature on dry seeds

Dry after-ripening

In his historical account of seed physiology from antiquity to the 20th century, Evenari (1984) considers that the Greek philosopher Theophrastus should be regarded as the father of seed physiology. Among the many aspects of seed biology known to Theophrastus more than 2000 years ago

was that germination requirements change during storage: 'Another thing which makes a difference as to the rapidity with which the seeds germinate is their age; for some herbs come up more quickly from fresh seeds ... , some come up more quickly from old seeds.'

During dry storage, seeds undergo physiological changes, which are often reflected in a decline in the level of innate dormancy. Accompanying these so-called dry after-ripening changes, germination requirements usually become less specific. For example, Derkx and Karssen (1993) reported that the sensitivity of *Arabidopsis thaliana* seeds to light and gibberellins increased as dormancy rapidly declined during dry storage. In a study of the effects of storage on the germination response to light and alternating temperatures in European populations of *Dactylis glomerata,* Probert *et al.* (1985b) showed that, while the relative effects of both factors did not change, the probability of seed germination in response to light and alternating temperatures increased linearly during storage at 15°C and 15% relative humidity (r.h.). In *D. glomerata,* the proportion of individuals that require light and/or alternating temperatures to trigger germination declines during dry storage. After prolonged storage, non-dormant seeds are capable of maximum germination, even in the dark, at constant temperatures.

For species adapted to areas of the world that experience rainfall throughout the year and soil types with high water retention, the ecological significance of changes in germination behaviour recorded during dry storage may be difficult to understand. However, for species adapted to regions of seasonal drought and dry soils – for example, winter annuals – physiological changes recorded during dry storage reflect a natural mechanism, which governs the timing of germination in the wild.

When seeds of winter annuals are in a state of conditional dormancy, shortly after dispersal, germination is characteristically restricted to a narrow range of low temperatures (Thompson, 1970a; Baskin and Baskin, 1971, 1972, 1978; Pemadasa and

Lovell, 1975). At this time of the year (late spring/early summer), soil temperatures are significantly higher than the maximum temperature favouring germination and therefore seeds do not germinate. During the subsequent warm, dry conditions of summer, after-ripening changes result in an increase in the maximum temperature for germination. During early autumn, falling soil temperatures overlap with the temperature range over which germination can occur and, providing other conditions necessary for germination are met (e.g. an adequate supply of water), germination will proceed, initially in the least dormant seeds. Laboratory studies of germination over a wide range of constant temperatures in seeds of the European sand-dune annual *Phleum arenarium* illustrate this trend (Fig. 11.1).

The rate of dry after-ripening in dry seeds is dependent on temperature and seed moisture content. In a detailed study of dormancy in rice, Roberts (1962) demonstrated a negative linear relationship between the logarithm of the period of dry storage required for 50% germination (mean

dormancy period) and temperature. This relationship is described by the equation:

$$\log d = K_d - C_d t$$

where d = the mean dormancy period, t = temperature, K_d = the intercept constant and C_d = the slope constant. Importantly, Roberts also demonstrated that the variation in the after-ripening period required to trigger germination between individual seeds in rice populations was normally distributed. This fact has led to the recent widespread use of population-based models to study seed dormancy (see later).

Subsequently, Roberts (1965) reported considerable variation in the depth of dormancy between rice cultivars indicated by differences in the intercept constant K_d but little change in the slope constant C_d (Fig. 11.2). Therefore, the relative effects of temperature on the rate of dry after-ripening (conveniently described by the temperature coefficient, Q_{10}) are constant within a species. The relatively high Q_{10} value of 3.1 calculated for rice, together with broadly similar values calculated for other grasses, e.g. 3.9 for barley (Roberts and Smith,

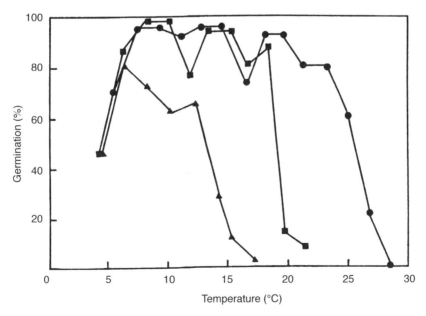

Fig. 11.1. Dry after-ripening in seeds of *Phleum arenarium*. Changes in the germination response to temperature during storage at 15°C and 15% r.h. Germination was tested on a thermogradient bar after 1 (▲), 6 (■) and 13 months (●) storage. The seeds were incubated on wetted filter-paper and received an 8-h photoperiod day^{-1}. (From R.J. Probert, unpublished data.)

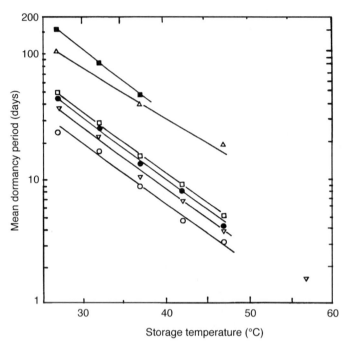

Fig. 11.2. The relationship between storage temperature and loss of seed dormancy during dry after-ripening in rice cultivars. The effect of storage temperature on mean dormancy period on cultivars: Masalacii A4 (■), Bayawuri (△), Lead 35 (□), India Pahil 46 (●), Nam Dawk Nai (▽), Mas 2401 (○). The anomalous regression line from Bayawuri is questionable since it is based on extrapolation from relatively few results. (Adapted from Roberts, 1988, after Roberts, 1965.)

1977), 2–3 for *D. glomerata* and *Festuca arundinacea* (Stoyanova and Kostov, 1983; Stoyanova *et al.*, 1984), implies a high activation energy for the controlling process. The acceleration of dry after-ripening in rice by high oxygen tensions suggests the operation of an oxidative process (Roberts, 1962). However, while there is some evidence of respiratory activity in dry seeds (Roberts and Smith, 1977), it is unlikely that oxidative electron transport and ATP synthesis are possible at the moisture contents optimal for dry after-ripening (Leopold *et al.*, 1988; Leopold and Vertucci, 1989).

The effect of seed water content on the rate of dry after-ripening has probably been underestimated, because of the dramatic effect of increasing moisture content on the rate of loss of viability in seeds. A number of studies have shown that, over a limited moisture range, there is a direct relation between the rate of after-ripening and seed moisture content. For example, Baskin and Baskin (1979) found an increased rate of

dry after-ripening in the winter annual *Draba verna* when seeds were stored at high, rather than low, r.h. In a detailed study of rice, Ellis *et al.* (1983) showed that increasing the moisture content (fresh weight basis) from 8 to 11% resulted in a 2.5-fold reduction in the storage period required for a given level of germination. Foley (1994), on the other hand, showed that there was an inverse relationship between the temperature and moisture content for dry after-ripening in seeds of *Avena fatua* (wild oat), such that lower moisture contents were needed for maximum after-ripening as temperature increased.

The activity of various physiological processes in seeds has been related to water binding (Vertucci and Leopold, 1986; Leopold and Vertucci, 1989; Vertucci and Roos, 1993; Sun *et al.*, 1997). Three critical regions of water binding have been identified in seeds, which correspond to the three parts of the reverse sigmoidal water sorption isotherms characteristic of desiccation-tolerant (orthodox) seeds (Fig. 11.3). Based on thermodynamic principles, it has been calculated that water is highly bound in region 1, weakly bound in region 2 and bound with negligible energy in region 3 (Vertucci and Leopold, 1984).

Using the same techniques, Leopold *et al.* (1988) have studied the relation between water status and dry after-ripening in red rice. After-ripening was most rapid at moisture contents between about 6 and 12% (fresh weight basis) and was severely restricted below 5%. Although evidence was presented that after-ripening was sharply reduced at moisture contents above 12%, the possibility that low germination may have been caused by a reduction in seed viability cannot be discounted from the published data. Using moisture isotherms constructed for red rice seeds at

2 and 22°C, Leopold *et al.* calculated water-binding enthalpies, using the Clausius– Clapeyron equation:

$$\Delta H = (R + T_1 \times T_2)/(T_2 - T_1) \times \ln (aw_1/aw_2)$$

where R = the gas constant; T_1 and T_2 = two temperatures (in this case 2 and 22°C) and aw_1 and aw_2 = the r.h. at the two temperatures for a given water content.

Calculated binding enthalpies indicated a region of high affinity at moisture contents below 10%, a region of lower binding affinity between 10 and about 20% and a region of negligible binding enthalpy at still higher moisture contents. These regions of water-binding enthalpy corresponded to the three distinct phases of the moisture isotherms constructed for seeds of red rice. Based on the range of moisture contents at which after-ripening was recorded, it may be concluded that dry

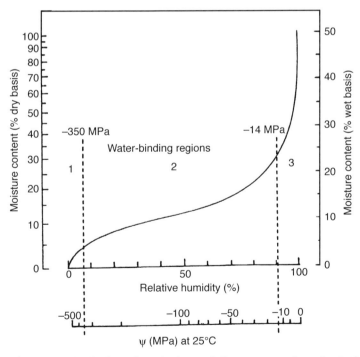

Fig. 11.3. Typical water sorption isotherm for orthodox seed. Data represent absorption isotherm for barley seeds at 25°C (adapted from Roberts and Ellis, 1989). Approximate water-potential limits for linear relations between water activity and seed longevity are shown in relation to the three regions of water binding (see text). Water potential (Ψ) expressed in megapascals (MPa) and relative humidity (r.h.) expressed as a proportion are related according to the equation $\Psi = (RT/\overline{V}_w) \ln$ r.h. where R = the gas constant (8.31 J K^{-1}), T = the absolute temperature and $\overline{V}_w$ = the partial molar volume of water (18 cm^3 mol^{-1}).

after-ripening in red rice occurred at moisture levels ranging from just below the
boundary between binding region 1 and 2
and over the lower part of region 2. It was
speculated that after-ripening may involve
some non-enzymatic oxidative reactions,
which are inhibited at lower moisture contents by rising free energy and at higher
moisture contents by oxidative metabolism. Optimum rates of after-ripening in
dormant seeds of A. fatua were also shown
to be associated with the second region of
water binding (Foley, 1994).

At the higher moisture levels favouring
dry after-ripening, i.e. in water-binding
region 2 and approaching water-binding
region 3, other important physiological
processes can occur, which could lead to
changes in germination requirements and
dormancy status. For example, the effects
of chilling (Vertucci and Leopold, 1986)
and light (via the photochemical conversion of phytochrome: Leopold and
Vertucci, 1989) can proceed at these moisture levels. These observations suggest that,
under certain conditions, it may be possible for seeds to enter secondary dormancy
(for example, as a result of an induced light
requirement) at the same time that primary
dormancy is being released by after-ripening. Indeed, Probert et al. (1985b) reported
a simultaneous release and induction of
dormancy when seeds of D. glomerata were
stored for several months over winter
under conditions of natural temperature
and r.h.

Progress in the modelling of seed dormancy loss under field conditions due to
dry after-ripening has been achieved
through the application of the hydrothermal time concept, first introduced by
Gummerson (1986) and developed further
by Bradford (1990, 1995, 1996, 1997).
Incubation time, temperature and water-
potential conditions are used to determine
progress towards germination according to
the equation:

$$\theta_{HT} = (T_i - T_b)[\Psi - \Psi_b(g)]t_g$$

where θ_{HT} is the hydrothermal time a seed
requires for germination (measured, for
example, in megapascal degree-days), T_i

and Ψ are, respectively, the temperature
and water potential of the incubation
medium, T_b is the base temperature for
germination, $\Psi_b(g)$ is the theoretical base
water potential for germination of the g
fraction of the population and t_g is the
germination time of the same fraction. As
with the distribution of thermal time
requirements, so the distribution of base
water potentials has been shown to be normal within seed populations, with mean
$\Psi_b(50)$ and standard deviation σ_{Ψ_b}.

Bradford (1997) suggested that seasonal or environmental effects on the
capacity for germination may act through
physiological shifts in the $\Psi_b(g)$ distribution, with variation in $\Psi_b(g)$ between seeds
providing differential sensitivity to local
conditions to ensure that there are both
opportunistic and conservative individuals
in the population.

Christensen et al. (1996) successfully
used the hydrothermal time concept to
model dry after-ripening in seeds of
Bromus tectorum (cheatgrass), an important winter-annual, alien weed of western
North America. It was found that variation
in the period of after-ripening, variation in
germination time courses during after-
ripening and the upward shift in the optimum germination temperature could all be
accounted for by varying the model parameter $\Psi_b(50)$. Whilst $\Psi_b(50)$ becomes more
negative as after-ripening progresses, other
parameters apparently change very little.

After-ripening in B. tectorum was further characterized by Bauer et al. (1998),
using a thermal time model for after-ripening, which assumed that the rate of change
in $\Psi_b(50)$ is a linear function of temperature above a base temperature according to
the equation:

$$\theta_{AT} = (T_s - T_1) t_{ar}$$

where θ_{AT} is the thermal time requirement
for after-ripening, T_s is the after-ripening
temperature, T_1 is the base temperature
below which after-ripening does not occur
and t_{ar} is the after-ripening time, i.e. the
time required for $\Psi_b(50)$ to change from its
initial value to a final value. As in any
thermal time model, θ_{AT} is a constant, so

that the time required for after-ripening decreases proportionately to the increase of the after-ripening temperature above the base temperature.

Bauer *et al.* (1998) showed that predictions of changes in $\Psi_b(50)$ for seeds held under field conditions were generally close to observed values, at least until seeds nearing the completion of after-ripening encountered significant rainfall events. Such important advances in the modelling of seed dormancy raise the prospect that very soon we shall be able explain much of the habitat-correlated variations in seed germination behaviour typical for natural populations of wild plant species (Allen and Meyer, 1998).

The changes in germination behaviour associated with dry after-ripening represent temperature- and moisture-dependent physiological changes occurring within seed embryos. Notwithstanding this, in numerous species – of which temperate grasses are a good example – seed covering structures greatly influence germination requirements and the expression of dormancy. In many cases, for example, removal or puncture of the covering structures can reduce or remove the requirement for dry after-ripening.

The seed-coat may influence germination and dormancy in a variety of ways (Bewley and Black, 1994; Baskin and Baskin, 1998). Perhaps the most familiar expression of physical dormancy is hard-seededness – a frequent condition, occurring in several important plant families. The term hard-seededness describes a dormancy mechanism entirely controlled by physical characteristics of the seed-coat, which prevents water uptake by otherwise non-dormant seed embryos. In certain ecosystems, natural temperature changes play an important role in the release of hard-seededness.

The role of temperature in the breakdown of hard-seededness

Dormancy controlled by the physical characteristics of a hard, water-impermeable seed-coat occurs in numerous species but is most common in members of the *Leguminosae, Malvaceae, Convolvulaceae, Chenopodiaceae, Cannaceae, Convallariaceae, Geraniaceae, Gramineae, Liliaceae* and *Solanaceae* (Ballard, 1973; Rolston, 1978; Bewley and Black, 1982).

The majority of experimental studies, however, have focused on species within the *Leguminosae* and *Malvaceae*, and the important structural and anatomical characteristics of the seed-coats of species within these families has been comprehensively covered elsewhere (Tran and Cavanagh, 1984; Egley, 1989; Kelly *et al.*, 1992). According to Tran and Cavanagh, impermeable seeds possess a specifically located structural weakness, which is the focus for hard-seed breakdown by natural agents (such as temperature). In the papilionoid and mimosoid legumes this point of weakness is located beneath the lens or strophiole, whereas in malvaceous species it is the region of chalazal discontinuity. In both cases, these regions are underlain by a single or double layer of thin-walled subpalisade cells, whose walls readily break up under stress. This palisade layer can then lift, exposing the underlying permeable tissue.

Dry heat and wide temperature fluctuations play an important role in hard-seed breakdown in species adapted to tropical, subtropical and Mediterranean ecosystems, which typically experience summer drought. In view of the considerable economic importance of legume species in the agricultural system of these areas and the high frequency of hard-seededness within the *Leguminosae*, most of the reports that appear in the literature relate to work on this family. Temperature mediates natural hard-seed breakdown in two principal ways:

1. Climatic, through the heating effect of solar radiation on the surface layers of dry soils (insolation), coupled with night-time cooling, resulting in a combination of exposure to high temperatures and wide temperature fluctuations (Quinlivan, 1966, 1971; Mott *et al.*, 1981; Taylor, 1981; McKeon and Mott, 1982; Russi, 1989; Russi *et al.*, 1992).

2. Fire-related, through the brief but intense heating of the surface layers of soil caused by man-made or natural fires (Floyd, 1966; Shea *et al.*, 1979; Pieterse and Cairns, 1986; Keeley, 1987; Thanos and Georghiou, 1988).

Climate effects

The rate of hard-seed breakdown during the dry season in populations of legume seeds is determined by the maximum temperature experienced and the amplitude of diurnal temperature fluctuations (Quinlivan, 1966; Mott *et al.*, 1981; McKeon and Mott, 1982; Russi, 1989). From combined laboratory and field experiments using the pasture legumes *Stylosanthes humilis* and *S. hamata* in the monsoon region of northern Australia, McKeon and Mott (1982) showed that maximum soil temperatures greater than 50–55°C were required for hard-seed breakdown. A more or less linear increase in the rate of hard-seed breakdown was observed over the temperature range

50–75°C and, for a given daily maximum temperature, a progressively greater number of days was required for a fixed reduction in hard-seededness. In a related publication, Mott *et al.* (1981) used these data to construct the predictive model:

$$y = a + (y_0 - a) \exp(-k\Sigma x)$$

where y = the hard-seed content (%); Σx = the accumulated degree-days of surface maximum temperature T_{max}, above a threshold T_0; y_0 = the initial hard-seed content (%); k = a rate constant (degree-days^{-1}) and a = the final (i.e. residual) hard-seed content (%). This model, which predicts an exponential increase in degree-days required for each successive reduction in hard-seededness by a fixed proportion, was successfully applied to field data gathered at three different subtropical and tropical locations in Australia, using seed populations of four different annual and perennial varieties of *Stylosanthes*.

Using seed populations of the annual legumes *Medicago rigidula* and *Trifolium stellatum* from northern Syria, Russi (1989)

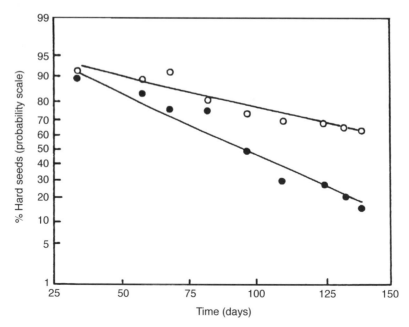

Fig. 11.4. The breakdown of hard-seededness in seeds of *Trifolium stellatum* (●) and *Medicago rigidula* (○) under natural conditions during the dry season in northern Syria (adapted from Russi, 1989).

quantified the relationship between hard-seed breakdown and duration of exposure to field conditions during the dry season, using probit analysis. In this case, a negative linear relationship between the probit of the percentage of hard seeds remaining and time was demonstrated for both species (Fig. 11.4). Interestingly, the regression slopes were significantly different, emphasizing species differences in seed-to-seed variability in hard-seededness. Russi's data also provide convincing evidence that the variation of hard-seededness within seed populations is normally distributed. This continuous variation in hard-seededness within seed populations, which is determined by both genetic and environmental factors (Rolston, 1978; Egley, 1989), has enabled the selective removal of hard-seededness from certain forage legumes, such as *Trifolium subterraneum* (Morley, 1958).

Although the majority of published evidence implicates climatic temperature effects in hard-seed breakdown in geographical regions that experience a hot dry season, there is no reason why temperature should not play a similar role in other ecosystems. Indeed, there is some evidence that increased temperature fluctuations on the soil surface following canopy clearance may be responsible for overcoming hard-seededness in successional species in humid tropical forests (Vázquez-Yanes and Orozco-Segovia, 1982, 1984). Moreover, when evidence that alternate freezing and thawing may increase the rate of hard-seed breakdown in temperate legumes is considered (Midgley, 1926; Pritchard *et al.*, 1988), it seems possible that temperature fluctuations involving freezing and thawing may play a role in the breakdown of hard-seededness in species adapted to cold climates.

Fire-related effects

In contrast to the gradual breakdown of hard-seededness that occurs in response to climatic temperature effects, the intense short-term heating effects caused by fire can lead to a rapid breakdown in hard-seededness in some species. Evidence to support this view is provided by an experiment conducted by Gill (1981), in which seeds of *Daviesia mimosoides* were either placed in grooves in dry clay–loam held in trays or sprinkled among litter, which was placed over the trays and ignited to simulate the effects of natural fire. When seeds were subsequently recovered and placed in Petri dishes with water, the seeds that had been placed in grooves germinated rapidly and to a high level, whereas untreated seeds showed no germination unless they were scarified. Seeds that were placed in the litter were killed. These results emphasize that a fine balance exists between seeds being killed if temperatures are too high and seeds being unaffected if temperatures are too low. In an earlier study of the effect of fire on weed seeds in the soil of wet sclerophyll forests in Australia, Floyd (1966) showed that the position of the seeds within the soil profile coupled with the intensity and duration of the fire are the critical determinants of whether or not hard-seed breakdown occurs. This point is further illustrated by the work of Pieterse and Cairns (1986), which showed that seeds of *Acacia longifolia* at a depth of 3 cm were unaffected by fire, at 2 cm a high proportion of seeds were rendered permeable without affecting viability, whereas at 1 cm all seeds were killed. In laboratory experiments, relatively brief exposures of a few minutes to temperatures of about 100°C have been shown to be effective in hard-seed breakdown of a number of species (Keeley, 1987; Munoz and Fuentes, 1989). Mott (1979) described a practical method for overcoming hard-seededness in bulk quantities of *Stylosanthes* seed, involving exposures of 15–30 s to temperatures of 140–150°C.

In a laboratory study of the effects of temperature on hard-seed breakdown in *Rhus javanica*, a pioneer tree species of temperate Japan, Washitani (1988) demonstrated a direct linear relation between the proportion of permeable seed (expressed in probits) and temperature. Therefore, the critical temperature for dormancy release between individual seeds was normally distributed. It was also shown that the proportion of seeds capable of germination was critically dependent on the duration of exposure at any given temperature. Germination curves describing the effect of heat-treatment duration were successfully modelled, using the assumption that the distribution of sensitivity to dormancy release and viability loss were both normal. Washitani concluded that, in the wild, dormancy release and viability retention in seeds of *R. javanica* would occur either as a result of several hours exposure to temperatures at about 60°C due to insolation or as a result of brief exposure to much higher temperatures during fires.

In Mediterranean ecosystems, regeneration and population dynamics of certain species is often dependent on the occurrence of natural fires. For example, in the Californian chaparral, certain woody *Ceanothus* species are unable to maintain their population size or colonize new areas in the absence of fire-stimulated regeneration from a soil reserve of hard-coated seed (Keeley, 1987).

From the available evidence, it may be concluded that, above a critical temperature, the rate of hard-seed breakdown is related to the product of exposure time and temperature. Under natural conditions, dormancy release in hard-seed populations occurs gradually, as a result of insolation during the dry season, or rapidly, as a result of brief exposure to intense heat during fires. The relative importance of these processes for different species is related to climatic and ecological factors, including agricultural practice.

Interestingly, in a study of the effects of fire on regeneration from buried seeds in 31 tropical tree species of the Amazon rainforest, temperatures within the range that stimulates germination of dry seeds in Mediterranean ecosystems were lethal to seeds of tropical tree species (Brinkmann and Vieira, 1971). Recolonization on slash-and-burn land in tropical rainforests is almost exclusively from airborne seed. The likely explanation for this difference is that the vulnerability of seeds to high temperature extremes is dependent on seed moisture status. In dry Mediterranean ecosystems, seeds in general and hard-seeded species in particular will be at a low moisture content and therefore relatively heat-resistant, whereas buried seeds of the vast majority of tropical rainforest species will be more or less fully imbibed and, as a result, susceptible to the damaging effects of high temperatures on cellular processes.

The effects of temperature on imbibed seeds

Chilling and seed dormancy

Although this section will be mainly concerned with a discussion of the effects of temperature on wet seeds, it is noteworthy that at least some of the physiological changes associated with temperature are possible in relatively dry seeds. For example, the promotive effects of chilling on the release of dormancy in apple (Vertucci and Leopold, 1986) and sitka spruce (Gosling and Rigg, 1990) can occur in seeds with a moisture content of less than 20% (fresh weight basis).

The stimulatory effects on germination of holding imbibed seeds at low temperatures has been recognized for centuries. In 1664, Evlyn recommended that, prior to the spring sowing of forestry species, such as *Acer* and *Fagus*, seeds should first be placed in moist sand or soil and held outdoors for the winter. In recent times, there have been countless reports of the effects of chilling on seed dormancy and these have been well covered in a number of reviews – for example, Stokes (1965), Lewak and Rudnicki (1977) and Nikolaeva (1977) – and general texts – for example, Bewley and Black (1982, 1994) and Mayer and Poljakoff-Mayber (1989).

The requirement for chilling, widespread amongst temperate species, represents a natural mechanism which ensures that germination occurs in the spring. For example, seeds of the annual aquatic grass *Zizania palustris*, which grows in shallow lakes in North America, are dormant at the time of dispersal in the autumn. During the winter, seeds buried in lake sediment are exposed to temperatures close to freezing. Under such conditions, dormancy steadily declines (Fig. 11.5) and high levels of germination are possible, coincident with rising water temperatures in the spring. Chilling represents a natural dormancy-breaking mechanism in many species, but the requirement can often be modified by other factors. Indeed, germination of freshly harvested *Z. palustris* seeds is possible if seeds are treated with a cocktail of alternative dormancy-breaking treatments, namely, removal of the palea and lemma and slitting of the pericarp above the embryo; incubation in a solution of gibberellins (GA$_{4+7}$) at 0.5 mM and in an alternating temperature regime of 25/10°C (8 h/16 h), with illumination during the warm phase. Moreover, the effect of this combination of treatments is enhanced when seeds are slightly dehydrated (Probert and Brierley, 1989). As in *Z. palustris*, there is often an interaction between the period of chilling required to overcome dormancy and the presence or absence of various seed covering struc-

tures. An excellent example of this is provided by the work of Visser (1954) on apple. In this case, more than 80 days of chilling at 3°C were required for maximum germination of intact seeds. Removal of the testa reduced this requirement by approximately 20 days and this was reduced still further by removal of the endosperm surrounding the embryo (Fig. 11.6).

In most species, a single period of chilling is sufficient to completely overcome dormancy. However, in some cases, a more complex situation exists, in which different organs of the embryo differ in their depth of dormancy. In such species (of which *Convallaria majalis* is an example), consecutive periods of chilling, warming and rechilling are required to overcome radicle and epicotyl dormancy (Stokes, 1965).

Optimum temperatures for the dormancy-breaking effect of chilling are generally close to 5°C (Bewley and Black, 1994). However, in a detailed study of dormancy in *Rumex obtusifolius* and *R. crispus*, Totterdell and Roberts (1979) reported that dormancy release was possible over a wide range of temperatures from 1.5 to 15°C. It

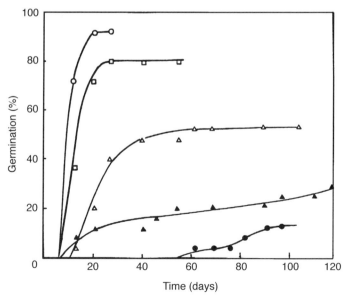

Fig. 11.5. Loss of dormancy in seeds of *Zizania palustris* during chilling in distilled water at 2°C. Germination was tested at 16°C in the light (12 h d^{-1}) immediately after harvest (●) and after 9 (▲), 13 (△), 20 (□) and 26 weeks' (○) chilling. (From Probert and Longley, 1989.)

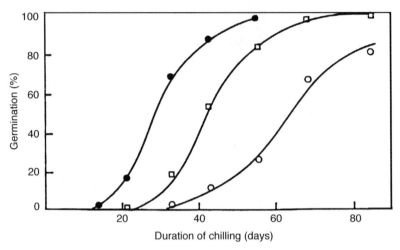

Fig. 11.6. Termination of apple-seed dormancy by chilling at 3°C. Germination of the following components was tested at 25°C at intervals during chilling: O, intact whole seeds; □, seeds with testa removed but endosperm intact; ●, isolated embryos. (Adapted from Stokes, 1965, after Visser, 1954.)

was argued that the reduced efficacy of warmer temperatures within this range to overcome dormancy resulted from a simultaneous, temperature-dependent, induction of secondary dormancy (Fig. 11.7). This concept has been developed further and successfully used to generate a descriptive model of seasonal changes in dormancy in buried weed-seed populations (Bouwmeester, 1990; Bouwmeester and Karssen, 1992) (see later).

The environmental control of germina-

(a)   (b)

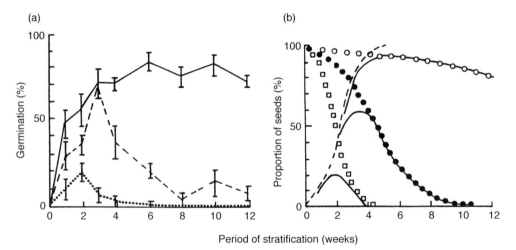

Period of stratification (weeks)

Fig. 11.7. (a) Germination of *Rumex obtusifolius* after 4 weeks at 25°C in the light following various periods of stratification at 1.5°C (—), 10°C (----) or 15°C (····); (b) Expected germination if loss of dormancy due to stratification (---) is independent of temperature over the range investigated, while the induction of dormancy increases with temperature so that the proportion of seeds with no induced dormancy is shown as follows: 1.5°C (O), 10°C (●), 15°C (□). Predicted germination is the product of (proportion of seeds with no induced dormancy) × (proportion of seeds that has lost innate dormancy through stratification) (—). It is assumed (arbitrarily) that seed-to-seed variability is similar for both processes with a coefficient of variation of 50%, and that log of the mean time taken to induce dormancy is linearly related to temperature. Note the similarity of the predicted curves to those in (a). (From Roberts, 1988, after Totterdell and Roberts, 1979.)

tion is often a complex process, involving the interaction of a number of factors, such as light, alternating temperatures, nitrate and desiccation. Several studies have shown that, in certain species, chilling reduces the requirement for other environmental factors (Roberts and Benjamin, 1979; Vincent and Roberts, 1979; Probert *et al.*, 1989). Van der Woude and Toole (1980) elegantly demonstrated a direct effect of chilling on phytochrome-controlled lettuce-seed germination. In this study, it was shown that a period of 6 h prechilling greatly enhanced the response of seeds to low, normally inhibitory levels of the active form of phytochrome (P_{fr}) established by far-red irradiation. Interestingly, the critical temperature, above which enhancement of P_{fr} action did not occur (13°C), was close to the threshold temperature for chilling effects identified by Totterdell and Roberts (1979).

Our understanding of the role of hormones in germination and dormancy has undoubtedly been advanced by the use of mutants. These studies reveal a central role for gibberellins in the germination process (Karssen *et al.*, 1987) and it has been shown that chilling results in an increase in responsivity to gibberellins (Karssen *et al.*, 1989; Hilhorst and Karssen, 1992; Derkx *et al.*, 1994). Hilhorst (1993, 1998) has developed a descriptive model for the effect of temperature on the release of seed dormancy, based on membrane-receptor occupancy theory. This model, also discussed by Vleeshouwers *et al.* (1995), assumes that the number of active membrane receptors for phytochrome and nitrate will increase or decrease as dormancy is released or induced by chilling. In a non-dormant seed, the activated receptor will bind P_{fr} and initiate a signal transduction chain, leading to GA synthesis and germination.

Chilling responses and climatic/genetic factors

Extensive studies by Meyer and co-workers on several Intermountain North American species has improved our understanding of

the relationship between seed germination patterns and climate-related features of habitat. The Intermountain region features habitats ranging from low-elevation warm deserts to high-elevation cold montane sites. Studies on several species adapted to these habitats (for a recent review, see Allen and Meyer, 1998) have revealed interesting habitat-correlated variation in seed germination patterns, including response to chilling. Populations from habitats with severe winters tended to be more dormant and required longer periods of chilling compared with populations from milder habitats. For spring-emerging populations, the rate of germination at low temperatures also varied, depending on whether the primary risk to seedling establishment was frost or drought. For example, time to 50% germination was negatively correlated with collection-site mean January temperature for three subspecies of *Artemisia tridentata* (big sagebrush) (Meyer and Monsen, 1992). Amongst three populations of the Mediterranean tree, *Pinus brutia*, Skordilis and Thanos (1995) also reported that the population adapted to the region experiencing the coldest winter climate had the highest level of dormancy and therefore chilling requirement.

Contrary to these findings however, Schütz and Milberg (1997) reported that seed populations of *Carex canescens* requiring prolonged periods of chilling were not associated with habitats experiencing severe winters. Indeed, in their study of populations from four regions of north-west Europe, the reverse was true. Schütz and Milberg argued that populations of *C. canescens* from areas experiencing predictably severe winters would require only short periods of chilling to ensure spring germination. Populations adapted to regions with unpredictable winters, on the other hand, would require a prolonged chilling requirement, coupled with an inability to germinate at low temperatures to prevent germination during mild intervals in otherwise cold winters. *Dioscorea* species from Japan, adapted to colder climatic regions, were also shown to be less dormant and therefore required less chilling than species adapted to warmer regions (Okagami and Kawai, 1982).

While it is clear that the dormancy-breaking effects of chilling are widespread, particularly among species adapted for spring germination, there is considerable evidence that low temperatures can also result in the induction of dormancy, particularly in species such as winter annuals, which are adapted for autumn germination (Baskin and Baskin, 1986, 1988). Even in species in which dormancy is released by chilling, extended incubation at low temperatures can lead to an induction of secondary dormancy (Willemsen, 1975; Bouwmeester, 1990). Even within a single seed population, genetic differences in the depth of dormancy between individual seeds can result in differential effects of chilling. For example, achenes of the ruderal annual *Ranunculus sceleratus* possess an absolute requirement for both light and alternating temperatures for germination (Probert *et al.*, 1987). Genetically distinct subpopulations differing in depth of dormancy were selected on the basis of the quantal response of a single population of individuals to daily temperature shifts. When these subpopulations were analysed for their sensitivity to an increasing period of chilling at 2°C, low-dormancy seed batches responded positively, with a marked decline in the requirements for light and alternating temperatures. High-dormancy batches, on the other hand, were much less responsive. Indeed, prolonged chilling of high-dormancy batches led to a reduction in both the rate and capacity of germination when seeds were tested in the presence of light and alternating temperatures, indicating that chilling led to the induction of secondary dormancy (Probert *et al.*, 1989). For colonizing species, genetic differences in the sensitivity of individuals within seed populations to seasonal temperature changes would maximize flexibility in the timing of emergence and may help to explain why certain species, such as *R. sceleratus*, possess both winter and summer annual life-forms.

Divergent responses to chilling amongst individual seeds in a population

were also reported in the short-lived perennial *Penstemon palmeri* (Meyer and Kitchen, 1992) and the same workers were able to demonstrate, through garden experiments, that habitat-related differences in response to chilling amongst populations of *Penstemon* were under genetic control (Meyer and Kitchen, 1994). Whilst there is clearly a strong genetic component to chilling responses, a recent study of between-year variation in chilling requirements in a single population of the temperate tree *Aesculus hippocastanum* (horse chestnut) also highlighted the importance of environmental factors (Pritchard *et al.*, 1996, 1999).

Recently, population-based models have been used to study aspects of seed dormancy, including response to chilling (Benech Arnold and Sanchez, 1995; Bradford, 1996; Pritchard *et al.*, 1996, 1999). In seeds of the temperate tree *A. hippocastanum*, for example, Pritchard and co-workers were able to demonstrate a negative linear relationship between the rate of dormancy loss and chilling temperature. As a result, the thermal time for dormancy loss ($\theta(D)$) could be described by the equation:

$$\theta(D) = [T_c(D) - T]\, t$$

where $T_c(D)$ is the ceiling temperature above which dormancy loss does not occur and t is the chilling time in days.

Observed between-year differences in the sensitivity of *A. hippocastanum* seed lots to chilling, reflected by differences in the standard deviation of the distribution of chilling periods required to overcome dormancy, were due to differences in $T_c(D)$. Seeds harvested in 1989, which were more responsive to chilling compared with seeds harvested in 1988, were shown to have a higher $T_c(D)$. Consequently, for a given chilling period, 1989 seeds would be expected to accumulate more chilling units (thermal time) compared with 1988 seeds.

In a subsequent analysis of the variation in sensitivity to chilling amongst seed lots harvested in 1988, 1990, 1991 and 1992, it was shown that the chilling period required to increase germination by 1 probit was negatively related to cumulative temperature above 0°C during a 7-week period before seed fall (Pritchard *et al.*, 1999). It was postulated that seed maturation temperature determines the value of the ceiling temperature for net dormancy loss $(T_c(D))$, such that warm seasons result in higher $T_c(D)$ values compared with cool seasons.

Warming and seed dormancy

In contrast to the immense literature on low-temperature effects, there have been far fewer reports of the effect of warm temperatures on changes in germination behaviour. However, it is clear that, like chilling, warm temperature preincubation of imbibed seeds can both release and induce dormancy. In some species – for example, *Spergula arvensis* – chilling and warming both release dormancy, although probably via a different mechanism (Karssen *et al.*, 1988). In many other species, however, chilling and warming have opposing effects. For example, in summer annuals in general, low winter temperatures release dormancy, whereas high summer temperatures induce dormancy (Baskin and Baskin, 1977, 1987a; Bouwmeester, 1990), while in winter annuals the reverse is true (Baskin and Baskin, 1978, 1986). In a recent study of germination in desert perennial shrub species from South Africa, Gutterman (1990) demonstrated that the effect of preincubating seeds at 45°C for 24 h depended on whether plants originated from areas receiving winter or summer rain. In the former, germination was adversely affected, whereas, in the latter, germination was stimulated by high-temperature treatment.

It is worthy of note that the promotive effects of high-temperature treatments are often rapid, and there have been numerous reports of a positive interaction between high-temperature treatments and the control of dormancy by other factors, notably light. For example, Taylorson and DiNola (1989) showed that the light requirement in *Echinochloa crus-galli* seeds was abolished

in a significant proportion of individuals (37%) following a 0.5 h treatment of imbibed seeds at 46°C. According to Taylorson and DiNola, such a response could have ecological consequences for field emergence.

Seasonal changes in germination response to temperature

The changing patterns of germination behaviour associated with the induction and release of dormancy in seeds was first discussed in terms of a general model by Vegis in 1964. Vegis proposed that, for all types of resting organs, including seeds, the temperature range favouring growth contracts as dormancy intensifies and expands as dormancy declines. Narrowing and widening of the temperature range for germination could occur by movement in the low- or high-temperature limit for germination, or both, depending on the species. At the time of maximum dormancy, seeds may be in a state of 'true dormancy', in which germination is prevented regardless of the environmental conditions; or in a state of 'relative' or 'conditional' dormancy, in which germination is prevented only under certain environmental conditions. Confirmation of the general rule formulated by Vegis has been derived from a number of detailed studies of the seasonal changes in germination behaviour of exhumed seeds (for general reviews, see Karssen, 1982; Baskin and Baskin, 1985a). For the most part, studies of seasonal patterns of germination behaviour involve the use of comparable methodology. Uniform samples of freshly harvested seeds or seeds previously held under conditions that minimize dormancy changes are usually placed in small sachets or envelopes. These are then buried in soil, either in unheated glasshouses or under natural conditions, to a specific depth and in such a way as to optimize contact with the surrounding soil. Under such conditions, seasonal changes in dormancy are possible but germination is usually prevented. At regular intervals, over periods usually of

1–3 years, samples of seeds are exhumed and tested for germination over a range of conditions. This approach facilitates an evaluation of the changes in germination response through time to a range of environmental factors, such as temperature, light, nitrate and desiccation.

Based on field observation and laboratory experiments, Went (1957) concluded that seasonal differences in the emergence of annual vegetation in the Californian deserts were mainly due to temperature differences after rain. When desert soils were watered in the laboratory at temperatures between 26 and 30°C, only summer annuals germinated, whereas, when the same soil was placed at 10°C, only winter annuals appeared. Since these important observations were made, a number of studies have demonstrated that, in strict summer and winter annuals, patterns of dormancy release and induction are seasonally distinct and out of phase by 6 months. Based on published evidence, Karssen et al. (1988) argued that, in general, dormancy is released during the season preceding the period with favourable conditions for seedling development and plant growth, whereas dormancy is induced in the season preceding the period with harmful conditions for plant survival. Karssen and co-workers concluded that patterns of dormancy are of high survival value to plants.

When seeds of summer annuals are dispersed in the autumn, they are either truly dormant or in a state of relative dormancy. If in a state of relative dormancy, germination is possible only over a restricted range of high temperatures. Since prevailing temperatures are below these values, germination is effectively prevented. During the winter, dormancy is released as a result of chilling (Fig. 11.8). Through time, the minimum temperature for germination extends to progressively lower temperatures. During early spring as soil temperatures begin to rise, the temperature range for germination and prevailing temperatures overlap and germination occurs, providing that the requirement for other factors, such as light, are also met.

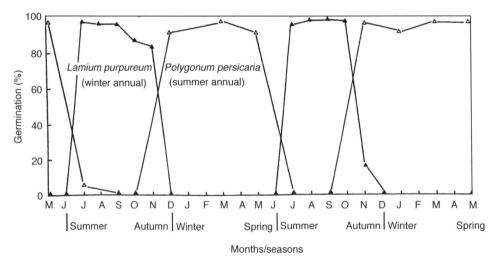

Fig. 11.8. Seasonal changes in the germination behaviour of buried seeds of the winter annual *Lamium purpureum* (▲) and summer annual *Polygonum persicaria* (△). Seeds of *L. purpureum* were buried in soil on 13 May 1980, 4 days after collection, at a depth of 7 cm in an unheated glasshouse at Lexington, Kentucky, USA. Germination of exhumed seeds was tested at 25/15°C (12 h/12 h) with a 14 h photoperiod. Seeds of *P. persicaria* were buried in sandy loam at a depth of approximately 10 cm, in December 1986, under field conditions at Wageningen, The Netherlands. Germination of exhumed seeds was tested at 20°C following a 15-min red-light irradiation. (Data for *L. purpureum* adapted from Baskin and Baskin, 1984, and *P. persicaria* from Bouwmeester, 1990.)

In strict summer annuals –for example, *Ambrosia artemisiifolia* (Baskin and Baskin, 1980) and *Polygonum persicaria* (Bouwmeester, 1990) – rising temperatures during late spring and summer result in the complete induction of secondary dormancy, indicated by a progressive increase in the minimum temperature for germination. Accordingly, germination does not generally occur during the summer and autumn (Fig. 11.9a). In some cases – for example, *Bidens polylepis* – seeds do not re-enter dormancy until late summer and consequently some germination may occur during the summer, in this case following disturbance by flooding (Baskin *ct al.*, 1995). Other species, which exhibit some of the general characteristics of summer annuals but are also capable of germination during the summer, including *Chenopodium album* (Baskin and Baskin, 1977; Bouwmeester, 1990), *Portulaca smallii* (Baskin and Baskin, 1987b) and *Spergula arvensis* (Karssen *et al.*, 1988; Bouwmeester, 1990), can be regarded as facultative summer annuals. It is interest-ing to note that, in contrast to Fig. 11.9a and b, in which T_{max} and T_{min}, respectively, are constant during the induction and release of dormancy, in *C. album* and *S. arvensis* expansion and contraction of the temperature range for germination occurs by movement of both the maximum and minimum temperature for germination (Bouwmeester, 1990). A downward shift in the minimum temperature for germination is by no means restricted to summer annuals. Using a thermal-time approach, Pritchard and co-workers (1999) recently reported a linear decline in the base temperature for germination when seeds of the temperate tree species *A. hippocastanum* were stratified in the laboratory at 6°C.

The dormancy patterns of winter annuals are exactly the reverse of those of summer annuals. When seeds are dispersed in late spring/early summer, they are either truly dormant or in a state of relative dormancy. If in a state of relative dormancy, germination is possible only over a restricted range of low temperatures. Since prevailing temperatures at this time of the

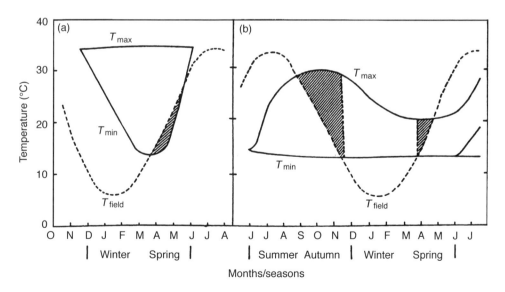

Fig. 11.9. Relationship between the field temperature and the changes in the range of temperatures over which germination can proceed. Solid line represent the maximum (T_{max}) and minimum (T_{min}) temperature at which germination is possible. The broken line indicates the mean daily maximum temperature in the field. In the hatched area the actual and the required temperature overlap. (a) Strict summer annual (data obtained from a study with *Ambrosia artemisiifolia*, Baskin and Baskin, 1980); (b) facultative winter annual (data obtained from a study with *Lamium amplexicaule*, Baskin and Baskin, 1981). (Adapted and redrawn from Bouwmeester, 1990, after Karssen, 1982.)

year are above these values, germination is prevented. During the summer, exposure to high temperatures results in dormancy release via the processes of dry after-ripening and warming (see previous sections; Fig. 11.8). Through time, the maximum temperature for germination extends to progressively warmer temperatures. During the autumn, as soil temperatures begin to fall, the temperature range for germination and prevailing temperatures overlap and germination occurs. In strict winter annuals – for example *Arabidopsis thaliana* (Baskin and Baskin, 1983) and *Lamium purpureum* (Baskin and Baskin, 1984) – falling temperatures during the late autumn and winter result in the complete induction of secondary dormancy, indicated by a progressive decline in the maximum temperature for germination. Accordingly, germination is precluded in the winter and spring. As with summer annuals, there are a number of species – for example, *Lamium amplexicaule* (Baskin and Baskin, 1981) and *Aphanes arvensis* (Roberts and Neilson, 1982) – that do not enter complete secondary dormancy (Fig. 11.9b). In these species, termed facultative winter annuals, seeds in a state of relative dormancy in the spring are capable of germination over a narrow range of temperatures. As soil temperatures rise, there is a brief overlap and germination occurs. Other species, which have been classed as facultative winter annuals and which share some of the characteristics of summer annuals, include *Capsella bursa-pastoris* (Baskin and Baskin, 1989) and *Sisymbrium officinale* (Bouwmeester, 1990).

In a comprehensive study of comparative germination ecophysiology in over 300 different species, including winter and summer annuals and monocarpic and polycarpic perennials, Baskin and Baskin (1988) showed that dormancy patterns were extremely variable among perennials. Nevertheless, it was evident that temperature played a pivotal role in the induction and release of dormancy in many cases and, in those species in which seasonal patterns could be identified, the temperature effects were identical to those operat-

ing in annual species and depended on whether emergence was programmed to occur in autumn or spring. However, some perennial species that develop persistent seed banks – for example, *Rumex crispus* (Baskin and Baskin, 1985b), *Cyperus odoratus* and *Penthorum sedoides* (Baskin et al., 1989) – do not exhibit cyclical patterns of dormancy controlled by seasonal changes in temperature. Instead, dormancy in these species appears to be simply controlled by an overriding requirement for light, coupled with a requirement for alternating temperatures.

Based on an original hypothesis of Totterdell and Roberts (1979), Bouwmeester (1990) and Bouwmeester and Karssen (1992) successfully developed a descriptive model of the seasonal changes in dormancy of four weedy species, using temperature-derived parameters only. Totterdell and Roberts had already demonstrated that the effect of chilling on dormancy release in *R. crispus* and *R. obtusifolius* depended on two subprocesses – dormancy relief and dormancy induction. For *Rumex* species, the effects of chilling on dormancy release were independent of temperature between 1.5 and 15°C. However, the actual response of seeds to preincubation within this range was dependent on the rate of dormancy induction, which was directly related to temperature over the same range (see previous section; Fig. 11.7). Based on this theory, Bouwmeester assumed that the process of dormancy relief was regulated by the period spent below a critical border temperature, which was quantified by the cold sum (C), and that the process of dormancy induction was regulated by the actual temperature over the same period quantified by the heat sum (H). Dormancy was therefore a function of both parameters:

$$D = f(C, H)$$

Using a quadratic function, which related expected germination to the cold and heat sum, germination temperature, composition of the germination medium and pretreatment temperature during a period prior to exhumation, equation parameters were selected that gave the best fit to germination data from laboratory experiments. Modelled in this way, changes in the minimum and maximum temperature for 50% germination were calculated and used to compare periods of predicted field emergence with changes in germination observed under natural conditions. For *Polygonum persicaria*, in which dormancy relief was optimal at temperatures just above freezing, there was good agreement between calculated periods of germination and actual results in the field (Fig. 11.10). However, for other species used in the study in which optimal temperatures for dormancy release were higher, the descriptive value of cold- and heat-sum parameters was weakened somewhat. Bouwmeester's results nonetheless provide confirmation of the dominant role of temperature in the regulation of seasonal patterns of dormancy. It is also evident that dormancy in seed populations is characterized by the range of temperatures over which germination can proceed. Temperature therefore has a dual effect. On the one hand, it regulates changes in dormancy and, on the other, it regulates the rate of germination when the actual temperature is within the germination temperature range.

Despite the dominant role of temperature in the control of seasonal patterns of dormancy in seeds, it is important to note that temperature does not act alone. Several studies have clearly demonstrated that other factors – for example, light, nitrate and desiccation – can all influence the expression of dormancy in seeds (Baskin and Baskin, 1980; Karssen, 1982; Karssen *et al.*, 1988; Bouwmeester, 1990). At any given time, the presence of one or more of these additional dormancy-breaking factors will result in an increase in the temperature range for germination.

The requirement for alternating temperatures

In numerous species, seed germination cannot occur or is, at best, severely reduced in constant-temperature environments. Steinbauer and Grigsby (1957) found that, of 85 species selected from 15 families, more than 80% showed higher germination at alternating temperatures compared with constant temperatures. In many cases, the sensitivity of seed populations to alternating temperatures may be

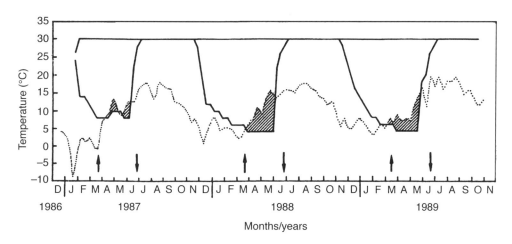

Fig. 11.10. Simulation of seasonal changes in the range of temperatures over which at least 50% of exhumed *Polygonum persicaria* seeds germinate. Solid lines represent maximum and minimum temperature required for 50% germination in water, calculated according to a descriptive model based on cold- and heat-sum parameters (see text). The dotted line indicates air temperature at 1.50 m recorded at a meteorological station in Wageningen, The Netherlands. Hatched areas indicate overlap of field temperature and germination temperature range. Arrows indicate the time when germination in Petri dishes outdoors actually increased above (↑) or decreased below 50% (↓). (Adapted from Bouwmeester, 1990.)

influenced by other environmental factors, particularly light (Toole and Koch, 1977; Roberts and Benjamin, 1979; Totterdell and Roberts, 1980; Probert *et al.*, 1986). Interaction between the active form of phytochrome (P_{fr}) and a requirement for alternating temperatures has been recognized since the early years of phytochrome study (Toole *et al.*, 1955). Interdependence between P_{fr} and the effects of brief high-temperature shifts have been shown in *Rumex* species (Taylorson and Hendricks, 1972; Takaki *et al.*, 1981, 1985; Hand *et al.*, 1982), and studies in our laboratory have demonstrated that the response to diurnal temperature cycles in seeds of *Dactylis glomerata* (Probert and Smith, 1986) and *Ranunculus sceleratus* (Probert *et al.*, 1987) is dependent on the presence of P_{fr}. Probert and Smith (1986) suggested that the capacity for dark germination at alternating temperatures may be explained by the sensitivity of some individuals to low levels of pre-existing P_{fr}.

Arising from detailed studies of dormancy in *Rumex* species, Roberts and Totterdell (1981) identified nine attributes of diurnal temperature cycles, each of which might be responsible for the stimulation of germination. It was also recognized that none of these characteristics could be altered experimentally without confounding it with a change in at least one other factor. Subsequently, Roberts and co-workers extended this list to 13 and classified the various attributes into primary, secondary and tertiary characteristics (Roberts, 1988; Murdoch *et al.*, 1989). Based on experiments on two contrasting species (*Chenopodium album* and *Panicum maximum*) and using a wide range of diurnal temperature cycles, Murdoch *et al.* (1989) developed a descriptive model relating germination (expressed in probits) to amplitude (the difference between the maximum and minimum temperature), mean temperature and thermoperiod (time in hours each day above the mean temperature). This model, which explained much of the variation in the germination response to alternating temperatures in both species, confirms published

evidence of the importance of these attributes and provides a valuable basis for the development of predictive models for the behaviour of seed populations under natural conditions.

A number of published reports have emphasized the importance and ecological significance of the effect of amplitude (Thompson *et al.*, 1977; Roberts and Totterdell, 1981; Thompson and Grime, 1983; Probert *et al.*, 1986, 1987). For several different species, Thompson and Grime (1983) found a more or less linear increase in germination with increasing amplitude. However, the distribution of sensitivity, i.e. the variation between individual seeds in the amplitude required to trigger germination, varied between species. For example, in *Urtica dioica* germination increased sharply with increasing amplitude, whereas in *Stellaria media* the increase in response was much more gradual.

There is an interesting and important relationship between the effects of amplitude, thermoperiod and number of temperature cycles required. Based on their studies on *Rumex*, Roberts and Totterdell (1981) stressed that, for maximum germination at increasingly large amplitudes, it was necessary to reduce the period spent at the warm phase of diurnal cycles. Providing this adjustment was made, the number of cycles required to trigger germination was lowered with increasing amplitude. These observations help to explain the connection between the effects of a series of diurnal cycles and the effects of single, brief, high-temperature shifts, shown to be effective in stimulating germination in a number of species sensitive to alternating temperatures and light (Takaki *et al.*, 1981; Probert, 1983; Taylorson and DiNola, 1989).

Evidence from studies on *D. glomerata* (Probert and Smith, 1986) and *R. sceleratus* (Probert *et al.*, 1987) showed that, in the presence of P_{fr}, variation between individuals in the number of diurnal temperature cycles required to trigger germination is normally distributed. In both species, a direct linear relation between germination percentage (converted to probits) and the logarithm of the number of diurnal cycles was demonstrated. Quantal analysis of dose–response experiments, in which the effects of various dormancy-modifying factors on the sensitivity of seed populations to diurnal temperature cycles may be analysed, is a useful approach in the understanding of the interactions between different environmental factors and the requirement for alternating temperatures.

Apart from the overriding influence of light (through phytochrome) on the response of seed populations to alternating temperatures, other factors, notably nitrate (Probert *et al.*, 1987), chilling (Probert *et al.*, 1989) and dry storage (Thompson and Grime, 1983; Probert *et al.*, 1986), tend to increase sensitivity. The proportion of individuals capable of germination at a given alternating temperature regime tends to increase in response to these additional factors. The influence of such factors may have important ecological consequences for the timing of emergence in some species. For example, the increased tendency toward germination in the dark (Probert *et al.*, 1987) and at reduced amplitudes (Benech Arnold *et al.*, 1990) as a result of exposure to low winter temperatures may help to synchronize germination in spring-germinating species.

Soil temperature measurements indicate that temperature fluctuations are greatest on or close to the surface of bare soil and in breaks in the vegetation canopy (Thompson *et al.*, 1977; Vázquez-Yanes and Orozco-Segovia, 1982; Probert, 1983; Van Assche and Vanlerberghe, 1989). It is therefore not surprising that the requirement for alternating temperatures and light represents an adaptation of small-seeded species that ensures that germination occurs close to the soil surface in vegetation gaps. Sensitivity to amplitude effectively prevents germination when seeds are buried below a few centimetres. Comparative studies have demonstrated that the requirement for alternating temperatures is extremely common in wetland species and in species of disturbed habitats (Thompson, 1974; Thompson *et al.*, 1977; Thompson and Grime, 1983). In temperate wetland ruderal annuals, such as *Rorippa islandica* and *R. sceleratus*, which typically colonize the bare mud at the edges of ponds and streams, the requirement for alternating temperatures serves to prevent germination when the soil is submerged in water. Conditions conducive to germination occur in the spring, when the falling water-table exposes the soil and increased solar radiation results in wide temperature fluctuations close to the soil surface.

Temperature and germination rate

As already discussed, natural temperature changes play a dominant role in controlling the temperature range over which germination may occur in seed populations of species that exhibit seasonal dormancy patterns. In addition to this effect, prevailing temperatures within the range over which germination can occur determine both the number of individuals capable of responding and, for each individual, the time taken to complete the germination process providing other conditions, such as water availability and light, are suitable.

For any non-dormant seed population, germination is possible only within well-defined temperature limits. The existence of three so-called cardinal temperatures, namely the maximum and minimum temperature beyond which germination is prevented, and the optimum temperature, which allows maximum germination in the shortest time, was first recognized by Sachs over a hundred years ago. Since then, there have been countless reports describing germination responses to temperature in many different species and ecotypes (Bewley and Black, 1994) and, during the last couple of decades, considerable progress has been made toward the quantification of germination responses to temperature and the development of predictive models.

Although earlier studies had demonstrated the linear relation between rate of seed germination (usually defined as the reciprocal of time taken to 50% germination) and temperature, Garcia-Huidobro *et al.* (1982) were the first to develop a model that incorporated the fact that different individuals in a seed population germinate at different rates. For the germination of different percentiles of a seed population of *Pennisetum typhoides* at suboptimal temperatures, Garcia-Huidobro *et al.* developed the equation:

$$1/t(G) = (T - T_b(G))/\theta(G) \qquad (11.1)$$

where T = the temperature; T_b = the base temperature at which the germination rate $(1/t)$ is zero; and θ = the thermal time in degree-days above T_b required to accumulate the given percentage germination (G).

Subsequently, Covell *et al.* (1986) showed that, within single seed populations of chickpea and soybean, base temperature did not vary and that the variation in thermal time was normal or log normal. Accordingly, the following equation was developed to describe the response to germination rate in seed populations at suboptimal constant temperatures:

$$1/t(G) = (T - T_b)/((\text{probit}(G) - K)\sigma) \qquad (11.2)$$

where σ = the standard deviation of the frequency distribution of thermal times and K = an intercept constant. It is noteworthy that, apart from the probit model, other distribution functions have been used successfully to describe variation in thermal-time characteristics in seed populations (Washitani, 1985; Benech Arnold *et al.*, 1990).

Equations similar to 11.1 and 11.2 have also been developed to describe germination at supra-optimal temperatures, where the rate of germination typically decreases linearly with increase in temperature and, for some species at least, thermal time is a constant, and seed-to-seed variation in germination rate is accounted for by a normal distribution in the ceiling (maximum) temperature (Roberts, 1988).

Returning to effects at suboptimal temperatures, where most attention has been focused; published reports have shown, not surprisingly, that base temperature varies between different species. For example, base temperature varied between 0 and 8.5°C for different grain legumes

(Covell *et al.*, 1986). Studies of intraspecific variation have shown no variation in base temperature in the case of chickpea (Ellis *et al.*, 1986) but significant variation between *Sorghum bicolor* genotypes (Harris *et al.*, 1987). At the population level, several studies suggest little or no variation in base temperature between individual seeds of non-dormant seed populations (Covell *et al.*, 1986; Ellis *et al.*, 1986; Benech Arnold *et al.*, 1990; Pritchard and Manger, 1990). Studies using onion (Ellis and Butcher, 1988) and barley (Ellis *et al.*, 1987) also suggest that, in the absence of dormancy, base temperature is not confounded with seed quality. Provided that base temperature does not vary within populations, Equation 11.2 may be used to screen populations or species for rate of germination at suboptimal temperatures. This approach would seem to be particularly useful in the selection of genotypes tolerant to cold seedbed temperatures (Covell *et al.*, 1986).

Temperature responses and geographical distribution

Numerous studies have been reported in which germination responses in partially dormant and non-dormant seed populations have been related to the geographical and ecological distribution of species and ecotypes. Since many of these earlier contributions were reviewed in some detail by Thompson (1981), the present discussion will be restricted to a few selected examples and more recent findings.

Intraspecific differences in germination requirements have been linked to the geographical distribution of a number of species – for example, *Tsuga canadensis* (Stearns and Olson, 1958), *Silene vulgaris* (Thompson, 1973b), *S. dioica* (Thompson, 1975), *Hyacinthoides non-scripta* (Thompson and Cox, 1978) and *D. glomerata* (Probert *et al.*, 1985a). When the germination response to alternating temperatures and light was compared in populations of *D. glomerata* collected from different parts of its European distribution, a distinct north–south trend was evident. While Mediterranean populations were relatively non-dormant, populations from northern Europe were characterized by a strong requirement for alternating temperatures and light in a high proportion of individuals. In contrast, when populations representing a range of ecological sites within the British Isles were compared, despite quantitative differences, the relative effects of light and alternating temperatures were remarkably consistent (Probert *et al.*, 1985a). Thompson's data on *Silene vulgaris* illustrate a similar trend. When populations were compared from a wide geographical range, marked differences in the temperature range for germination were noted, but, when populations all originating from one geographical zone (in this case, Scandinavia) were compared, differences were negligible (Thompson, 1973b). However, in a study of 19 populations of the arable weed *Agrostemma githago* from throughout its European distribution, Thompson (1973c) found only minor differences in the temperature ranges for germination, which were all characteristic of Mediterranean species of *Caryophyllaceae*. Thompson concluded that this species originated in the Mediterranean basin and, despite its spread throughout Europe as an arable weed, its germination character had remained more or less stable since its association with Neolithic people.

Intraspecific differences in germination response to temperature have also been related to altitudinal distribution. For example, Linington *et al.* (1979) reported that populations of the perennial forage grass *Festuca pratensis* var. *apennina* from lower altitudes had lower optima than populations from higher altitudes.

Not surprisingly, a number of studies have shown that interspecific differences in germination responses to temperature may be related to ecological and geographical factors. In some instances, contrasting temperature requirements for germination associated with different life-cycle strategies have been identified in different species adapted to the same ecological habitat. This has been recorded in certain arid and semi-arid regions where there is a sharp delineation of growing seasons due to seasonal rainfall (Went, 1957; Mott, 1972). In the *Acacia* shrubland of the Murchison region of Western Australia, there are two major wet periods of the year, in the summer and winter, which are also characterized by large differences in mean monthly temperature. When soil from this region was incubated in the laboratory, Mott (1972) found that low temperatures typical of winter stimulated germination of mainly forbes, whereas high temperatures stimulated predominantly grasses. Examination of the temperature range for germination of non-dormant populations of the dominant species of the winter and summer flora revealed that the summer grass *Aristida contorta* had a much higher optimum temperature for germination compared with the two winter *Compositae*, *Helipterum craspedioides* and *Helichrysum cassinianum* (Fig. 11.11).

Intraspecific variation in germination responses to temperature in European species of *Caryophyllaceae* has been studied in detail by Thompson (1970b, c, 1973a, c). In these studies, mainly non-dormant seed populations were tested over a wide range of constant temperatures and then compared by constructing germination 'character curves', in which the maximum and minimum temperature for 50% germination recorded on successive days was plotted against time. Although marked differences in the temperature range for germination were recorded and some generalizations were possible (for example, temperature minima and maxima for germination were consistently lower in Mediterranean species compared with species with more northerly distributions), definite trends were difficult to identify.

Thompson's approach has undoubtedly yielded a great deal of useful information on the relationship between germination responses to temperature and the ecological and geographical distribution of certain species and ecotypes. Unfortunately, one of the problems of a broad approach using seed populations previously stored for variable periods of

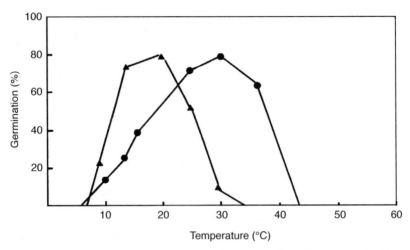

Fig. 11.11. A comparison of the temperature ranges for germination of non-dormant seeds of the summer grass *Aristida contorta* (●), and winter composite *Helichrysum cassinianum* (▲) from the Murchison region of Western Australia (adapted from Mott and Groves, 1981, after Mott, 1972).

time is that, unless experiments are conducted at intervals during storage, it may be difficult to assess whether the measured temperature response is stable and hence whether valid comparisons between populations can be made. An additional drawback is that, if single experiments are conducted on after-ripened seed, potentially important differences that may have been present at the time of collection and different trends of change during storage may be missed.

Even when experiments are conducted on freshly harvested seeds, extreme caution is needed in the interpretation of similarities or differences in germination response between seed populations. For example, in a recent study of four annual weed species, Andersson and Milberg (1998) reported a random pattern of variation in dormancy between populations, years and mother plants in laboratory tests on freshly harvested seeds. Reflecting the divergent responses of winter and summer annuals to chilling, the same authors were also able to demonstrate that, in some cases, meaningful differences in germination behaviour may not appear until seeds have been chilled. In others, differences present in freshly harvested seeds may disappear following chilling (Milberg and Andersson, 1998).

Using an experimental approach specifically designed to detect meaningful differences in seed germination, it is interesting to note that Baskin and co-workers (Baskin and Baskin, 1988; Baskin et al., 1991; Buchele et al., 1991; Walck et al., 1997) were unable to demonstrate unique germination characteristics in several endemic species compared with their widely distributed relatives. Clearly, whatever the reasons for their endemism, it was apparently not the result of particular germination characteristics.

Conclusions

Two different mechanisms of temperature-dependent dormancy release occur in dry seeds. These are mainly apparent in species adapted to survive seasonal drought. First, temperature-dependent dry after-ripening results in dormancy release during artificial dry storage or under natural conditions during the dry season. Secondly, temperature controls the rate of hard-seed breakdown in seeds that possess non-dormant embryos but are enclosed in a hard water-impermeable seed-coat. Not surprisingly, the critical temperatures required and the relationship between the rate of dormancy release and temperature differ in each case. In dry after-ripening, there is an exponential increase in the rate of dormancy release with increasing temperature, with optima in the region of 40–50°C. Interestingly, these temperatures are around the minimum for hard-seed breakdown and, above these values, the rate of dormancy removal increases more or less linearly with increasing temperature.

Quite separately from the effects of temperature on the rate of dormancy release in dry seeds, temperature also determines the rate of germination itself in fully hydrated non-dormant seeds. The well-described positive and negative linear relations between the rate of germination and temperature over the sub- and supra-optimal ranges, respectively, and the more recently discovered normal or log-normal distribution of thermal-time requirements between individual seeds within populations have facilitated the development of a number of useful predictive models. Recently, considerable progress has been made in the application of such population-based models to the behaviour of dormant seed populations. Driven partly by the economic need to predict the emergence of weed seeds, these approaches have improved our understanding of the behaviour of natural seed populations.

Seasonal temperature changes regulate the dormancy status of natural seed populations in countless species. For temperate annuals in particular, seeds are programmed to germinate at a particular time of the year. This is achieved by cyclical changes in dormancy status, regulated by seasonal changes in temperature. Induction or release of dormancy can be measured by the contraction or expansion of the temperature limits for seed germination, coupled with changes in sensitivity to germination-regulating factors, such as light. Our understanding of such behaviour has been advanced by extensive studies undertaken during the last 30 years in the USA and Europe, in which buried seed populations are carefully monitored.

Too frequently in the past, bold statements on the ecological significance of germination responses at the species level were made when laboratory experiments had been conducted on a single seed lot after storage for an unspecified period. Now we are much more aware of the extreme plasticity of germination responses and the need to use a wide variety of test conditions in the laboratory. The importance of 'ground-truthing' laboratory data with field observations is also now widely recognized, and useful practical guidelines for the setting up of meaningful experiments have been published (Cohn, 1996; Baskin and Baskin, 1998).

Despite this greater awareness, it would appear that many seed ecologists still neglect the likely importance of seed moisture status in the expression of seed dormancy. Desiccation of previously imbibed seeds to low moisture levels can reduce dormancy and as a result alter germination responses to environmental factors, including temperatures (Karssen, 1982; Karssen *et al.*, 1988; Bouwmeester, 1990). Moreover, even partial desiccation of freshly harvested seeds can result in an increase in sensitivity to dormancy-regulating factors (Kermode *et al.*, 1989; Probert and Brierley, 1989; Aldridge, 1991). These findings may have important practical implications for studies directed toward an understanding of ecological aspects of the environmental control of germination. In seed biology research, it is common practice for seeds to be dried to a low moisture content following harvest, in order to retain seed viability. Unlike seeds of crop species, which have been selected for shatter resistance, it is becoming increasingly clear that seeds of wild plants are frequently dispersed when the moisture content is still high (Hay, 1997). Indeed, in many cases, it is questionable whether the moisture content under natural conditions ever drops to the low levels used in artificial storage. Further work is clearly needed to determine the extent of changes in dormancy and germination behaviour associated with artificial drying treatments.

References

Aldridge, C.D. (1991) Physiological studies on desiccation intolerance in propagules of aquatic grasses. Unpublished PhD thesis, University of Reading, UK.

Allen, P.S. and Meyer, S.E. (1998) Ecological aspects of seed dormancy loss. *Seed Science Research* 8, 183–191.

Andersson, L. and Milberg, P. (1998) Variation in seed dormancy among mother plants, populations and years of seed collection. *Seed Science Research* 8, 29–38.

Angevine, M.W. and Chabot, B.F. (1979) Seed germination syndromes in higher plants. In: Solbrig, O.T., Jain, S., Johnson, G.B. and Raven, P.H. (eds) *Topics in Plant Population Biology*. Columbia University Press, New York, pp. 188–206.

Ballard, L.A.T. (1973) Physical barriers to germination. *Seed Science and Technology* 1, 285–303.

Baskin, C.C. and Baskin, J.M. (1988) Germination ecophysiology of herbaceous plant species in a temperate region. *American Journal of Botany* 75, 286–305.

Baskin, C.C. and Baskin, J.M. (1998) *Seeds: Ecology, Biogeography, and Evolution of Dormancy and Germination*. Academic Press, New York.

Baskin, C.C., Baskin, J.M. and Chester, E.W. (1995) Role of temperature in the germination ecology of the summer annual *Bidens polylepis* Blake (Asteraceae). *Bulletin of the Torrey Botanical Club* 122, 275–281.

Baskin, J.M. and Baskin, C.C. (1971) Germination ecology of *Phacelia dubia* var. *dubia* in Tennessee glade. *American Journal of Botany* 58, 98–104.

Baskin, J.M. and Baskin, C.C. (1972) Physiological ecology of germination of *Viola rafinesquii*. *American Journal of Botany* 59, 981–988.

Baskin, J.M. and Baskin, C.C. (1977) Role of temperature in the germination ecology of three summer annual weeds. *Oecologia* 30, 377–382.

Baskin, J.M. and Baskin, C.C. (1978) Temperature requirements for afterripening of seeds of a winter annual induced into secondary dormancy by low winter temperatures. *Bulletin of the Torrey Botanical Club* 105, 104–107.

Baskin, J.M. and Baskin, C.C. (1979) Effect of relative humidity on afterripening and viability in seeds of the winter annual *Draba verna*. *Botanical Gazette* 140, 284–287.

Baskin, J.M. and Baskin, C.C. (1980) Ecophysiology of secondary dormancy in seeds of *Ambrosia artemisiifolia*. *Ecology* 61, 475–480.

Baskin, J.M. and Baskin, C.C. (1981) Seasonal changes in the germination responses of buried *Lamium amplexicaule* seeds. *Weed Research* 21, 299–306.

Baskin, J.M. and Baskin, C.C. (1983) Seasonal changes in the germination responses of buried seeds of *Arabidopsis thaliana* and ecological interpretation. *Botanical Gazette* 144, 540–543.

Baskin, J.M. and Baskin, C.C. (1984) Role of temperature in regulating timing of germination in soil seed reserves of *Lamium purpureum* L. *Weed Research* 24, 341–349.

Baskin, J.M. and Baskin, C.C. (1985a) The annual dormancy cycle in buried weed seeds: a continuum. *Bioscience* 35, 492–498.

Baskin, J.M. and Baskin, C.C. (1985b) Does seed dormancy play a role in the germination ecology of *Rumex crispus*? *Weed Science* 33, 340–343.

Baskin, J.M. and Baskin, C.C. (1986) Temperature requirements for after-ripening in seeds of nine winter annuals. *Weed Research* 26, 375–380.

Baskin, J.M. and Baskin, C.C. (1987a) Temperature requirement for after-ripening in buried seeds of four summer annual weeds. *Weed Research* 27, 385–389.

Baskin, J.M. and Baskin, C.C. (1987b) Seasonal changes in germination responses of buried seeds of *Portulaca smallii*. *Bulletin of the Torrey Botanical Club* 114, 169–172.

Baskin, J.M. and Baskin, C.C. (1989) Germination response of buried seeds of *Capsella bursa-pastoris* exposed to seasonal temperature changes. *Weed Research* 29, 205–212.

Baskin, J.M., Baskin, C.C. and Spooner, D.M. (1989) Role of temperature, light and date seeds were exhumed from soil on germination of four wetland perennials. *Aquatic Botany* 35, 387–394.

Baskin, J.M., Baskin, C.C., Parr, P.D. and Cunningham, M. (1991) Seed germination ecology of the rare hemiparasite *Tomanthera auriculata* (Scrophulariaceae). *Castanea* 56, 51–58.

Bauer, M.C., Meyer, S.E. and Allen, P.S. (1998) A simulation model to predict seed dormancy loss in the field for *Bromus tectorum* L. *Journal of Experimental Botany* 49, 1235–1244.

Benech Arnold, R.L. and Sanchez, R.A. (1995) Modelling weed seed germination. In: Kigel, J. and Galili, G. (eds) *Seed Development and Germination*. Marcel Dekker, New York, pp. 545–566.

Benech Arnold, R.L., Ghersa, C.M., Sanchez, R.A. and Insausti, P. (1990) Temperature effects on dormancy release and germination rate in *Sorghum halepense* L. Pers. seeds: a quantitative analysis. *Weed Research* 30, 81–90.

Bewley, J.D. and Black, M. (1982) *Physiology and Biochemistry of Seeds in Relation to Germination*. Vol. 2. *Viability, Dormancy and Environmental Control*. Springer-Verlag, Berlin.

Bewley, J.D. and Black, M. (1994) *Seeds Physiology of Development and Germination*. Plenum Press, New York.

Bouwmeester, H.J. (1990) The effect of environmental conditions on the seasonal dormancy pattern and germination of weed seeds. Unpublished PhD thesis, Agricultural University, Wageningen, The Netherlands.

Bouwmeester, H.J. and Karssen, C.M. (1992) The dual role of temperature in the regulation of the seasonal changes in dormancy and germination of seeds of *Polygonum persicaria* L. *Oecologia* 90, 88–94.

Bradford, K.J. (1990) A water relations analysis of seed germination rates. *Plant Physiology* 94, 840–849.

Bradford, K.J. (1995) Water relations in seed germination. In: Kigel, J. and Galili, G. (eds) *Seed Development and Germination*. Marcel Dekker, New York, pp. 351–396.

Bradford, K.J. (1996) Population-based models describing seed dormancy behaviour: implications for experimental design and interpretation. In: Lang, G.A. (ed.) *Plant Dormancy: Physiology, Biochemistry and Molecular Biology*. CAB International, Wallingford, UK, pp. 313–339.

Bradford, K.J. (1997) The hydro-time concept in seed germination and dormancy. In Ellis, R.H., Black, M., Murdoch, A.J. and Hong, T.D. (eds) *Basic and Applied Aspects of Seed Biology. Proceedings of the Fifth International Workshop on Seeds, Reading, 1995*. Kluwer Academic Publishers, Dordrecht, pp. 349–360.

Brinkmann, W.L.F. and Vieira, A.N. (1971) The effect of burning on germination of seeds at different soil depth of various tropical tree species. *Turrialba* 21, 77–82.

Buchele, D.E., Baskin, J.M. and Baskin, C.C. (1991) Ecology of the endangered species *Solidago shortii*. III. Seed germination ecology. *Bulletin of the Torrey Botanical Club* 118, 288–291.

Christensen, M., Meyer, S.E. and Allen, P.S. (1996) A hydrothermal time model of seed after-ripening in *Bromus tectorum* L. *Seed Science Research* 6, 155–163.

Cohn, M.A. (1996) Operational and philosophical decisions in seed dormancy research. *Seed Science Research* 6, 147–153.

Covell, S., Ellis, R.H., Roberts, E.H. and Summerfield, R.J. (1986) The influence of temperature on seed germination rate in grain legumes. 1. A comparison of chickpea, lentil, soyabean and cowpea at constant temperatures. *Journal of Experimental Botany* 37, 705–715.

Derkx, M.P.M. and Karssen, C.M. (1993) Variability in light, gibberellin and nitrate requirement of *Arabidopsis thaliana* seeds due to harvest time and conditions of dry storage. *Journal of Plant Physiology* 141, 574–582.

Derkx, M.P.M., Vermeer, E. and Karssen, C.M. (1994) Gibberellins in seeds of *Arabidopsis thaliana*: biological activities, identification and effects of light and chilling on endogenous levels. *Plant Growth Regulation* 15, 223–234.

Egley, G.H. (1989) Water-impermeable seed coverings as barriers to germination. In: Taylorson, R.B. (ed.) *Recent Advances in the Development and Germination of Seeds*. Plenum Press, New York, pp. 207–223.

Egley, G.H. and Duke, S.O. (1985) Physiology of weed seed germination and dormancy. In: Duke, S.O. (ed.) *Weed Physiology*, Vol. I. *Reproduction and Ecophysiology*. CRC Press, Boca Raton, Florida, pp. 27–64.

Ellis, R.H. and Butcher, P.D. (1988) The effects of priming and 'natural' differences in quality amongst onion seed lots on the response of the rate of germination to temperature and the identification of characteristics under genotypic control. *Journal of Experimental Botany* 39, 935–950.

Ellis, R.H., Hong, T.D. and Roberts, E.H. (1983) Procedure for the safe removal of dormancy from rice seed. *Seed Science and Technology* 11, 77–112.

Ellis, R.H., Covell, S., Roberts, E.H. and Summerfield, R.J. (1986) The influence of temperature on seed germination in grain legumes. II. Intraspecific variation in chickpea (*Cicer arietinum* L.) at constant temperatures. *Journal of Experimental Botany* 37, 1503–1515.

Ellis, R.H., Hong, T.D. and Roberts, E.H. (1987) Comparison of cumulative germination and rate of germination of dormant and aged barley seed lots at different constant temperatures. *Seed Science and Technology* 15, 717–727.

Evenari, M. (1984) Seed physiology: its history from antiquity to the beginning of the 20th century. *Botanical Review* 50, 119–142.

Evlyn, J. (1664) *A Discourse of Forest Trees and the Propagation of Timber*. J. Martyn and J. Allestry, Printer to the Royal Society, London.

Fenner, M. (1985) *Seed Ecology*. Chapman and Hall, London.

Fenner, M. (ed.) (1992) *Seeds: The Ecology of Regeneration in Plant Communities*. CAB International, Wallingford, UK.

Floyd, A.G. (1966) Effect of fire upon weed seeds in wet sclerophyll forests of northern New South Wales. *Australian Journal of Botany* 14, 243–256.

Foley, M.E. (1994) Temperature and water status of seed affect after-ripening in wild oat (*Avena fatua*). *Weed Science* 42, 200–204.

Garcia-Huidobro, J., Monteith, J.L. and Squire, G.R. (1982) Time, temperature and germination of pearl millet (*Pennisetum typhoides* S. and H.). 1. Constant temperature. *Journal of Experimental Botany* 33, 288–296.

Gill, A.M. (1981) Coping with fire. In: Pate, J.S. and McComb, A.J. (eds) *The Biology of Australian Plants*. University of Western Australia Press, Nedlands, pp. 65–87.

Gosling, P.G. and Rigg, P. (1990) The effect of moisture content and prechill duration on the efficiency of dormancy breakage in Sitka spruce (*Picea sitchenisis*) seed. *Seed Science and Technology* 18, 337–343.

Grime, J.P. (1979) *Plant Strategies and Vegetation Processes*. Wiley, Chichester.

Gummerson, R.J. (1986) The effect of constant temperatures and osmotic potentials on the germination of sugar beet. *Journal of Experimental Botany* 37, 729–741.

Gutterman, Y. (1990) Do germination mechanisms differ in plants originating in deserts receiving winter or summer rain? *Israel Journal of Botany* 39, 355–372.

Gutterman, Y. (1993) *Seed Germination in Desert Plants*. Springer-Verlag, Berlin.

Hand, D.J., Craig, G., Takaki, M. and Kendrick, R.E. (1982) Interaction of light and temperature on seed germination of *Rumex obtusifolius* L. *Planta* 156, 457–460.

Harper, J.L. (1977) *Population Biology of Plants*. Academic Press, London.

Harris, D., Hamdi, Q.A. and Terry, A.C. (1987) Germination and emergence of *Sorghum bicolor*: genotypic and environmentally induced variation in the response to temperature and depth of sowing. *Plant, Cell and Environment* 10, 501–508.

Hay, F.R. (1997) The development of seed longevity in wild plant species. Unpublished PhD thesis, Kings College, University of London.

Hegarty, T.W. (1973) Temperature relations of germination in the field. In: Heydecker, W. (ed.) *Seed Ecology*. Butterworths, London, pp. 411–432.

Heydecker, W. (ed.) (1973) *Seed Ecology*. Butterworths, London.

Hilhorst, H.W.M. (1993) New aspects of seed dormancy. In: Côme, D. and Corbineau, F. (eds) *Fourth International Workshop on Seeds. Basic and Applied Aspects of Seed Biology*. ASFIS, Paris, pp. 571–579.

Hilhorst, H.W.M. (1998) The regulation of secondary dormancy: the membrane hypothesis revisited. *Seed Science Research* 8, 77–90.

Hilhorst, H.W.M. and Karssen, C.M. (1992) Seed dormancy and germination: the role of abscisic acid and gibberellins and the importance of hormone mutants. *Plant Growth Regulation* 11, 225–238.

Karssen, C.M. (1982) Seasonal patterns of dormancy in weed seeds. In: Khan, A.A. (ed.) *The Physiology and Biochemistry of Seed Development, Dormancy and Germination*. Elsevier Biomedical Press, Amsterdam, pp. 243–270.

Karssen, C.M., Groot, S.P.C. and Koornneef, M. (1987) Hormone mutants and seed dormancy in *Arabidopsis* and Tomato. In: Thomas, H. and Grierson, D. (eds) *Development Mutants in Higher Plants*. SEB Seminar Series 32, Cambridge University Press, Cambridge, pp. 119–133.

Karssen, C.M., Derkx, M.P.M. and Post, B.J. (1988) Study of seasonal variation in dormancy of *Spergula arvensis* L. seeds in a condensed annual temperature cycle. *Weed Research* 28, 449–457.

Karssen, C.M., Zagorski, S., Kepczynski, J. and Groot, S.P.C. (1989) Key role for endogenous gibberellins in the control of seed germination. *Annals of Botany* 63, 71–80.

Keeley, F.E. (1987) Role of fire in seed germination of woody taxa in California chaparral. *Ecology* 68, 434–443.

Kelly, K.M., Van Staden, J. and Bell, W.E. (1992) Seed coat structure and dormancy. *Plant Growth Regulation* 11, 201–209.

Kermode, A.R., Dumbroff, E.B. and Bewley, J.D. (1989) The role of maturation drying in the transition from seed development to germination VII. Effects of partial and complete desiccation on abscisic acid levels and sensitivity in *Ricinus communis* L. seeds. *Journal of Experimental Biology* 40, 303–313.

Kigel, J. and Galili, G. (1995) *Seed Development and Germination*. Marcel Dekker, New York.

Koller, D. (1972) Environmental control of seed germination. In: Kozlowski, T.T. (ed.) *Seed Biology*, Vol. II. Academic Press, London, pp. 1–101.

Leopold, A.C. and Vertucci, C.W. (1989) Moisture as a regulator of physiological reaction in seeds. In: Stanwood, P.C. and McDonald, M.B. (eds) *Seed Moisture*. CSSA Special Publication No. 14, Crop Science Society of America, Madison, Wisconsin, USA, pp. 51–67.

Leopold, A.C., Glenister, R. and Cohn, M.A. (1988) Relationship between water content and after-ripening in red rice. *Physiologia Plantarum* 74, 659–662.

Lewak, S. and Rudnicki, R.M. (1977) Afterripening in cold-requiring seeds. In: Khan, A.A. (ed.) *The Physiology and Biochemistry of Seed Dormancy and Germination*. North Holland Publishing, Amsterdam, pp. 193–217.

Linington, S., Bean, E.W. and Tyler, B.F. (1979) The effects of temperature upon seed germination in *Festuca pratensis* var. *apennina*. *Journal of Applied Ecology* 16, 933–938.

McKeon, G.M. and Mott, J.J. (1982) The effect of temperature on the field softening of hard seed of *Stylosanthes humilis* and *S. hamata* in a dry monsoonal climate. *Australian Journal of Agricultural Research* 33, 75–85.

Mayer, A.M. and Poljakoff-Mayber, A. (1989) *The Germination of Seeds*, 4th edn. Pergamon Press, Oxford.

Meyer, S.E. and Kitchen, S.G. (1992) Cyclic seed dormancy in the short-lived perennial *Penstemon palmeri*. *Journal of Ecology* 80, 115–122.

Meyer, S.E. and Kitchen, S.G. (1994) Habitat-correlated variation in seed germination response to chilling in *Penstemon* section *glabri* (Scrophulariaceae). *American Midland Naturalist* 132, 349–365.

Meyer, S.E. and Monsen, S.B. (1992) Big sagebrush germination patterns: subspecies and population differences. *Journal of Range Management* 45, 87–93.

Midgley, A.R. (1926) Effect of alternate freezing and thawing on the impermeability of alfalfa and dodder seeds. *Journal of American Society of Agronomy* 18, 1087–1098.

Milberg, P. and Andersson, L. (1998) Does cold stratification level out differences in seed germinability between populations? *Plant Ecology* 134, 225–234.

Morley, F.H.W. (1958) The inheritance and ecological significance of seed dormancy in subterranean clover (*Trifolium subterraneum* L.). *Australian Journal of Biological Sciences* 11, 261–274.

Mott, J.J. (1972) Germination studies on some annual species from an arid region of Western Australia. *Journal of Ecology* 60, 293–304.

Mott, J.J. (1979) High temperature contact treatment of hard seed in *Stylosanthes*. *Australian Journal of Agricultural Research* 30, 847–854.

Mott, J.J. and Groves, R.H. (1981) Germination strategies. In: Pate, J.S. and McComb, A.J. (eds) *The Biology of Australian Plants*. University of Western Australia Press, Nedlands, Western Australia, pp. 307–341.

Mott, J.J., McKeon, G.M., Gardener, C.J. and Mannetje, L. (1981) Geographic variation in the reduction of hard seed content of *Stylosanthes* seeds in the tropics and subtropics of Northern Australia. *Australian Journal of Agricultural Research* 32, 861–869.

Munoz, M.R. and Fuentes, E.R. (1989) Does fire induce shrub germination in the Chilean matorral? *Oikos* 56, 177–181.

Murdoch, A.J., Roberts, E.H. and Goedert, C.O. (1989) A model for germination responses to alternating temperatures. *Annals of Botany* 63, 97–111.

Nikolaeva, M.G. (1977) Factors controlling the seed dormancy pattern. In: Khan, A.A. (ed.) *The Physiology and Biochemistry of Seed Dormancy and Germination*. North Holland Publishing, Amsterdam, pp. 51–74.

Okagami, N. and Kawai, M. (1982) Dormancy in *Dioscorea*: differences in temperature responses in seed germination among six Japanese species. *Botanical Magazine* 95, 155–166.

Pemadasa, M.A. and Lovell, P.H. (1975) Factors controlling germination of some dune annuals. *Journal of Ecology* 63, 41–59.

Pieterse, P.J. and Cairns, A.L.P. (1986) The effect of fire on an *Acacia longifolia* seed bank in the south western Cape. *South African Journal of Botany* 52, 233–236.

Pritchard, H.W. and Manger, K.R. (1990) Quantal response of fruit and seed germination rate in *Quercus robur* L. and *Castanea sativa* Mill. to constant temperatures and photon dose. *Journal of Experimental Botany* 41, 1549–1557.

Pritchard, H.W., Manger, K.R. and Prendergast, F.G. (1988) Changes in *Trifolium arvense* seed quality following alternating temperature treatment using liquid nitrogen. *Annals of Botany* 62, 1–11.

Pritchard, H.W., Tompsett, P.B. and Manger, K.R. (1996) Development of a thermal time model for the quantification of dormancy loss in *Aesculus hippocastanum* seeds. *Seed Science Research*. 6, 127–135.

Pritchard, H.W., Steadman, K.J., Nash, J.V. and Jones, C. (1999) Kinetics of dormancy release and the high temperature germination response in *Aesculus hippocastanum* seeds. *Journal of Experimental Botany* 50, 1507–1514.

Probert, R.J. (1983) Germination studies in European populations of *Dactylis glomerata* L. Unpublished PhD thesis, Council for National Academic Awards.

Probert, R.J. (1992) The role of temperature in germination ecophysiology. In: Fenner, M. (ed.) *Seeds: The Ecology of Regeneration in Plant Communities*. CAB International, Wallingford, UK, pp. 285–325.

Probert, R.J. and Brierley, E.R. (1989) Desiccation intolerance in seeds of *Zizania palustris* is not related to developmental age or the duration of post-harvest storage. *Annals of Botany* 64, 669–674.

Probert, R.J. and Longley, P.L. (1989) Recalcitrant seed storage physiology in three aquatic grasses. (*Zizania palustria, Spartina anglica* and *Porteresia coarctata*). *Annals of Botany* 63, 53–63.

Probert, R.J. and Smith, R.D. (1986) The joint action of phytochrome and alternating temperatures in the control of seed germination in *Dactylis glomerata*. *Physiologia Plantarum* 67, 299–304.

Probert, R.J., Smith, R.D. and Birch, P. (1985a) Germination responses to light and alternating temperatures in European populations of *Dactylis glomerata* L. I. Variability in relation to origin. *New Phytologist* 99, 305–316.

Probert, R.J., Smith, R.D. and Birch, P. (1985b) Germination responses to light and alternating temperatures in European populations of *Dactylis glomerata* L. IV. The effects of storage. *New Phytologist* 101, 521–529.

Probert, R.J., Smith, R.D. and Birch, P. (1986) Germination responses to light and alternating temperatures in European populations of *Dactylis glomerata* L. V. The principal components of the alternating temperature requirement. *New Phytologist* 102, 133–142.

Probert, R.J., Gajjar, K.H. and Haslam, I.K. (1987) The interactive effects of phytochrome, nitrate and thiourea on the germination response to alternating temperatures in seeds of *Ranunculus sceleratus* L.: a quantal approach. *Journal of Experimental Botany* 38, 1012–1025.

Probert, R.J., Dickie, J.B. and Hart, M.R. (1989) Analysis of the effect of cold stratification on the germination response to light and alternating temperatures using selected seed populations of *Ranunculus sceleratus* L. *Journal of Experimental Botany* 40, 293–301.

Quinlivan, B.J. (1966) The relationship between temperature fluctuations and the softening of hard seeds of some legume species. *Australian Journal of Agricultural Research* 17, 625–631.

Quinlivan, B.J. (1971) Seed coat impermeability in legumes. *Journal of the Australian Institute of Agricultural Science* 37, 283–295.

Roberts, E.H. (1962) Dormancy in rice seed. III. The influence of temperature, moisture and gaseous environment. *Journal of Experimental Botany* 13, 75–94.

Roberts, E.H. (1965) Dormancy in rice seed. IV. Varietal responses to storage and germination temperatures. *Journal of Experimental Botany* 16, 341–349.

Roberts, E.H. (1972) Dormancy: a factor affecting seed survival in the soil. In: Roberts, E.H. (ed.) *Viability of Seeds*. Chapman and Hall, London, pp. 321–359.

Roberts, E.H. (1988) Temperature and seed germination. In: Long, S.P. and Woodward, F.I. (eds) *Plants and Temperature*. Symposia of the Society of Experimental Biology, Company of Biologists, Cambridge, pp. 109–132.

Roberts, E.H. and Benjamin, S.K. (1979) The interactions of light, nitrate and alternating temperature on the germination of *Chenopodium album, Capsella bursa-pastoris* and *Poa annua* before and after chilling. *Seed Science and Technology* 7, 379–392.

Roberts, E.H. and Ellis, R.H. (1989) Water and seed survival. *Annals of Botany* 63, 39–52.

Roberts, E.H. and Smith, R.D. (1977) Dormancy and the pentose phosphate pathway. In: Khan, A.A. (ed.) *The Physiology and Biochemistry of Seed Dormancy and Germination*. North Holland Publishing, Amsterdam, pp. 385–411.

Roberts, E.H. and Totterdell, S. (1981) Seed dormancy in *Rumex* sp. in response to environmental factors. *Plant, Cell and Environment* 4, 97–106.

Roberts, H.A. and Neilson, J.E. (1982) Seasonal changes in the temperature requirements for germination of buried seeds of *Aphanes arvensis*. *New Phytologist* 92, 159–166.

Rolston, M.P. (1978) Water-impermeable seed dormancy. *Botanical Review* 44, 365–396.

Russi, L. (1989) Ecological and physiological aspects of seeds of annual grasslands in a Mediterranean environment. Unpublished PhD thesis, University of Reading, UK.

Russi, L., Cocks, P.S. and Roberts, E.H. (1992) Hard-seededness and seed bank dynamics of six pasture legumes. *Seed Science Research* 2, 231–241.

Schütz, W. and Milberg, P. (1997) Seed dormancy in *Carex canescens*: regional differences and ecological consequences. *Oikos* 78, 420–428.

Shea, S.R., McCormick, J. and Portlock, C.C. (1979) The effect of fires on regeneration of leguminous species in the northern jarrah (*Eucalyptus marginata* Sm.) forest of Western Australia. *Australian Journal of Ecology* 4, 195–205.

Simon, E.W. (1979) Seed germination at low temperatures. In: Lyons, J.M., Graham, D. and Raison, J.K. (eds) *Low Temperature Stress in Crop Plants: the Role of the Membrane*. Academic Press, New York, pp. 37–45.

Simpson, G.M. (1990) *Seed Dormancy in Grasses*. Cambridge University Press, Cambridge.

Skordilis, A. and Thanos, C.A. (1995) Seed stratification and germination strategy in the Mediterranean pines *Pinus brutia*, and *P. halepensis*. *Seed Science Research* 5, 151–160.

Stearns, F. and Olson, J. (1958) Interactions of photoperiod and temperature affecting seed germination in *Tsuga canadensis*. *American Journal of Botany* 45, 53–58.

Steinbauer, G.P. and Grigsby, B. (1957) Interactions of temperature, light and moistening agent in the germination of weed seeds. *Weeds* 5, 681–688.

Stokes, P. (1965) Temperature and seed dormancy. In: Ruhland, W. (ed.) *Encyclopaedia of Plant Physiology*, Vol. XV/2. Springer-Verlag, Berlin, pp. 746–803.

Stoyanova, S. and Kostov, K. (1983) Effect of seed storage temperature on the post harvest dormancy period of *Dactylis glomerata* L. *Rastenier dui Nauki* 20, 94–100.

Stoyanova, S., Kostov, K. and Angelova, A. (1984) Influence of storage temperature on the post-harvest dormancy period of *Festuca arundinacea* Schreb. *Plant Science (Sofia)* 21, 99–108.

Sun, W.Q., Koh, D.C.Y. and Ong, C.M. (1997) Correlation of modified water sorption properties with the decline of storage stability of osmotically-primed seeds of *Vigna radiata* (L.) Wilczek. *Seed Science Research* 7, 391–397.

Takaki, M., Kendrick, R.E. and Dietrich, S.M.C. (1981) Interaction of light and temperature on the germination of *Rumex obtusifolius* L. *Planta* 152, 209–214.

Takaki, M., Heeringa, G.H., Cone, J.W. and Kendrick, R.E. (1985) Analysis of the effect of light and temperature on the fluence response curves for germination of *Rumex obtusifolius*. *Plant Physiology* 77, 731–734.

Taylor, G.B. (1981) Effect of constant temperature treatments followed by fluctuating temperatures on the softening of hard seeds of *Trifolium subterraneum* L. *Australian Journal of Plant Physiology* 8, 547–555.

Taylorson, R.B. and DiNola, L. (1989) Increased phytochrome responsiveness and a high temperature transition in barnyard grass (*Echinochloa crus-galli*) seed dormancy. *Weed Science* 37, 335–338.

Taylorson, R.B. and Hendricks, S.B. (1972) Phytochrome control of germination of *Rumex crispus* L. seeds induced by temperature shifts. *Plant Physiology* 50, 645–648.

Thanos, C.A. and Georghiou, K. (1988) Ecophysiology of fire-stimulated seed germination in *Cistus incanus* ssp. *creticus* (L.) Heywood and *C. salvifolius* L. *Plant, Cell and Environment* 11, 841–849.

Thompson, K. and Grime, J.P. (1983) A comparative study of germination responses to diurnally-fluctuating temperatures. *Journal of Applied Ecology* 20, 141–156.

Thompson, K., Grime, J.P. and Mason, G. (1977) Seed germination in response to diurnal fluctuations in temperature. *Nature* 267, 147–148.

Thompson, P.A. (1970a) Changes in germination responses of *Silene secundiflora* in relation to the climate of its habitat. *Physiologia Plantarum* 23, 739–746.

Thompson, P.A. (1970b) A comparison of the germination character of species of Caryophyllaceae collected in Central Germany. *Journal of Ecology* 58, 699–711.

Thompson, P.A. (1970c) Germination of species of Caryophyllaceae in relation to the geographical distribution in Europe. *Annals of Botany* 34, 427–449.

Thompson, P.A. (1973a) Geographical adaptation of seeds. In: Heydecker, W. (ed.) *Seed Ecology*. Butterworths, London, pp. 31–58.

Thompson, P.A. (1973b) Seed germination in relation to ecological and geographical distribution. In: Heywood, V.H. (ed.) *Taxonomy and Ecology*. Academic Press, London, pp. 93–119.

Thompson, P.A. (1973c) Effects of cultivation on the germination character of the corn cockle (*Agrostemma githago* L.). *Annals of Botany* 37, 133–154.

Thompson, P.A. (1974) Effects of fluctuating temperatures on germination. *Journal of Experimental Botany* 25, 164–175.

Thompson, P.A. (1975) Characterisation of the germination responses of *Silene dioica* (L.) Clairv. populations from Europe. *Annals of Botany* 39, 1–19.

Thompson, P.A. (1981) Ecological aspects of seed germination. In: Thompson, J.R. (ed.) *Advances in Research and Technology of Seeds*, Part 6. Centre for Agricultural Publishing and Documentation, Wageningen, pp. 9–42.

Thompson, P.A. and Cox, S.A. (1978) Germination of the bluebell *Hyacynthoides non-scripta* (L.) Chouard in relation to its distribution and habitat. *Annals of Botany* 42, 51–62.

Toole, E.H., Toole, V.K., Borthwick, H.A. and Hendricks, S.B. (1955) Interaction of temperature and light in germination of seeds. *Plant Physiology* 30, 473–478.

Toole, V.K. and Koch, E.J. (1977) Light and temperature controls of dormancy and germination in bentgrass seeds. *Crop Science* 17, 806–810.

Totterdell, S. and Roberts, E.H. (1979) Effects of low temperatures on the loss of innate dormancy, and the development of induced dormancy in seeds of *Rumex obtusifolius* and *Rumex crispus* L. *Plant, Cell and Environment* 2, 131–137.

Totterdell, S. and Roberts, E.H. (1980) Characteristics of alternating temperatures which stimulate loss of dormancy in seeds of *Rumex obtusifolius* L. and *R. crispus* L. *Plant, Cell and Environment* 3, 3–12.

Tran, V.N. and Cavanagh, A.K. (1984) Structural aspects of dormancy. In: Murray, D.R. (ed.) *Seed Physiology*, Vol. 2. *Germination and Reserve Mobilisation*. Academic Press, Sydney, Australia, pp. 1–44.

Van Assche, J.A. and Vanlerberghe, K.A. (1989) The role of temperature on the dormancy cycle of seeds of *Rumex obtusifolius* L. *Functional Ecology* 3, 107–115.

Van der Woude, W.J. and Toole, V.K. (1980) Studies of the mechanism of enhancement of phytochrome-dependent lettuce seed germination by pre-chilling. *Plant Physiology* 66, 220–224.

Vázquez-Yanes, C. and Orozco-Segovia, A. (1982) Seed germination of a tropical rain forest pioneer tree (*Heliocarpus donnell smithii*) in response to diurnal fluctuations of temperature. *Physiologia Plantarum* 56, 295–298.

Vázquez-Yanes, C. and Orozco-Segovia, A. (1984) Ecophysiology of seed germination in the tropical humid forests of the world: a review. In: Medina, E., Mooney, H.A. and Vázquez-Yanes, C. (eds) *Physiological Ecology of Plants of the Wet Tropics*. Junk, The Hague, pp. 37–50.

Vegis, A. (1964) Dormancy in higher plants. *Annual Review of Plant Physiology* 15, 185–224.

Vertucci, C.W. and Leopold, A.C. (1984) Bound water in soyabean seed and its relation to respiration and imbibitional damage. *Plant Physiology* 75, 114–117.

Vertucci, C.W. and Leopold, A.C. (1986) Physiological activities associated with hydration level in seeds. In: Leopold, A.C. (ed.) *Membranes Metabolism and Dry Organisms*. Comstock Publishing Associates, Cornell University Press, New York, pp. 35–49.

Vertucci, C.W. and Roos, E. (1993) Theoretical basis of protocols for seed storage II. The influence of temperature on optimum moisture levels. *Seed Science Research* 3, 201–213.

Vincent, E.M. and Roberts, E.H. (1979) The influence of chilling, light and nitrate on the germination of dormant seeds of common weed species. *Seed Science and Technology* 7, 3–14.

Visser, T. (1954) After-ripening and germination of apple seeds in relation to the seed coats. *Proceedings Koniklijke Nederlandse akademie van Wetenschappen* 57, 175–185.

Vleeshouwers, L.M., Bouwmeester, H.J. and Karssen, C.M. (1995) Redefining seed dormancy: an attempt to integrate physiology and ecology. *Journal of Ecology* 83, 1031–1037.

Walck, J.L., Baskin, C.C. and Baskin, J.M. (1997) Comparative achene germination requirements of the rockhouse endemic *Ageratina luciae-brauniae* and its widespread close relative *A. altissima* (Asteraceae). *The American Midland Naturalist* 137, 1–12.

Washitani, I. (1985) Germination rate dependency on temperature of *Geranium carolinianum* seeds. *Journal of Experimental Botany* 36, 330–337.

Washitani, I. (1988) Effects of high temperatures on the permeability and germination of the hard seeds of *Rhus javanica* L. *Annals of Botany* 62, 13–16.

Went, F.W. (1957) *Experimental Control of Plant Growth*. Chronica Botanica, Waltham, Massachusetts, pp. 248–251.

Willemsen, R.W. (1975) Effect of stratification temperature and germination temperature on germination and the induction of secondary dormancy in common ragweed seeds. *American Journal of Botany* 62, 1–5.

Chapter 12

Effect of Chemical Environment on Seed Germination

Henk W.M. Hilhorst and Cees M. Karssen

Laboratory of Plant Physiology, Wageningen Agricultural University, Arboretumlaan 4, Wageningen, The Netherlands

Introduction

Soil is the natural physical and chemical environment of most seeds. Essentially, soil is a three-phase system consisting of solids, liquids and gases in varying proportions. In most soils, the solids are predominantly mineral, derived from rock materials. Direct chemical effects of minerals in soil on the germination of seeds are not known. Minerals inhibit germination non-specifically when they occur in high concentrations in soils. The effects of high salinity can be either osmotic or toxic. Soil may also contain organic matter. The amount of organic matter is determined partly by the rate at which fresh plant residues are added and partly by the rate at which they are decomposed by the microflora and fauna. Primary soil particles of different sizes are mixed in various proportions to give recognizable textural classes of soils, e.g. sand loams, sands, etc. (Currie, 1973). The matrix formed by such a mixture has a fundamental pore size, which reflects the proportions of the ingredients. Thus clay soils have many small pores; sands have fewer but larger pores. Soil structure depends on the primary composition of soil, on the interaction between different soil components and, in arable land, on the cultivation methods.

Therefore, structure forms a heterogeneous pattern, which may show large open crevices in dry summers but may collapse under the impact of rain or pressure. Freshly ploughed fields contain clods, which, in turn, are broken down to become the crumbs of the seedbed during subsequent cultivation. Soil structure is important to germinating seeds because it determines the distribution and availability of water, solutes and gases. The seed–soil contact and the process of water uptake have been described in detail by Hadas (1982).

The natural chemical environment of seeds is composed of the liquid and gas phases of soil. The dissolved substances in soil that may affect germination are either inorganic or organic. Most inorganic ions do not have any specific effect on seed germination (Egley and Duke, 1985). Nitrate ions and, to a certain extent, ammonium ions are the notable exceptions.

Organic substances in soil that influence germination partly originate from the material in the direct vicinity of the seeds, and are often neighbouring seeds, fruits or other maternal tissues. Mostly the substances are added to the soil solution by leakage or secretion from living underground plant organs, by leaking from leaves or by decomposition of such plant organs. An inhibitory role is claimed for several of these compounds, often in an allelopathic fashion. Promotive action is well documented for the chemical signals from roots that stimulate germination of parasitic angiosperms (Visser, 1989).

The gas phase of soil contains the usual components of the atmosphere. The modification in its composition by the respiratory activity of soil organisms or by high water saturation may have either profound promotive or inhibitory effects on germination. Also several volatile compounds that occur in soil, such as ethylene or products of anaerobic metabolism, influence germination.

Nitrate

Soil nitrate and dormancy

Nitrate is the major naturally occurring inorganic soil component that stimulates seed germination. This feature of the nitrate ion has been known for a long time. Lehmann (1909) was the first to report its promotive action. Since then, numerous wild species, both monocots and dicots, have been found to be sensitive to the stimulatory action of nitrate (Roberts and Smith, 1977; Bewley and Black, 1982). In

addition, the germination of fern spores may also be promoted by nitrate (Scheuerlein *et al.*, 1985). Since nitrate is central to the nitrogen cycle, most soil types contain nitrate, often at levels within the range of concentrations that are effective in laboratory germination tests. However, the ecological significance of nitrate for germination in the field cannot be treated in isolation from other biotic and abiotic factors, such as water, temperature, light, responsiveness of seeds, other chemical soil constituents, etc. Indeed, interactions between such factors and nitrate have been described for a large number of species (e.g. Vincent and Roberts, 1977). To complicate matters even more, soil nitrate levels, interactions of factors and responsiveness of seeds are all dynamic, showing fluctuations over shorter or longer periods. In this limited survey, we shall make an attempt to assess the importance of nitrate and interacting factors for the survival of weed species. Moreover, a possible mechanism of nitrate action in seed germination will be discussed in some detail.

Seeds of many weed species undergo annual cycles of dormancy (for reviews, see Karssen, 1982; Baskin and Baskin, 1998). After shedding, seeds often possess primary dormancy (Fig. 12.1). Primarily dormant seeds, which are either dry or

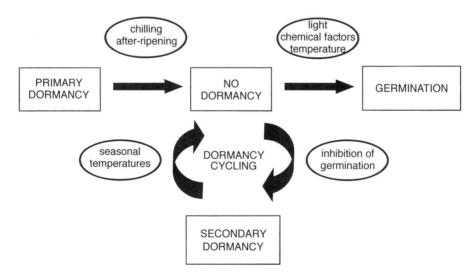

Fig. 12.1. Schematic presentation of changes in dormancy and germination.

imbibed, require exposure to a certain temperature for a certain period to relieve dormancy. Only then will seeds become sensitive to factors that stimulate germination. However, germination will only occur when the complete set of required factors is present. If this is not the case, germination will be inhibited. When this state is extended for longer periods, the seeds may enter a new state of dormancy, called secondary dormancy. Again, secondary dormancy may be broken under suitable temperature conditions. It has been shown convincingly that this cycle depends on seasonal fluctuations of the soil temperature (Bouwmeester and Karssen, 1993). Although nitrate has often been regarded as a breaker of dormancy, the scheme of Fig. 12.1 implies that it is a stimulator of germination. Thus, the process of dormancy relief can be separated from the germination event (Hilhorst and Karssen, 1990). Hence, modification of (depth of) dormancy is a modification of germinability, rather than of germination.

In the field situation, a seed bank, consisting of seeds that undergo annual dormancy cycling, will produce a (sub)population of non-dormant seeds at a specific period. Depending on their specific requirements and the actual environmental conditions, these seeds will either germinate or return to the dormant state. If soil nitrate levels also fluctuate in a seasonal pattern, it may be speculated that this fluctuation interferes with the seasonal pattern of weed emergence. From the literature, it appears that there is no general seasonal pattern of nitrate levels in different soil types (e.g. Rice, 1983; Runge, 1983). The rate of mineralization of nitrogen depends on temperature. Hence, a higher production of both nitrate and ammonium can be expected during the growing season in summer. Obviously, consumption also increases during this period (Runge, 1983). Other reports show high nitrate levels during winter and early spring, followed by a steady decline, due to consumption and possibly inhibition of nitrification by root systems of established plants (Rice, 1983; Obermann, 1985). In addition, a number of

site effects may strongly influence nitrate and ammonium levels. These site effects may be soil type, soil pH, moisture content, burial depth, disturbance of soil, cultivation practice, etc. Therefore, we can only conclude that soil nitrate is not part of the mechanism by which seeds sense the time of the year. As mentioned earlier, temperature is the most likely candidate for this role. However, as will be discussed later, sensing of nitrate may play a role when small subpopulations of seeds require information on the local growth conditions.

Uptake of nitrate

Seeds may take up nitrate during development on the mother plant or, after maturation and shedding, directly from the soil nitrate pool. For species such as *Sisymbrium officinale* and *Chenopodium album*, it has been shown that the nitrate content of seeds on the mother plant is directly related to soil nitrate levels (Saini *et al.*, 1985a, b; Bouwmeester *et al.*, 1994; Fig. 12.2). However, other species, such as *Polygonum persicaria* and *P. lapatifolia*, did not show this relationship. In these species, seed nitrate levels were generally much lower. The amount of nitrate that will reach the nitrate pools in plants depends on several factors, both in the soil and in the plant. One of the most important factors is the plant's capacity to assimilate nitrate. Other known factors are the level of soil ammonium, water status of the soil, volume of the root system, competition of neighbouring plants, efficiency of the uptake system, etc. This all makes predictions of nitrate levels in seeds on the mother plant very complicated. Therefore, assessment of the ecological significance of a possible indirect sensing of the nitrogen status of the soil by seeds on the plant is equally complicated.

When a mature, dry seed is shed, it comes into direct contact with the soil (micro)environment. For uptake of nitrate from the soil, several requirements have to be fulfilled. Nitrate must be available in a

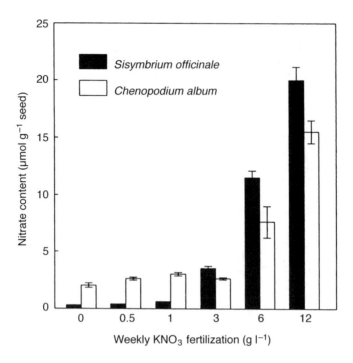

Fig. 12.2. The effect of potassium nitrate fertilization of *Chenopodium album* and *Sisymbrium officinale* plants on nitrate content of produced seeds. The plants were cultivated in plots in the open field. Nitrate contents are means of duplicates ± SE. (After Bouwmeester, 1990.)

freely diffusible form. This implies that sufficient water must be present. Uptake of nitrate by seeds of *Sinapis arvensis* is a function of both the nitrate and the water content of the soil (Fig. 12.3). It can be concluded that nitrate uptake is maximal at an optimal combination of nitrate and water content of the soil. Apparently, suboptimal moisture contents will negatively influence diffusion of nitrate, whereas supra-optimal water contents will dilute the available nitrate to levels below the physiological threshold for germination.

Nitrate and seed germination

The stimulation of seed germination by nitrate has been the subject of many studies. In general, germination is stimulated within a range of 0–0.05 M nitrate. The nitrate concentration in the soil solution also fluctuates within this range (Young and Aldag, 1982). This might be an indica-

tion for an ecological role of nitrate in the seedbed. Supra-optimal nitrate concentrations inhibit germination. It is not known whether this is a toxic consequence, a specific inhibition or an osmotic effect.

A positive relationship between endogenous nitrate content and germination in water has been found for *C. album* (Saini *et al.*, 1985b) and *Sisimbrium officinale*. Figure 12.4 clearly shows the absolute dependency of seeds of the latter species on the presence of nitrate. Seed nitrate levels below approximately 0.1 μmol g^{-1} are too low for any seed of these populations to germinate. Moreover, since the seed populations of this experiment were of differing origins and years of harvest, it may be concluded that the nitrate content of seeds is the major limiting factor in germination. However, most nitrate-dependent seeds are light-requiring. This leads us to conclude that a positive interaction between light and nitrate is the principal limitation to germination.

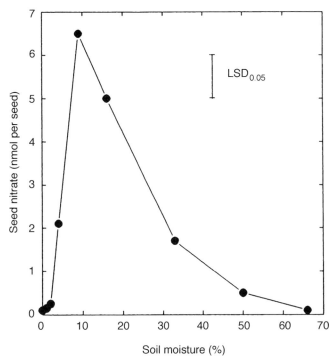

Fig. 12.3. Nitrate uptake by seeds of *Sinapis arvensis* incubated in a soil containing 26 mg nitrate N kg^{-1} and different water contents. Moisture content expressed as percentage by weight. LSD$_{0.05}$, least significant difference at 5% probability. (From Goudey *et al.*, 1988.)

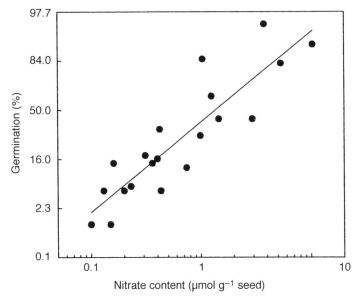

Fig. 12.4. Plot of nitrate concentration (log scale) against germination (probit scale) of seeds from 20 different seed lots of *Sisymbrium officinale*. Seed lots were from plants grown on hydroculture or from plants harvested in the field in different years. Seeds were preincubated for 48 h at 15°C and germinated at 24°C after red-light irradiation. (From Bouwmeester *et al.*, 1994; Hilhorst and Karssen, 1990.)

The interaction between light and nitrate

Promotion of germination by light is mediated by the plant pigment phytochrome. This pigment occurs in two forms: a stable inactive form with an absorption maximum at 660 nm (P_r) and an unstable form with an absorption maximum at 730 nm (P_{fr}), which is physiologically active (for extensive reviews, see Bewley and Black, 1982; Frankland and Taylorson, 1983). The interactive effect of light and nitrate on germination has been demonstrated in many weed species (e.g. Vincent and Roberts, 1977; Williams, 1983). In seeds *of Avena fatua* (Hilton, 1984) and *S. officinale* (Hilhorst and Karssen, 1990), this interaction has been studied in more detail. In both species, the effectiveness of nitrate depended on the level of P_{fr}. In *S. officinale*, the dependency on light is absolute. It has been shown that seeds may contain P_{fr} at the moment of shedding. This pre-existing phytochrome may persist in the dry seeds for long periods (Cone and Kendrick, 1985). If the amount of pre-existing P_{fr} is above the threshold level and if sufficient endogenous nitrate is present, seeds may germinate in the dark. It is not known whether this is the case for all dark-germinating seeds, but it is a possibility that is often overlooked. Burial experiments with *S. officinale* have shown that dormancy relief may be accompanied by nitrate-stimulated dark germination. This is most probably the result of a hypersensitivity to pre-existing phytochrome (Derkx and Karssen, 1993).

A flush of germination of weeds in a crop field after disturbance of the soil indicates that light may be the limiting factor for germination in nitrogen-rich soils. It has been shown that, during soil disturbance, very short exposure times (even less than a second) may induce germination. Germination may proceed even when seeds are returned beneath the soil surface to depths of up to 10 cm (Hartmann and Nezadal, 1990). However, it has been shown that soil disturbance may also result in considerable release of nitrate ions from the soil. This phenomenon is probably a significant factor in the promotion of weeds, such as *A. fatua*, in North America, where summer fallowing is commonly practised (Simpson, 1990). As shown in *S. officinale* and *Arabidopsis thaliana* (Hilhorst and Karssen, 1988), the effect of P_{fr} and nitrate on seed germination may be reciprocal. This implies that the light requirement of seeds in a soil with low nitrate is higher than that of seeds in nitrate-rich soils. Whether this has any ecological meaning remains to be shown.

In summary, seeds are equipped with a sensing mechanism for light and nitrate. This mechanism seems to be 'tuned' in such a way that (non-dormant) seeds will only germinate when the combination of P_{fr} and nitrate levels in the seed is adequate. To some extent, the two factors may replace each other.

Any assessment of the ecological significance of this sensing mechanism in this perspective can only be based on teleological considerations. Obviously, a plant that will eventually grow from a seed requires light and nitrogen for optimal development. However, a relationship between the light and nitrate levels that promote seed germination and the levels required by the growing plant for these factors has yet to be shown.

A more feasible explanation for the sensing mechanism is the ability of seeds to detect local disturbances of light and nitrate levels in the immediate environment. Seeds that are shaded by foliage receive light with a lower ratio of red to far-red light than direct sunlight. This is due to the leaf chlorophyll, which absorbs red light more effectively than far-red light. Consequently, the levels of P_{fr} that are established in the shaded seeds will be lower and more often below the threshold levels for germination (Frankland and Taylorson, 1983). In this way, germination beneath established plants may be prevented and competition avoided. Similarly, established plants may lower the nitrate content of the soil around their root systems. Nitrate is consumed and nitrification may be inhibited (Rice, 1983). As a result, the seeds in the immediate environment are depleted of nitrate and germination probability will be reduced (Pons, 1989).

In addition, soil temperature and diurnal temperature fluctuations are strongly influenced by the vegetation. Diurnal soil-temperature fluctuations are much less pronounced beneath an established sward than in soils with no vegetation. Many weed species require diurnal temperature fluctuations for successful germination (Thompson *et al.*, 1977). Again this may be a mechanism to detect the proximity of competitors.

In conclusion, indications are present for at least three environmental factors that can be accurately sensed by seeds: light, nitrate and temperature (fluctuations). Strong interactions occur among these factors (Vincent and Roberts, 1977; Probert *et al.*, 1987; Hilhorst and Karssen, 1990). This provides the seed with a powerful and sensitive device for sensing major elements of its physical and chemical environment. Evidently, this has potential survival value for the species.

Mechanism of nitrate action

In the past, various hypotheses have been proposed to account for the action of nitrate in seed germination. These include activation of the pentose phosphate pathway (Hendricks and Taylorson, 1975; Roberts and Smith, 1977), stimulation of oxygen uptake (Hilton and Thomas, 1986) and action as a co-factor of phytochrome (Hilhorst and Karssen, 1988; Hilhorst, 1990a, b). The pentose-shunt hypothesis has been criticized extensively (Bewley and Black, 1982). Experiments designed to test the hypothesis gave ambiguous results and a clear role for the pentose phosphate pathway in dormancy breaking has yet to be shown (Cohn, 1987). An increased uptake of oxygen by seeds of several weed species after nitrate application was attributed to breakage of dormancy (Hilton and Thomas, 1986). However, critical evaluation of these experiments led to the conclusion that the increased oxygen uptake was the result and not the cause of the onset of germination. Thus, it was concluded that the effect of nitrate was on germination and not on oxygen uptake (Hilhorst and Karssen, 1990). Moreover, breakage of dormancy in *S. officinale* by lowered temperature, but without allowing seeds to germinate (by depriving seeds of light and nitrate), was not accompanied by an increase in oxygen uptake, indicating that there was no induction of general metabolic activity (Derkx *et al.*, 1993).

A model that integrates the effects of temperature, light and nitrate on germination and breakage of dormancy has been proposed recently (Hilhorst, 1990a, b; Hilhorst *et al.*, 1996). In this model, these three factors interact on a common reaction site, the plasma membrane. A number of arguments add to the suggestion that membranes are involved in the regulation of dormancy (Hilhorst, 1998). There is evidence that phytochrome binds to a component located in a membrane. Membrane fractions to which phytochrome binds have been isolated (Gallagher *et al.*, 1988). Furthermore, factors known to influence membrane properties, such as anaesthetics, may influence phytochrome-induced responses (Van der Woude, 1985; Taylorson, 1988). For the third factor, nitrate, direct binding to the phytochrome receptor protein has been suggested (Hilhorst, 1990a, b). This was based on observed shifts to higher concentrations of nitrate and red-light fluence values during induction of secondary dormancy in seeds of *Sisimbrium officinale*. The magnitude of these shifts was similar for the responses to both nitrate and light. Similar observations were made during burial experiments with *S. officinale*, indicating that the natural dormancy cycling of this species is also accompanied by a changing sensitivity to light and nitrate (Derkx and Karssen, 1993; Hilhorst, 1993; Hilhorst *et al.*, 1996). Other studies also point to a very close action of P_{fr} and nitrate in seeds (De Petter *et al.*, 1985) and fern spores (Haas and Scheuerlein, 1990).

In the model of Hilhorst (Fig. 12.5), the transduction chain leading to germination begins at a membrane-bound (receptor)

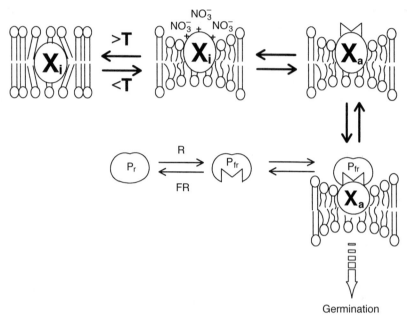

Germination

Fig. 12.5. Pictorial model for the regulation of dormancy and germination. For explanation, see text. X_i and X_a are inactive and active receptors X, respectively. P_r and P_{fr} are inactive and active forms of phytochrome, respectively. R, red light; FR, far red; T, threshold temperature for membrane transition. The interrupted arrow denotes multiple steps. (Based on Hilhorst, 1993.)

protein, which becomes exposed as a result of a temperature-induced change in membrane fluidity. This facilitates binding of nitrate, resulting in an increased affinity of the protein for P_{fr}. Upon binding of P_{fr}, the receptor–protein complex is activated and induces the signal-transduction pathway, ultimately leading to germination. Although this model is highly speculative, it may be an option for future research. Its merit is that it integrates three major environmental factors in the germination mechanism. Moreover, if it is assumed that synthesis and breakdown of the receptor protein or its accessibility to nitrate and phytochrome parallel breakage and induction of dormancy (or may even be the underlying mechanism), the model also accounts for a clear distinction between dormancy and germination, as discussed earlier.

Interaction of nitrate and ammonium

Ammonium is the other major inorganic nitrogen source in the soil. Its effect on

germination will be treated only briefly, because the number of species that respond to ammonium is only a fraction of the number of nitrate-responsive species. For example, Roberts and Smith (1977) have listed 60 species with a positive nitrate response, but only three responded to reduced nitrogen sources. This differing response to inorganic nitrogen sources was explained by assuming that induction of germination required an electron acceptor. However, in *S. officinale*, nitrate does not act by virtue of its capacity to accept electrons (Hilhorst and Karssen, 1989). Therefore, other differences in properties, such as charge, size and structure, may also be responsible for the notably lower activity of the ammonium ion.

A combination of ammonium and nitrate, however, seems to be more effective than either factor alone (Karssen and de Vries, 1983; Goudey *et al.*, 1986). We are not aware of any reports dealing with the mechanism of action of ammonium in imbibed seeds. Treatment of dry *A. fatua* seeds with ammonia gas resulted in increased germination but also in an increase in electrolyte leakage. This may be an indication that the site of action is located on the plasma membrane (Cairns and de Villiers, 1986).

Organic chemicals

Soils contain numerous organic compounds, both volatile and non-volatile. These are often the products of decaying plant and animal remains and the accompanying microorganisms. Also living plants produce a vast array of organics, usually in their root exudates. These compounds have the potential to inhibit or stimulate germination.

Organic inhibitors

Burial of seeds in soil inhibits germination of a great number of species. The absence of light and diurnally fluctuating temperature are important reasons for this inhibition. However, the germination of light-independent seeds, which are capable of germination in darkness, is often also inhibited during burial (Holm, 1972; Frankland, 1977). It has often been postulated that chemical inhibition plays a major role in the regulation of germination in soil. The study of germination-inhibiting compounds has received much attention. We shall restrict ourselves in this chapter to those organic inhibitors that are released in soil and inhibit germination in the natural environment.

The so-called allelopathic substances or allelochemicals form an important group of natural inhibitors. Allelopathy generally refers to detrimental effects of higher plants of one species – the donor – on the germination, growth or development of plants of another species – the recipient (Putman, 1985). In a few cases, allelochemicals may be stimulatory at lower concentrations. There is extensive evidence that allelopathy may contribute to interactive relationships in plants, such as those which determine species composition in community structure, species replacement in succession, etc. We shall confine ourselves to the effects of allelochemicals on seeds. Reviews on allelopathy have been published by Rice (1983), Putman (1985) and Baskin and Baskin (1998).

Allelopathic inhibition of seed germination can play a major role in the regulation of plant succession. Rice (1983) describes in detail the roles of allelopathy in the development of vegetation in old fields that have been abandoned for cultivation because of low fertility. Succession in such fields included four stages: (i) pioneer weed (2–3 years); (ii) annual grass (9–13 years); (iii) perennial bunch grass (for 30 years); and (iv) true prairie grasses. The weed stage, which was dominated by robust plants, such as *Helianthus annuus, C. album, Sorghum halepense* and others, was rapidly replaced by small annual grass species, such as *Aristida oligantha*. It was found that the pioneer weeds eliminated themselves through the production of toxins that inhibited germination of their own seeds (autotoxicity) and those of accompanying weeds. Germination of *A. oligantha*, a grass that often followed the weeds in succession, was not inhibited. Inhibitors were detected in extracts from decaying material from, for example, *S. halepense and H. annuus*. Experiments with soil samples showed that the toxins also leach into soil from living sunflower organs. However, there was greater toxicity from accumulated debris.

Several allelochemicals have been chemically identified. For instance, Numata (1982) identified *cis*-dehydroxymatricaria ester (*cis*-DME) as the active principle in the underground organs of *Solidago altissima*. This weed succeeds the pioneer *Ambrosia artemisiifolia* in old-field succession. The compound formed 2.5% of the dry weight of plant material and actively inhibited germination of the *Ambrosia* seeds. A soil block 10 cm in depth from the rhizosphere of *Solidago* contained 5 p.p.m. *cis*-DME, a level sufficiently high to inhibit germination. The compound survived in soil for several months without decomposition by microorganisms.

The allelochemicals not only leach from leaves, stems and roots but also from seeds. *Parthenium hysterophorus,* an aggressive weed on the American continent, contains two major water-soluble sesquiterpene lactones (up to 8% of its dry weight): parthenin and coronopilin (Picman and Picman, 1984). Although achenes contain lower amounts than leaves and stems, the concentrations are sufficiently high to inhibit seed germination. Autotoxicity has also been shown for seeds. Germination rates of the achenes of *P. hysterophorus* increased with decreasing achene density and increasing washing periods preceding germination. The toxins were mainly located in the coat of the achenes. The inhibitor acts as a kind of rain-gauge, which determines that germination will only occur when sufficient inhibitor has been washed out. These germination inhibitors may also be effective in the formation of zones of inhibition, which may deter competitors.

Seeds of *P. hysterophorus* contain the phenolic compounds caffeic, vanillic, *p*-coumaric, anisic, *p*-hydroxybenzoic, chlorogenic and ferulic acids. These compounds are also known for their allelopathic activities (Rice, 1983). However, removal of phenolic compounds from the extracts showed that in *P. hysterophorus* sesquiterpene lactones are the major compounds responsible for the inhibition of germination. Williams and Hoogland (1982) also expressed some doubt about the role of phenolic compounds as allelopathic agents in the control of germination. Phenolic compounds may have other functions, however. An interesting suggestion is that phenolics in seed-coats help to prevent seed decay by inhibition of microbial attack (Rice, 1983). Côme and Tissaoui (1973) suggested that, in apple seeds, phenolic compounds interact in the testa with oxygen and therefore lower the concentrations inside the testa. Restricted oxygen levels are related to the induction of secondary dormancy.

Organic promoters

In order to germinate, the seeds of most angiospermous root parasites require chemical stimulants emanating from their hosts' roots. Several of these species heavily parasitize cultivated crops and cause severe damage. The most critical phase in the development of these host-specific root parasites is the early and unambiguous recognition of the correct host. The chemical communication between parasite and host is basic to the recognition. The physiology of the root parasite seed is complicated (Joel *et al.*, 1995). Newly shed seeds of root parasites possess primary dormancy and require a period of dry after-ripening to relieve dormancy. When brought into contact with water, a so-called conditioning period of several days is required, during which the seed becomes responsive to the chemical stimulant from the host. The transition to the responsive state can be very abrupt. Contact with the host stimulant induces germination. The conditioning period has been the subject of a number of physiological and biochemical studies. Conditioning is characterized by a specific pattern of respiration and synthesis of DNA, proteins and hormones, such as gibberellins. Temperature and light modify this pattern. If a chemical stimulus is absent after conditioning, the seeds may enter a state of secondary dormancy. It has been shown that seeds of parasitic weeds may pass through dormancy cycles in the soil and in this way remain viable for long periods.

The host-dependent germination of many root-parasitic weeds has been investigated extensively (for references, see Visser, 1989). Little is known, however, about the actual host-liberated chemical stimulants of germination. Chemical stimulants are produced not only by true host plants but also by false hosts, i.e. plants that cannot be parasitized. One of the first stimulants that were identified was named strigol. It stimulates the germination of *Striga asiatica* seeds at extremely low concentrations (Cook *et al.*, 1972). Strigol was isolated from the false host *Gossypium hir-*

sutum (cotton). A germination stimulant for *Striga* has also been isolated from glandular hairs in the root-hair zone of a host, *Sorghum bicolor.* The compound was identified as a labile hydroquinone that triggered the germination of *Striga asiatica* seed at 10^{-7} M and was called sorgoleone (Chang *et al.*, 1986). The relatively low sensitivity of the *Striga* seeds to this compound has been interpreted as a mechanism to ensure that the seeds do not germinate at too great a distance from the host root. We know relatively little about the chemical stimulation of germination in root parasites. It is difficult to understand why certain root parasites have a rather wide host range, while others exhibit an extremely narrow range. It is even more difficult to understand with our currently limited knowledge why false hosts also produce germination stimulants but are not infected. More definite information is needed on the chemical nature of other germination stimulants and their dose–response reactions.

Attempts to control these weeds have largely concentrated on compounds that will stimulate germination in the absence of a host. This phenomenon has been exploited in the case of ethylene, which promotes the germination of *S. asiatica* (Eplee, 1975). When injected into soil, ethylene will cause suicidal germination, because the seedling must attach to the host root within a few days in order to survive.

A wide variety of relatively simple organic substances are capable of breaking dormancy and/or stimulating germination. These compounds include alcohols, ketones, aldehydes and weak organic acids (Cohn, 1997). Although the stimulating effects of these compounds on the germination of many species has been demonstrated only in the laboratory, they should be mentioned here, since many of them can be found in the soil. The extensive dose–response analyses, with red rice (*Oryza sativa*) as the model species, by Cohn and co-workers (Cohn, 1987, 1989; Cohn *et al.*, 1989) have shown that the concentrations required to break dormancy

may vary by as much as five orders of magnitude among the different compounds. Another important observation was that the concentration optima for the response could be very narrow. This implies that the efficacy of organic compounds in the soil is highly concentration-dependent. This explains why both promotive and inhibitory action of similar compounds on similar species have been reported. The fact that seeds respond to these organic substances may have ecological significance. However, attributing specific functions can, as yet, only be speculative.

Some organic compounds are deliberately applied to arable soils to induce (suicidal) germination of weeds (Taylorson, 1987) as a means of weed management and control. These compounds also include pesticides and herbicides (Baskin and Baskin, 1998).

Gases

The gaseous phase of soil occupies those pores that are not already filled with water. The amount of air space can be controlled only by managing the soil structure and water content. Movement of gases through soil is primarily by molecular diffusion. The diffusion pattern may be simple or complex, depending on where the seed lies within the soil structure. When a crop seed lies in the intercrumb pores in a newly prepared seedbed, it is necessary to consider only diffusion through the macropores. The water held in a cultivated soil at field capacity does little to restrict gas diffusion through the soil in bulk. But, if the seed lies inside the crumb matrix, the diffusion coefficient will be restricted by bonding between particles in the crumb and by water in the crumb pores. At field capacity, when the crumb is saturated, gas diffusion inside the crumb is entirely in solution, giving a drastic increase in diffusion resistance of four orders of magnitude (Currie, 1973). In waterlogged soils and especially in heavy soils, the oxygen content of the gaseous phase may drop considerably below that in normal air. The gas phase is also influenced by the presence of vegetation. Roots of plants will take up oxygen and produce carbon dioxide (CO_2), changing the balance between the gases. In soils having a high organic content and containing an active microflora, the balance may shift in a similar way. Although oxygen levels in soils rarely drop below 19% and CO_2 levels rarely exceed 1%, much greater extremes may occur at microsites, such as those adjacent to plant roots or decaying organic matter, and in soil of flooded areas (Roberts, 1972; Egley and Duke, 1985). In addition to oxygen, CO_2 and nitrogen, soils may contain several other gases and volatile compounds. Those gaseous compounds are mostly due to anaerobic conditions and the activity of microorganisms. Soils may contain methane, hydrogen sulphide, hydrogen, nitrous oxide and small amounts of carbon monoxide, ethylene and ammonia (Mayer and Poljakoff-Mayber, 1989).

Oxygen

In general, germination and early seedling growth require oxygen at atmospheric levels. Oxygen uptake usually rises rapidly during the first hours of imbibition, followed by adenosine triphosphate (ATP) formation. The rapid rise in fresh weight and respiration during imbibition is, in most species, followed by a lag phase in both the uptake of water and oxygen until radicle protrusion. Oxygen diffusion can be strongly limited during the lag phase, because, in a fully hydrated seed, oxygen diffusion is limited by oxygen solubility in water. Moreover, oxygen is often utilized in coats and endosperm in non-respiratory reactions. Therefore oxygen concentrations at the level of meristems in the embryonic axis may be rather low. In some seeds, the period of tolerance to anaerobic conditions is considerably extended. This is particularly true for seeds that germinate under water, and also for the many terrestrial plants that germinate under water. The genus Echinochloa contains some of the well-studied flooding-tolerant species. Seeds of Echinochloa crus-galli germinate

well under either anaerobic or aerobic conditions. This ability undoubtedly contributes to its seriousness as a water weed in rice. The seeds of Echinochloa species are all tolerant to the ethanol produced during anaerobic germination. In Echinochloa species, the tolerance to anaerobiosis is not related to dormancy breaking. However, in several species, incubation under anaerobic conditions delays the induction of secondary dormancy, in contrast to seeds incubated under aerobic conditions (Le Deunf, 1973; Karssen, 1980/81). Seeds of Viola species and Veronica hederifolia were induced to germinate in 100% nitrogen or 2% oxygen, but not in 8% oxygen (Longchamp and Gora, 1979). On the other hand, oxygen was necessary for the release of dormancy during chilling of Ambrosia artemisiifolia (Brennan et al., 1978) and during warm-temperature after-ripening of Avena fatua and other cereal seeds (Simmonds and Simpson, 1971).

In summary, the response to oxygen is highly variable. Obviously, this is due to soil characteristics, such as soil type, water content and burial depth, and to oxygen availability to the embryo, which largely depends on the structural and chemical properties of the seed-coat and other embryo-surrounding tissues.

Carbon dioxide

As with oxygen, the level of CO_2 in the soil depends on depth, temperature, moisture, porosity and amount of biotic activity. Carbon dioxide levels increase with depth, ranging from 1 dm^3 m^{-3} in the top 10 cm of soil to 80 dm^3 m^{-3} at 50 cm (Egley, 1984). Carbon dioxide levels may increase five to tenfold when biological activity in the soil peaks in spring and summer. In general, soil CO_2 levels are greatly influenced by moisture content, due to the restricted diffusion of gases (Yabuki and Kitaya, 1984, as cited in Yoshioka et al., 1998). In addition, the respiratory activity of microorganisms and actively growing plant roots and CO_2 evolving from decaying plant material increase with soil moisture content.

In general, CO_2 levels of 20–50 dm³ m⁻³ stimulate germination (for references, see Baskin and Baskin, 1998). This is a higher level than is usually found in the soil top layer. However, it has been shown that rainfall may cause an immediate increase in CO_2 levels. The CO_2 concentration at a depth of 3 cm rose from 8 to 30 dm³ m⁻³ within hours after rainfall. It was clearly demonstrated that this rise in CO_2 levels, rather than moisture content, light, ethylene or nitrate, caused the intermittent flushes of germination of *E. crus-galli* seeds (Yoshioka *et al.*, 1998).

Since levels of CO_2 in the soil are generally below the levels that stimulate germination, it is not likely that the gas plays a significant role in changing the states of dormancy (Egley and Duke, 1985; Baskin and Baskin, 1998).

Ethylene

Ethylene is a common constituent of the soil atmosphere. This gas can be produced by both aerobic microorganisms and anaerobic bacteria (*Clostridium*) (Pazout *et al.*, 1981), as well as by plant roots (Stumpff and Johnson, 1987). Ethylene concentrations of several parts per million have been recorded in soils (Smith and Restall, 1971). However, the concentration of ethylene may vary between different microenvironments, especially at sites where ethylene-producing microorganisms flourish. Ethylene can have a strong influence on plant growth (Arshad and Frankenberger, 1990a, b). For several weed species, a promotive or inhibitory effect of ethylene on seed germination has been reported (e.g. Olatoye and Hall, 1973). These authors demonstrated a positive interaction between the effects of ethylene and light and, for *Sinapis arvensis*, between ethylene and nitrate. In a study of the effect of ethylene on seed germination of 43 species (ten grasses and 33 broad-leaved weeds), germination in nine species was promoted by ethylene in concentrations between 0 and 100 p.p.m. Germination was inhibited in two species, while the other species were not affected (Taylorson, 1979). Interestingly, the nine species that responded positively to ethylene are now known to be responsive to nitrate as well. This may also be indicative of an interaction between ethylene and nitrate. The interaction between ethylene and nitrate has been examined in more detail in *C. album* (Saini *et al.*, 1985a, b; Saini and Spencer, 1986), *Portulaca oleracea* (Egley, 1984) and *A. fatua* (Saini *et al.*, 1985c). In *C. album*, a clear synergism between the effects of ethylene and nitrate was found. This was shown for both endogenous and applied nitrate. In *A. fatua*, however, the effects were additive. In seeds of *P. oleracea*, the effects were also additive, but seeds had to be preincubated in nitrate before the positive interaction with ethylene could be detected. Also interactions between ethylene and CO_2 have been reported. It appears that elevated levels may enhance the effect of ethylene on germination (e.g. Esashi *et al.*, 1988).

The mechanism of action of ethylene in seed germination still awaits elucidation. Roles for ethylene in respiration and in alteration of protein body membranes have been proposed (reviewed in Ross, 1984). Assessment of the ecological significance of soil ethylene for germination of seeds in the field is difficult, since ethylene can both promote and inhibit germination. As to the inhibitory properties of ethylene, this again may be an adaptation to avoiding competition with established plants. Ethylene concentrations are highest in the rhizosphere, because it is rich in microorganisms that are capable of synthesizing ethylene. Moreover, production of ethylene by microorganisms can be enhanced by compounds in root exudates (Arshad and Frankenberger, 1990a, b).

Conclusions

To guarantee successful emergence of seedlings in the field, germination has to occur at the proper time and place. Seeds receive information about the succession of seasons through fluctuations in temperature. In arid and semi-arid zones, the timing of precipitation adds important information. Seeds receive information about their depth in the soil and neighbouring vegetation through the dependency of the germination process on light and fluctuating diurnal temperatures. Therefore, germination of many seed species often only occurs at or close to the surface of the soil and in vegetation gaps.

The chemical environment provides seeds with information about the quality of their environment with respect to suitability for growth. In general, chemical factors that promote germination are also beneficial for emergence and seedling growth. The dependence of many species on nitrate for germination is a clear example of this rule. The presence of high soil nitrate levels may even stimulate the germination of the next generation of seeds, via the accumulation of nitrate during seed formation. The dependence of parasitic seeds on chemical promoters excreted by the host plant illustrates the parallelism between the stimulation of germination and seedling growth. Seedlings of the parasite also depend fully on host factors. Similarly, allelopathic substances in the soil, which inhibit germination, are generally deleterious to seedling growth.

However, most conclusions about the role of the chemical environment of seeds can only be preliminary. The list of active promoters and inhibitors is by no means complete. Our knowledge about allelopathic compounds is still very limited. Undoubtedly, many more compounds will be isolated and characterized. Also the information about mutual interactions of the different compounds in the chemical environment is very restricted. The physiological and biochemical mechanisms underlying promotion and inhibition by these chemical factors are largely unknown. The physiological action of nitrate is so far the best explored. Structure–activity studies of stimulants of parasitic weeds and of some organic dormancy-breaking compounds are making steady progress. However, the actual mechanisms of action on the molecular level have yet to be elucidated.

The chemical environment of seeds needs better exploration. A more detailed knowledge will undoubtedly stimulate a better understanding of the environmental control of germination and emergence. This may ultimately lead to the development of alternative methods for weed control and management.

References

Arshad, M. and Frankenberger, W.T., Jr (1990a) Ethylene accumulation in soil in response to organic amendments. *Soil Science Society of America Journal* 54, 1026–1031.

Arshad, M. and Frankenberger, W.T., Jr (1990b) Production and stability of ethylene in soil. *Biology and Fertility of Soils* 10, 29–34.

Baskin, C.C. and Baskin, J.M. (1998) *Seeds: Ecology, Biogeography, and Evolution of Dormancy and Germination*. Academic Press, San Diego, 666 pp.

Bewley, J.D. and Black, M. (1982) *Physiology and Biochemistry of Seeds*, Vol. 2: *Viability, Dormancy and Environmental Control*. Springer-Verlag, Berlin.

Bouwmeester, H.J. (1990) The effect of environmental conditions on the seasonal dormancy pattern and germination of weed seeds. Unpublished PhD thesis, Agricultural University, Wageningen, The Netherlands.

Bouwmeester, H.J. and Karssen, C.M. (1993) Annual changes in dormancy and germination in seeds of *Sisymbrium officinale* (L.) Scop. *New Phytologist* 124, 179–191.

Bouwmeester, H.J., Derks, L., Keizer, J.J. and Karssen, C.M. (1994) Effects of endogenous nitrate content of *Sisymbrium officinale* seeds on germination and dormancy. *Acta Botanica Neerlandica* 43, 39–50.

Brennan, T., Willemsen, R., Rudd, T. and Frenkel, C. (1978) Interaction of oxygen and ethylene in the release of ragweed seeds from dormancy. *Botanical Gazette* 139, 46–52.

Cairns, A.L.P. and de Villiers, O.T. (1986) Breaking seed dormancy of *Avena fatua L.* seed by treatment with ammonia. *Weed Research* 26, 191–197.

Chang, M., Netzly, O.H., Butler, L.G. and Lynn, D.C. (1986) Chemical regulation of distance: characterization of the first natural host germination stimulant for *Striga asiatica*. *Journal of the American Chemical Society* 108, 7858–7860.

Cohn, M.A. (1987) Mechanisms of physiological seed dormancy. In: Frasier, G.W. and Evans, R.A. (eds) *Seed and Seedbed Ecology of Rangeland Plants*. USDA-ARS, Washington, DC, pp. 14–20.

Cohn, M.A. (1989) Factors influencing the efficacy of dormancy-breaking chemicals. In: Taylorson, R.B. (ed.) *Recent Advances in the Development and Germination of Seeds*. Plenum Press, New York, pp. 261–267.

Cohn, M.A. (1997) QSAR modelling of dormancy-breaking chemicals. In: Ellis, R.H., Black, M., Murdoch, A.J. and Hong. T.D. (eds) *Basic and Applied Aspects of Seed Biology*. Kluwer Academic Publishers, Dordrecht, pp. 289–295.

Cohn, M.A., Jones, K.L., Chiles, L.A. and Church, D.F. (1989) Seed dormancy in red rice. VII. Structure–activity studies of germination stimulants. *Plant Physiology* 89, 879–882.

Côme, D. and Tissaoui, T. (1973) Interrelated effects of imbibition, temperature and oxygen on seed germination. In: Heydecker, W. (ed.) *Seed Ecology*. Butterworths, London, pp. 157–168.

Cone, J.W. and Kendrick, R.E. (1985) Fluence–response curves and action spectra for promotion and inhibition of seed germination in wildtype and long-hypocotyl mutants of *Arabidopsis thaliana* L. *Planta* 163, 43–54.

Cook, C.E., Whichard, L.P., 'Wall, M.E., Egley, G.H., Coggan, P., Luhan, B.A. and McPahil, A.T. (1972) Germination stimulants. 11. The structure of strigol – a potent seed germination stimulant for witchweed (*Striga lutea* Lour.). *Journal of the American Chemical Society* 94, 6198—6199.

Currie, J.A. (1973) The seed–soil system. In: Heydecker, W. (ed.) *Seed Ecology*. Butterworths, London, pp. 463–480.

De Petter, E., Van Wiemeersch, L., Rethy, R., Dedonder, A., Fredericq, H. and de Greef, J. (1985) Probit analysis of low and very-low fluence-responses of phytochrome-controlled *Kalanchoë blossfeldiana* seed germination. *Photochemistry and Photobiology* 42, 697–703.

Derkx, M.P.M. and Karssen, C.M. (1993) Changing sensitivity to light and nitrate but not to gibberellins regulates seasonal dormancy patterns in *Sisymbrium officinale* seeds. *Plant, Cell and Environment* 16, 469–479.

Derkx, M.P.M., Smidt, W.J., VanDerPlas, L.H.W. and Karssen, C.M. (1993) Changes in dormancy of *Sisymbrium officinale* seeds do not depend on changes in respiratory activity. *Physiologia Plantarum* 89, 707–718.

Egley, G.H. (1984) Ethylene, nitrate and nitrite interactions in the promotion of dark germination of common purslane seeds. *Annals of Botany* 53, 833–840.

Egley, G.H. and Duke, S.O. (1985) Physiology of weed seed dormancy and germination. In: Duke, S.O. (ed.) *Weed Physiology*, Vol. 1, *Reproduction and Ecophysiology*. CRC Press, Boca Raton, Florida, pp. 27–64.

Eplee, R.E. (1975) Ethylene, a witch weed seed germination stimulant. *Weed Science* 23, 433–436.

Esashi, Y., Kawabe, K., Isuzugawa, K. and Ishizawa, K. (1988) Interrelations between carbon dioxide and ethylene on the stimulation of cocklebur seed germination. *Plant Physiology* 86, 39–43.

Frankland, B. (1977) Phytochrome control of seed germination in relation to the light environment. In: Smith, H. (ed.) *Light and Plant Development*. Butterworths, London, pp. 477–491.

Frankland, B. and Taylorson, R.B. (1983) Light control of seed germination. In: Shropshire, W. and Mohr, H. (eds) *Photomorphogenesis*. Encyclopedia of Plant Physiology, New Series, Vol. 16A, Springer-Verlag, Berlin, pp. 428–456.

Gallagher, S., Short, T.W., Ray, P.M., Pratt, L.H. and Briggs, W.R. (1988) Light-mediated changes in two proteins found associated with plasma membrane fractions from pea stem sections. *Proceedings of the National Academy of Sciences (USA)* 85, 8003–8007.

Goudey, J.S., Saini, H.S. and Spencer, M.S. (1986) Seed germination of wild mustard (*Sinapis arvensis*): factors required to break primary dormancy. *Canadian Journal of Botany* 65, 849–852.

Goudey, J.S., Saini, H.S. and Spencer, M.S. (1988) Role of nitrate in regulating germination of *Sinapis arvensis* L. (wild mustard). *Plant, Cell and Environment* 11, 9–12.

Haas, C.J. and Scheuerlein, R. (1990) Phase-specific effects of nitrate on phytochrome mediated germination in spores of *Dryopteris filix-mas* L. *Photochemistry and Photobiology* 52, 67–72.

Hadas, A. (1982) Seed–soil contact and germination. In: Khan, A.A. (ed.) *The Physiology and Biochemistry of Seed Development, Dormancy and Germination*. Elsevier Biomedical Press, Amsterdam, pp. 507–527.

Hartmann, K.M. and Nezadal, W. (1990) Photocontrol of weeds without herbicides. *Naturwissenschaften* 77, 158–163.

Hendricks, S.B. and Taylorson, R.B. (1975) Breaking of seed dormancy by catalase inhibition. *Proceedings of the National Academy of Sciences (USA)* 72, 306–309.

Hilhorst, H.W.M. (1990a) Dose–response analysis of factors involved in germination and secondary dormancy of seeds of *Sisymbrium officinale*. I. Phytochrome. *Plant Physiology* 94, 1090–1095.

Hilhorst, H.W.M. (1990b) Dose–response analysis of factors involved in germination and secondary dormancy of seeds of *Sisymbrium officinale*. II. Nitrate. *Plant Physiology* 94, 1096–1102.

Hilhorst, H.W.M. (1993) New aspects of seed dormancy. In: Côme, D. and Corbineau, F. (eds) *Fourth International Workshop on Seeds. Basic and Applied Aspects of Seed Biology*. ASFIS, Paris, pp. 571–579.

Hilhorst, H.W.M. (1998) The regulation of secondary dormancy: the membrane hypothesis revisited. *Seed Science Research* 8, 77–90.

Hilhorst, H.W.M. and Karssen, C.M. (1988) Dual effect of light on the gibberellin and nitrate-stimulated seed germination of *Sisymbrium officinale and Arabidopsis thaliana*. *Plant Physiology* 86, 591–597.

Hilhorst, H.W.M. and Karssen, C.M. (1989) Nitrate reductase independent stimulation of seed germination in *Sisymbrium officinale* L. (hedge mustard) by light and nitrate. *Annals of Botany* 63, 131–137.

Hilhorst, H.W.M. and Karssen, C.M. (1990) The role of light and nitrate in seed germination. In: Taylorson, R.B. (ed.) *Recent Advances in the Development and Germination of Seeds*. NATO-ASI Series A 187, Plenum Press, New York, pp. 191–206.

Hilhorst, H.W.M., Derkx, M.P.M. and Karssen, C.M. (1996) An integrating model for seed dormancy cycling: characterization of reversible sensitivity. In: Lang, G.A. (ed.) *Plant Dormancy*. CAB International, Wallingford, UK, pp. 341–360.

Hilton, J.R. (1984) The influence of light and potassium nitrate on the dormancy and germination of *Avena fatua* L. (wild oat) and its ecological significance. *New Phytologist* 96, 31–34.

Hilton, J.R. and Thomas, J.A. (1986) Regulation of pregermination, rates of respiration in seeds of various seed species by potassium nitrate. *Journal of Experimental Botany* 37, 1516–1524.

Holm, R.E. (1972) Volatile metabolites controlling germination in buried weed seeds. *Plant Physiology* 50, 293–297.

Joel, D.M., Steffens, J.C. and Matthews, D.E. (1995) Germination of weedy root parasites. In: Kigel, J., Negbi, M. and Gallili, G. (eds) *Seed Development and Germination*. Marcel Dekker, New York, pp. 567–596.

Karssen, C.M. (1980/81) Environmental conditions and endogenous mechanisms involved in secondary dormancy of seeds. *Israel Journal of Botany* 29, 45–64.

Karssen, C.M. (1982) Seasonal patterns of dormancy in weed seeds. In: Khan, A.A. (ed.) *The Physiology and Biochemistry of Seed Development, Dormancy and Germination*. Elsevier Biomedical Press, Amsterdam, pp. 243–270.

Karssen, C.M. and de Vries, B. (1983) Regulation of dormancy and germination by nitrogenous compounds in the seeds of *Sisymbrium officinale* L. (hedge mustard). *Aspects of Applied Biology* 4, 47–54.

Le Deunf, Y. (1973) Interactions entre l'oxygène et la lumière dans la germination et l'induction d'une dormance secondaire chez les sernences de *Rumex crispus* L. *Comptes Rendues des Académie des Sciences Série D* 276, 2381–2384.

Lehmann, E. (1909) Zur Keimungsphysiologie und -biologie von *Ranunculus sceleratus* L. und einigen anderen Samen. *Berichte der Deutsche Botanische Gesellschaft* 27, 476–494.

Longchamp, J.P. and Gora, M. (1979) Influences of oxygen deficiencies on the germination of weed seeds. *Oecologia Plantarum* 14, 121–126.

Mayer, A.M. and Poljakoff-Mayber, A. (1989) *The Germination of Seeds*, 4th edn. Pergamon Press, Oxford.

Numata, M. (1982) Weed-ecological approaches to allelopathy. In: Holzner, W. and Numata, N. (eds) *Biology and Ecology of Weeds*. Junk, The Hague, pp. 169–173.

Obermann, P. (1985) Die Belastung des Grundwassers aus landwirtschaftlicher Nutzung nach heutigem Kenntnisstand. In: Nieder, H. (ed.) *Nitrat im Grundwasser*. VCH Verlagsgesellschaft, Weinheim, pp. 53–64.

Olatoye, S.T. and Hall, M.A. (1973) Interaction of ethylene and light on dormant weed seeds. In: Heydecker, W. (ed.) *Seed Ecology*. Butterworths, London, pp. 233–249.

Pazout, J., Wurst, M. and Vancura, V. (1981) Effect of aeration on ethylene production by soil bacteria and soil samples cultivated in a closed system. *Plant and Soil* 62, 431–437.

Picman, J. and Picman, A.K. (1984) Autotoxicity in *Parthenium hysterophorus* and its possible role in control of germination. *Biochemical Systematics and Ecology* 12, 287–292.

Pons, T.L. (1989) Breaking of seed dormancy as a gap detection mechanism. *Annals of Botany* 63, 139–143.

Probert, R.J., Gajjar, K.H. and Haslarn, I.K. (1987) The interactive effects of phytochrome, nitrate and thiourea on the germination response to alternating temperatures in seeds of R*anunculus sceleratus* L.: a quantal approach. *Journal of Experimental Botany* 38, 1012–1025.

Putman, A.R. (1985) Weed alelopathy. In: Duke, S.O. (ed.) *Weed Physiology*, Vol. 1. *Reproduction and Ecophysiology*. CRC Press, Boca Raton, Florida, pp. 131–155.

Rice, E.L. (1983) *Allelopathy*, 2nd edn. Academic Press, New York.

Roberts, E.H. (1972) Dormancy: a factor affecting seed survival in the soil. In: Roberts, E.H. (ed.) *Viability of Seeds*. Chapman and Hall, London, pp. 321–359.

Roberts, E.H. and Smith, R.D. (1977) Dormancy and the pentose phosphate pathway. In: Khan, A.A. (ed.) *The Physiology and Biochemistry of Seed Dormancy and Germination*. Elsevier Biomedical Press, Amsterdam, pp. 385–411.

Ross, J.D. (1984) Metabolic aspects of dormancy. In: Murray, D.R. (ed.) *Seed Physiology*, Vol. 2. *Germination and Reserve Mobilisation*. Academic Press, London, pp. 45–75.

Runge, M. (1983) Physiology and ecology of nitrogen nutrition. In: Lange, O.L., Nobel, P.S., Osmond, C.B. and Ziegier, H. (eds) *Physiological Plant Ecology*, Vol. III. Encyclopedia of Plant Physiology, New Series, Vol. 12C, Springer-Verlag, Berlin, pp. 163–200.

Saini, H.S. and Spencer, M.S. (1986) Manipulation of seed nitrate content modulates the dormancy breaking effect of ethylene on *Chenopodium album* seed. *Canadian Journal of Botany* 65, 876–878.

Saini, H.S., Bassi, P.K. and Spencer, M.S. (1985a) Seed germination in *Chenopodium album* L.: relationships between nitrate and the effects of plant hormones. *Plant Physiology* 77, 940–943.

Saini, H.S., Bassi, P.K. and Spencer, M.S. (1985b) Seed germination of *Chenopadium album* L.: further evidence for the dependence of the effects of growth regulators on nitrate availability. *Plant, Cell and Environment* 8, 707–711.

Saini, H.S., Bassi, P.K. and Spencer, M.S. (1985c) Interactions among ethephon, nitrate and after-ripening in the release of dormancy of wild oat (*Avena fatua*) seed. *Weed Science* 34, 43–47.

Scheuerlein, R., Mader, U. and Haupt, W. (1985) Phytochrome mediated fernspore germination: sensitivity to red light is increased by NH_4 NO_3. In: *Book of Abstracts, European Symposium on Photomorphogenesis in Plants, 1985, Wageningen*, abstract 118.

Simmonds, J.A. and Simpson, C.M. (1971) Increased participation of pentose phosphate in response to afterripening and gibberellic acid treatment in caryopses of *Avena fatua*. *Canadian Journal of Botany* 49, 1833–1840.

Simpson, C.M. (1990) *Seed Dormancy in Grasses*. Cambridge University Press, Cambridge.

Smith, K.A. and Restall, S.W.F. (1971) The occurrence of ethylene in anaerobic soil. *Journal of Soil Science* 22, 430–436.

Stumpff, N. and Johnson, J.D. (1987) Ethylene production by loblolly pine seedlings associated with water stress. *Physiologia Plantarum* 69, 167–172.

Taylorson, R.B. (1979) Response of weed seeds to ethylene and related hydrocarbons. *Weed Science* 27, 7–10.

Taylorson, R.B. (1987) Environmental and chemical manipulation of weed seed dormancy. *Review of Weed Sciences* 3, 135–154.

Taylorson, R.B. (1988) Anaesthetic enhancement of *Echinochloa crus-galli* (L.) Beauv. seed germination: possible membrane involvement. *Journal of Experimental Botany* 39, 50–58.

Thompson, K., Grime, J.P. and Mason, G. (1977) Seed germination in response to diurnal fluctuations of temperature. *Nature* 267, 147–149.

Van der Woude, W.J. (1985) A dimeric mechanism for the action of phytochrome: evidence from photothermal interactions in lettuce seed germination. *Photochemistry and Photobiology* 58, 686–692.

Vincent, E.M. and Roberts, E.H. (1977) The interactions of light, nitrate and alternating temperatures in promoting the germination of dormant seeds of common weed species. *Seed Science and Technology* 6, 659–670.

Visser, J.H. (1989) Germination requirements of some root-parasite flowering plants. *Naturwissenschaften* 76, 253–261.

Williams, E.D. (1983) Effects of temperature, light, nitrate and pre-chilling on seed germination of grassland plants. *Annals of Applied Biology* 103, 161–172.

Williams, R.D. and Hoogland, R.E. (1982) The effects of naturally occurring phenolic compounds on seed germination. *Weed Science* 30, 206–210.

Yabuki, K. and Kitaya, Y. (1984) Studies on the control of gaseous environment in the rhizosphere. (1) Changes in CO_2 concentration and gaseous diffusion coefficient in soils after irrigation. *Journal of Agricultural Meteorology* 40, 1–7.

Yoshioka, T., Satoh, S. and Yamasue, Y. (1998) Effects of increased concentration of soil CO_2 on intermittent flushes of seed germination in *Echinochloa crus-galli* var. *crus-galli*. *Plant, Cell and Environment* 21, 1301–1306.

Young, J.L. and Aldag, R.W. (1982) Organic forms of nitrogen soils. In: Stevenson, F.J. (ed.) *Nitrogen in Agricultural Soils*. American Society of Agronomy, Crop Science Society of America, Soil Science Society of America, Madison, USA, pp. 43–66.

Chapter 13
Role of Fire in Regeneration from Seed

Jon E. Keeley[1] and C.J. Fotheringham[2]

[1]US Geological Survey Biological Resources Division, Western Ecological Research Center, Sequoia-Kings Canyon Field Station, Three Rivers, California, USA; [2]Organismic Biology, Ecology and Evolution, University of California, Los Angeles, California, USA

Introduction

Fire is a disturbance factor in ecosystems worldwide and affects the reproduction of many plant species. For some species, it is just one of several disturbances that trigger seed germination and subsequent seedling recruitment, whereas in other 'fire-dependent' species, fire may be required for seedling recruitment. Fire may trigger seed regeneration directly, through the opening of serotinous fruits or cones or by inducing the germination of dormant soil-stored seed banks. Fire may also indirectly initiate seedling recruitment by opening gaps in closed vegetation, thus providing conditions suitable for colonization. There is a multitude of mechanisms for capitalizing upon such disturbances and the particular mode is a function of fire regime, climate, growth form, phylogeny and biogeography.

Postfire environments

Fire causes a multitude of changes in the environment that enhance site quality for seedling recruitment and thus provide the selective impetus for fire-dependent germination. Perhaps foremost is the removal of vegetation (formation of gaps), resulting in

an increase in irradiance at ground (seedling) level and a reduction in competition for water. In addition, fire accelerates the mineralization of organic matter, making inorganic nutrients more readily available (e.g. Dunn and DeBano, 1977; Wells *et al.*, 1979). Fire reduces herbivore populations by direct mortality and indirectly by opening the habitat and making herbivores more vulnerable to predators (Quinn, 1994). The pulse of recruitment that potentially satiates predators (O'Dowd and Gill, 1984) also reduces seedling predation. Other advantages include soil sterilization, which alters microbial populations and reduces pathogens (Fletcher, 1910; Sabiiti and Wein, 1987; Wicklow, 1988).

Costs associated with recruitment in burned sites include delayed reproduction for the many species that accumulate dormant seed banks between fires (e.g. Gadgil and Bossert, 1970). Also, on sites following high-intensity fires, the postfire seedbed may create pH and osmotic conditions unfavourable for the germination of some species (Henig-Sever *et al.*, 1996; Ne'eman *et al.*, 1999). In addition, postfire gaps may be drought-prone, as the increased exposure may lead to elevated evaporation and thus reduced moisture availability at shallower depths, where germination occurs.

Direct effects of fire on germination

Many species accumulate seed banks that are triggered to germinate by fire. These seed banks develop by the accumulation over 1 or more years of dormant or quiescent seeds. Distinguishing between dormancy and quiescence is in part a matter of semantics. For example, some restrict the term dormancy to just those seeds that fail to germinate following imbibition, whereas germination inhibited by seed-coat characteristics (e.g. restricting entry of water or oxygen or release of inhibitors) is considered quiescence (see Murdoch and Ellis, Chapter 8, this volume). Others consider both of these conditions dormancy and distinguish them as 'endogenous' versus 'exogenous' dormancy (e.g. Baskin and Baskin, 1998). We shall follow the latter convention and reserve the term quiescence for seeds held within serotinous cones or fruits.

Heat-induced release of canopy-stored seed banks

Serotiny is the delayed opening of fruits or cones and is often interpreted as an adaptation for timing seed dispersal to favourable seedbed conditions. Some desert species control seed dispersal with serotinous fruits that open following rainfall, whereas, in many fire-prone ecosystems, serotiny cues dispersal to the postfire environment. It is widespread in northern hemisphere coniferous genera, such as *Cupressus* and *Pinus* (Wolf, 1948; Keeley and Zedler, 1998). In the southern hemisphere, serotiny is uncommon in coniferous trees but widespread in shrubby angiosperms, and is particularly well represented in the *Proteaceae* and *Myrtaceae* of Mediterranean-climate South Africa and Western Australia (Lamont *et al.*, 1991).

In all of these taxa, cones or fruits will open and disperse seeds within days of being scorched by fire, resulting in a pulse of seedling recruitment that generates stands of even-aged cohorts. Some species, such as the serotinous *Pinus brutia*, have

seeds with endogenous dormancy, which is overcome by cold stratification following release from the cone (Skordilis and Thanos, 1995). In the absence of fire, cones may open due to death of the branch, heating from solar irradiance or other causes, but in communities with closed canopies this generally does not lead to successful recruitment, although on more open sites it may. There are interesting intraspecific patterns of varying degrees of serotiny and consequent seedling recruitment, which generate even-aged or uneven-aged populations, possibly tied to fire-return intervals (reviewed for *Pinus* in Keeley and Zedler, 1998). Retention of seeds is also highly variable across species, ranging from a few months (e.g. *Pinus muricata* cones remain closed through the autumn fire season and disperse seeds in winter) to many years or decades.

The distribution of serotiny is fairly predictable, as illustrated by the genus *Pinus*. Many pines inhabit fire-prone environments, but only a small subset of these species are serotinous, specifically those on sites subject to high-intensity/severity 'stand-replacing' fire regimes (Fig. 13.1).

Heat-shock-stimulated germination of soil-stored seed banks

Many species that disperse seeds at maturity have innate barriers to germination and thus accumulate dormant seed banks, which are triggered to germinate by fire-related cues, such as heat shock (e.g. *Ceanothus crassifolius* in Table 13.1). This is widespread in many fire-prone ecosystems and is common in the *Fabaceae*, *Rhamnaceae*, *Malvaceae*, *Sterculiaceae*, *Cistaceae* and *Convolvulaceae* (e.g. Ballard, 1981; Barro and Poth, 1988; Trabaud and Oustric, 1989; Keeley, 1991, 1995; Kilian and Cowling, 1992; Thanos *et al.*, 1992; Bell *et al.*, 1993; Cocks and Stock, 1997; Keeley and Bond, 1997; Herranz *et al.*, 1998).

Heat-stimulated seeds exhibit exogenous dormancy imposed by a dense palisade tissue beneath the generally smooth and highly cutinized testa. This barrier to

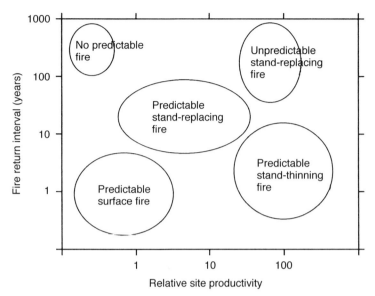

Fig. 13.1. Fire regimes generated by patterns of site productivity and fire recurrence interval (redrawn from Keeley and Zedler, 1998).

water entry inhibits seeds from imbibing water and has led to the term 'hard-seeded'. Fire disrupts the water-imperme-able tissues, allowing imbibition, which typically leads to germination. While it is generally thought that heat directly dis-rupts the palisade tissues, in certain cases it appears to be tied to fire-triggered desic-cation of the seed-coat (Brits *et al.*, 1993). In some hard-seeded species the barrier to germination may not be lack of imbibition but embryo anoxia, due to seed-coat restriction of oxygen (Brits *et al.*, 1995). While the breaking of the barrier to water or oxygen entry may suffice to germinate many species, for others the exogenous dormancy is coupled with endogenous dor-mancy and thus germination occurs only when heat shock is coupled with other environmental cues, such as light (Keeley, 1987; Bell, 1994) or cold stratification (Keeley, 1991).

Hard-seeded species with postfire seedling recruitment appear to differ sub-stantially in the intensity and duration of heat shock most stimulatory to germina-tion. Some exhibit optimal germination after brief bursts of high temperatures (e.g. 5 min at 105°C), whereas others have

higher germination after a long duration at lower temperature (e.g. 1 h at 70°C). The same applies to lethal temperature regimes, e.g. large-seeded species survive a short duration at high temperature but are killed by a long duration at lower temperature, whereas very small seeds exhibit the oppo-site pattern (Keeley *et al.*, 1985). Soil mois-ture during heating also diminishes heat tolerance (Westermeier, 1978; Parker, 1987). Such differences in stimulation/tol-erance regimes may explain some of the variance in microhabitat segregation of postfire floras (e.g. Davis *et al.*, 1989). In addition, variations in heating may affect subsequent seedling growth (Hanley and Fenner, 1998).

Since many species are stimulated by a long duration at 70–80°C, they are not strictly tied to postfire environments – such conditions may be encountered by seeds exposed to direct sun-rays on open sites. Thus, heat-shock-stimulated germina-tion does not limit recruitment to burned sites; rather, such species can establish in gaps created by other types of disturbance as well. Also, since unburned landscapes often comprise a heterogeneous collection of suitable and unsuitable recruitment sites, it is not surprising that most heat-stimulated species have polymorphic seed pools (Keeley, 1991, 1995). Thus, while the bulk of the seed bank may be deeply dormant, a portion may germinate readily and establish in the absence of fire, for reasons elaborated upon by Westoby (1981).

Heat-stimulated species are common in all Mediterranean-climate shrublands, where they are regularly exposed to intense stand-replacing fires (Keeley, 1991, 1995; Bell *et al.*, 1993; Arianoutsou and Thanos, 1996). Outside these ecosystems, heat-stimulated germination and postfire recruitment are not common but are present in species from a diversity of families and vegetation types. For example, it is known from temperate forests and heathlands, in species of *Anacardiaceae* (Marks, 1979; Washitani, 1988), *Convolvulaceae* (McCormac and Windus, 1993), *Ericaceae* (Jaynes, 1968), *Fabaceae* (Cushwa *et al.*, 1968; Martin *et al.*, 1975; Grigore and Tramer, 1996), *Geraniaceae* (Abrams and Dickmann, 1984), *Hypericaceae* (Mallik and Gimingham, 1985), *Restionaceae* (Musil and de Witt, 1991) and *Malvaceae* (Baskin and Baskin, 1997). In grasslands, it is apparently present in both temperate (McCormac and Windus, 1993) and tropical legumes (Sabiiti and Wein, 1987).

Smoke- and charred-wood-stimulated germination of soil-stored seed banks

It has only recently become evident that, in some fire-prone environments, the majority of species that recruit after fires lack heat-stimulated germination, but that chemical products of biomass combustion trigger germination (Table 13.1). This response is present in hundreds of species from the Mediterranean-climate ecosystems of California, South Africa and Western Australia (Jefferey *et al.*, 1988; Keeley, 1991, 1995; Brown, 1993a; Dixon *et al.*, 1995; Roche *et al.*, 1997; Keeley and Fotheringham, 1998b). Preliminary surveys in the other two Mediterranean-climate regions – central Chile and the Mediterranean basin – have thus far failed

to find clear-cut examples of charred-wood- or smoke-stimulated germination (Keeley and Keeley, 1999; J.E. Keeley, C.J. Fotheringham and W.J. Bond, unpublished data). This may be linked with natural lightning fires being less predictable in these two regions. While knowledge of charred-wood- or smoke-stimulated germination is relatively recent, the phenomenon was apparently recognized centuries ago by early Americans (Indians) in the western USA, who routinely sowed tobacco seeds in postfire ash beds (Harrington, 1932). At least one of these species, *Nicotiana attenuata*, is known to be smoke-stimulated (Baldwin *et al.*, 1994).

Charred-wood- or smoke-stimulated germination is found in a wide diversity of families, although it is largely lacking in the families noted for heat-stimulated germination. In Mediterranean-climate regions of California, it is common in members of the *Hydrophyllaceae*, *Papaveraceae*, *Polemoniaceae* and *Scrophulariaceae* – families with centres of radiation in western North America. It is also known from families with more cosmopolitan distributions – *Asteraceae*, *Boraginaceae*, *Brassicaceae*, *Caryophyllaceae*, *Lamiaceae*, *Loasaceae*, *Onagraceae* and *Solanaceae*. In the southern hemisphere, it is known from some of the same families, e.g. *Asteraceae* and *Scrophulariaceae*, but is most common in other families – the *Dilleniaceae*, *Epacridaceae*, *Ericaceae*, *Goodeniaceae*, *Haemodoraceae*, *Myrtaceae*, *Poaceae*, *Proteaceae*, *Restionaceae*, *Rutaceae* and *Thymelaceae* (Brown, 1993a; Baxter *et al.*, 1994; Dixon *et al.*, 1995; van Staden *et al.*, 1995; Enright *et al.*, 1997; Keeley and Bond, 1997; Roche *et al.*, 1997). Smoke-stimulated germination has also been demonstrated for species from both tropical and temperate grasslands (Baxter *et al.*, 1994; Read and Bellairs, 1999).

Charred wood and smoke are equally effective in triggering germination of a wide range of species (Brown, 1993b; Keeley and Fotheringham, 1998b). The combustion products that trigger germination are transferred to seeds as vapour or liquid and this may occur directly from smoke or be secondarily transferred from soil particles (Fig. 13.2). Heating of wood products is sufficient to release active agents (Keeley and Pizzorno, 1986) and the same applies to heating of (organic matter in) soil (Keeley and Nitzberg, 1984). Thus, the conclusion that 'chemical constituents from smoke do not appear to provide a stimulus separate from the effects of heat' (Enright *et al.*, 1997), which was based upon heat treatments of soil, should be re-evaluated.

Two factors make it critical that experimental determination of smoke-stimulated germination include a dosage–response curve: species differ both in level of smoke products stimulatory to germination and in the level that is lethal to seeds (e.g. Table 13.1). This is well illustrated by two California chaparral species with deeply dormant seed banks. Only 1 min of direct exposure to smoke will induce 100% germination of the annual *Emmenanthe penduliflora*, but fails to stimulate the perennial *Romneya coulteri*. On the other

hand, if seeds are sown in soil exposed to smoke for 30 s, both species germinate completely. However, germination of *Emmenanthe* is inhibited by soils exposed to a longer duration (30 min exposure is lethal), whereas *Romneya* germinates well in soils exposed to 30 min of smoke treatment (Keeley and Fotheringham, 1998b).

Considering the postfire increase in soil nitrate levels (Christensen, 1973), it is worth considering the role of this ion. Pons (1989b) showed that nitrate-triggered germination is a potential cue for germination on open sites for weedy species, particularly light-inhibited weeds (Hilhorst and Karssen, 1989). Thanos and Rundel (1995) suggested that nitrate was the 'trigger' that cued species to respond to postfire conditions in California chaparral. However, the nitrate ion alone has been shown to be insufficient in triggering germination (Table 13.1), whereas nitrogen dioxide (at levels found in biomass smoke) is highly effective for some species (Fig. 13.3).

Seed-coats of smoke-stimulated species

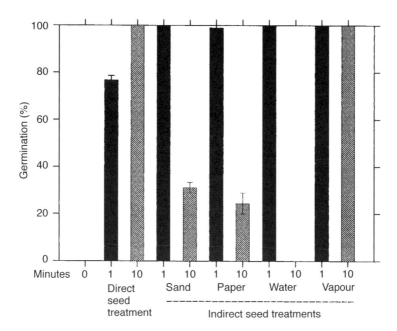

Fig. 13.2. Germination of *Emmenanthe penduliflora* for control (0) and smoke treatments of 1- or 10-min exposures for direct treatment (smoke-treated seeds incubated on non-treated filter-paper) and indirect treatments (untreated seeds incubated on smoke-treated sand or filter-paper or untreated seeds incubated with smoke water or exposed to gases emitted by smoke-treated filter-paper) (redrawn from Keeley and Fotheringham, 1997).

Table 13.1. California chaparral species illustrating the range of variation in germination response (mean percentage germination + SE, $n = 3$ replicates of 30 seeds) (data from Keeley and Fotheringham, 1998b, and unpublished).

Species	Family	Percentage germination												
		Control	70°C 1 h	105°C 5 min	115°C 5 min	CW[a]	Smoke (min) 5	15	Scar	Scar + GA[b]	GA[b]	Nitrate[c] (mol m^{-3}) 1	10	100
Ceanothus crassifolius	Rhamnaceae	0	50	87	47	0	0	0	98	99	0	0	0	0
Emmenanthe penduliflora	Hydrophyllaceae	0	0	0	0	80	100	15	100	100	0	0	2	1
Phacelia grandiflora	Hydrophyllaceae	1	0	0	0	44	59	66	99	95	0	0	4	0
Romneya coulteri	Papaveraceae	0	0	0	0	46	74	99	19	100	0	0	0	0
Silene multinervia	Caryophyllaceae	8	7	4	8	84	98	92	98	95	29	15	47	9
Dicentra chrysantha	Papaveraceae	0	0	0	0	0	0	0	0	0	0	0	0	0

[a]CW = 10% aqueous leachate from charred wood.
[b]GA = gibberellic acid (GA$_3$) highest germination with 1, 5 or 10 mmol m^{-3}.
[c]KNO$_3$ in distilled water, pH ~6.

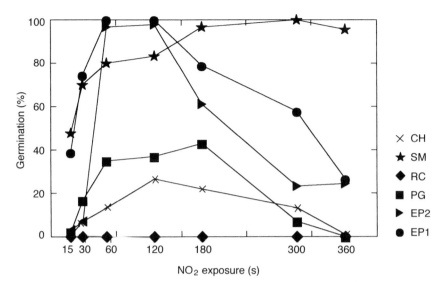

Fig. 13.3. Germination response to nitrogen dioxide (7.7×10^3 mg m^{-3}) at different durations of exposure applied to dry seeds for smoke-stimulated California chaparral species: *Caulanthus heterophyllus* (CH), *Silene multinervia* (SM), *Romneya coulteri* (RC), *Phacelia grandiflora* (PG) and two populations of *Emmenanthe penduliflora* (EP1 and EP2) (redrawn from Keeley and Fotheringham, 1998b).

are structurally quite different from those of heat-stimulated species. Outer seed-coats are highly sculptured and, in all cases so far examined, they lack a dense palisade layer. Commonly, the outer coat comprises loosely packed tissues with a subdermal semi-permeable cuticle. Most of the species so far tested fully imbibe water during dormancy, indicating endogenous dormancy (Keeley and Fotheringham, 1998b).

Mechanism of smoke- and charred-wood-stimulated germination

It is unknown what chain of reactions triggers germination of smoke-stimulated species and there is every likelihood that the mechanism is not the same in all species. One of the best-studied species is *E. penduliflora* and it provides a useful model system for investigating smoke-induced germination because of the following:

1. Historically, it was the species first identified to have charred-wood-stimulated germination (Wicklow, 1977) and its ger-

mination behaviour has been well characterized (Jones and Schlesinger, 1980; Keeley and Nitzberg, 1984; Keeley *et al.*, 1985; Keeley, 1987, 1991; Keeley and Fotheringham, 1997, 1998a, b).
2. This species has deeply dormant seed pools, often with zero germination under 'control' conditions, and this is consistent across most chaparral populations, unlike some other smoke-stimulated species, where a sizeable fraction of the seed bank germinates in the absence of fire-related cues.
3. Germination is light-neutral and most populations are cold-stratification-dependent.

Emmenanthe germinates in response to either charred wood, smoke or aqueous extracts or vapours from either of these combustion products (Fig. 13.2). Under the proper concentrations, these fractions generate 100% germination of otherwise completely dormant seed pools. Chemicals eliminated as germination cues include nitrate ion, nitrous oxide, carbon dioxide, ethylene and methane (Keeley and Fotheringham, 1997, 1998a). Some compounds, e.g. carbon monoxide, sulphuric acid and hydrogen peroxide, are capable of stimulating substantial germination, but the strongest and most consistent response is to nitrogen oxides (Fig. 13.3). Several characteristics suggest that this may be the active component of smoke-stimulated germination. Nitrogen dioxide is capable of stimulating complete germination at the levels present in biomass smoke and, like smoke, it can stimulate germination by direct exposure or indirectly by binding to soil particles and being transferred in water or vapours (Keeley and Fotheringham, 1997).

Dormant *Emmenanthe* seeds readily imbibe water. However, while the seed-coat layers are permeable to water and solutes, there is a semi-permeable subdermal cuticle, which allows water to pass but blocks larger-molecular-weight solutes, e.g. eosin and other dyes (Keeley and Fotheringham, 1997, 1998a, b). Smoke or low levels (7.7×10^3 mg m^{-3}) of nitrogen dioxide change the characteristics of this membrane and allow diffusion of solutes that would otherwise be blocked. It is unknown whether or not this change in membrane diffusivity has any role in the germination of this species, but one characteristic suggests it might: seeds germinate completely following seed-coat (including the subdermal cuticle) scarification (Table 13.1). However, scarification-induced germination is widespread in many species, and may not reflect on the smoke-induced germination mechanism. It has been posited (K. Bradford, 1997, personal communication) that scarification-induced germination is a 'wound response', widely selected for because of the inevitable microbial infection to be expected once the integrity of the seed-coat is broken. Further evidence that the basis for scarification- and smoke-induced germination may differ is the fact that in *Emmenanthe* a gibberellic acid inhibitor prevents germination of scarified seeds but not of smoke-treated seeds (Keeley and Fotheringham, 1998a).

In characterizing the germination response of *Emmenanthe*, it has been shown that germination is strongly dependent upon oxidizing agents or a combination of acidity and certain anions, such as nitrate or sulphate (however, these ions alone are ineffective). Many characteristics of *Emmenanthe* germination fit Cohn's model of dormancy-breaking behaviour of weak acids, which increase membrane permeability in the disassociated form (Cohn *et al.*, 1983, 1987; Cohn and Castle, 1984; Cohn and Hughes, 1986; Cohn, 1989, 1996). However, alternatives have been proposed (e.g. Hendricks and Taylorson, 1972; Keeley and Fotheringham, 1998a; Raven and Yin, 1998).

It is likely that the mechanism of smoke-stimulated germination is different between species, which should be expected, considering its wide phylogenetic distribution. Some smoke-stimulated chaparral species share many of the same germination responses with *Emmenanthe*, including response to nitrogen dioxide, but not all (Fig. 13.3). Two other smoke-stimulated species are worth noting, because they illustrate some of the variation in a single community. The response of *Romneya coulteri* to smoke is not unlike that of *Emmenanthe*, but *Romneya* fails to respond to nitrogen dioxide (although it is stimulated by nitrite ions), to seed-coat scarification or to treatment by strong acids and oxidizing agents (Keeley and Fotheringham, 1998b). *Dicentra chrysantha* fails to germinate under all conditions (Table 13.1), unless seeds are first pretreated by a long period (*c.* 1 year) of soil burial. Following burial, seeds are still dormant but are now highly sensitized and will germinate in response to smoke (Keeley and Fotheringham, 1998b). Ageing has been shown to increase the response to smoke in a great many Australian species (Roche *et al.*, 1997). However, in the case of *Dicentra*, ageing *per se* is insufficient, since shelf-stored seeds fail to respond to smoke, suggesting that interactions with the soil environment are necessary. It is apparent that smoke-stimulated seeds have different barriers to germination, which require a particular order of environmental cues. It is conceivable that such differences in response may affect postfire community structure, both spatially and temporally.

Further variation is evident in the response of *Nicotiana attenuata,* an annual widespread in scrub and woodlands throughout arid parts of western North America. As with chaparral species so far studied, *Nicotiana* seeds freely imbibe water in the dormant state and thus have endogenous dormancy (Baldwin *et al.,* 1994). It has been proposed that germination in this species is not tied to nitrogenous compounds, because germination is triggered by treatment with pyrolysis products of α-cellulose (Baldwin *et al.,* 1994). Whether or not nitrogen is involved remains to be determined, since these experiments were all conducted with ~10 mol m^{-3} KNO$_3$ in the incubation medium of controls and treatments. Previous studies on other smoke-stimulated species have shown that, while nitrate in water fails to stimulate germination (e.g. Table 13.1), under acidic conditions, either from buffers or added pyrolysis products, nitrate is stimulatory to germination (Keeley and Fotheringham, 1998a), suggesting that the combination of protons and the nitrate anion were involved in the germination response. *Nicotiana* requires further work before a role for nitrogenous compounds can be ruled out.

Induced or enforced dormancy in soil-stored seed banks

As discussed above, many species have innate seed dormancy, which is overcome by fire-related cues, such as heat and charred wood. However, some species in fire-prone environments do not exhibit innate dormancy at the time of dispersal and will germinate readily under laboratory conditions, but develop dormancy under field conditions. For example, California chaparral annuals – *Cryptantha* spp., *Nicotiana* spp., *Papaver californica* and others – collected from recent burn sites have non-dormant seeds, which germinate readily upon wetting (Keeley, 1987, 1991; Keeley and Fotheringham, 1998b). In the field, however, populations of these species are abundant only immediately

after fire, arising from dormant seed banks. Their persistence is closely tied to the post-fire environment and they decline in subsequent years, some disappearing entirely by the third postfire year (J.E. Keeley, unpublished data). This is circumstantial evidence that dormancy is induced or enforced by environmental factors developing after fire; Harper (1977) distinguishes induced dormancy as an acquired inability to germinate and enforced dormancy as imposed by an environmental constraint, such as lack of light or the presence of an inhibitor. One of the earliest cases studied was that of chemical inhibition (allelopathy) of herbaceous species by chemicals leached from the shrub canopy, putatively selected to inhibit competitors (Muller *et al.,* 1964; McPherson and Muller, 1969; cf. Keeley and Keeley, 1989). Alternatively, inhibition of germination may represent induced dormancy and result from an evolved sensitivity to compounds indicative of unfavourable shrub-dominated environments (Koller, 1972; Angevine and Chabot, 1979; Keeley, 1991); such compounds are perhaps best described as 'infochemicals' (Smith and van Staden, 1995). Preston and Baldwin (1999) have recently provided field evidence consistent with the idea of negative effectors enforcing dormancy on *N. attenuata,* and they appear to be associated with specific shrub species (Baldwin and Morse, 1994).

Studies of chemical inhibition (allelopathy) have proposed that postfire germination is triggered by high-temperature destruction of inhibitory compounds in the soil environment (Muller *et al.,* 1964; McPherson and Muller, 1969). This, of course, implies enforced dormancy, although it is possible that the chemical environment in unburned soils induced dormancy by converting the seeds to a state where germination required heat or smoke.

In California chaparral, there is evidence that species most closely tied to postfire environments – i.e. strict 'fire-following species' – have innate dormancy, and germination is not chemically inhibited by the prefire soil environment, whereas the more 'opportunistic' species, arising under a variety of disturbances, have dormancy that is induced by chemicals in the soil environment. Evidence of this is in the 'allelopathy' experiments reported by Christensen and Muller (1975). They found that several postfire endemics (e.g. *Allophyllum glutinosum, Emmenanthe penduliflora* and *Lotus scoparius*) were insensitive to shrub leachates, even when primed to germinate by scarification (Table 13.1). In contrast, several of the opportunistic species (e.g. *Centaurea melitensis, Cryptantha intermedia* and *Lactuca serriola*) were highly sensitive to shrub leachates (Table 13.2).

Other examples of induced or enforced dormancy include the highly opportunistic coastal sage-scrub vegetation in California. These subligneous shrubs readily recruit in a variety of disturbances (DeSimone and Zedler, 1999). Dominants, such as *Artemisia californica, Salvia mellifera* and *Mimulus aurantiacus*, have seeds that exhibit a marked interaction between light and charred wood. Seeds have a positive photoblastic response – germinating readily in the light – but are dormant in the dark, thus cueing germination to gaps but blocking germination of buried seeds. However, following fire, this dark-induced dormancy is broken by chemicals leached from charred wood (Keeley, 1991). As with the opportunistic *N. attenuata* (Preston and Baldwin, 1999), these subshrubs appear to be responding to both positive and negative effectors.

Ecological distribution of fire-dependent regeneration

Fire-dependent seed release from serotinous cones/fruits or smoke/heat-triggered germination is largely concentrated in

Table 13.2. Effect of leaf leachate from the dominant chaparral shrub, *Adenostoma fasciculatum*, on germination of herbaceous species, including: (i) 'opportunistic' species (common in many types of disturbance); and (ii) 'postfire endemics' (closely associated with fire; *Convolvulus cyclostegius* and *Lotus scoparius* are heat-stimulated and the rest are smoke-stimulated). Seedcoats of these postfire endemics were scarified to overcome innate dormancy. (Based on data from Christensen and Muller, 1975.)

Species	Percentage germination (mean ± SD)	
	Controls	Leachate
Opportunistic species		
Bromus rigidus (Poaceae)	94 ± 6	97 ± 4
Centaurea melitensis (Asteraceae)	93 ± 5	0 ± 0
Cryptantha intermedia (Boraginaceae)	93 ± 7	3 ± 6
Erigeron divergens (Asteraceae)	94 ± 1	0 ± 0
Lactuca serriola (Asteraceae)	94 ± 1	37 ± 11
Postfire endemics		
Allophyllum glutinosum (Polemoniaceae)	90 ± 0	83 ± 0
Convolvulus cyclostegius (Convolvulaceae)	92 ± 2	90 ± 6
Emmenanthe penduliflora (Hydrophyllaceae)	80 ± 14	73 ± 7
Eucrypta chrysanthemifolia (Hydrophyllaceae)	95 ± 2	91 ± 9
Lotus scoparius (Fabaceae)	96 ± 2	95 ± 4
Phacelia grandiflora (Hydrophyllaceae)	6 ± 5	4 ± 3

ecosystems prone to stand-replacing fire regimes (Fig. 13.1). Characteristics in common with such systems include predictable but infrequent ignitions and moderate site productivity, sufficient to accumulate woody fuels but insufficient for arborescent growth forms capable of 'outgrowing' fire effects (e.g. high canopy, self-pruning, thick bark, etc.). These are mostly Mediterranean-climate shrublands, where the winter/spring rainfall occurs under mild temperatures, conducive to moderate biomass production, but the severe summer drought limits the stature of the vegetation, maintaining a dense flammable vegetation capable of fire spread through the canopy and over large areas. Other vegetation types conducive to stand-replacing fire regimes, such as densely stocked *Pinus contorta* forests at mid-elevations in the Rocky Mountains and high-latitude small-stature *Pinus banksiana* forests, are also dominated by postfire recruitment from serotinous cones. These forests, however, generally lack an understorey vegetation with dormant seed banks that cue seed recruitment to postfire conditions. Most grasslands are subject to frequent stand-replacing fires, but, with a few exceptions, these species are not known to establish dormant seed banks dependent upon fire for recruitment.

Risks associated with fire-dependent recruitment

In woody plants, vegetative regeneration from underground stems or roots is often associated with postfire regeneration, and this provides a measure of protection in the event of seed recruitment failure following fire. However, with the exception of certain gymnosperms, there is little evidence that such resprouting has evolved in response to fire, as it is almost ubiquitous in woody dicotyledonous plants (Keeley, 1981; Keeley and Zedler, 1998). In this light, it is particularly interesting that, in the most fire-prone habitats (i.e. shrublands of California, South Africa and Western Australia), shrub species in a number of genera have evolutionarily lost the ability to resprout and thus are dependent upon seedling recruitment. These obligate-seeding species are vulnerable to localized extirpation if fire regimes fall outside their range of tolerance, in terms of time to maturity, adult longevity and seed-bank persistence (Parker and Kelly, 1989). They

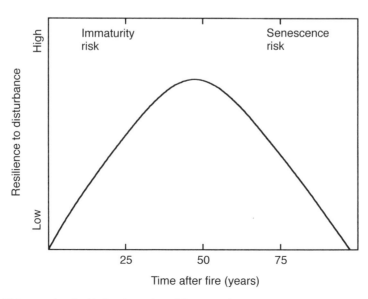

Fig. 13.4. Risks associated with fire-dependent obligate seed regeneration.

face an 'immaturity risk' (Fig. 13.4) if fires are too frequent, i.e. occur before sufficient seed banks have established. On the other hand, if the fire return interval exceeds the lifespan of the shrubs and their seed bank, the population faces a 'senescence risk'. At present, humans contribute sufficiently to the fire regime in most shrubland regions (e.g. Keeley et al., 1999b) for the senescence risk to be minimal (cf. Bond, 1980); however, such is not the case for the immaturity risk. There are a number of well-documented cases where human intervention in fire regimes has increased fire frequency sufficiently to threaten fire-dependent species (van Wilgen and Kruger, 1981; Zedler et al., 1983; Haidinger and Keeley, 1993; Keeley et al., 1999a).

Another risk is the potential loss of inclusive fitness due to delayed reproduction (e.g. Schaffer, 1974). Various arguments have been proposed to account for the selective factors involved in the evolution of dormancy (Cohen, 1968). Ellner's (1986) hypothesized parent–offspring conflict in seed germination would be interesting to examine in these fire-type species. Germination of heat-shock-stimulated seeds is controlled by maternal seed-coat tissues, but smoke-stimulated species appear to have endogenous (embryonic) control of germination. We posit that due to the time-varying nature of fire-prone ecosystems, rapid germination is avoided, because the chances for successful recruitment are nil in competition with the dominant vegetation.

Life-history syndromes

Fire-dependent recruitment appears to be tied to other life-history characteristics, including tolerance to drought stress and shading (Keeley, 1998). Dormant soil-stored seed banks triggered to germinate by smoke/charred wood (and heat shock) are not randomly distributed across growth forms, but rather are common in annuals and shrubby perennials, uncommon in trees and rare in herbaceous perennials (Keeley, 1991, 1995; Keeley and Bond, 1997). In California chaparral, species with

dormant seed banks have markedly smaller seeds than those lacking dormancy (Keeley, 1991), a phenomenon observed in other vegetation types as well (Rees, 1996). Species with fire-dependent recruitment are predominantly passive seed dispersers – seeds are dispersed in time rather than in space because, when fires occur, they generate gaps of extraordinary size. Indeed, animal-dispersed seeds in most Mediterranean-climate shrublands lack seed dormancy and animals play a crucial role in targeting seeds to localized safe sites present during the fire-free interval (Keeley, 1991, 1992b, 1995).

Phylogenetic distribution of fire-dependent regeneration

Serotiny versus soil-borne seed banks

Distribution of canopy- versus soil-stored seeds differs globally. In the northern hemisphere, serotiny is largely restricted to conifer trees – primarily Pinus and Cupressus – and soil-stored seed banks are widespread in angiosperm shrubs. Hypotheses for why gymnosperms have evolved to adopt serotiny rather than soil seed banks include: (i) lack of genetic potential for the evolution of seed-coat/embryo characteristics that enforce dormancy and cue germination to heat or smoke; (ii) selection for serotiny is linked to other cone characteristics that reduce predation; and (iii) serotinous conifer habitats have moderate productivity and long intervals between fires, resulting in forest-floor fuel loads that essentially sterilize soil-stored seeds when a fire occurs (Thomas and Wein, 1994; Ne'eman, 1997; Keeley and Zedler, 1998). Concomitantly, lack of serotiny in northern-hemisphere angiosperms may be tied to fuel structure in these lower-stature shrubs, which leads to fire intensities sufficient to decimate above-ground seed banks.

In the southern hemisphere, particularly in the South African and Western Australian Mediterranean-climate shrublands, serotiny is widespread in angiosperm shrubs (Lamont *et al.*, 1991). This is in sharp contrast to the lack of serotiny in California chaparral and we hypothesize that the infertile soils typical of these southern-hemisphere sites reduce production and consequently reduce fire intensity (van Wilgen and van Hensbergen, 1992), thus making canopy seed storage less likely to succumb to fire than would be the case in California chaparral. Also, the nutrient-poor soils of these regions place a premium on nutrient-rich seeds and thus another selective advantage of canopy over soil storage may be predator avoidance (Keeley, 1992a).

Heat- and smoke-stimulated germination

No detailed phylogenetic studies of fire-stimulated germination are available. However, some generalizations are reasonable. Based on the taxonomic distribution of smoke- versus heat-stimulated germination, it is clear that there is a strong phylogenetic component to these germination responses. For example, countless numbers of species with heat-stimulated germination are known worldwide for the *Fabaceae* and yet no example has been reported for smoke-stimulated germination in this family. It appears that this generalization applies equally well to other families with heat-stimulated germination, such as the *Rhamnaceae* and *Malvaceae*. These families are also tied together in terms of their seed-coat structure – a relatively conservative trait (Atwater, 1980).

Smoke-stimulated germination in distantly related families in the northern and southern hemispheres is reasonably interpreted as convergent evolution, although

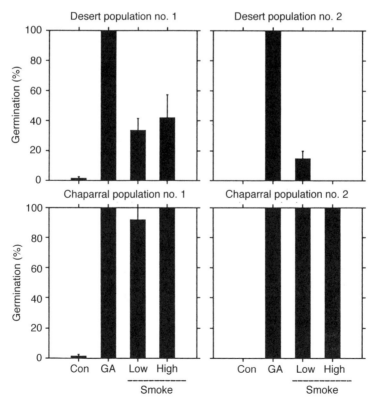

Fig. 13.5. Germination response for *Emmenanthe* populations from desert and chaparral; Con, control; GA, gibberellic acid (GA$_3$); low smoke, half sheet of a 10 min smoked filter-paper, high smoke, whole smoked filter-paper (redrawn from Fotheringham, 1999). GA$_3$ germination demonstrates high viability in all populations. (Where columns are absent, germination was zero.)

within family lineages it may have arisen one or more times.

Recent studies have focused on widespread species distributed in both fire-prone chaparral and non-fire-prone desert habitats (Fotheringham, 1999). Of particular interest is the observation that desert populations of smoke-stimulated chaparral species have a low level of smoke-triggered germination (Fig. 13.5). Smoke-stimulated germination has also been demonstrated in arid karoo species of South Africa, where fires are rare, and it has been suggested that this observation calls into question the adaptive role of smoke-stimulated germination in fire-prone habitats (Pierce *et al.*, 1995). There are alternative explanations. In some chaparral species, it is apparent that nitrogen oxides in smoke are the chemical cue for germination. Biotic production of nitrogen oxides are additional sources and may even be more important triggers in recent burns or even in non-fire-prone environments. In deserts, such cues would be associated with resource islands in and around shrubs. Consistent with this notion is the fact that there is significant microhabitat segregation – smoke (or nitrogen oxide)-stimulated germination is commonly found in species growing in association with shrubs (Fotheringham, 1999).

Indirect effects of fire on seed regeneration

Light

European heathland species accumulate substantial seed banks, which contribute to postfire recruitment, and yet there is little evidence that they are directly stimulated by heat or fire-related chemicals (Hobbs *et al.*, 1984; Mallik *et al.*, 1984; Mallik and Gimingham, 1985). Two of the more common heath species, *Calluna vulgaris* and *Erica tetralix*, have seeds that are light-stimulated and it is proposed that burning, as well as other disturbances, open the site sufficiently for light to trigger germination (Pons, 1989a).

Alternating temperatures

Some species accumulate seed banks with innate dormancy that is broken by the more extreme diurnal fluctuation in soil temperatures on sites opened up by fire (Brits, 1986; Pierce and Moll, 1994). These are typically short-lived seed banks, which also recruit on other open sites and are not strictly tied to fire.

Colonization

In many conifer forests, trees and shrubs are highly dependent upon disturbances such as fire to produce gaps suitable for seedling recruitment. Natural lightning-ignited fires are frequent in these forests and, because of the high site productivity, woody fuels lead to a mixed-intensity stand-thinning fire regime (Fig. 13.1). While much of the forest remains intact, gaps of the order of 10^2–10^4 m^2 are common. Species such as *Pinus ponderosa* often require these gaps for successful seedling recruitment, but, because they do not maintain dormant seed banks, they also require a seed source within dispersal distance. Mature trees typically survive these fires, because site productivity is sufficient to promote rapid growth, which maintains the canopy of many trees above the flame length, self-pruning of dead branches limits fire spread into the canopy and thick bark protects stems at flame level (Keeley and Zedler, 1998).

Other species, such as *Epilobium angustifolium* (Solbreck and Andersson, 1987) or *Betula* spp. (Hobbs *et al.*, 1984), have long-distance wind-dispersed seeds capable of colonizing distant burned sites. Such species typically have no dormancy and establish rapidly upon encountering suitable growing conditions (Romme *et al.*, 1995). Success of these colonizing species is controlled by burn severity, which affects gap size and competition, and by fire size, which affects proximity of seed sources (Turner *et al.*, 1997). These species commonly recruit on burned sites but they are capable of capitalizing on other disturbances as well.

Fire-induced flowering

A wide variety of species exhibit varying degrees of flowering restriction to postfire sites. The most extreme case is that of the South African lily *Cyrtanthus ventricosus,* which has only ever been observed to flower after fire. Generally, it flowers within a week following fire, regardless of the season (Le Maitre and Brown, 1992), and flowering is triggered by smoke (Keeley, 1993). Other geophytes in several fire-prone habitats show a marked flowering response in the first growing season after fire, tied to increased light or nutrients or alternating soil temperatures. The demography of such species has not been studied in much detail, but seedling recruitment appears to be restricted to subsequent postfire years. The typical pattern is resprouting from buried vegetative structures that survive fire and flowering and dispersing seeds in the first postfire year, followed by massive seedling recruitment in the second year. This pattern is evident in many geophytes and also perennial grasses in grasslands and savannahs, as well as subshrubs in coastal scrub vegetation.

Bamboo fire cycle

It is a widespread characteristic in long-lived woody bamboos to delay flowering and seed production for many decades and then to synchronize flowering and fruiting at the landscape scale, followed by massive mortality of bamboo clones. There are numerous reports in the literature of a linkage between bamboo mortality, wildfires and seedling recruitment, e.g. 'fierce forest fires followed the death of the bamboo to be accompanied almost immediately thereafter by copious natural regeneration' (Kadambi, 1949). It has been hypothesized that delayed reproduction, mast flowering and semelparity are a character syndrome selected to promote catastrophic wildfires, which open forest canopies and maintain the early successional stages necessary for seedling success and long-term persistence of clones (Keeley and Bond, 1999). In essence, mast mortality has been selected as a niche construction mechanism (e.g. Odling-Smee *et al.*, 1996) that increases the likelihood of wildfires and promotes successful recruitment.

References

Abrams, M.D. and Dickmann, D.I. (1984) Apparent heat stimulation of buried seeds of *Geranium bicknellii* on jack pine sites in northern lower Michigan. *Michigan Botanist* 23, 81–88.

Angevine, M.W. and Chabot, B.F. (1979) Seed germination syndromes in higher plants. In: Solbrig, O.T., Jain, S., Johnson, G.B. and Raven, P.H. (eds) *Topics in Plant Population Biology*. Columbia University Press, New York, pp. 188–206.

Arianoutsou, M. and Thanos, C.A. (1996) Legumes in the fire-prone Mediterranean region: an example from Greece. *International Journal of Wildland Fire* 6, 77–82.

Atwater, B.R. (1980) Germination, dormancy and morphology of the seeds of herbaceous ornamental plants. *Seed Science and Technology* 8, 523–573.

Baldwin, I.T. and Morse, L. (1994) Up in smoke. 2. Germination of *Nicotiana attenuata* in response to smoke-derived cues and nutrients in burned and unburned soils. *Journal of Chemical Ecology* 20, 2373–2392.

Baldwin, I.T., Stasza Kozinski, L. and Davidson, R. (1994) Up in smoke. 1. Smoke-derived germination cues for postfire annual, *Nicotiana attenuata* Torr ex Watson. *Journal of Chemical Ecology* 20, 2345–2372.

Ballard, L.A.T. (1981) Physical barriers to germination. *Seed Science and Technology* 1, 285–303.

Barro, S.C. and Poth, M. (1988) Seeding and sprouting *Ceanothus* species: their germination responses to heat. In: di Castri, F., Floret, C., Rambal, S. and Roy, J. (eds) *Time Scales and Water Stress. Proceedings of the 5th International Conference on Mediterranean Ecosystems (MEDECOS V)*. International Union of Biological Sciences, Paris, pp. 155–158.

Baskin, C.C. and Baskin, J.M. (1998) *Seeds. Ecology, Biogeography, and Evolution of Dormancy and Germination*. Academic Press, San Diego.

Baskin, J.M. and Baskin, C.C. (1997) Methods of breaking seed dormancy in the endangered species *Iliamna corei* (Sherff) Sherr (Malvaceae), with special attention to heating. *Natural Areas Journal* 17, 313–323.

Baxter, B.J.M., Staden, J.V., Granger, J.E. and Brown, N.A.C. (1994) Plant-derived smoke and smoke extracts stimulate seed germination of the fire climax grass *Themeda triandra. Environmental and Experimental Botany* 34, 217–223.

Bell, D.T. (1994) Interaction of fire, temperature and light in the germination response of 16 species from the *Eucalyptus marginata* forest of south-western Western Australia. *Australian Journal of Botany* 42, 501–509.

Bell, D.T., Plummer, J.A. and Taylor, S.K. (1993) Seed germination ecology in southwestern Western Australia. *Botanical Review* 59, 24–73.

Bond, W.J. (1980) Fire and senescent fynbos in the Swartberg, Southern Cape. *South African Forestry Journal* 114, 68–71.

Brits, G.J. (1986) Influence of fluctuating temperatures and H_2O_2 treatment on the germination of *Leucospermum cordifolium* and *Serruria florida* (Proteaceae) seeds. *South African Journal of Botany* 123, 286–290.

Brits, G.J., Calitz, F.J., Brown, N.A.C. and Manning, J.C. (1993) Desiccation as the active principle in heat-stimulated seed germination of *Leucospermum* R.Br. (Proteaceae) in fynbos. *New Phytologist* 125, 397–403.

Brits, G.J., Cutting, J.G.M., Brown, N.A.C. and van Staden, J. (1995) Environmental and hormonal regulation of seed dormancy and germination in Cape fynbos *Leucospermum* R.Br. (Proteaceae) species: a working model. *Plant Growth Regulation* 17, 181–193.

Brown, N.A.C. (1993a) Promotion of germination of fynbos seeds by plant-derived smoke. *New Phytologist* 123, 185–192.

Brown, N.A.C. (1993b) Seed germination in the fynbos fire ephemeral, *Syncarpha vestita* (L.) B.Nord. is promoted by smoke, aqueous extracts of smoke and charred wood derived from burning the ericoid-leaved shrub, *Passerina vulgaris* Thoday. *International Journal of Wildland Fire* 3, 203–206.

Christensen, N.L. (1973) Fire and the nitrogen cycle in California chaparral. *Science* 181, 66–68.

Christensen, N.L. and Muller, C.H. (1975) Effects of fire on factors controlling plant growth in *Adenostoma* chaparral. *Ecological Monographs* 45, 29–55.

Cocks, M.P. and Stock, W.D. (1997) Heat stimulated germination in relation to seed characteristics in fynbos legumes of the western Cape Province, South Africa. *South African Journal of Botany* 63, 129–132.

Cohen, D. (1968) A general model of optimal reproduction in a randomly varying environment. *Journal of Ecology* 56, 219–228.

Cohn, M.A. (1989) Factors influencing the efficacy of dormancy-breaking chemicals. In: Taylorson, R.B. (ed.) *Recent Advances in the Development and Germination of Seeds*. Plenum Press, New York, pp. 261–267.

Cohn, M.A. (1996) Chemical mechanisms of breaking seed dormancy. *Seed Science Research* 6, 95–99.

Cohn, M.A. and Castle, L. (1984) Dormancy in red rice. IV. Response of unimbibed and imbibing seeds to nitrogen dioxide. *Plant Physiology* 60, 552–556.

Cohn, M.A. and Hughes, J.A. (1986) Seed dormancy in red rice. VI. Response to azide, hydroxylamine, and cyanide. *Plant Physiology* 80, 531–533.

Cohn, M.A., Butera, D.L. and Hughes, J.A. (1983) Seed dormancy in red rice. III. Response to nitrite, nitrate, and ammonium ions. *Plant Physiology* 73, 381–384.

Cohn, M.A., Chiles, L.A., Hughes, J.A. and Boullion, K.J. (1987) Seed dormancy in red rice. VI. Monocarboxylic acids: a new class of pH-dependent germination stimulants. *Plant Physiology* 84, 716–719.

Cushwa, C.T., Martin, R.E. and Miller, R.T. (1968) The effects of fire on seed germination. *Journal of Range Management* 21, 250–254.

Davis, F.W., Borchert, M.I. and Odion, D.C. (1989) Establishment of microscale vegetation pattern in maritime chaparral after fire. *Vegetatio* 84, 53–67.

DeSimone, S.A. and Zedler, P.H. (1999) Shrub seedling recruitment in unburned Californian coastal sage scrub and adjacent grassland. *Ecology* 80, 2018–2032.

Dixon, K.W., Roche, S. and Pate, J.S. (1995) The promotive effect of smoke derived from burnt native vegetation on seed germination of Western Australian plants. *Oecologia* 101, 185–192.

Dunn, P.H. and DeBano, L.F. (1977) Fire's effect on biological and chemical properties of chaparral soils. In: Mooney, H.A. and Conrad, C.E. (eds) *Proceedings of the Symposium on Environmental Consequences of Fire and Fuel Management in Mediterranean Ecosystems*. USDA Forest Service, Washington, DC, pp. 75–84.

Ellner, S. (1986) Germination dimorphisms and parent–offspring conflict in seed germination. *Journal of Theoretical Biology* 123, 173–185.

Enright, N.J., Goldblum, D., Ata, P. and Ashton, D.H. (1997) The independent effects of heat, smoke and ash on emergence of seedlings from the soil seed bank of a healthy *Eucalyptus* woodland in Grampians (Gariwerd) National Park, western Victoria. *Australian Journal of Ecology* 22, 81–88.

Fletcher, F. (1910) Effect of previous heating of the soil on the growth of plants and the germination of seeds. *Cairo Scientific Journal* 4, 81–86.

Fotheringham, C.J. (1999) Spatial and temporal factors influencing desert annual seed germination behavior. Unpublished MS thesis, California State University, Los Angeles.

Gadgil, M. and Bossert, W.H. (1970) Life historical consequences of natural selection. *American Naturalist* 104, 1–24.

Grigore, M.T. and Tramer, E.J. (1996) The short-term effect of fire on *Lupinus perennis* (L.). *Natural Areas Journal* 16, 41–48.

Haidinger, T.L. and Keeley, J.E. (1993) Role of high fire frequency in destruction of mixed chaparral. *Madroño* 40, 141–147.

Hanley, M.E. and Fenner, M. (1998) Pre-germination temperature and the survivorship and onward growth of Mediterranean fire-following plant species. *Acta Oecologica* 19, 181–187.

Harper, J.L. (1977) *Population Biology of Plants.* Academic Press, New York.

Harrington, J.P. (1932) *Tobacco among the Karok Indians of California.* Bureau of American Ethnology Bulletin 94, Smithsonian Institution, Washington, DC.

Hendricks, S.B. and Taylorson, R.B. (1972) Promotion of seed germination by nitrates and cyanides. *Nature* 237, 169–170.

Henig-Sever, N., Eshel, A. and Ne'eman, G. (1996) pH and osmotic potential of pine ash as post-fire germination inhibitors. *Physiologia Plantarum* 96, 71–76.

Herranz, J.M., Gerrandis, P. and Martinez-Sanchez, J.J. (1998) Influence of heat on seed germination of seven Mediterranean Leguminosae species. *Plant Ecology* 136, 95–103.

Hilhorst, H.W.M. and Karssen, C.M. (1989) The role of light and nitrate in seed germination. In: Taylorson, R.B. (ed.) *Recent Advances in the Development and Germination of Seeds.* Plenum Press, New York, pp. 191–205.

Hobbs, R.J., Mallik, A.U. and Gimingham, C.H. (1984) Studies on fire in Scottish heathland communities. III. Vital attributes of the species. *Journal of Ecology* 72, 963–976.

Jaynes, R.A. (1968) Breaking seed dormancy of *Kalmia hirsuta* with high temperatures. *Ecology* 49, 1196–1198.

Jefferey, D.J., Holmes, P.M. and Rebelo, A.G. (1988) Effects of dry heat on seed germination in selected indigenous and alien legume species in South Africa. *South African Journal of Botany* 54, 28–34.

Jones, C.S. and Schlesinger, W.H. (1980) *Emmenanthe penduliflora* (Hydrophyllaceae): further consideration of germination response. *Madroño* 27, 122–125.

Kadambi, K. (1949) On the ecology and silviculture of *Dendrocalamus strictus* in the bamboo forests of Bhadrvati Division, Mysore State, and comparative notes on the species *Bambusa arundinacea, Ochlandra travancorica, Oxytenanthera monostigma* and *O. stocksii. Indian Forester* 75, 289–299.

Keeley, J.E. (1981) Reproductive cycles and fire regimes. In: Mooney, H.A., Bonnicksen, T.M., Christensen, N.L., Lotan, J.E. and Reiners, W.A. (eds) *Proceedings of the Conference Fire Regimes and Ecosystem Properties.* General Technical Report WO-26, USDA Forest Service, Washington, DC, pp. 231–277.

Keeley, J.E. (1987) Role of fire in seed germination of woody taxa in California chaparral. *Ecology* 68, 434–443.

Keeley, J.E. (1991) Seed germination and life history syndromes in the California chaparral. *Botanical Review* 57, 81–116.

Keeley, J.E. (1992a) A Californian's view of fynbos. In: Cowling, R.M. (ed.) *The Ecology of Fynbos.* Oxford University Press, Cape Town, pp. 372–388.

Keeley, J.E. (1992b) Temporal and spatial dispersal syndromes. In: Thanos, C.A. (ed.) *MEDECOS VI. Proceedings of the 6th International Conference on Mediterranean Climate Ecosystems, Plant–animal Interactions in Mediterranean-type ecosystems.* University of Athens, Athens, Greece, pp. 251–256.

Keeley, J.E. (1993) Smoke-induced flowering in the fire-lily *Cyrtanthus ventricosus. South African Journal of Botany* 59, 638.

Keeley, J.E. (1995) Seed germination patterns in fire-prone Mediterranean-climate regions. In: Arroyo, M.T.K., Zedler, P.H. and Fox, M.D. (eds) *Ecology and Biogeography of Mediterranean Ecosystems in Chile, California and Australia.* Academic Press, San Diego, pp. 239–273.

Keeley, J.E. (1998) Coupling demography, physiology and evolution in chaparral shrubs. In: Rundel, P.W., Montenegro, G. and Jaksic, F.M. (eds) *Landscape Diversity and Biodiversity in Mediterranean-type Ecosystems.* Springer, New York, pp. 257–264.

Keeley, J.E. and Bond, W.J. (1997) Convergent seed germination in South African fynbos and Californian chaparral. *Plant Ecology* 133, 153–167.

Keeley, J.E. and Bond, W.J. (1999) Mast flowering and semelparity in bamboo: the fire cycle hypothesis. *American Naturalist* 154, 383–391.

Keeley, J.E. and Fotheringham, C.J. (1997) Trace gas emissions in smoke-induced seed germination. *Science* 276, 1248–1250.

Keeley, J.E. and Fotheringham, C.J. (1998a) Mechanism of smoke-induced seed germination in a postfire chaparral annual. *Journal of Ecology* 86, 27–36.

Keeley, J.E. and Fotheringham, C.J. (1998b) Smoke-induced seed germination in California chaparral. *Ecology* 79, 2320–2336.

Keeley, J.E. and Keeley, M.B. (1999) Role of charred wood, heat-shock, and light in germination of postfire *Phrygana* species from the eastern Mediterranean Basin. *Israel Journal of Plant Sciences* 47, 11–16.

Keeley, J.E. and Keeley, S.C. (1989) Allelopathy and the fire induced herb cycle. In: Keeley, S.C. (ed.) *The California Chaparral: Paradigms Reexamined*. Natural History Museum of Los Angeles County, Los Angeles, pp. 65–72.

Keeley, J.E. and Nitzberg, M.E. (1984) The role of charred wood in the germination of the chaparral herbs *Emmenanthe penduliflora* (Hydrophyllaceae) and *Eriophyllum confertiflorum* (Asteraceae). *Madroño* 31, 208–218.

Keeley, J.E. and Zedler, P.H. (1998) Evolution of life histories in *Pinus*. In: Richardson, D.M. (ed.) *Ecology and Biogeography of Pines*. Cambridge University Press, Cambridge, UK, pp. 219–251.

Keeley, J.E., Morton, B.A., Pedrosa, A. and Trotter, P. (1985) Role of allelopathy, heat, and charred wood in the germination of chaparral herbs and suffrutescents. *Journal of Ecology* 73, 445–458.

Keeley, J.E., Ne'eman, G. and Fotheringham, C.J. (1999a) Immaturity risk in a fire dependent pine. *Journal of Mediterranean Ecology* 1, 41–47.

Keeley, J.E., Fotheringham, C.J. and Morais, M. (1999b) Reexamining fire suppression impacts on brushland fire regimes. *Science* 284, 1829–1832.

Keeley, S.C. and Pizzorno, M. (1986) Charred wood stimulated germination of two fire-following herbs of the California chaparral and the role of hemicellulose. *American Journal of Botany* 73, 1289–1297.

Kilian, D. and Cowling, R.M. (1992) Comparative seed biology and co-existence of two fynbos shrub species. *Journal of Vegetation Science* 3, 637–646.

Koller, D. (1972) Environmental control of seed germination. In: Kozlowski, T.T. (ed.) *Seed Biology*, Vol. II. Academic Press, New York, pp. 1–107.

Lamont, B.B., Maitre, D.C.L., Cowling, R.M. and Enright, N.J. (1991) Canopy seed storage in woody plants. *Botanical Review* 57, 277–317.

Le Maitre, D.C. and Brown, P.J. (1992) Life cycles and fire-stimulated flowering in geophytes. In: van Wilgen, B.W., Richardson, D.M., Kruger, F.J. and van Hensbergen, H.J. (eds) *Fire in South African Mountain Fynbos*. Springer-Verlag, Berlin, pp. 145–160.

McCormac, J.S. and Windus, J.L. (1993) Fire and *Cuscuta glomerata* Choisy in Ohio: a connection? *Rhodora* 95, 158–165.

McPherson, J.K. and Muller, C.H. (1969) Allelopathic effects of *Adenostoma fasciculatum*, 'chamise,' in the California chaparral. *Ecological Monographs* 39, 177–198.

Mallik, A.U. and Gimingham, C.H. (1985) Ecological effects of heather burning. II. Effects of seed germination and vegetative regeneration. *Journal of Ecology* 73, 633–644.

Mallik, A.U., Hobbs, R.J. and Legg, C.J. (1984) Seed dynamics in *Calluna–Arctostaphylos* heath in north-eastern Scotland. *Journal of Ecology* 72, 855–871.

Marks, P.L. (1979) Apparent fire-stimulated germination of *Rhus typhina* seeds. *Bulletin of the Torrey Botanical Club* 106, 41–42.

Martin, R.E., Miller, R.L. and Cushwa, C. (1975) Germination response of legume seeds subjected to moist and dry heat. *Ecology* 56, 1441–1445.

Muller, C.H., Muller, W.H. and Haines, B.L. (1964) Volatile growth inhibitors produced by aromatic shrubs. *Science* 143, 471–473.

Musil, C.F. and de Witt, D.M. (1991) Heat-stimulated germination in two Restionaceae species. *South African Journal of Botany* 57, 175–176.

Ne'eman, G. (1997) Regeneration of natural pine forest – review of work done after the 1989 fire in Mount Carmel, Israel. *International Journal of Wildland Fire* 7, 295–306.

Ne'eman, G., Henig-Sever, N. and Eshel, A. (1999) Regulation of the germination of *Rhus coriaria*, a post-fire pioneer, by heat, ash, pH, water potential and ethylene. *Physiologia Plantarum* 106, 47–52.

Odling-Smee, F.J., Laland, K.N. and Feldman, M.W. (1996) Niche construction. *American Naturalist* 147, 641–648.

O'Dowd, D.J. and Gill, A.M. (1984) Predator satiation and site alteration following fire: mass reproduction of alpine ash (*Eucalyptus delegatensis*) in southeastern Australia. *Ecology* 65, 1052–1066.

Parker, V.T. (1987) Effects of wet-season management burns on chaparral vegetation: implications for rare species. In: Elias, T.S. (ed.) *Conservation and Management of Rare and Endangered Plants.* California Native Plant Society, Sacramento, pp. 233–237.

Parker, V.T. and Kelly, V.R. (1989) Seed banks in California chaparral and other Mediterranean climate shrublands. In: Leck, M.A., Parker, V.T. and Simpson, R.L. (eds) *Ecology of Soil Seed Banks.* Academic Press, New York, pp. 231–255.

Pierce, S.M. and Moll, E.J. (1994) Germination ecology of six shrubs in fire-prone Cape fynbos. *Vegetatio* 110, 25–41.

Pierce, S.M., Esler, K. and Cowling, R.M. (1995) Smoke-induced germination of succulents (Mesembryanthemaceae) from fire-prone and fire-free habitats in South Africa. *Oecologia* 102, 520–522.

Pons, T.L. (1989a) Dormancy and germination of *Calluna vulgaris* (L.) Hull and *Erica tetralix* L. seeds. *Acta Oecologica* 10, 35–43.

Pons, T.L. (1989b) Breaking of seed dormancy by nitrate as a gap detection mechanism. *Annals of Botany* 63, 139–143.

Preston, C.A. and Baldwin, I.T. (1999) Positive and negative signals regulate germination in the post-fire annual, *Nicotiana attenuata. Ecology* 80, 481–494.

Quinn, R.D. (1994) Animals, fire, and vertebrate herbivory in Californian chaparral and other Mediterranean-type ecosystems. In: Moreno, J.M. and Oechel, W.C. (eds) *The Role of Fire in Mediterranean-type Ecosystems.* Springer-Verlag, New York, pp. 46–77.

Raven, J.A. and Yin, Z.H. (1998) The past, present and future of nitrogenous compounds in the atmosphere, and their interactions with plants. *New Phytologist* 139, 205–219.

Read, T.R. and Bellairs, S.M. (1999) Smoke affects the germination of native grasses of New South Wales. *Australian Journal of Botany* 47, 563–576.

Rees, M. (1996) Evolutionary ecology of seed dormancy and seed size. *Philosophical Transactions of the Royal Society of London, B* 351, 1299–1308.

Roche, S., Dixon, K.W. and Pate, J.S. (1997) Seed ageing and smoke: partner cues in the amelioration of seed dormancy in selected Australian native species. *Australian Journal of Botany* 45, 783–815.

Romme, W.H., Bohland, L., Persichetty, C. and Caruso, T. (1995) Germination ecology of some common forest herbs in the Yellowstone National Park, Wyoming, USA. *Arctic and Alpine Research* 27, 407–412.

Sabiiti, E.N. and Wein, R.W. (1987) Fire and *Acacia* seeds: a hypothesis of colonization success. *Journal of Ecology* 74, 937–946.

Schaffer, W.M. (1974) Optimal reproductive effort in fluctuating environments. *American Naturalist* 103, 783–790.

Skordilis, A. and Thanos, C.A. (1995) Seed stratification and germination strategy in the Mediterranean pines *Pinus brutia* and *P. halepensis. Seed Science Research* 5, 151–160.

Smith, M.T. and van Staden, J. (1995) Infochemicals: the seed–fungus–root continuum. *Environmental and Experimental Biology* 35, 113–123.

Solbreck, C. and Andersson, D. (1987) Vertical distribution of fireweed, *Epilobium angustifolium*, seeds in the air. *Canadian Journal of Botany* 65, 2177–2178.

Thanos, C.A. and Rundel, P.W. (1995) Fire-followers in chaparral: nitrogenous compounds trigger seed germination. *Journal of Ecology* 83, 207–216.

Thanos, C.A., Georghiou, K., Kadis, C. and Pantazi, C. (1992) Cistaceae: a plant family with hard seeds. *Israel Journal of Botany* 41, 251–263.

Thomas, P.A. and Wein, R.W. (1994) Amelioration of wood ash toxicity and jack pine establishment. *Canadian Journal of Forest Research* 24, 748–755.

Trabaud, L. and Oustric, J. (1989) Heat requirements for seed germination of three *Cistus* species in the garrigue of southern France. *Flora* 183, 321–325.

Turner, M.G., Romme, W.H., Gardner, R.H. and Hargrove, W.W. (1997) Effects of fire size and pattern on early succession in Yellowstone National Park. *Ecological Monographs* 67, 411–433.

van Staden, J., Drewes, F.E. and Jager, A.K. (1995) The search for germination stimulants in plant-derived smoke extracts. *South African Journal of Botany* 61, 260–263.

van Wilgen, B.W. and Kruger, F.J. (1981) Observations on the effects of fire in mountain fynbos at Zachariashoek, Paarl. *Journal of South African Botany* 47, 195–212.

van Wilgen, B.W. and van Hensbergen, H.J. (1992) Fuel properties of vegetation in Swartboskloof. In: van Wilgen, B.W., Richardson, D.M., Kruger, F.J. and van Hensbergen, H.J. (eds) *Fire in South African Mountain Fynbos.* Springer-Verlag, Berlin, pp. 37–53.

Washitani, I. (1988) Effects of high temperatures on the permeability and germinability of the hard seeds of *Rhus javanica* L. *Annals of Botany* 62, 13–16.

Wells, C.G., DeBano, L.F., Lewis, C.E., Fredriksen, R.L., Franklin, E.C., Froelich, R.C. and Dunn, P.H. (1979) *Effects of Fire on Soil: a State-of-knowledge Review.* General Technical Report WO-7.1, USDA Forest Service, Washington, DC. 134 pp.

Westermeier, L.J. (1978) Effects of dry and moist heat shocks on seed viability and germination of *Lotus strigosus* and *Lupinus excubitus* var. *hallii*. Unpublished MA thesis, California State University, Fullerton.

Westoby, M. (1981) How diversified seed germination behaviour is selected. *American Naturalist* 118, 882–885.

Wicklow, D.T. (1977) Germination response in *Emmenanthe penduliflora* (Hydrophyllaceae). *Ecology* 58, 201–205.

Wicklow, D.T. (1988) Parallels in the development of post-fire fungal and herb communities. *Proceedings of the Royal Society of Edinburgh* 94B, 87–95.

Wolf, C.B. (1948) Taxonomic and distributional studies of the New World cypresses. *Aliso* 1, 1–250, 325–444.

Zedler, P.H., Gautier, C.R. and McMaster, G.S. (1983) Vegetation change in response to extreme events: the effect of a short interval between fires in California chaparral and coastal scrub. *Ecology* 64, 809–818.

Chapter 14
Ecology of Seedling Regeneration

Kaoru Kitajima[1] and Michael Fenner[2]

[1]*Botany Department, University of Florida, Gainesville, Florida, USA;* [2]*School of Biological Sciences, University of Southampton, Southampton, UK*

Introduction

Seedlings represent the final stage in the process of regeneration from seed. The period between seed germination and the establishment of an independent juvenile plant is one of the most vulnerable in the life cycle. The newly emerged seedling no longer has the ability to withstand the adverse conditions tolerated by the ungerminated seed, but does not yet have the physical robustness it will acquire with age. Mortality in these early stages can be due to a wide range of biotic and abiotic factors, which vary from place to place and from one year to the next (Mack and Pyke, 1984).

Definition of a seedling

The seedling phase can be said to begin with the protrusion of the radicle from the seed (i.e. the externally visible consequences of the internal processes of germination), but there is no agreement as to when the young plant ceases to be a seedling. Often, any small plant grown from a single seed is called a seedling, without any clear indication of the stage of development reached. One possible defini-tion of a seedling is a young plant still dependent on the food reserves stored in the seed. Once dependence is transferred to external sources, the plant can be con-sidered to have passed the seedling stage. However, this definition has certain disad-vantages. The transference of dependence from internal to external sources of miner-als takes place gradually, with no very clear-cut end-point. Moreover, the shift in dependence takes place at different rates for different elements (Krigel, 1967; Fenner, 1986; Fenner and Lee, 1989). The definition would need to specify which particular nutrient was to be monitored. Even if a seedling is continuing to use its stored resources, there is no reason to sup-pose that it is dependent on them. Some tropical tree species drop their cotyledons (with stored nutrients) without apparently making much use of them (Ng, 1978); other species, such as oaks, can lose them through predation without major ill effects (Bossema, 1979; Sonesson, 1994).

Although development of seedling autotrophy is gradual, it is possible to iden-tify the time when seedlings start to utilize external resources by comparing growth curves under contrasting resource avail-ability (Kitajima, 1996b; Fig. 14.1). Growth curves should be identical under

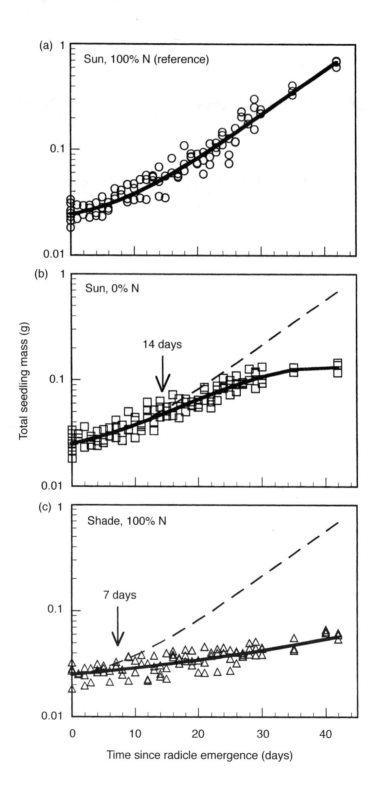

contrasting resource regimes as long as seedlings are completely dependent on the seed reserves for their resource demands. As soon as seedlings start to depend on external supplies of resources, relative growth rate (RGR) under limited resource availability (e.g. limited supply of soil nitrogen or light energy) becomes significantly lower than RGR under optimal resource supply. Kitajima (1992a) found that the start of dependency on external supply of resources occurs earlier for light (Fig. 14.1c) than for soil nitrogen (Fig. 14.1b) in three tropical woody species. For a time, the seedlings are dependent on both seed reserves and external supplies, with an increasing degree of dependence on the latter. As long as the external supplies of resources are adequate and balanced, seedlings grow exponentially until ontogenetic constraints become important. But, if the external supply of, for example, a particular mineral is severely limited, the growth of the seedling becomes negligible after the exhaustion of that element in the seed reserves (Fig. 14.2).

Another possible cut-off point for defining the end of the seedling phase has been proposed by Sattin and Sartorato (1997). Young plants typically follow a sigmoidal growth curve with three phases: a period of exponential growth (with a constant RGR), a period of linear growth (with a declining RGR) and a period of declining absolute growth as the curve approaches the final asymptote. Sattin and Sartorato (1997) propose that the seedling phase be deemed to terminate at the end of the period of exponential growth.

Seed-reserve utilization

In general, seed size is positively correlated with tolerance of seedlings to various abiotic stress factors, including shade, drought, fire or freezing (Westoby *et al.*, 1996). Burke and Grime (1996) found that large-seeded species establish more readily over a wider range of conditions, while smaller-seeded species are more dependent on disturbance. However, how a large seed enhances survivorship of seedlings may not be understood completely without knowing how seed reserves are utilized during germination and seedling establishment. Three important factors that characterize seed-reserve utilization are: (i) the quantity of seed reserves; (ii) the duration of seed-reserve dependency; and (iii) cotyledon functional morphology.

Seed-reserve size

The concentration of individual mineral nutrient reserves (nitrogen, phosphorus, potassium, sulphur, etc.) tends to have a much smaller coefficient of variation than seed mass within and among species (Pate *et al.*, 1985; Oladokun, 1989; Grubb and Coomes, 1997; Grubb and Burslem, 1998; Grubb *et al.*, 1998). As a result, how much of a given resource is available in the seed

---

Fig. 14.1. (*Opposite*) Seedling growth of a tropical tree, *Tabebuia rosea* (*Bignoniaceae*), with foliaceous epigeal cotyledons, under three contrasting conditions. At radicle emergence (day 0), seedlings were transplanted individually to pots filled with washed sand. Nitrogen availability was controlled by flushing the pot daily with 20% strength Johnson solution (100% N = complete nutrient solution containing 2.2 mM NO_3^- and 0.4 mM NH_4^+; 0% N = complete except for nitrogen). Light availability was controlled by a shade cloth (sun = 27% full sun or 6.8 mol photons day^{-1}; shade = 1.2% full sun or 0.3 mol photons day^{-1}). Three plants per treatment were harvested every day for 30 days, then at day 35 and day 42. The third-order polynomial was fitted to log_{10} (biomass) under each growth condition. The mean seed mass was 25.6 mg. (a) Growth of seedlings given sufficient external supply of light and nitrogen exhibited a slight lag during the seed-reserve dependency period (0 – c.14 days), followed by a constant relative growth rate (RGR) (= linear growth in semilog scale). This growth curve is indicated in the other two graphs as a broken line to serve as a reference. (b) Growth of plants without a nitrogen supply significantly dropped below the reference at 14 days after radicle emergence (no longer overlapping the 95% confidence limits). Thus, after 14 days, the seedling demand for nitrogen cannot be met by seed reserves alone. (c) Growth of plants under limited light supply significantly dropped below the reference at 7 days. Thus, after 7 days the seedling's demand for energy cannot be met by the seed reserves alone. (Adapted from Kitajima, 1992a.)

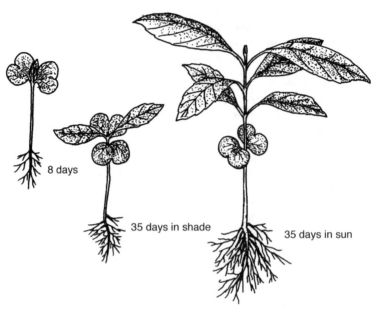

Fig. 14.2. Typical *Tabebuia rosea* seedlings at 8 days (under all growth conditions) and at 35 days after radicle emergence (in shade and sun treatments with 100% N supply). The height and leaf area of 35-day-old seedlings raised under sun + 0% N were similar to those of 35-day-old seedlings grown in shade, except that the leaves and cotyledons were thicker and yellowish due to nitrogen deficiency.

is largely a function of the total seed mass, excluding the seed-coat (i.e. the embryo-cum-endosperm fraction of Grubb and Coomes, 1997). Concentrations of most of the major mineral elements (with the exception of potassium) are usually higher in the embryo-cum-endosperm fraction than in vegetative tissues, while the seed-coat has much lower concentrations (Fenner, 1983; Grubb and Coomes, 1997). Water is generally not a stored reserve, with the possible exception of recalcitrant seeds, which lose viability when relative water content drops below 16–30% of fresh weight (Garwood and Lighton, 1990; Vázquez-Yanes and Orozco-Segovia, 1996). Initially, high water content may be important for large seeds, as large seeds have slower imbibition rates, due to their lower surface : volume ratio than smaller seeds (Kikuzawa and Koyama, 1999). As water uptake is partly determined by the contact between the seed and the soil, the coarseness of the soil topography is an important factor influencing germination (Chambers, 1995).

There appears to be no general relationship between the effective seed nutrient supply and the balance of minerals in the soil environment likely to be encountered by the seedling. Most seeds seem to behave in a broadly similar way, regardless of phylogeny or habitat. In general, species from poor soils have similar concentrations of minerals in their seeds to those from fertile soils (Lee and Fenner, 1989). However, some degree of mineral nutrient enrichment in seeds may occur in seeds of species adapted to very infertile soils. Tall trees in Amazonian caatinga have smaller seeds with higher concentrations of phosphorus and magnesium than tall trees on fertile soils in other lowland rainforest sites (Grubb and Coomes, 1997).

In terms of energy storage, theoretical calculations of biosynthetic pathways suggest that total seedling biomass synthesized from seed reserves alone varies among species from 50 to 80% of the initial seed mass (Penning de Vries and Van Laar, 1977; Kitajima, 1992a). Part of this interspecific variation is explainable by the lipid contents in seeds. Seeds of some taxa may store an exceptionally large quantity of lipids (Vaughan, 1970; Barclay and Earl, 1974). Lipids contain approximately twice the level of chemical bonding energy per gram as starch, the typical energy reserve compound in vegetative tissues (Williams et al., 1987). Levin (1974) recorded higher seed lipid contents in forest species than in species from open habitats, and hypothesized that increased lipid contents may be advantageous in the energy-limited forest understorey. However, energy retrieval from beta-oxidation of lipids in plants is not as efficient as from starch (Chapin, 1989). For example, a 100% increase in energy content per gram of seed tissue by packing lipids may result in only a 40% increase in terms of initial seedling mass (Kitajima, 1992a). Thus, increasing lipid content in seeds is less economical for the parent plant than simply increasing seed mass, unless there is a strong simultaneous selection for large energy reserves, which enhance establishment, and for small-seededness, which enhances dispersal.

The size a seedling can attain in a given time may be a crucial trait determining its fitness. Large seeds give rise to large seedlings. Size is presumed to be an advantage for establishment, as a bigger seed can penetrate from deeper layers of soil, exploit a larger volume of soil more quickly and compete for light more effectively than a small seedling. Mitchell and Allsopp (1984) argue that the large-seeded Australian Hakea sericea is able to invade the phosphorus-deficient soils of the South African fynbos because of the greater absolute phosphorus content of its seeds relative to that in the smaller-seeded native species. However, small seeds may at least partially compensate by their greater concentration of minerals and the higher RGR and larger relative root ratios of their seedlings (Fenner, 1983). These features may not always provide sufficient compensation, especially in poor soils. The reason why well-dispersed small-seeded legumes are seldom found as pioneers in primary succession may be that the seeds may not contain enough nitrogen capital for nodule

formation before the seedling becomes nitrogen-independent through symbiotic fixation (Grubb, 1986).

Duration of seed-reserve dependency

The early seedling stage is developmentally and physiologically dynamic. During germination, a developing embryo is completely dependent on maternally derived nutrients stored in seeds. During the period that the seedling is developing organs for autotrophic resource acquisition (such as foliaceous cotyledons and leaves, as well as roots for water and mineral uptake), it depends largely on seed reserves. Even after seed reserves are exhausted (or are no longer available, due to the abscission of cotyledons), a significant proportion of the seedling biomass may be attributable to maternal provisions in seeds, rather than to the resources autotrophically acquired by the seedling itself. Hence, seedling size is strongly correlated with seed size when the first leaf becomes fully expanded (Howe and Richter, 1982; Kitajima, 1992a; Cornelissen et al., 1998). Interspecific variation in growth rates may dampen this relationship with time. The size of shade-suppressed tree seedlings, however, may be a function of seed size more than a year after germination (Augspurger, 1984b; Ernst, 1988; Osunkoya et al., 1994; Canham et al., 1999).

The minerals stored in seeds are markedly unbalanced, in the sense that the supply of certain elements becomes insufficient to meet the seedling's requirements long before the others. This can be shown by growing seedlings in nutrient solutions with one element missing. This forces the seedling to utilize its internal reserves of the missing element, and so the maximum size attained by the seedling is a measure of the effective internal supply of that element. Experiments on a range of species (Fenner, 1986; Fenner and Lee, 1989; Hanley and Fenner, 1997) show that certain elements (notably nitrogen) are depleted very quickly, leaving the seedling dependent on external supplies within a few days of germination. Other elements (e.g. sulphur) are relatively well supplied by the seed, and the seedling can live off its seed store for a considerable period, provided that the other elements are supplied externally. Figure 14.1b illustrates the considerable degree of seedling development that can be achieved without an external supply of nitrogen in a tropical-forest species.

Proteaceae, a successful family in fire-frequented sites on infertile soils in Australia, includes members with unusually high concentrations of nitrogen and phosphorus (Pate *et al.*, 1985; Stock *et al.*, 1990; Grubb *et al.*, 1998). Nitrogen and phosphorus contents in the whole plants of these species do not increase in the first 80–120 days after germination, indicating an unbalanced enrichment of seeds with these nutrients, which complements the relatively high availability of calcium, magnesium and potassium in the soil after fire (Stock *et al.*, 1990). Among three *Bignoniaceae* species in Panama, seedlings of *Callichlamys latifolius*, with a high nitrogen concentration in seeds (8.2% of dry mass, excluding seed-coat), show a longer independence of external resources than the other two species, which have much lower nitrogen concentrations (3.7 and 4.3%) (Kitajima, 1996a). High nitrogen storage in seeds may lower the seedlings' energy demands for nitrate reduction, and hence may be advantageous for seedlings living near the photosynthetic light compensation point in a deeply shaded understorey (Burslem *et al.*, 1995). Poor soil species may compensate to some degree by having larger seeds, thereby increasing the absolute amount of nutrients available (Lee and Fenner, 1989; Maranon and Grubb, 1993; Milberg and Lamont, 1997).

Cotyledon functional morphology

There is a great diversity of cotyledon types, especially among tropical tree species (Duke, 1965; Bokdam, 1977; Ng, 1978; de Vogel, 1980; Hladik and Miquel, 1990; Garwood, 1996; Kitajima, 1996a).

Epigeal and hypogeal germination refers to whether cotyledons are raised above the ground or remain at or below the ground level during germination. Within each type, there is variation in the degree of cotyledon exposure and thickness, which is related to the primary function of mature cotyledons (Duke, 1965; Garwood, 1996). Reserve-type cotyledons either store nutrients within their tissues or absorb nutrients stored in endosperm during germination (Garwood, 1996). Many reserve-type cotyledons remain inside the seed-coat (cryptocotylar seedlings), while others grow free of the seed-coat and become green (phanerocotylar seedlings). Among phanerocotylar species with green cotyledons of various thicknesses, light-saturated photosynthetic rates per gram of cotyledon tissue are inversely correlated with cotyledon thickness (Kitajima, 1992b). Cotyledons thinner than 1 mm (foliaceous cotyledons) have high net photosynthetic rates, while cotyledons thicker than 1 mm have photosynthetic rates that are just enough to balance their respiration rates (Lovell and Moore, 1970; Kitajima, 1992b).

The duration of seed-reserve dependency is related to the position and the functional morphology of the cotyledons. Species with epigeal foliaceous cotyledons invest a large proportion of seed reserves into development of cotyledons as persistent photosynthetic organs, which supply the energy necessary for the development of the true leaves (Lovell and Moore, 1971; Marshall and Kozlowski, 1974; Ampofo *et al.*, 1976; Mulligan and Patrick, 1985). Seedlings with foliaceous photosynthetic cotyledons start to use light to meet their energy demands earlier than species with storage cotyledons (Marshall and Kozlowski, 1974; Kitajima, 1996a). In contrast, fleshy, epigeal, green cotyledons, such as those of beans, have a much shorter lifespan (Lovell and Moore, 1971; Garwood, 1983, 1996). Net photosynthetic rates of such cotyledons are not enough to make a significant contribution to seedling growth, and true leaves are constructed with stored reserves in seeds (Lovell and Moore, 1971). Nevertheless, cotyledon photosynthesis in thick green cotyledons can at least pay for the large metabolic cost of the rapid export of reserves from cotyledons (Marshall and Kozlowski, 1974; Kitajima, 1992b).

Reserve-type cotyledons are generally associated with large seed size (Ng, 1978; Hladik and Miquel, 1990; Garwood, 1996; Kitajima, 1996a). They may remain attached to the seedling axis for various periods (Garwood, 1996). Some persistent cotyledons may not be used continuously for growth after expansion of the initial set of leaves, after which removal of cotyledons has a negligible effect on seedling growth (Sonesson, 1994). However, such reserves may aid rapid resprouting if the above-ground portion is damaged by herbivory (Harms and Dalling, 1997; Harms *et al.*, 1997; Hoshizaki *et al.*, 1997). Even in species that do not maintain persistent cotyledon reserves, maternally derived tissues may remain of key importance for a prolonged period, especially in competitive or stressful environments, where growth rates are suppressed. It appears that seedlings of large-seeded species develop a proportionally large storage pith in the stem and main root (Castro-Diez *et al.*, 1998). Some 2-week-old seedlings of *Anacardium excelsum* manage to resprout leaves using storage in the main stem after complete removal of green-storage cotyledons and leaves (K. Kitajima, unpublished data).

These early-seedling morphological traits appear to be conservative through evolutionary time. Within a genus, there is generally only one type of functional morphology, and it is uncommon to find more than two types within a family (Garwood, 1996). Thus, the rate of evolutionary change in cotyledon functional morphology appears to lag behind evolutionary changes in seed mass, which can be found at the species and genus levels. In broad taxonomic comparisons, small-seeded species tend to have photosynthetic cotyledons, while large-seeded species tend to have non-photosynthetic cotyledons (Hladik and Miquel, 1990; Kitajima, 1992a, 1996a; Garwood, 1996). As a result,

a comparison of seed size across a wide variety of taxa is also a comparison of seedling functional morphology, which may act as a constraint on selection for larger or smaller seed size. For a review of the broad significance of seed size, see Leishman *et al.*, Chapter 2, this volume.

Allocation-based trade-offs in seedlings

Developing seedlings face trade-offs in resource allocation to growth, defence against tissue loss, and storage. Growth rates of older and completely autotrophic plants may be negatively correlated with allocation to chemical and physical defences (Bloom *et al.*, 1985; Coley, 1988). Hence, an important source of variation in seedling-biomass conversion efficiency from seed to seedling is the difference in types of tissue created. For example, lignin confers greater structural and chemical defensive properties than cellulose alone. The biosynthetic cost of lignin, however, is about twice as high as that of cellulose (Williams *et al.*, 1987). Hence, the unit of seedling biomass that can be converted from a unit of seed reserve will be less in cases where the lignin content of the seedling tissue is high. All else being equal, synthesis of other defensive compounds (such as tannins, phenolics and alkaloids) should also result in lower seedling biomass per gram of seed mass.

Tissue created cheaply has a shorter duration, partly because it is faster to pay back for its construction (the innate optimization criterion: Williams *et al.*, 1989), but more importantly because it suffers greater physical damage and loss to consumers (Coley *et al.*, 1985). Leaf mass per unit area, leaf toughness, leaf chemical defences and longer leaf lifespan are all correlated with each other (Koike, 1988; Lei and Lechowicz, 1990; Reich *et al.*, 1991). Carbon allocations to defence and storage both result in higher survivorship of organs and individuals, at the cost of less carbon available for immediate growth (Bloom *et al.*, 1985). Allocation to defence enhances survival by preventing tissue loss, which is especially costly in a resource-poor environment (Coley et al., 1985).

Carbon storage is important for recovery from tissue loss and survival when net photosynthetic rate is negative due to environmental stresses. Energy stored in cotyledons has been shown to be important for recovery from herbivory (Armstrong and Westoby, 1993; Harms and Dalling, 1997; Hoshizaki et al., 1997) and frost damage (Aizen and Woodcock, 1996). Likewise, carbon storage in the non-cotyledonous tissues of young seedlings (such as stems and tap roots) contributes to survival. As plants do not store large amounts of lipids in vegetative tissues, total non-structural carbohydrates (TNC, total of starch and sugars) provides an adequate measure of energy reserve in non-seed tissues (Chapin et al., 1990). In a study of 2-year-old seedlings of four temperate deciduous trees, defoliation treatments resulted in refoliation responses, which then caused a decrease in TNC pool size and a decrease in survival through the next year (Canham et al., 1999). McPherson and Williams (1998) found that a high level of TNC in the stem of juvenile cabbage palms (26–54% of dry mass in intact plants) enabled them to recover and survive three sequential defoliation treatments at 7-week intervals, while the TNC level declined with the number of defoliation treatments. How ontogenetic development and availability of external resources affect the storage dynamics of seedlings is an important area of investigation, in order to understand the allocation-based trade-offs in seedlings (Bryant and Julkunentiitto, 1995).

Once a seedling becomes dependent on light as the main source of energy, fast growth is achieved by a large carbon allocation to leaf area increase, which gives a compounded return of future carbon income (Bloom et al., 1985). Hence, there is generally a strong positive correlation between leaf-area ratio (LAR, ratio of leaf area to whole plant mass) and RGR (biomass growth rate per gram biomass per day) of seedlings (Poorter and Lambers,

1991; Walters et al., 1993a, b; Kitajima, 1994; Cornelissen et al., 1996, 1998; Reich et al., 1998; Veneklaas and Poorter, 1998). Leaf morphological traits, including specific leaf area (leaf area per unit leaf mass) and its components, explain a large portion of interspecific variation in LAR (Kitajima, 1996b; Wright and Westoby, 1999). Root respiration rates and the ratio of leaf mass to supportive-structure mass are also important components of RGR in woody species (Cornelissen et al., 1998; Reich et al., 1998). These allocation and morphological traits may be selected in suites according to the life history of the species, and together they exhibit negative and positive interspecific correlation with the seed mass. The negative correlation between RGR and seed mass found in interspecific comparisons of a variety of species is an integrative result of these associations of morphological and allocational traits.

Factors affecting seedling survival and growth

Vulnerability of seedlings

The early seedling stage is demographically dynamic (Harper, 1980; Silvertown and Dickie, 1981; Fenner, 1987). The high mortality at the seedling stage may act as a strong selective filter on seed traits, as well as seedling traits. Hence, the adaptive significance of various seed traits is intimately linked to successful seedling establishment. Seed traits that enhance dispersal lessen sibling competition among seedlings, as well as density- and distance-dependent mortality. Traits that can be interpreted as 'gap-detection mechanisms' in seeds restrict germination in time and space, enhancing the likelihood of seedling survival and growth (see Bullock, Chapter 16, this volume).

Seedling survival studies of various plants have often, but not always, identified continuously decreasing rates of mortality, which can be depicted as decreasing steepness of the survivorship curve (log (fraction surviving + 1) plotted against time since germination: Li *et al.*, 1996). Thus, it is common to see a decline in mortality rates with size and age of seedlings (De Steven, 1991; Alvarez-Buylla and Martinez-Ramos, 1992; Sato *et al.*, 1994; Tanouchi *et al.*, 1994). The high mortality of young seedlings reflects their small size and the soft and palatable tissue required for their rapid development (Fenner, 1987; Hanley *et al.*, 1996b). The disadvantage of small-ness of seedling reflects a trade-off relation-ship with dispersal distance, such that smaller and better-dispersed seeds may have a better chance of escaping density-dependent mortality agents, while larger (and less well-dispersed) seeds have resources to cope with abiotic stresses (Howe *et al.*, 1985). The causes of death may be ascertained and quantified by direct observation, as attempted in *Bromus tectorum* (Mack and Pyke, 1984). Various hazard agents may operate simultaneously or sequentially. For example, seedlings of a neotropical tree, *Tachigalia versicolor*, suf-fer a high mortality rate from mammalian herbivory during the first 2–4 weeks after germination, while damping-off pathogens become the main source of mortality at 4–8 weeks of age (Kitajima and Augspurger, 1989). This is perhaps related to an increase in tissue toughness and loss of food reserves in cotyledons with time. Such observation alone entails many technical problems in ascertaining the causes of mortality, while experimental approaches test the importance of individ-ual mortality factors (Fenner, 1987).

Abiotic factors

Ecologists have long been attracted to the idea that seedling tolerance of various abi-otic factors may explain species specializa-tion along environmental gradients. Abiotic stresses, such as sudden freezing and severe drought, may kill seedlings directly. Many abiotic stress factors, such as shade, excess light, heat, water stress and flooding, may not kill seedlings imme-diately, but may lower their tolerance to biotic mortality agents, such as herbivores

and pathogens (Augspurger, 1984a; Augspurger and Kelly, 1984). Mortality of both shade-tolerant and intolerant trees is greater in shade than in light gaps (Grime and Jeffrey, 1965; Augspurger, 1984a), per-haps because the less favourable carbon balance in deep shade makes it harder to maintain defensive traits and recover from tissue loss to pests.

Conversely, biotic factors often cause abiotic stress factors. Most importantly, competition with neighbouring plants, which deprives emerging seedlings of light and soil resources and raises seedling mor-tality (Fenner, 1978a; Denslow *et al.*, 1991; Aguilera and Lauenroth, 1993; Gordon and Rice, 1993). Seedlings often suffer severely from physical damage, as young seedlings are easily uprooted or snapped (Kitajima and Augspurger, 1989; Clark and Clark, 1991). Adult neighbours of similar life-form competitively affect seedling sur-vivorship, perhaps due to their similar resource requirements (Leishman, 1999). Trampling and soil digging by non-herbivo-rous birds and mammals often cause such physical damage (Gillham, 1956; Maesako, 1997). Fallen branches from the overstorey, soil shifting by freeze–thawing actions, flooding and landslide also result in severe physical damage and seedling death (Cook, 1979; Shibata and Nakashizuka, 1995). Standardized assays to examine the overall intensity of physical disturbance with arti-ficial seedling models often show large dif-ferences between habitat types (Clark and Clark, 1989; McCarthy and Facelli, 1990; Mack, 1998).

Burial

Burial is the first abiotic stress factor that a seedling has to cope with before emerging from the soil and litter layer. Large-seeded species are better equipped to emerge from a deeper layer of soil and litter (Carter and Grace, 1986; Cheplick and Quinn, 1987; Molofsky and Augspurger, 1992; Seiwa and Kikuzawa, 1996). Light transmission through the litter layer is virtually nil (Vázquez-Yanes *et al.*, 1990). Grubb and

Metcalfe (1996) argue that some small-seeded species have adapted to survive in shade by specialization to the bare soil of steep topographies. Many small-seeded pioneer species remain ungerminated unless they are close to the surface of the bare ground (Putz, 1983; Vázquez-Yanes and Orozco-Segovia, 1992; Dalling et al., 1995). Nevertheless, small seeds may germinate and exhaust their energy reserves before emergence from soil and litter (Fenner, 1995). Such 'wasted' germination is hard to quantify in relation to the other possible reasons for failed emergence from the soil. Small seedlings that manage to emerge through a thick litter layer via etiolation may suffer greater mortality (Goldberg and Werner, 1983; Collins, 1990; Molofsky and Augspurger, 1992; Smith and Capelle, 1992; Dalling, 1995; Seiwa and Kikuzawa, 1996), but this is not always the case (Cheplick and Quinn, 1987; Kitajima and Tilman, 1996; Hoffmann, 1997). In contrast, litter cover protects seedlings of large-seeded species, such as oaks, from desiccation damage (Seiwa and Kikuzawa, 1996). Although litter may have both negative and positive effects, a meta-analysis of 32 published studies found that the overall effects of litter are negative on both seed germination and seedling establishment (Xiong and Nilsson, 1999). In trees, seedlings of large-seeded species also tend to have a greater biomass allocation to tap roots than small-seeded species, which facilitates penetration through the thick litter layer (Kohyama and Grubb, 1994). Fenner (1983), however, found that, among temperate herbaceous plants, seedlings of smaller-seeded species, whose germination is likely to be inhibited under the litter, had higher root-weight ratios.

Gaps in vegetation

A common requirement for seedling establishment is the absence of competition from other species within the immediate vicinity. Breaks in the continuous vegetation cover provide localized and temporary release from competition, and are probably a prerequisite for the successful regeneration of many species. Gaps are continually created in vegetation by both biotic and abiotic agents. Examples of gaps caused by animals include molehills (Inouye et al., 1987), anthills (McGinley et al., 1994), rabbit scrapes, hoofprints and worm-casts. Abiotic agents of gap creation include windthrow of trees (Whitmore, 1978), fire (Menges and Hawkes, 1998), lightning strike (Sousa and Mitchell, 1999), frost-heave and flood deposition. External agents are not essential for gap creation; they may arise through purely internal vegetation processes, such as the death of individual plants or fallen trees or branches (Fenner, 1978b; Whitmore, 1978). Each cause creates a gap with unique qualities, and many studies indicate that gap type is important in favouring regeneration of particular species (see Bullock, Chapter 16, this volume).

The survival and growth of seedlings, especially those of small-seeded species, are often correlated with the gap size in the vegetation (Fenner, 1978a, b; Gross and Werner, 1982). Clear evidence comes from experimental studies comparing seedling performance in gaps of different sizes (McConnaughay and Bazzaz, 1987; Bullock et al., 1995; Gray and Spies, 1996). Gap size can also affect the chances of a seedling being grazed (Hanley et al., 1996a, b). Although there appears to be a limit to how finely the regeneration niche may be divided along a gap-size gradient in a tropical forest, species differ in minimum gap size required for successful regeneration (Denslow et al., 1998; Hubbell et al., 1999). A gap created by a fallen tree creates a particularly complex set of heterogeneous microhabitats, favouring distinct groups of species at the root, trunk and crown (Denslow, 1980; Nuñez-Farfán and Dirzo, 1988). There is a large environmental gradient from the centre to the edge of a given gap (Chazdon, 1992; Brown, 1993; Sipe and Bazzaz, 1994), and within-gap environmental heterogeneity may be more important to small seedlings than overall gap size per se (Brown and Whitmore, 1992; Whitmore and Brown, 1996). One species may be favoured in the centre, whereas another survives better near the margins. Gap shape, which determines the ratio of margin to area, and gap orientation, which determines the timing of high light availability, may be important too.

Gap creation exhibits a certain seasonality (Brokaw, 1982). Records of similar gaps created at different times of the year show differences in colonization (Hobbs and Mooney, 1985). Stochastic factors, such as the vicinity of potential seed sources in time and space, influence the species composition of seedlings colonizing a given gap (Dalling *et al.*, 1998a, b). Using mathematical equations to simulate seed shadow, it is possible to predict spatial patterns of seedling recruitment in relation to gap dynamics in the community (Ribbens *et al.*, 1994). Even without detailed knowledge of seed-shadow distribution, it is possible to show that certain types of gaps favour the establishment of certain species from seeds. Practical examples of this include the preparation of seedbeds by foresters to encourage natural regeneration (Prevost, 1996) and the expansion of gaps to favour regeneration of commercial timber species with relatively high light demands (Mostacedo and Fredericksen, 1999).

Shade avoidance and tolerance

Many small-seeded species avoid germination in shade by means of a suite of traits that can be interpreted as gap-detection mechanisms (Swain and Whitmore, 1988; Horvitz and Schemske, 1994; Vázquez-Yanes and Orozco-Segovia, 1994; Dalling *et al.*, 1998a). Once germinated, seedlings may employ strategies to avoid shade temporally and spatially. Small-seeded species in temperate grasslands tend to germinate in the autumn, possibly reducing the likelihood of shading by neighbours (Silvertown, 1981). Small-seeded tree species in temperate forests may germinate and grow before the overstorey canopy fully develops (Seiwa, 1998), although such temporal escapes are less likely to be successful in evergreen forests. Stem elongation in response to a low red : far-red ratio of light may help avoid shading in conditions where there is a steep vertical gradient of light due to shade from the surrounding vegetation, as in herbaceous communities and within a tree-fall gap (Smith, 1982). The generally higher relative growth rate of small-seeded species (Shipley and Peters, 1980; Fenner, 1983; Cornelissen *et al.*, 1998; Reich *et al.*, 1998) may help small seedlings to reach the canopy or keep up with the growing height of vegetation following a disturbance. The responsiveness of seedlings to the red : far-red ratio depends on both phylogeny and the regeneration niche of the species (Morgan and Smith, 1979; Corré, 1983; Kitajima, 1994, Lee *et al.*, 1996, 1997). Rapid seedling growth may allow rapid graduation from the vulnerable small sizes. However, rapid development and growth require lower tissue density, which makes seedlings more vulnerable to herbivores and pathogens (Kitajima, 1994; Ryser, 1996). Rapid tissue growth also requires higher hydraulic conductivity, which may make seedlings more vulnerable to freeze-induced embolism and potential death (Castro-Diez *et al.*, 1998).

While small-seeded species tend to have mechanisms to avoid or escape shade, large-seeded species tend to exhibit shade tolerance (Grime and Jeffrey, 1965; Foster and Janson, 1985; Osunkoya *et al.*, 1993, 1994; Leishman and Westoby, 1994a; Osunkoya, 1996; Saverimuttu and Westoby, 1996; Reich *et al.*, 1998). The physiological requirement for survival in shade is the maintenance of a positive, but not necessarily large, carbon balance over time. The carbon balance in shade is a function of photosynthetic income minus tissue loss to consumers (herbivores and pathogens) and inherent tissue senescence. Within a given species, the greater the shade, the lower the photosynthetic income and survival. However, the efficiency of photosynthetic light utilization alone does not explain interspecific variation in shade tolerance of tree seedlings (Walters *et al.*, 1993a; Kitajima, 1994; Veneklaas and Poorter, 1998). Rather, higher survivorship of large-seeded species in shade is based on the combination of biomass allocation patterns and tissue densities that confer resistance to tissue loss from herbivores and pathogens (Kitajima, 1996b; Reich et al., 1998). The maintenance of carbon storage is also important for survival in the shade, enabling seedlings to resprout and refoliate following tissue loss (Canham et al., 1999). Otherwise, the process of recovery may be extremely slow in the shade.

Flood

Species zonation along hydrological gradients may be explained by seed and seedling tolerance of flood and soil anoxia. Seed buoyancy and flood tolerance, rather than drought tolerance, appear to be the key traits that explain the segregation of two large-seeded trees, Mora excelsum and M. gonggrijpii, along a hydrological gradient in the Guyana rainforest (Steege, 1994). Likewise, buoyancy of seeds and propagules is important in wetland Asclepias (Edwards et al., 1994) and many mangrove-forest species (Rabinowitz, 1978; McGuinness, 1997). Lack of oxygen and increased levels of potentially phytotoxic compounds in flooded soils impair physiological function and the growth of roots (Kozlowski et al., 1991; McKee, 1995). Large seeds and seedlings may enhance flood tolerance of seedlings by prevention of complete submersion and better anchorage against being washed away. Parkia discolor, a species restricted to seasonally flooded forests, has larger seeds and tolerates longer periods of flooding than P. pendula, a widely distributed species in upland forests (tierra-firme) in Brazil (Scarano and Crawford, 1992). Tree species frequent along rivers and on wetland differ in their degree of flood tolerance, reflecting differences in the physiological and anatomical responses of seedlings to prolonged flooding (Davidson, 1985; Topa, 1986; Topa and McLeod, 1986; Sena Gomes and Kozlowski, 1988; Kozlowski et al., 1991; Terazawa and Kikuzawa, 1994).

Fire

Small seedlings generally do not tolerate fire. Where fire is frequent but of low intensity, regeneration by ramets (i.e. resprouts) appears to be a successful strategy, while regeneration by genets (i.e. seedlings) is advantageous where adults are unlikely to survive infrequent and intense fire (Hoffmann, 1997; Kruger et al., 1997). Seeds of some fire adapted species need to be buried sufficiently near the surface to receive heat to break dormancy (Auld, 1986). Serotinous (late-to-open) cones and fruits store seeds in an above-ground reserve, from which they are released in response to intense heat (Kozlowski et al., 1991). In Florida, where most communities, including wetlands, are either fire-adapted or fire-dependent (Myers and Ewel, 1990; Menges and Hawkes, 1998), pine species exhibit contrasting seedling regeneration strategies. Seedlings of Pinus palustris, found in well-drained and frequently burned sites, protect the apical meristem inside thick layers of scales and dense leaf clusters, which resemble a bunch grass. Juveniles remain in this 'grass' stage for years and can survive repeated, frequent, low-intensity fires. In contrast, P. serotina, dominant in wet and infrequently burned sites, maintains aboveground seed banks in serotinous cones and also resprouts from the base of the main trunk after fire (Godfrey, 1988). Generalization on fire-dependent systems, however, requires caution. Postfire seedling densities differ greatly between climate regions, more so than among soil types within a climatic region (Carrington and Keeley, 1999; see also Keeley and Fotheringham, Chapter 13, this volume).

Drought

Drought may be fatal to germinating seeds and small seedlings (Miles, 1972; Maruta, 1976; Veenendaal et al., 1996a). Microsites that prevent desiccation are safe sites for seedling establishment of dominant grasses in the short-grass prairies (Fowler, 1986).

Seed germination in seasonally dry tropical forests exhibits high seasonality, peaking during the early rainy season (Garwood, 1982; Veenendaal et al., 1996a). Many tropical species whose seeds disperse at the end of the wet season exhibit a strong innate dormancy, lasting 3–6 months, regardless of the watering regime during this period (Garwood, 1989). In an African tropical dry forest, small rain episodes at the end of the dry season do not trigger germination; this has the effect of preventing seedlings from being exposed to subsequent drought stress (Veenendaal et al., 1996a). Likewise, desert annuals typically germinate only after a heavy rain above the species-specific threshold (Baskin and Baskin, 1998).

Larger seeds and seedlings are advantageous for seedling survival in dry soil in some studies (Baker, 1972), but not in others (Mazer, 1989). Leishman and Westoby (1994b) found large seed advantages in their glasshouse study but not in the field. The genetic background of the maternal parent also affects the outcome of competition with neighbours for water (Rice et al., 1993). Seedlings established in tree-fall gaps are larger than those established in the shaded understorey, and hence survive seasonal drought better, despite higher evapotranspirational demands in gaps (Howe, 1990; Veenendaal et al., 1996b). Another reason why gaps are advantageous for seedling survival through the dry season is because soil water potentials are often more favourable without larger competitors (Becker et al., 1988; Veenendaal et al., 1996b). Within light gaps, the earlier the seed germination, the larger the seedling and the higher the survival during the dry season (Garwood, 1982). In tropical dry forest, which experiences severe seasonal drought, species specialized in disturbed habitats have a lower root : shoot ratio, a higher specific root length and more branched roots than species from undisturbed habitats (Huante et al., 1992). The herring-bone pattern of the roots of the latter type of species may be an adaptive morphology to probe down to a deeper layer of soil.

Seedlings exhibit anatomical and physiological acclimatization to drought stress. Moderate drought stress decreases the root hydraulic conductance of wild olive seedlings by 16–66% compared with unstressed plants, but the root hydraulic conductance recovers after rewatering (Lo Gullo et al., 1998). Severe drought stress, in contrast, results in root anatomical changes, such as a thicker and well-suberized exodermis and a multilayer endodermis. Hydraulic conductance of stems and roots increases in proportion to leaf-area increase. When standardized by either leaf area or dry mass, fast-growing pioneers have significantly higher specific hydraulic conductivity than slow-growing shade-tolerant species (Tyree et al., 1998).

Biotic factors affecting seedling survival

Predation of seedlings

Predation is one of the greatest hazards facing newly germinated seedlings. For example, mollusc grazing has been shown to have a major effect on recruitment and eventual species composition (Hanley et al., 1995a). Meristematic damage as well as continuous loss of leaves to herbivory decreases seedling survivorship (Clark and Clark, 1985). Below-ground herbivores are also important mortality agents (Inouye et al., 1987; Gange et al., 1991). Predator-exclusion experiments have demonstrated the importance of vertebrate grazers, especially of rodents, in various communities (Sork, 1987; Howe, 1990; Osunkoya et al., 1992; Molofsky and Fisher, 1993; Canham et al., 1994; Terborgh and Wright, 1994; Asquith et al., 1997; Edwards and Crawley, 1999). Invertebrates, such as molluscs (Hanley et al., 1996a, b), crabs (Smith, 1987; Osborne and Smith, 1990; McGuinness, 1997) and insects (Maron, 1997), affect seedling survival and the distribution of species. The relative importance of these seedling predators differs among communities and microhabitats. In a study of seedling herbivory in temperate grassland, rodents and molluscs (but not arthropods) were found to be important mortality agents (Hulme, 1994). The likelihood of a seedling being grazed by molluscs is influenced by several factors, including gap size and season (Hanley et al., 1996a). In Amazonian lowland forests, leaf-cutter ants influence seedling recruitment sites (Vasconcelos and Cherrett, 1997). Selective grazing on seedlings by insects (Brown and Gange, 1999) or molluscs (Hanley et al., 1995a; Fenner et al., 1999) can have a long-term influence on species composition.

The absolute size of a seedling may be the determining factor in deciding whether or not it is killed by predation. Grazing by molluscs preferentially kills the smaller (younger) seedlings of herbaceous species (Fenner, 1987; Hanley et al., 1995b), at least in monoculture stands. Molluscs can kill a seedling with the removal of a very small amount of material (e.g. from the hypocotyl: Dirzo and Harper, 1980), whereas a similar bite would have a negligible effect on a bigger seedling. Defence compounds in seedlings appear to be temporary in many cases. The concentration of hydroxamic acids (known to have toxic and anti-feedant effects against insects) is high in newly germinated seeds of wheat (Thackray et al., 1990) and maize (Klun and Robinson, 1969), but the levels decline in the first 8 days after germination. A study comparing the palatability of seedlings and adults of herbaceous plants found that, in the majority of cases, the seedlings were actually more palatable than the adults (Fenner et al., 1999). Seedlings with similar leaf defence chemicals to that of the parent may be attacked more by pests and selected against, favouring genetically dissimilar seedlings. This hypothesis explains why the leaf chemical composition of seedlings is different from that of the parent tree in several tropical species (Langenheim and Stubblebine, 1983; Sanchez-Hidalgo et al., 1999).

Community factors, such as seedling density and identity of nearest neighbours (Hanley, 1995, 1998), may play a part in determining the vulnerability of seedlings to grazing. Janzen (1970) and Connell (1971) put forward a model relating recruitment to distance from the parent tree. Predators attracted to the parent tree or the high densities of seeds or seedlings may prevent seedling regeneration in the immediate vicinity of the parent. This Janzen–Connell model is a specific case of density-dependent mortality, which would serve to increase spacing among conspecific trees in mature tropical forests (Wills et al., 1997). Survival of the tropical tree *Dipteryx panamensis* is much lower under the parent crown (12%) than under nonparent trees (33%) (De Steven and Putz, 1984). Seedling survival of the neotropical nutmeg *Virola nobilis* increases with distance from the parent tree (Howe et al., 1985). However, spatial patterns of mortality as predicted by the Janzen–Connell model may not always be detectable for seedlings (Clark and Clark, 1984; Augspurger and Kitajima, 1992; Condit et al., 1992; Burkey, 1994; Cintra, 1997). The overall consequences may be more complex for seedlings than for seeds, as various predators and pathogens respond to seedling density in different manners. The response of vertebrate predators may be further influenced by the abundance of alternative food sources and their territorial and foraging behaviours. At a high density, some predators may be satiated (Augspurger and Kitajima, 1992; Forget, 1993), resulting in a higher overall survival near the parent, and this pattern may be further modified by other mortality agents that attack seedlings at later stages.

Disease

Soil-borne pathogens may cause high mortality in a density-dependent manner. Crowded seedlings suffer disproportionately greater mortality (Burdon and Chilvers, 1976a; Augspurger, 1983). Damping off, or rotting of seedlings at the soil level, is caused by soil-borne pathogens, including bacteria (Pedersen et al., 1999) and anamorphic fungi in *Ascomycota* (e.g. *Fusarium*), *Basidio-*

mycota (e.g. *Rhizoctonia*) and *Oomycota* (e.g. *Pythium* and *Phytophthora*) (Garrett, 1970). Many of these soil-borne damping-off pathogens can infect multiple host species and genotypes, although hosts differ in their susceptibility (Burdon and Chilvers, 1976b). The age structure of the seedling population, reflecting its germination chronology, also affects damping-off mortality (Neher *et al.*, 1987). The infectivity and virulence of pathogens may be influenced by the density of dormant spores, the epidemic history of the site and the environmental conditions that enhance pathogen dispersal and survival, as well as host vigour (Lockwood, 1986). Many dormant spores of soil-borne fungi germinate under a similar hydrological regime to that which promotes germination of host-plant seeds, or are stimulated to germinate by the root exudates of the potential hosts (Lockwood, 1986). Zoospores of some soil-borne pathogens, such as *Pythium* and *Phytophthora*, exhibit negative geotropism and swim up to the surface of the flooded soil, penetrate down the soil stratum as floodwater recedes and then exhibit chemotactic and electrotactic attraction to the host root tips (Cameron and Carlie, 1977; Deacon, 1988; Morris and Gow, 1993). Consequently, slope orientation and hydrology may be important factors in the spatial spread of damping-off epidemics within a seedling cohort.

Damping-off mortality is generally higher in shade than in gaps (Grime and Jeffrey, 1965; Augspurger, 1983, 1984a; Augspurger and Kelly, 1984), either because host seedling vigour is greater in gaps or because the environment is more favourable for the survival of pathogens. Regardless of the cause, this underlines the fact that shade tolerance is an ability not merely to maintain positive net carbon gain but also to survive pathogen attacks. Indeed, in aseptic culture, seedlings survive for a long time, even when light availability is marginally above the photosynthetic light compensation point (Vaartaja, 1962). Various means of defence against soil-borne pathogens include the isolation and premature death of infected

tissues and the induction of defensive chemicals. Infection by bacteria and zoosporitic fungi, such as *Pythium* and *Phytophthora*, is limited to the root tips. The programmed death and dispersion of the border cells that constitute the root cap may have an important defensive role (Hawkes, 1998). Exudates of border cells act as decoys that lure potentially dangerous soil-borne organisms and then mucilage surrounding border cells swells and mechanically repels border cells from each other and away from the root apical meristem. Modification of microbiota and soil pH in the rhizosphere by exudates from the roots may favour beneficial fungi, which may compete with pathogens for resources released by the host (Hendelsman and Stabb, 1996). Seedling survival is influenced by a complex of ecological interactions between pathogenic and non-pathogenic competitors in the soil.

Mycorrhizae

The establishment of mycorrhizae is of critical importance for the uptake of mineral nutrients from the soil, especially in infertile soil and in competition with neighbours (Janos, 1980a, 1983; Alexander, 1989; Read, 1991). The absence of the appropriate fungus in the soil is known to be a factor in influencing the course of succession in tropical forests (Janos, 1980b) and may prevent the invasion of savannah by forest trees through the inability of seedlings to establish (Bowman and Panton, 1993). Mycorrhiza formation is thought mainly to facilitate the uptake of phosphorus (e.g. Colpaert *et al.*, 1999). This element is relatively immobile in the soil, and a newly germinated seedling with a limited rootstock may have difficulty in obtaining an adequate supply without the aid of the symbiotic fungus. The fact that phosphorus is often limiting for seedlings is shown by the positive response obtained when it is added as a fertilizer (Brandon *et al.*, 1997). Species of arbuscular mycorrhizal fungi in the *Zygomycota* form associations with a wide range of host plants, including most herbaceous plants and many dicotyledonous trees. In arbuscular mycorrhizae, the fungus is not normally species-specific with regard to the host. In contrast, trees and shrubs in certain families, such as *Pinaceae*, *Fagaceae* and *Dipterocarpaceae*, form ectomycorrhizal associations with species-specific fungi in *Basidiomycota*. For tiny seedlings of orchids, mycorrhizae are essential as a means of acquiring mineral nutrients; for seedlings of parasitic orchids, the mycorrhizae enable the plant to obtain all necessary resources throughout its life. Ericaceous mycorrhizae are considered to be the reason for dominance of this family in arctic and alpine environments with low pH and nitrogen availability (Read, 1991).

Colonization by mutualistic fungi may protect host roots from pathogenic organisms. However, these benefits come with a cost of carbon allocation to mycorrhizae. Although, in the long term, the benefit of improved soil nutrient availability is likely to enhance growth and survival, the carbon drain by mycorrhizae often lowers the growth rates and survival of seedlings, at least in the short term (McGee, 1990; Lovelock *et al.*, 1996, 1997). The relative benefit of mycorrhizae, however, should depend on the host demand for soil resources relative to photosynthates (Lovelock *et al.*, 1997). For example, the relative growth enhancement by arbuscular mycorrhizae was shown to be greater on nutrient-poor ant mounds than on fertile ant mounds created by different ant species, perhaps because of the carbon cost (McGinley *et al.*, 1994).

The timing and degree of infection of seedling roots by mycorrhizal fungi depend on the availability of fungal spores, the proximity to the infected roots of neighbours and the seedling demand for soil mineral resources, which develops as seed-stored reserves become insufficient. Low carbohydrate levels in the roots of shaded seedlings may be a regulatory mechanism to limit mycorrhizal development (Sasaki and Ng, 1981). However, when spores are available, mycorrhizal infection may start while cotyledons are still attached (Janos, 1980a; Herrera *et al.*, 1992) or as early as 20 days after germination (Lee and Alexander, 1996). Indeed, it is probable that the formation of mycorrhizae is the norm for most seedlings. Gay *et al.* (1982) found that seedlings of a wide range of herbaceous species in chalk grassland became infected with arbuscular mycorrhizae within 2 weeks of germination. Cotyledons have been shown to supply the carbohydrates necessary for the establishment of ectomycorrhizae in seedlings of the tropical woody legume *Afzelia africana* (Bâ *et al.*, 1994). Cotyledon retention may be longer for infected seedlings, probably because mineral nutrient reserves in cotyledons are withdrawn more slowly in mycorrhizal seedlings (Janos, 1980a).

Mycorrhizae modify competitive interaction of seedlings with neighbours (Moora and Zobel, 1996, 1998; Hartnett and Wilson, 1999; Marler *et al.*, 1999). Although a given species of arbuscular mycorrhizal fungus may successfully establish mycorrhizae with various hosts, its infectivity and effectiveness differ among host species (Sylvia *et al.*, 1993; Allen *et al.*, 1995; van der Heijden *et al.*, 1998). The effect of mycorrhizae on seedling emergence differs among different host species (Hartnett *et al.*, 1994). Mycorrhizae may enhance the dominance of a certain species (Hartnett and Wilson, 1999). On the other hand, a smaller, competitively inferior, seedling may benefit from hyphal connection to a larger partner, which may take a larger share in feeding the fungus with carbon (Moora and Zobel, 1996).

There are many unexplored questions on the role of mycorrhizae in spatial patterns of seedling establishment. For example, do seedlings benefit from being close to an adult plant of the same genotype or to an individual of another species that is likely to have mutualistic fungi in the roots? Experimental tests, as well as a detailed cost–benefit analysis of mycorrhizae in seedlings would advance our understanding of plant community dynamics.

Seedling banks

Regeneration of many perennial (especially woody) species depends on multiple cohorts of suppressed seedlings that await release by gap creation. This is a more common strategy in mature forests than the alternative strategies of soil seed banks and repeated seeding for recruitment (Canham, 1985; Brokaw and Scheiner, 1989; see also Grime and Hillier, Chapter 15, this volume). The relative importance of the seedling bank may be identified in the demographic matrix as high survival at the seedling stage and a slow transition probability to a larger size class (Silvertown et al., 1996). Mechanistic understanding of this important strategy for perennial plants requires a knowledge of morphological and physiological responses of seedlings to temporally heterogeneous environments.

Ontogenetic drift

Seedlings exhibit rapid ontogenetic change in biomass allocation patterns during and after the seed-reserve dependency period. During the seed dependency period, seedlings may lose weight. This phase may be prolonged in large-seeded species in a resource-poor environment. Once seed-derived reserves are exhausted, however, the seedling needs to achieve a positive carbon gain. Acclimatization of seedlings to light and nutrient availability through the early stages of ontogenetic development is important in balancing carbon budgets. Many good examples of the phenotypic plasticity in biomass allocation and the physiology of seedlings are found in studies of seedling acclimatization to contrasting light environments (for example, Langenheim et al., 1984; Fetcher et al., 1987; Walters et al., 1993a; Kitajima, 1994). Below-ground competition and resource availability likewise affect seedling morphology (McConnaughay and Coleman, 1999). Seedlings are particularly responsive to herbivory in studies of induction of chemical defences (Karban and Baldwin, 1997).

When seedlings of comparable ages are compared, apparent acclimatization responses may merely reflect ontogenetic or size differences (Rice and Bazzaz, 1989; McConnaughay and Coleman, 1999). In favourable environments, the time necessary to achieve a certain developmental stage (e.g. expansion of the third true leaf), as well as to achieve a certain biomass (e.g. ten times the seed mass), is shortened. Successive leaves increase in area and thickness until they achieve relatively constant dimensions. This ontogenetic increase in size of the individual leaf is accompanied by a proportional increase in stem diameter, according to species-specific allometric relationships (Brouat et al., 1998). In general, as seedlings grow, proportionally more biomass is allocated to structural support (Kohyama, 1983; Walters et al., 1993b; McConnaughay and Coleman, 1999), resulting in lower LAR for larger seedlings. Thus, differences in leaf morphologies and LAR of 2-month-old seedlings grown in sunny and shaded environments reflect differences due to ontogenic stages, as well as environments. Allometric analysis and standardization by developmental stages are powerful tools to tease out the seedling's acclimatization responses from ontogenic development. For example, in a study of three annuals, apparent differences in LAR for the same-aged seedlings at different levels of water availability disappear in all species when ontogenetic drift is accounted for by comparison of allometric relationships (McConnaughay and Coleman, 1999).

Comparisons of seedling leaves with adult leaves demonstrates ontogenetic shifts in many physiological and morphological traits. Seedling leaves are much smaller and thinner than adult leaves (King, 1999), and often do not achieve as high a photosynthetic rate per unit leaf area as sapling and adult leaves, even when they develop in the full sun (Koike, 1988). Seedling leaves are often more palatable than adult leaves, especially where the latter are unpalatable (Fenner et al., 1999). The leaf longevity of rapidly growing seedlings may be shorter or longer than that of adults. For temperate deciduous trees, mean leaf longevity is shorter for seedlings than for adults (Kikuzawa and Ackerly, 1999), but, in shade-suppressed saplings of evergreen species, leaf longevity appears to be longer than for adults (Coley, 1983). In a dry environment, shallow-rooted seedlings may experience a greater degree of drought than adults. In hemi-epiphytes, the seedlings (which are entirely epiphytic) exhibit very different physiological and morphological traits from those of the ground-rooted adults (Holbrook and Putz, 1996).

Phenotypic plasticity of seedlings

A relatively high degree of phenotypic plasticity is probably to be expected in seedlings as a consequence of the fact that the causes of seedling mortality change from season to season and from place to place. Demographic studies, such as those by Sharitz and McCormick (1973) and Mack and Pyke (1984), indicate that, within a single species, the selection pressures vary constantly. This may prevent the development of highly specialized adaptations for specific mortality factors, such as drought or shade. A more effective strategy for coping with constantly varying selective pressures would be a wide flexibility in phenotypic responses. Features such as hypocotyl extension and root/shoot ratio show a considerable degree of flexibility (Fenner, 1987; Lee *et al.*, 1996, 1997).

Seedlings that are initially acclimatized to a certain environmental condition need to acclimatize continuously to changing environments. This is a particularly important issue in the release of shade-suppressed seedlings in the forest understorey (Whitmore, 1978; Newell *et al.*, 1993). A sudden and large increase in light intensity may cause temporary or permanent photoinhibition to existing seedling leaves (Langenheim *et al.*, 1984; Turner and Newton, 1990; Kamaluddin and Grace, 1992a; Lovelock *et al.*, 1994; Castro *et al.*,

1995). Both the existing leaves and the leaves developing after increase in light availability acclimatize, with higher photosynthetic capacity and conductance, greater amount of photoprotective pigments and greater leaf mass per area. In contrast, the degree of phenotypic plasticity is much more limited in existing leaves (Turnbull *et al.*, 1993). These leaf-level responses to opening in the canopy are accompanied by the whole-plant level responses of increased hydraulic conductivity of stems, higher root:shoot ratio, greater internodal length and faster height growth rates (Popma and Bongers, 1988). The degree of phenotypic plasticity is often greater for pioneers and light-demanding species than for climax and shade-tolerant species (Strauss-Debenedetti and Bazzaz, 1996; Veenendaal *et al.*, 1996c; Zipperlen and Press, 1996; Veneklaas and Poorter, 1998; Valladares *et al.*, 2000).

A large and long-lasting gap may be sufficient to promote a seedling to a larger size class and possibly a reproductive size. More often, however, a seedling of a forest tree species experiences repeated episodes of small opening and closure of the canopy (Canham, 1985). Existing leaves reduce dark respiration and photosynthetic capacity quickly in response to shading (Kamaluddin and Grace, 1992b; Turnbull *et al.*, 1993). Adjusting to reshading may be more stressful than adjustment to opening, because limited carbon income makes it harder to pay for the production of new leaves. This possible cost of acclimatizing back to a shadier environment may be a possible reason why some species take a 'pessimistic strategy' of limited phenotypic plasticity, while others may take an 'optimistic strategy' of strong response in leaf turnover rates and shoot growth rates to light-gap opening (Kohyama, 1987). Phenotypic plasticity of established seedlings is an important modifier of the spatial and temporal patterns of regeneration, which are primarily set by where and when seeds fall and survive to seedlings.

References

Aguilera, M.O. and Lauenroth, W.K. (1993) Seedling establishment in adult neighbourhoods: intraspecific constraints in the regeneration of the bunchgrass *Bouteloua gracilis*. *Journal of Ecology* 81, 253–261.

Aizen, M.A. and Woodcock, H. (1996) Effects of acorn size on seedling survival and growth in *Quercus rubra* following simulated spring freeze. *Canadian Journal of Botany* 74, 308–314.

Alexander, I. (1989) Mycorrhizas in tropical forests. In: Proctor, J. (ed.) *Mineral Nutrients in Tropical Forest and Savanna Ecosystems*, Vol. 9. Blackwell Scientific Publications, Oxford, pp. 169–188.

Allen, E.B., Allen, M.F., Helm, D.J., Trappe, J.M., Molina, R. and Rincon, E. (1995) Patterns and regulation of mycorrhizal plant and fungal diversity. *Plant and Soil* 170, 47–62.

Alvarez-Buylla, E.R. and Martinez-Ramos, M. (1992) Demography and allometry of *Cecropia obtusifolia*, a neotropical pioneer tree: an evaluation of the climax–pioneer paradigm for tropical rain forests. *Journal of Ecology* 80, 275–290.

Ampofo, S.T., Moore, K.G. and Lovell, P.H. (1976) Cotyledon photosynthesis during seedling development in *Acer*. *New Phytologist* 76, 41–52.

Armstrong, D.P. and Westoby, M. (1993) Seedlings from large seeds tolerate defoliation better: a test using phylogenetically independent contrasts. *Ecology* 74, 1092–1116.

Asquith, N.M., Wright, S.J. and Clauss, M.J. (1997) Does mammal community composition control recruitment in neotropical forests? Evidence from Panama. *Ecology* 78, 941–946.

Augspurger, C.K. (1983) Seed dispersal of the tropical tree, *Platypodium elegans*, and the escape of its seedlings from fungal pathogens. *Journal of Ecology* 71, 759–771.

Augspurger, C.K. (1984a) Seedling survival of tropical tree species: interactions of dispersal distance, light gaps, and pathogens. *Ecology* 65, 1705–1712.

Augspurger, C.K. (1984b) Light requirements of neotropical tree seedlings: a comparative study of growth and survival. *Journal of Ecology* 72, 777–795.

Augspurger, C.K. and Kelly, C.K. (1984) Pathogen mortality of tropical tree seedlings: experimental studies of the effects of dispersal distance, seedling density, and light conditions. *Oecologia* 61, 211–217.

Augspurger, C.K. and Kitajima, K. (1992) Experimental studies of seedling recruitment from contrasting seed distributions. *Ecology* 73, 1270–1284.

Auld, T.D. (1986) Population dynamics of the shrub *Acacia suaveolens* Sm. Willd.: fire and the transition to seedlings. *Australian Journal of Ecology* 11, 373–386.

Bâ, A.M., Garbaye, J., Martin, F. and Dexheimer, J. (1994) Root soluble carbohydrates of *Afzelia africana* Sm.: seedlings and modifications of mycorrhiza establishment in response to the excision of cotyledons. *Mycorrhiza* 4, 269–275.

Baker, H.G. (1972) Seed weight in relation to environmental condition in California. *Ecology* 53, 997–1010.

Barclay, A.S. and Earl, F.R. (1974) Chemical analyses of seeds III. Oil and protein content of 1253 species. *Economic Botany* 28, 178–236.

Baskin, C.C. and Baskin, J.M. (1998) *Seeds: Ecology, Biogeography and Evolution of Dormancy and Germination*. Academic Press, San Diego.

Becker, P., Rabenold, P.E., Idol, J.R. and Smith, A.P. (1988) Water potential gradients for gaps and slopes in a Panamanian tropical moist forest's dry season. *Journal of Tropical Ecology* 4, 173–184.

Bloom, A.J., Chapin, F.S. and Mooney, H.A. (1985) Resource limitation in plants: an economic analogy. *Annual Reviews of Ecology and Systematics* 16, 363–392.

Bokdam, J. (1977) Seedling morphology of some African Sapotaceae and its taxonomical significance. *Medel Lanbouwhogeschool* 77, 1–84.

Bossema, I. (1979). Jays and oaks: an eco-ethological study of a symbiosis. *Behaviour* 70, 1–117.

Bowman, D.M.J.S. and Panton, W.J. (1993) Factors that control monsoon-rainforest seedling establishment and growth in north Australian *Eucalyptus* savannah. *Journal of Ecology* 81, 297–304.

Brandon, N.J., Shelton, H.M. and Peck, D.M. (1997) Factors affecting the early growth of *Leucaena leucocephala*. 2. Importance of arbuscular mycorrhizal fungi, grass competition and phosphorus application on yield and nodulation of *Leucaena* in pots. *Australian Journal of Experimental Agriculture* 37, 35–43.

Brokaw, N.V.L. (1982) Treefalls: frequency, timing, and consequences. In: Leigh, E.G., Jr, Rand, A.S. and Windsor, D.M. (eds) *The Ecology of a Tropical Forest: Seasonal Rhythms and Long-Term Changes*. Smithsonian Institution Press, Washington, DC, pp. 101–108.

Brokaw, N.V.L. and Scheiner, S. (1989) Species composition in gaps and structure of a tropical forest. *Ecology* 70, 538–540.

Brouat, C., Gibernau, M., Amsellem, L. and McKey, D. (1998) Corner's rules revisited: ontogenetic and interspecific patterns in leaf–stem allometry. *New Phytologist* 139, 459–470.

Brown, N. (1993) The implications of climate and gap microclimate for seedling growth conditions in a Bornean lowland rain forest. *Journal of Tropical Ecology* 9, 153–168.

Brown, N.D. and Whitmore, T.C. (1992) Do dipterocarp seedlings really partition tropical rain forest gaps? *Philosophical Transactions of the Royal Society of London Series B – Biological Sciences* 335, 369–378.

Brown, V.K. and Gange, A.C. (1999). Plant diversity in successional grassland: how is it modified by foliar insect herbivory? In: Kratochwil, A. (ed.) *Biodiversity in Ecosystems*. Kluwer, Dordrecht, pp. 133–146.

Bryant, J.P. and Julkunentiitto, R. (1995) Ontogenic development of chemical defense by seedling resin in birch: energy-cost of defense production. *Journal of Chemical Ecology* 21, 883–896.

Bullock, J.M., Hill, B.C., Silvertown, J. and Sutton, M. (1995) Gap colonization as a source of grassland community change: effects of gap size and grazing on the rate and mode of colonization by different species. *Oikos* 72, 273–282.

Burdon, J.J. and Chilvers, G.A. (1976a) The effect of clumped planting patterns on epidemics of damping-off disease in cress seedlings. *Oecologia* 23, 17–29.

Burdon, J.J. and Chilvers, G.A. (1976b) Epidemiology of *Pythium* induced damping-off in mixed species seedling stands. *Annals of Applied Biology* 82, 233–240.

Burke, M.J.W. and Grime, J.P. (1996) An experimental study of plant community invasibility. *Ecology* 77, 776–790.

Burkey, T.V. (1994) Tropical tree species diversity: a test of the Janzen–Connell model. *Oecologia* 97, 533–540.

Burslem, D.F.R.P., Grubb, P.J. and Turner, I.M. (1995) Responses to nutrient addition among shade-tolerant tree seedlings of lowland tropical rain forest in Singapore. *Journal of Ecology* 83, 113–122.

Cameron, G.W. and Carlie, M.J. (1977) Negative geotaxis of zoospores of the fungus *Phytophthora*. *Journal of General Microbiology* 98, 599–602.

Canham, C.D. (1985) Suppression and release during canopy recruitment in *Acer saccharum*. *Bulletin of Torrey Botanical Club* 112, 134–145.

Canham, C.D., McAninch, J.B. and Wood, D.M. (1994) Effects of the frequency, timing, and intensity of simulated browsing on growth and mortality of tree seedlings. *Canadian Journal of Forest Research* 24, 817–825.

Canham, C.D., Kobe, R.K., Latty, E.F. and Chazdon, R.L. (1999) Interspecific and intraspecific variation in tree seedling survival: effects of allocation to roots versus carbohydrate reserves. *Oecologia* 121, 1–11.

Carrington, M.E. and Keeley, J.E. (1999) Comparison of post-fire seedling establishment between scrub communities in Mediterranean and non-Mediterranean climate ecosystems. *Journal of Ecology* 87, 1025–1036.

Carter, M.F. and Grace, J.B. (1986) Relative effects of *Justicia americana* litter on germination, seedlings, and established plants of *Polygonum lapathifolium*. *Aquatic Botany* 23, 341–349.

Castro, Y., Fetcher, N. and Fernandez, D.S. (1995) Chronic photoinhibition in seedlings of tropical trees. *Physiologia Plantarum* 94, 560–565.

Castro-Diez, P., Puyravaud, J.P., Cornelissen, J.H.C. and Villar-Salvador, P. (1998) Stem anatomy and relative growth rate in seedlings of a wide range of woody plant species and types. *Oecologia* 116, 57–66.

Chambers, J.C. (1995) Relationships between seed fates and seedling establishment in an alpine ecosystem. *Ecology* 76, 2124–2133.

Chapin, F.S. (1989) The cost of tundra plant structures: evaluation of concepts and currencies. *American Naturalist* 133, 1–19.

Chapin, F.S., Schulze, E.-D. and Mooney, H.A. (1990) The ecology and economics of storage in plants. *Annual Review Ecology and Systematics* 21, 432–447.

Chazdon, R.L. (1992) Photosynthetic plasticity of two rain forest shrubs across natural gap transect. *Oecologia* 92, 586–595.

Cheplick, G.P. and Quinn, J.A. (1987) The role of seed depth, litter and fire in the seedling establishment of amphicarpic peanutgrass (*Amphicarpum pushii*). *Oecologia* 73, 459–464.

Cintra, R. (1997) A test of the Janzen–Connell model with two common tree species in Amazonian forest. *Journal of Tropical Ecology* 13, 641–658.

Clark, D.A. and Clark, D.B. (1984) Spacing dynamics of a tropical rain forest tree: evaluation of the Janzen–Connell model. *American Naturalist* 124, 769–788.

Clark, D.B. and Clark, D.A. (1985) Seedling dynamics of a tropical tree: impacts of herbivory and meristem damage. *Ecology* 66, 1884–1892.

Clark, D.B. and Clark, D.A. (1989) The role of physical damage in the seedling mortality regime of a neotropical rain forest. *Oikos* 55, 225–230.

Clark, D.B. and Clark, D.A. (1991) The impact of physical damage on canopy tree regeneration in tropical rain forest. *Journal of Ecology* 79, 447–457.

Coley, P.D. (1983) Herbivory and defensive characteristics of tree species in a lowland tropical forest. *Ecological Monographs* 53, 209–233.

Coley, P.D. (1988) Effects of plant growth rate and leaf lifetime on the amount and type of anti-herbivore defense. *Oecologia* 74, 531–536.

Coley, P.D., Bryant, J.P. and Chapin, F.S. (1985) Resource availability and plant anti-herbivore defense. *Science* 230, 895–899.

Collins, S.L. (1990) Habitat relationships and survivorship of tree seedlings in hemlock–hardwood forest. *Canadian Journal of Botany* 68, 790–797.

Colpaert, J.V., van Tichelen, K.K., van Assche, J.A. and van Laere, A. (1999). Short-term phosphorus uptake rates in mycorrhizal and non-mycorrhizal roots of intact *Pinus sylvestris* seedlings. *New Phytologist* 143, 589–597.

Condit, R., Hubbell, S.P. and Foster, R.B. (1992) Recruitment near conspecific adults and the maintenance of tree and shrub diversity in a neotropical forest. *American Naturalist* 140, 261–286.

Connell, J.H. (1971) On the role of natural enemies in preventing competitive exclusion in some marine animals and in rain forest trees. In: Boer, P.J.D. and Gradwell, G.R. (eds) *Dynamics of Populations*. Pudoc, Wageningen, pp. 290–310.

Cook, R.E. (1979) Patterns of juvenile mortality and recruitment in plants. In: Solbrig, T.O., Jain, S., Johnson, B.G. and Raven, H.P. (eds) *Topics in Plant Population Biology*. Columbia Universtiy Press, New York, pp. 207–231.

Cornelissen, J.H.C., Castro-Diez, P. and Hunt, R. (1996) Seedling growth, allocation and leaf attributes in a wide range of woody plant species and types. *Journal of Ecology* 84, 755–765.

Cornelissen, J.H.C., Castro-Diez, P. and Carnelli, A.L. (1998) Variation in relative growth rate among woody species. In: Lambers, H., Poorter, H. and Van Vuuren, M.M.I. (eds) *Inherent Variation in Plant Growth*. Backhuys Publishers, Leiden, pp. 363–392.

Corré, W.J. (1983) Growth and morphogenesis of sun and shade plants II. The influence of light quality. *Acta Botanica Neerlandica* 32, 185–202.

Dalling, J.W. (1995) The effect of litter and soil disturbance on seed germination in upper montain rain forest, Jamaica. *Caribbean Journal of Science* 31, 223–229.

Dalling, J.W., Swaine, M.D. and Garwood, N.C. (1995) Effect of soil depth on seedling emergence in tropical soil seed-bank investigations. *Functional Ecology* 9, 119–121.

Dalling, J.W., Hubbell, S.P. and Silvera, K. (1998a) Seed dispersal, seedling establishment and gap partitioning among tropical pioneer trees. *Journal of Ecology* 86, 674–689.

Dalling, J.W., Swaine, M.D. and Garwood, N.C. (1998b) Dispersal patterns and seed bank dynamics of two pioneer tree species in moist tropical forest, Panama. *Ecology* 79, 564–578.

Davidson, E.M. (1985) The effect of waterlogging on seedlings of *Eucalyptus marginata*. *New Phytologist* 101, 743–753.

Deacon, J.W. (1988) Behavioral responses of fungal zoospores. *Microbiological Science* 5, 249–252.

Denslow, J. (1980) Gap partitioning among tropical rainforest trees. *Biotropica* 12, 47–55.

Denslow, J.S., Newell, E. and Ellison, A. (1991) The effect of understory palms and cyclanths on the growth and survival of *Inga* seedlings. *Biotropica* 23, 225–234.

Denslow, J.S., Ellison, A.M. and Sanford, R.E. (1998) Treefall gap size effects on above- and below-ground processes in a tropical wet forest. *Journal of Ecology* 86, 597–609.

De Steven, D. (1991) Experiments on mechanisms of tree establishment in old-field succession: seedling survival and growth. *Ecology* 72, 1076–1088.

De Steven, D. (1994) Tropical tree seedling dynamics: recruitment patterns and their population consequences for three canopy species in Panama. *Journal of Tropical Ecology* 10, 369–383.

De Steven, D. and Putz, F.E. (1984) Impact of mammals on early recruitment of a tropical canopy tree, *Dipteryx panamensis,* in Panama. *Oikos* 43, 207–216.

de Vogel, E.F. (1980) *Seedlings of Dicotyledons*. Centre for Agricultural Publishing and Documentation, Pudoc, Wageningen, The Netherlands.

Dirzo, R. and Harper, J.L. (1980) Experimental studies on plant–slug interactions.II. The effect of grazing by slugs on high density monocultures of *Capsella bursa-pastoris* and *Poa annua*. *Journal of Ecology* 68, 999–1011.

Duke, J.A. (1965) Keys for the identification of seedlings of some prominent woody species in eight forest types in Puerto Rico. *Annals of Missouri Botanical Garden* 52, 314–350.

Edwards, A.L., Wyatt, R. and Sharitz, R.R. (1994) Seed buoyancy and viability of the wetland milkweed *Asclepias perennis* and an upland milkweed, *Asclepias exaltata*. *Bulletin of the Torrey Botanical Club* 121, 160–169.

Edwards, G.R. and Crawley, M.J. (1999) Rodent seed predation and seedling recruitment in mesic grassland. *Oecologia* 118, 288–296.

Ernst, W.H.O. (1988) Seed and seedling ecology of *Bachystegia spiciformis*, a predominant tree component in miombo woodlands in south central Africa. *Forest Ecology and Management* 25, 195–210.

Fenner, M. (1978a) Susceptibility to shade in seedlings of colonizing and closed turf species. *New Phytologist* 81, 739–744.

Fenner, M. (1978b) A comparison of the abilities of colonizers and closed-turf species to establish from seed in artificial swards. *Journal of Ecology* 66, 953–963.

Fenner, M. (1983) Relationships between seed weight, ash content and seedling growth in twenty-four species of Compositae. *New Phytologist* 95, 697–706.

Fenner, M. (1986) A bioassay to determine the limiting minerals for seeds from nutrient-deprived *Senecio vulgaris* plants. *Journal of Ecology* 74, 497–505.

Fenner, M. (1987) Seedlings. *New Phytologist* 106 (Suppl.), 35–47.

Fenner, M. (1995) Ecology of seed banks. In: Kigel, J. and Galili, G. (eds) *Seed Development and Germination*. Marcel Dekker, New York, pp. 507–528.

Fenner, M. and Lee, W.G. (1989) Growth of seedlings of pasture grasses and legumes deprived of single mineral nutrients. *Journal of Applied Ecology* 26, 223–232.

Fenner, M., Hanley, M.E. and Lawrence, R. (1999) Comparison of seedling and adult palatability in annual and perennial plants. *Functional Ecology* 13, 546–551.

Fetcher, N., Oberbauer, S.F., Rojas, G. and Strain, B.R. (1987) Efectos del régimende luz sobre la fotosintesis y el crecimiento en plántulas de árboles de un bosque lluvioso tropical de Costa Rica. *Revista de Biologia Tropical* 35, 97–110.

Forget, P.M. (1993) Post-dispersal predation and scatterhoarding of *Dipteryx panamensis* (Papilionaceae) seeds by rodents in Panama. *Oecologia* 94, 255–261.

Foster, S.A. and Janson, C.H. (1985) The relationship between seed size and establishment conditions in tropical woody plants. *Ecology* 66, 773–780.

Fowler, N.L. (1986) Microsite requirements for germination and establishment of three grass species. *American Midland Naturalist* 115, 131–145.

Gange, A.C., Brown, V.K. and Farmer, L.M. (1991) Mechanisms of seedling mortality by subterranean insect herbivores. *Oecologia* 88, 228–232.

Garrett, S.D. (1970) *Pathogenic Root Infecting Fungi*. Cambridge University Press, Cambridge.

Garwood, N.C. (1982) Seasonal rhythm of seed germination in a semi-deciduous tropical forest. In: Leigh, E.G., Rand, A.S. and Windsor, D.M. (eds) *The Ecology of a Tropical Forest: Seasonal Rhythms and Long-Term Changes*. Smithsonian Institution Press, Washington, DC, pp. 173–185.

Garwood, N.C. (1983) Seed germination in a seasonal tropical forest in Panama: a community study. *Ecological Monographs* 53, 159–181.

Garwood, N.C. (1989) Tropical soil seed banks: a review. In: Leck, M.A., Simpson, R.L. and Parker, V.T. (eds) *Ecology of Seed Banks*. Academic Press, San Diego, pp. 149–209.

Garwood, N.C. (1996) Functional morphology of tropical tree seedlings. In: Swaine, M.D. (ed.) *The Ecology of Tropical Forest Tree Seedlings*. Parthenon, Carnforth, pp. 59–129.

Garwood, N.C. and Lighton, J.R.B. (1990) Physiological ecology of seed respiration in some tropical species. *New Phytologist* 115, 549–558.

Gay, P.E., Grubb, P.J. and Hudson, H.J. (1982) Seasonal changes in the concentration of nitrogen, phosphorus and potassium, and in the density of mycorrhiza in biennial and matrix-forming perennial species of closed chalkland turf. *Journal of Ecology* 70, 571–593.

Gillham, M.E. (1956) Ecology of the Pembrokeshire islands. IV. Effects of grazing on the vegetation. *Journal of Ecology* 41, 84–99.

Godfrey, R.K. (1988) *Trees, Shrubs, and Woody Vines of Northern Florida and Adjacent Georgia and Alabama*. University of Georgia Press, Athens.

Goldberg, D.E. and Werner, P.A. (1983) The effects of size of opening in vegetation and litter cover on seedling establishment of goldenrods (*Solidago* spp.). *Oecologia* 60, 149–155.

Gordon, D.R. and Rice, K.J. (1993) Competitive effects of grassland annuals on soil-water and blue oak (*Quercus douglasii*) seedlings. *Ecology* 74, 68–82.

Gray, A.N. and Spies, T.A. (1996) Gap size, within-gap position and canopy structure effects on conifer seedling establishment. *Journal of Ecology* 84, 635–645.

Grime, J.P. and Jeffrey, D.W. (1965) Seedling establishment in vertical gradients of sunlight. *Journal of Ecology* 53, 621–642.

Gross, K.L. and Werner, P.A. (1982) Colonizing abilities of 'biennial' plants species in relation to ground cover: implications for their distribution in a successional sere. *Ecology* 63, 921–931.

Grubb, P.J. (1986) The ecology of establishment. In: Bradshaw, A.D. and Chadwick, M.J. (eds) *Ecology and Design in Landscape*. Blackwell Scientific Publications, Oxford, pp. 83–97.

Grubb, P.J. and Burslem, D.F.R.P. (1998) Mineral nutrient concentrations as a function of seed size within seed crops: implications for competition among seedlings and defence against herbivory. *Journal of Tropical Ecology* 14, 177–185.

Grubb, P.J. and Coomes, D.A. (1997) Seed mass and nutrient content in nutrient-starved tropical rainforest in Venezuela. *Seed Science Research* 7, 269–280.

Grubb, P.J. and Metcalfe, D.J. (1996) Adaptation and inertia in the Australian tropical lowland rainforest flora: contradictory trends in intergeneric and intrageneric comparisons of seed size in relation to light demand. *Functional Ecology* 10, 512–520.

Grubb, P.J., Metcalfe, D.J., Grubb, E.A.A. and Jones, G.D. (1998) Nitrogen-richness and protection of seeds in Australian tropical rainforest: a test of plant defence theory. *Oikos* 82, 467–482.

Hanley, M.E. (1995) The influence of molluscan herbivory on seedling regeneration in grassland. Unpublished PhD thesis, University of Southampton.

Hanley, M.E. (1998) Seedling herbivory, community composition and plant life history traits. *Perspectives in Plant Ecology, Evolution and Systematics* 1, 191–205.

Hanley, M.E. and Fenner, M. (1997) Seedling growth of four fire-following Mediterranean plant species deprived of single mineral nutrients. *Functional Ecology* 11, 398–405.

Hanley, M.E., Fenner, M. and Edwards, P.J. (1995a) An experimental field study of the effects of mollusc grazing on seedling recruitment and survival in grassland. *Journal of Ecology* 83, 621–627.

Hanley, M.E., Fenner, M. and Edwards, P.J. (1995b) The effect of seedling age on the likelihood of herbivory by the slug *Deroceras reticulatum*. *Functional Ecology* 9, 745–759.

Hanley, M.E., Fenner, M. and Edwards, P.J. (1996a) Mollusc grazing and seedling survivorship of four common grassland plant species: the role of gap size, species and season. *Acta Oecologica* 17, 331–341.

Hanley, M.E., Fenner, M. and Edwards, P.J. (1996b) The effect of mollusc-grazing on seedling recruitment in artificially created grassland gaps. *Oecologia* 106, 240–246.

Harms, K.E. and Dalling, J.W. (1997) Damage and herbivory tolerance through resprouting as an advantage of large seed size in tropical trees and lianas. *Journal of Tropical Ecology* 13, 617–621.

Harms, K.E., Dalling, J.W. and Aizprua, R. (1997) Regeneration from cotyledons in *Gustavia superba* (Lecythidaceae). *Biotropica* 29, 234–237.

Harper, J.L. (1980) Plant demography and ecological theory. *Oikos* 35, 244–253.

Hartnett, D.C. and Wilson, G.W.T. (1999) Mycorrhizae influence plant community structure and diversity in tallgrass prairie. *Ecology* 80, 1187–1195.

Hartnett, D.C., Samenus, R.J., Fischer, L.E. and Hetrick, B.A.D. (1994) Plant demographic responses to mycorrhizal symbiosis in tallgrass prairie. *Oecologia* 99, 21–26.

Hawkes, M.C. (1998) Function of root border cells in plant health: pioneers in the rhizosphere. *Annual Review of Phytopathology* 36, 311–327.

Hendelsman, J. and Stabb, E.V. (1996) Biocontrol of soilborne plant pathogens. *Plant Cell* 8, 1855–1869.

Herrera, R.A., Capote, R.P., Menédez, L. and Rodríguez, M.E. (1992) Silvigenesis stages and the role of mycorrhiza in natural regeneration in Sierra del Rosario, Cuba. In: Gomea-Pompa, A., Whitmore, T.C. and Hadley, M. (eds) *Rain Forest Regeneration and Management*. Parthenon, Carnforth, pp. 211–221.

Hladik, A. and Miquel, S. (1990) Seedling types and plant establishment in an African rain forest. In: Bawa, K.S. and Hadley, M. (eds) *Reproductive Ecology of Tropical Forest Plants*. Parthenon, Carnforth, pp. 261–282.

Hobbs, R.J. and Mooney, H.A. (1985) Community and population dynamics of serpentine grassland annuals in relation to gopher disturbance. *Oecologia* 67, 342–351.

Hoffmann, W.A. (1997) The effects of fire and cover on seedling establishment in a neotropical savannah. *Journal of Ecology* 84, 383–393.

Holbrook, M. and Putz, F. (1996) Ecophysiology of tropical vines and hemi-epiphytes: plants that climb up and plants that climb down. In: Mulkey, S.S., Chazdon, R.L. and Smith, A.P. (eds) *Tropical Forest Plant Ecophysiology*. Chapman and Hall, New York, pp. 363–394.

Horvitz, C.C. and Schemske, D.W. (1994) Effects of dispersers, gaps, and predators on dormancy and seedling emergence in a tropical herb. *Ecology* 75, 1949–1958.

Hoshizaki, K., Suzuki, W. and Sasaki, S. (1997) Impacts of secondary seed dispersal and herbivory on seedling survival in *Aesculus turbinata*. *Journal of Vegetation Science* 8, 735–742.

Howe, H.F. (1990) Survival and growth of juvenile *Virola surinamensis* in Panama: effects of herbivory and canopy closure. *Journal of Tropical Ecology* 6, 259–280.

Howe, H.F. and Richter, W.M. (1982) Effects of seed size on seedling size in *Virola surinamensis*: a within and between tree analysis. *Oecologia* 53, 347–351.

Howe, H.F., Schupp, E.W. and Westley, L.C. (1985) Early consequences of seed dispersal for a neotropical tree (*Virola surinamensis*). *Ecology* 66, 781–791.

Huante, P., Rincon, E. and Gavito, M. (1992) Root system analysis of seedlings of seven tree species from a tropical dry forest in Mexico. *Trees – Structure and Function* 6, 77–82.

Hubbell, S.P., Foster, R.B., O'Brien, S.T., Harms, K.E., Condit, R., Wechsler, B., Wright, S.J. and de Lao, S.L. (1999) Light-gap disturbances, recruitment limitation, and tree diversity in a neotropical forest. *Science* 283, 554–557.

Hulme, P.E. (1994) Seedling herbivory in grassland: relative impact of vertebrate and invertebrate herbivores. *Journal of Ecology* 82, 873–880.

Inouye, R.S., Huntly, N.J., Tilman, D. and Tester, J.R. (1987) Pocket gophers (*Geomys bursarius*), vegetation, and soil nitrogen along a successional sere in east central Minnesota. *Oecologia* 72, 178–184.

Janos, D.P. (1980a) Vesicular-arbuscular mycorrhizae affect lowland tropical rain forest plant growth. *Ecology* 61, 151–162.

Janos, D.P. (1980b) Mycorrhizae influence tropical succession. *Biotropica* 12, 56–64.

Janos, D.P. (1983) Tropical mycorrhizas, nutrient cycles and plant growth. In: Sutton, S.L., Whitmore, T.C. and Chadwick, A.C. (eds) *Tropical Rain Forest: Ecology and Management.* Blackwell Scientific Publications, Oxford, pp. 327–345.

Janzen, D.H. (1970) Herbivores and the number of tree species in tropical forests. *American Naturalist* 104, 501–528.

Kamaluddin, M. and Grace, J. (1992a) Photoinhibition and light acclimation in seedlings of *Bischofia javanica*, a tropical forest tree from Asia. *Annals of Botany* 69, 47–52.

Kamaluddin, M. and Grace, J. (1992b) Acclimation in seedlings of a tropical tree, *Bischofia javanica*, following a stepwise reduction in light. *Annals of Botany* 69, 557–562.

Karban, I. and Baldwin, I.T. (1997) *Induced Responses to Herbivory.* University of Chicago Press, Chicago.

Kikuzawa, K. and Ackerly, D. (1999) Significance of leaf longevity in plants. *Plant Species Biology* 14, 39–45.

Kikuzawa, K. and Koyama, H. (1999) Scaling of soil water absorption by seeds: an experiment using seed analogues. *Seed Science Research* 9, 171–178.

King, D.A. (1999) Juvenile foliage and the scaling of tree proportions, with emphasis on *Eucalyptus. Ecology* 80, 1944–1954.

Kitajima, K. (1992a) The importance of cotyledon functional morphology and patterns of seed reserve utilization for the physiological ecology of neotropcial tree seedlings. Unpublished PhD thesis, University of Illinois.

Kitajima, K. (1992b) Relationship between photosynthesis and thickness of cotyledons for tropical tree species. *Functional Ecology* 6, 582–589.

Kitajima, K. (1994) Relative importance of photosynthetic traits and allocation patterns as correlates of seedling shade tolerance of 13 tropical trees. *Oecologia* 98, 419–428.

Kitajima, K. (1996a) Cotyledon functional morphology, seed reserve utilization, and regeneration niches of tropical tree seedlings. In: Swaine, M.D. (ed.) *The Ecology of Tropical Forest Tree Seedlings.* UNESCO, Paris, pp. 193–208.

Kitajima, K. (1996b) Ecophysiology of tropical tree seedlings. In: Mulkey, S.S., Chazdon, R.L. and Smith, A.P. (eds) *Tropical Forest Plant Ecophysiology.* Chapman and Hall, New York, pp. 559–596.

Kitajima, K. and Augspurger, C.K. (1989) Seed and seedling ecology of a monocarpic tropical tree, *Tachigalia versicolor. Ecology* 70, 1102–1114.

Kitajima, K. and Tilman, D. (1996) Seed banks and seedling establishment on an experimental productivity gradient. *Oikos* 76, 381–391.

Klun, J.A. and Robinson, J.F. (1969) Concentrations of two 1,4-benzoxazinones in dent corn at various stages of development of the plant and its relation to resistance of the host plant to the European corn borer. *Journal of Economic Entomology* 62, 214–220.

Kohyama, T. (1983) Seedling stage of two subalpine *Abies* species in distinction from sapling stage: a matter-economic analysis. *Botanical Magazine Tokyo* 49–65.

Kohyama, T. (1987) Significance of architecture and allometry in saplings. *Functional Ecology* 1, 399–404.

Kohyama, T. and Grubb, P.J. (1994) Below- and above-ground allometries of shade-tolerant seedlings in a Japanese warm-temperate rainforest. *Functional Ecology* 8, 229–236.

Koike, T. (1988) Leaf structure and photosynthetic performance as related to the forest succession of deciduous broad-leaved trees. *Plant Species Biology* 3, 77–87.

Kozlowski, T.T., Kramer, P.J. and Pallardy, S.G. (1991) *The Physiological Ecology of Woody Plants.* Academic Press, San Diego.

Krigel, I. (1967) The early requirement for plant nutrients by subterranean clover seedlings (*Trifolium subterraneum*). *Australian Journal of Agricultural Research* 18, 879–886.

Kruger, L.M., Midgley, J.J. and Cowling, R.M. (1997) Resprouters *vs.* reseeders in South African forest trees: a model based on forest canopy height. *Functional Ecology* 11, 101–105.

Langenheim, J.H. and Stubblebine, W.H. (1983) Variation in leaf resin composition between parent tree and progeny in *Hymenaea*: implications for herbivory in the humid tropics. *Biochemistry Systematics and Ecology* 11, 97–106.

Langenheim, J.H., Osmond, C.B., Brooks, A. and Ferra, P.J. (1984) Photosynthetic responses to light in seedlings of selected Amazonian and Australian rainforest tree species. *Oecologia* 63, 215–224.

Lee, D.W., Baskaran, K., Mansor, M., Mohamad, H. and Yap, S.K. (1996) Irradiance and spectral quality affect Asian tropical rain forest tree seedling development. *Ecology* 77, 568–580.

Lee, D.W., Oberbauer, S.F., Krishnapilay, B., Mansor, M., Mohamad, H. and Yap, S.K. (1997) Effects of irradiance and spectral quality on seedling development of two southeast Asian *Hopea* species. *Oecologia* 11, 1–9.

Lee, S.S. and Alexander, I.J. (1996) The dynamics of ectomycorrhizal infection of *Shorea leprosula* seedlings in Malaysian rain forests. *New Phytologist* 132, 297–305.

Lee, W.G. and Fenner, M. (1989) Mineral nutrient allocation in seeds and shoots of twelve *Chionochloa* species in relation to soil fertility. *Journal of Ecology* 77, 704–716.

Lei, T.T. and Lechowicz, M.J. (1990) Shade adaptation and shade tolerance in saplings of three *Acer* species from eastern North America. *Oecologia* 84, 224–228.

Leishman, M.R. (1999) How well do plant traits correlate with establishment ability? Evidence from a study of 16 calcareous grassland species. *New Phytologist* 141, 487–496.

Leishman, M.R. and Westoby, M. (1994a) The role of large seed size in shaded conditions: experimental evidence. *Functional Ecology* 8, 205–214.

Leishman, M.R. and Westoby, M. (1994b) The role of seed size in seedling establishment in dry soil conditions: experimental evidence from semi-arid species. *Journal of Ecology* 82, 249–258.

Levin, D.A. (1974) The oil content of seeds: an ecological perspective. *American Naturalist* 108, 193–206.

Li, M., Lieberman, M. and Lieberman, D. (1996) Seedling demography in undisturbed tropical wet forest in Costa Rica. In: Swain, M.D. (ed.) *The Ecology of Tropical Forest Tree Seedlings.* Parthenon, Carnforth, pp. 285–314.

Lockwood, J.L. (1986) Soilborne plant pathogens: concepts and connections. *Phytopathology* 76, 20–27.

Lo Gullo, M.A., Nardini, A., Salleo, S. and Tyree, M.T. (1998) Changes in root hydraulic conductance (K_R) of *Olea oleaster* seedlings following drought stress and irrigation. *New Phytologist* 140, 25–31.

Lovell, P.H. and Moore, K.G. (1970) A comparative study of cotyledons as assimilatory organs. *Journal of Experimental Botany* 21, 1017–1030.

Lovell, P.H. and Moore, K.G. (1971) A comparative study of the role of the cotyledons in seedling development. *Journal of Experimental Botany* 22, 153–162.

Lovelock, C.E., Jebb, M. and Osmond, C.B. (1994) Photoinhibition and recovery in tropical plant species: response to disturbance. *Oecologia* 97, 297–307.

Lovelock, C.E., Kyllo, D. and Winter, K. (1996) Growth responses to versicular-arbuscular mycorrhizae and elevated CO_2 in seedlings of a tropical tree, *Beilschmiedia pendula. Functional Ecology* 10, 662–667.

Lovelock, C.E., Kyllo, D., Popp, M., Isopp, H., Virgo, A. and Winter, K. (1997) Symbiotic vesicular-arbuscular mycorrhizae influence maximum rates of photosynthesis in tropical tree seedlings grown under elevated CO_2. *Australian Journal of Plant Physiology* 24, 185–194.

McCarthy, B.C. and Facelli, J.M. (1990) Microdisturbances in oldfields and forests: implications for woody seedling establishment. *Oikos* 58, 55–60.

McConnaughay, K.D.M. and Bazzaz, F.A. (1987) The relationships between gap size and performance of several colonizing annuals. *Ecology* 68, 411–416.

McConnaughay, K.D.M. and Coleman, J.S. (1999) Biomass allocation in plants: ontogeny or optimality? A test along three resource gradients. *Ecology* 80, 2581–2593.

McGee, P.A. (1990) Survival and growth of seedlings of coachwood (*Ceratopetalum apetalum*): effects of shade, mycorrhizas and a companion plant. *Australian Journal of Botany* 38, 583–592.

McGinley, M.A., Dhillion, S.S. and Neumann, J.C. (1994) Environmental heterogeneity and seedling establishment: ant plant–microbe interactions. *Functional Ecology* 8, 607–615.

McGuinness, K.A. (1997) Dispersal, establishment and survival of *Ceriops tagal* propagules in a north Australian mangrove forest. *Oecologia* 109, 80–87.

McKee, K.L. (1995) Seedling recruitment patterns in a Belizean mangrove forest: effects of establishment ability and physico-chemical factors. *Oecologia* 101, 448–460.

McPherson, K. and Williams, K. (1998) The role of carbohydrate reserves in the growth, resilience, and persistence of cabbage palm seedlings (*Sabal palmetto*). *Oecologia* 117, 460–468.

Mack, A.L. (1998) The potential impact of small-scale physical disturbance on seedlings in a Papuan rainforest. *Biotropica* 30, 547–552.

Mack, R.N. and Pyke, D.A. (1984) The demography of *Bromus tectorum*: the role of microclimate, grazing and disease. *Journal of Ecology* 72, 731–748.

Maesako, Y. (1997) Effects of streaked shearwaters (*Calonectris leucomelas*) burrowing on the lucido-phyllous forest in Ohshima Island, Japan. *Vegetation Science* 14, 61–74.

Maranon, T. and Grubb, P.J. (1993) Physiological basis and ecological significance of the seed size and relative growth rate relationship in Mediterranean annuals. *Functional Ecology* 7, 591–599.

Marler, M.J., Zabinski, C.A. and Callaway, R.M. (1999) Mycorrhizae indirectly enhance competitive effects of an invasive forb on a native bunchgrass. *Ecology* 80, 1180–1186.

Maron, J.L. (1997) Interspecific competition and insect herbivory reduce bush lupine (*Lupinus arboreus*) seedling survival. *Oecologia* 110, 284–290.

Marshall, P.E. and Kozlowski, T.T. (1974) Photosynthetic activity of cotyledons and foliage leaves of young angiosperm seedlings. *Canadian Journal of Botany* 52, 2023–2032.

Maruta, E. (1976) Seedling establishment of *Polygonum cuspidatum* on Mt. Fuji. *Japanese Journal of Ecology* 26, 101–105.

Mazer, S.J. (1989) Ecological, taxonomic, and life-history correlates of seed mass among Indiana dune angiosperms. *Ecological Monographs* 59, 153–175.

Menges, E.S. and Hawkes, C.V. (1998) Interactive effects of fire and microhabitat on plants of Florida scrub. *Ecological Applications* 8, 935–946.

Milberg, P. and Lamont, B.B. (1997) Seed/cotyledon size and content play a major role in early performance of species on nutrient-poor soils. *New Phytologist* 137, 665–672.

Miles, J. (1972) Early mortality and survival of self-sown seedlings in Glenfeshie, Inverness-shire. *Journal of Ecology* 61, 93–98

Mitchell, D.T. and Allsopp, N. (1984) Changes in the phosphorus composition of *Hakea sericea* (Proteacea) during germination under low phosphorus conditions. *New Phytologist* 96, 239–247.

Molofsky, J. and Augspurger, C.K. (1992) The effect of leaf litter on early seedling establishment in a tropical forest. *Ecology* 73, 68–77.

Molofsky, J. and Fisher, B.L. (1993) Habitat and predation effects on seedling survival and growth in shade-tolerant tropical trees. *Ecology* 74, 261–264.

Moora, M. and Zobel, M. (1996) Effect of arbuscular mycorrhiza on inter- and intraspecific competition of two grassland species. *Oecologia* 108, 79–84.

Moora, M. and Zobel, M. (1998) Can arbuscular mycorrhiza change the effect of root competition between conspecific plants of different ages? *Canadian Journal of Botany* 76, 613–619.

Morgan, D.C. and Smith, H. (1979) A systematic relationship between phytochrome-controlled development and species habitat, for plants grown in simulated natural radiation. *Planta* 145, 253–258.

Morris, B.M. and Gow, N.A.R. (1993) Mechanism of electrotaxis of zoospores of phytopathogenic fungi. *Phytopathology* 83, 877–882.

Mostacedo, C.B. and Fredericksen, T.S. (1999) Regeneration status of important tropical forest tree species in Bolivia: assessment and recommendations. *Forest Ecology and Management* 124, 263–273.

Mulligan, D.R. and Patrick, J.W. (1985) Carbon and phosphorus assimilation and deployment in *Eucalyptus pilularis* Smith seedlings with special reference to the role of cotyledons. *Australian Journal of Botany* 33, 485–496.

Myers, R.L. and Ewel, J.J. (1990) *Ecosystems of Florida*. University of Central Florida, Orlando.

Neher, D.A., Augspurger, C.K. and Wilkinson, H.T. (1987) Influence of age structure of plant populations on damping-off epidemics. *Oecologia* 74, 419–424.

Newell, E.A., McDonald, E.P., Strain, B.R. and Denslow, J.S. (1993) Photosynthetic responses of *Miconia* species to canopy openings in a lowland tropical rain forest *Oecologia* 94, 49–56.

Ng, F.S.P. (1978) Strategies of establishment in Malayan forest trees. In: Tomlinson, P.B. and Zimmerman, M.H. (eds) *Tropical Trees as Living Systems*. Cambridge University Press, Cambridge, pp. 129–162.

Núñez-Farfán, J. and Dirzo, R. (1988) Within-gap spatial heterogeneity and seedling performance in a Mexican tropical forest. *Oikos* 51, 274–284.

Oladokun, M.A.O. (1989) Nut weight and nutrient contents of *Cola acuminata* and *C. nitida* (Sterculiaceae). *Economic Botany* 43, 17–22.

Osborne, K. and Smith, T.J. (1990) Differential predation on mangrove propagules in open and closed canopy forest habitats. *Vegetatio* 89, 1–6.

Osunkoya, O.O. (1996) Light requirements for regeneration in tropical forest plants: taxon-level and ecological attribute effects. *Australian Journal of Ecology* 21, 429–441.

Osunkoya, O.O., Ash, J.E., Hopkins, M.S. and Graham, A.W. (1992) Factors affecting survival of tree seedlings in North Queensland rainforests. *Oecologia* 91, 569–578.

Osunkoya, O.O., Ash, J.E., Graham, A.W. and Hopkins, M.S. (1993) Growth of tree seedlings in tropical rain forests of North Queensland, Australia. *Journal of Tropical Ecology* 9, 1–18.

Osunkoya, O.O., Ash, J.E., Hopkins, M.S. and Graham, A.W. (1994) Influence of seed size and seedling ecological attributes on shade-tolerance of rain forest tree species in Northern Queensland. *Journal of Ecology* 82, 149–163.

Pate, J.S., Rasins, E., Rullo, J. and Kuo, J. (1985) Seed nutrient reserves of Proteaceae with special reference to protein bodies and their inclusions. *Annals of Botany* 57, 747–770.

Pedersen, E.A., Reddy, M.S. and Chakravarty, P. (1999) Effect of three species of bacteria on damping-off, root rot development, and ectomycorrhizal colonization of lodgepole pine and white spruce seedlings. *European Journal of Forest Pathology* 29, 123–134.

Penning de Vries, F.W.T. and Van Laar, H.H. (1977) Substrate utilization in germinating seeds. In: Landsberg, J.J. and Cutting, C.V. (eds) *Environmental Effects on Crop Physiology.* Academic Press, London, pp. 217–228.

Poorter, H. and Lambers, H. (1991) Is interspecific variation in relative growth rate positively correlated with biomass allocation to the leaves? *American Naturalist* 138, 1264–1268.

Popma, J. and Bongers, F. (1988) The effect of canopy gaps on growth and morphology of seedlings of rain forest species. *Oecologia* 75, 625–632.

Prevost, M. (1996) Effects of scarification on soil properties and natural seeding of a black spruce stand in the Quebec boreal forest. *Canadian Journal of Forestry Research* 26, 72–86.

Putz, F.E. (1983) Treefall pits and mounds, buried seeds, and the importance of soil disturbance to pioneer trees on Barro Colorado Island, Panama. *Ecology* 64, 1069–1074.

Rabinowitz, D. (1978) Mortality and initial propagule size in mangrove seedlings in Panama. *Journal of Ecology* 66, 45–51.

Read, D.J. (1991) Mycorrhizas in ecosystems: nature's response to the 'law of minimum'. In: Hawksworth, D.L. (ed.) *Frontiers in Mycology.* CAB International, Wallingford, UK, pp. 101–130.

Reich, P.B., Uhl, C., Walters, M.B. and Ellsworth, D.S. (1991) Leaf lifespan as a determinant of leaf structure and function among 23 Amazonian tree species. *Oecologia* 86, 16–24.

Reich, P.B., Tjoelker, M.G., Walters, M.B., Vanderklein, D.W. and Bushena, C. (1998) Close association of RGR, leaf and root morphology, seed mass and shade tolerance in seedlings of nine boreal tree species grown in high and low light. *Functional Ecology* 12, 327–338.

Ribbens, E., Silander, J.A. and Pacala, S.W. (1994) Seedling recruitment in forests: calibrating models to predict patterns of tree seedling dispersion. *Ecology* 75, 1794–1806.

Rice, K.J., Gordon, D.R., Hardison, J.L. and Welker, J.M. (1993) Phenotypic variation in seedlings of a keystone tree species (*Quercus douglasii*): the interactive effects of acorn source and competitive environment. *Oecologia* 96, 537–547.

Rice, S.A. and Bazzaz, F.A. (1989) Quantification of plasticity of plant traits in response to light intensity: comparing phenotypes at a common weight. *Oecologia* 78, 502–507.

Ryser, P. (1996) The importance of tissue density for growth and life span of leaves and roots: a comparison of five ecologically constrasting grasses. *Functional Ecology* 10, 717–723.

Sánchez-Hidalgo, M.E., Martínez-Ramos, M. and Espinosa-García, J. (1999) Chemical differentiation between leaves of seedlings and spatially close adult trees from the tropical rain-forest species *Nectandra ambigens* (Lauraceae): an alternative test of the Janzen–Connell model. *Functional Ecology* 13, 725–732.

Sasaki, S. and Ng, F.S.P. (1981) Physiological studies on germination and seedling development in *Intsia palembanica* (Merbau). *Malay Forester* 44, 43–59.

Sato, T., Tanouchi, H. and Takeshita, K. (1994) Initial regenerative processes of *Distylium racemosum* and *Persea thunbergii* in an evergreen broad-leaved forest. *Journal of Plant Research* 107, 331–337.

Sattin, M. and Sartorato, I. (1997) Role of seedling growth on weed–crop competition. In: Praczyk, T. (ed.) *Proceedings of the 10th EWRS Symposium, Poznan, Poland.* European Weed Research Society, Poznan, pp. 3–12.

Saverimuttu, T. and Westoby, M. (1996) Seedling longevity under deep shade in relation to seed size. *Journal of Ecology* 84, 681–689.

Scarano, F.R. and Crawford, R.M.M. (1992) Ontogeny and the concept of anoxia-tolerance: the case of the Amazonian leguminous tree *Parkia pendula*. *Journal of Tropical Ecology* 8, 349–352.

Seiwa, K. (1998) Advantages of early germination for growth and survival of seedlings of *Acer mono* under different overstorey phenologies in deciduous broad-leaved forests. *Journal of Ecology* 86, 219–228.

Seiwa, K. and Kikuzawa, K. (1996) Importance of seed size for the establishment of seedlings of five deciduous broad-leaved tree species. *Vegetatio* 123, 51–64.

Sena Gomes, A.R. and Kozlowski, T.T. (1988) Physiological and growth responses to flooding of seedlings of *Hevea braziliensis*. *Biotropica* 20, 286–293.

Sharitz, R.R. and McCormick, J.F. (1973). Population dynamics of two competing annual specoes. *Ecology* 54, 723–740.

Shibata, M. and Nakashizuka, T. (1995) Seed and seedling demography of four co-occurring *Carpinus* species in a temperate deciduous forest. *Ecology* 76, 1099–1108.

Shipley, B. and Peters, R.H. (1980) The allometry of seed weight and seedling relative growth rate. *Functional Ecology* 4, 523–529.

Silvertown, J.W. (1981) Seed size, life span, and germination date as coadapted features of plant life history. *American Naturalist* 118, 860–864.

Silvertown, J.W. and. Dickie, J.B. (1981) Seedling survivorship in natural populations of nine perennial chalk grassland plants. *New Phytologist* 88, 555–558.

Silvertown, J.W., Franco, M. and Menges, E. (1996) Interpretation of elasticity matrices as an aid to the management of plant populations for conservation. *Conservation Biology* 10, 591–597.

Sipe, T.W. and Bazzaz, F.A. (1994) Gap partitioning among maples (*Acer*) in central New England: shoot architecture and photosynthesis. *Ecology* 75, 2318–2332.

Smith, H. (1982) Light quality, photoperception, and plant strategy. *Annual Review of Plant Physiology* 33, 481–518.

Smith, M. and Capelle, J. (1992) Effects of soil surface microtopography and litter cover on germination, growth and biomass production of chicory (*Cichorium intybus* L). *American Midland Naturalist* 128, 246–253.

Smith, T.J. (1987) Seed predation in relation to tree dominance and distribtuion in mangrove forests. *Ecology* 68, 266–273.

Sonesson, L.K. (1994) Growth and survival after cotyledon removal in *Quercus robur* seedlings, grown in different natural soil types. *Oikos* 69, 65–70.

Sork, V.L. (1987) Effects of predation and light on seedling establishment in *Gustavia superba*. *Ecology* 68, 1341–1350.

Sousa, W.P. and Mitchell, B.J. (1999) The effect of seed predators on plant distributions: is there a general pattern in mangroves? *Oikos* 86, 55–66.

Steege, H.T. (1994) Flooding and drought tolerance in seeds and seedlings of two *Mora* species segregated along a soil hydrological gradient in the tropical rain forest of Guyana. *Oecologia* 100, 356–367.

Stock, W.D., Pate, J.S. and Delfs, J. (1990) Influence of seed size and quality on seedling development under low nutrient conditions in five Australian and South African members of the Proteaceae. *Journal of Ecology* 78, 1005–1020.

Strauss-Debenedetti, S. and Bazzaz, F.A. (1996) Photosynthetic characteristics of plants along successional gradients. In: Mulkey, S.S., Chazdon, R.L. and Smith, A.P. (eds) *Tropical Forest Plant Ecophysiology*. Chapman and Hall, New York, pp. 162–186.

Swain, M.D. and Whitmore, T.C. (1988) On the definition of ecological species groups in tropical rain forest. *Vegetatio* 75, 81–86.

Sylvia, D.M., Wilson, D.O., Graham, J.H., Madox, J.J., Millner, P., Morton, J.B., Skipper, H.D., Wright, S.F. and Jarstfer, A.G. (1993) Evaluation of vesicular-arbuscular mycorrhizal fungi in diverse plants and soils. *Soil Biology and Biochemistry* 25, 705–713.

Tanouchi, H., Sato, T. and Takeshita, K. (1994) Comparative-studies on acorn and seedling dynamics of four *Quercus* species in an evergreen broad-leaved forest. *Journal of Plant Research* 107, 153–159.

Terazawa, K. and Kikuzawa, K. (1994) Effects of flooding on leaf dynamics and other seedling responses in flood-tolerant *Alnus japonica* and flood-intolerant *Betula platyphylla* var *japonica*. *Tree Physiology* 14, 251–261.

Terborgh, J. and Wright, S.J. (1994) Effects of mammalian herbivores on plant recruitment in two neotropical forests. *Ecology* 75, 1829–1833.

Thackray, D.J., Wratten, S.W., Edwards, P.J. and Niemeyer, H.M. (1990) Resistence to the aphids *Sitobion avenae* and *Rhopalosiphum padi* in Gramineae in relation to hydroxamic acid levels. *Annals of Applied Biology* 116, 573–583.

Topa, M. (1986) Aerenchyma and lenticel formation in pine sedlings: a possible avoidance mechanism to anaerobic growth conditions. *Physiologia Plantarum* 68, 540–550.

Topa, M.A. and McLeod, K.W. (1986) Responses of *Pinus clausa, Pinus serotina* and *Pinus taeda* seedlings to anaerobic solution culture. I. Changes in growth and root morphology. *Physiologia Plantarum* 68, 523–531.

Turnbull, M.H., Doley, D. and Yates, D.J. (1993) The dynamics of photosynthetic acclimation to changes in light quantity and quality in three Australian rainforest tree species. *Oecologia* 94, 218–228.

Turner, I.M. and Newton, A.C. (1990) The initial responses of some tropical rain forest tree seedlings to a large gap environment. *Journal of Applied Ecology* 27, 605–608.

Tyree, M.T., Velez, V. and Dalling, J.W. (1998) Growth dynamics of root and shoot hydraulic conductance in seedlings of five neotropical tree species: scaling to show possible adaptation to differing light regimes. *Oecologia* 114, 293–298.

Vaartaja, O. (1962) The relationship of fungi to survival of shaded tree seedlings. *Ecology* 43, 547–549.

Valladares, F., Wright, S.J., Lasso, E., Kitajima, K. and Pearcy, R.W. (2000) Plastic phenotypic response to light of 16 congeneric shrubs from a Panamanian rain forest. *Ecology* 81, 1925–1936.

van der Heijden, M.G.A., Klironomos, J.N., Ursic, M., Moutoglis, P., Streitwolf-Engel, R., Boller, T., Wiemken, A. and Sanders, I.R. (1998) Mycorrhizal fungal diversity determines plant biodiversity, ecosystem variability and productivity. *Nature* 396, 69–72.

Vasconcelos, H.L. and Cherrett, J.M. (1997) Leaf-cutting ants and early forest regeneration in central Amazonia: effects of herbivory on tree seedling establishment. *Journal of Tropical Ecology* 13, 357–370.

Vaughan, J.G. (1970) *The Structure and Utilization of Oil Seeds*. Chapman and Hall, London.

Vázquez-Yanes, C. and Orozco-Segovia, A. (1992) Effects of litter from a tropical rainforest on tree seed germination and establishment under controlled conditions. *Tree Physiology* 11, 391–400.

Vázquez-Yanes, C. and Orozco-Segovia, A. (1994) Signals for seeds to sense and respond to gaps. In: Caldwell, M.M. and Pearcy, R.W. (eds) *Exploitation of Environmental Heterogeneity by Plants*. Academic Press, New York, pp. 209–236.

Vázquez-Yanes, C. and Orozco-Segovia, A. (1996) Physiological ecology of seed dormancy and longevity. In: Mulkey, S.S., Chazdon, R.L. and Smith, A.P. (eds) *Tropical Forest Plant Ecophysiology*. Chapman and Hall, New York, pp. 535–558.

Vázquez-Yanes, C., Orozco-Segovia, A., Rincon, E., Sanchez-Coronado, M.E., Huante, P., Toledo, J.R. and Barradas, V.L. (1990) Light beneath the litter in a tropical forest: effect on seed germination. *Ecology* 71, 1952–1958.

Veenendaal, E.M., Ernst, W.H.O. and Modise, G.S. (1996a) Effect of seasonal rainfall pattern on seedling emergence and establishment of grasses in a savannah in south-eastern Botswana. *Journal of Arid Environments* 32, 305–317.

Veenendaal, E.M., Swaine, M.D., Agyeman, V.K., Blay, D., Avebresse, I.K. and Mullins, C.E. (1996b) Differences in plant and soil water relations in and around a forest gap in West Africa during the dry season may influence seedling establishment and survival. *Journal of Ecology* 84, 83–90.

Veenendaal, E.M., Swaine, M.D., Lecha, R.T., Wals, M.F., Abebrese, I.K. and Owusuafriyie, K. (1996c) Responses of West African forest tree seedlings to irradiance and soil fertility. *Functional Ecology* 10, 501–511.

Veneklaas, E.J. and Poorter, L. (1998) Growth and carbon partitioning of tropical tree seedlings in contrasting light environment. In: Lambers, H., Poorter, L. and Van Vuuren, M.M.I. (eds) *Inherent Variation in Plant Growth: Physiological Mechanisms and Ecological Consequences*. Bunkhuys Publishers, Leiden, pp. 337–361.

Walters, M.B., Kruger, E.L. and Reich, P.B. (1993a) Growth, biomass distribution and CO_2 exchange of northern hardwood seedlings in high and low light: relationships with successional status and shade tolerance. *Oecologia* 94, 7–16.

Walters, M.B., Kruger, E.L. and Reich, P.B. (1993b) Relative growth rate in relation to physiological and morphological traits for northern hardwood tree seedlings: species, light environment and ontogenetic considerations. *Oecologia* 96, 219–231.

Westoby, M., Leishman, M.R. and Lord, J. (1996) Comparative ecology of seed size and dispersal. *Philosophical Transactions of the Royal Society of London, Series B (Biological Sciences)* 351, 1309–1317.

Whitmore, T.C. (1978) Gaps in the forest canopy. In: Thomlinson, P.B. and Zimmerman, M.H. (eds) *Tropical Trees as Living Systems*. Cambridge University Press, Cambridge, pp. 639–655.

Whitmore, T.C. and Brown, N.D. (1996) Dipterocarp seedling growth in rain forest canopy gaps during six and a half years. *Philosophical Transactions of the Royal Society of London Series B (Biological Sciences)* 351, 1195–1203.

Williams, K., Percival, F., Merino, J. and Mooney, H.A. (1987) Estimation of tissue construction cost from heat of combustion and organic nitrogen content. *Plant Cell and Environment* 10, 725–734.

Williams, K., Field, C.B. and Mooney, H.A. (1989) Relationships among leaf construction cost, leaf longevity, and light environment in rain-forest plants of the genus *Piper*. *American Naturalist* 133, 198–211.

Wills, C., Condit, R., Foster, R.B. and Hubbell, S.P. (1997) Strong density- and diversity-related effects help to maintain tree species diversity in a neotropical forest. *Proceedings of the National Academy of Sciences of the USA* 94, 1252–1257.

Wright, I.J. and Westoby, M. (1999) Differences in seedling growth behaviour among species: trait correlations across species, and trait shifts along nutrient compared to rainfall gradients. *Journal of Ecology* 87, 85–97.

Xiong, S. and Nilsson, C. (1999) The effects of plant litter on vegetation: a meta-analysis. *Journal of Ecology* 87, 984–994.

Zipperlen, S.W. and Press, M.C. (1996) Photosynthesis in relation to growth and seedling ecology of two dipterocarp rain forest tree species. *Journal of Ecology* 84, 863–876.

Chapter 15

The Contribution of Seedling Regeneration to the Structure and Dynamics of Plant Communities, Ecosystems and Larger Units of the Landscape

J. Philip Grime and Susan H. Hillier

Unit of Comparative Plant Ecology, Department of Animal and Plant Sciences, University of Sheffield, Sheffield, UK

Introduction

In terrestrial vegetation, seeds and seedlings are implicated in various ecological phenomena. These extend beyond population processes (persistence, dispersal, genetic variability and change) to influences upon the distribution, dynamics and diversity of much larger units (communities, ecosystems, landscapes, local floras). With increasing evidence of regional and global impacts of land-use and climate change upon vegetation, there is a need to model and predict regenerative processes at scales that are compatible with those employed by geographers, climatologists and landscape ecologists. This chapter reviews progress in the development of the theoretical framework and database required to meet this objective and provides examples of the study of regenerative patterns in relatively large units of vegetation. In the period that has elapsed since the appearance of the first edition of this book, seedling establishment has assumed a central place in the very active worldwide debate concerning the consequences for ecosystem function of declining plant species diversity. Accordingly, this chapter has been expanded to review the hypothesis that our concerns about continuing losses in the species richness of plant communities should focus upon their long-term consequences for the recruitment of vegetation dominants. It is suggested that dangers to ecosystem viability and their utility to humans are much more likely to arise from future failures in regenerative processes than from any immediate losses of ecosystem functions.

Theory

Functional types

It is now widely recognized that, for dynamic models of vegetation to achieve a wide applicability, it is necessary to classify plants into functional types. Some ecologists (e.g. Grubb, 1985) have argued that the number of types should be large, while others (e.g. Grime, 1985) have advocated a relatively simple classification. However, as Keddy (1989) has pointed out, the argument between Grubb and Grime arises from different research objectives. A careful reading of Grubb (1985) reveals a primary concern to explain the detailed ecology of individual plant populations. In contrast, Grime (1985) follows the approach of Noble and Slatyer (1979) in seeking to recognize the minimal typology sufficient to capture the essential functioning of communities. In this chapter, we explore the hypothesis that many of the dynamic features of terrestrial vegetation can be analysed by reference to a subset of functional types implicated in regeneration.

Functional types in the regenerative phase

From the contributions of many ecologists and evolutionary biologists, including Stebbins (1951, 1974), Sagar and Harper (1961), Wilbur *et al.* (1974), Grubb (1977),

Gill (1978) and Grime (1979), there has been gradual recognition of the advantage to be gained by separate analysis of the strategies exhibited by organisms during the regenerative phase of the life history. This acknowledges the principle that the microhabitats exploited by juveniles may be quite different from those experienced by the parents. In addition, many juveniles have characteristics, such as small size, high potential mobility or capacity for dormancy, which not only expose them to additional hazards but also confer a rich potential for adaptive specialization.

For plants, theories of ecological specialization in the regenerative phase have been formalized as a set of five regenerative strategies, four of which describe mechanisms apparently maximizing the potential for successful seedling establishment in specific circumstances (Table 15.1). Within the landscape, the success of one seed regenerative type rather than another may be determined by the predictability of disturbance events in space and/or time (Table 15.1, Fig. 15.1). This is consistent with the notion that natural selection has generated recurring sets of juvenile traits,

Table 15.1. Five regenerative strategies of widespread occurrence in terrestrial vegetation (from Grime, 1989).

Strategy		Functional characteristics	Conditions under which strategy appears to enjoy a selective advantage
Vegetative expansion	(V)	New shoots vegetative in origin and remaining attached to parent plant until well established	Productive or unproductive habitats subject to low intensities of disturbance
Seasonal regeneration	(S)	Independent offspring (seeds or vegetative propagules) produced in a single cohort	Habitats subjected to seasonally predictable disturbance by climate or biotic factors
Persistent seed or spore bank	(B_s)	Viable but dormant seeds or spores present throughout the year, some persisting more than 12 months	Habitats subjected to temporally unpredictable disturbance
Numerous widely dispersed seeds or spores	(W)	Offspring numerous and exceedingly buoyant in air; widely dispersed and often of limited persistence	Habitats subjected to spatially unpredictable disturbance or relatively inaccessible (cliffs, walls, tree trunks, etc.)
Persistent juveniles	(B_{sd})	Offspring derived from an independent propagule but seedling or sporeling capable of long-term persistence in a juvenile state	Unproductive habitats subjected to low intensities of disturbance

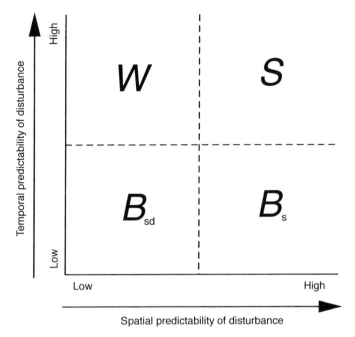

Fig. 15.1. Theoretical relationship between the predictability of disturbance events in space and/or time and strategies of regeneration from seed. *W*, widely dispersed seeds; B_{sd}, persistent juveniles; *S*, seasonal regeneration; B_s, persistent seed bank.

which confer either stress tolerance (seedlings that persist in unfavourable conditions, such as beneath a closed canopy) or, more commonly, the capacity for stress avoidance (gap exploitation by various mechanisms of dormancy or dispersal). In vascular plants, these ideas are supplemented by seed-bank models describing four major seasonal patterns of dormancy and persistence in the soil (Thompson and Grime, 1979). Several of these regenerative functional types have numerous close analogues in the regenerative biology of heterotrophs (Grime, 1979).

Regenerative strategies and vegetation dynamics

Regenerative functional types have been examined hitherto independently of the broader characteristics of plant community dynamics. However, the apparent 'uncoupling' between established and regenerative phases in the life history of organisms need

not hinder attempts to analyse the contribution of particular regenerative strategies to processes of vegetation change. As a starting-point, we re-examine in Fig. 15.2 various familiar successional phenomena portrayed in terms of Grime's triangular array of strategies in the established phase (Grime, 1974, 1987). In Fig. 15.2k–o, an attempt has been made to identify the role played by each of the regenerative strategies of Table 15.1 in each of the successional processes described in Fig. 15.2f–j.

A major contribution is predicted for *W* (the production of numerous widely dispersed seeds or spores) during the early stages of both secondary (Fig. 15.2k and l) and primary (Fig. 15.2m) succession. This prediction, supported by a wealth of empirical evidence (Ridley, 1930; van der Pijl, 1972), rests on the assumption that a considerable selective advantage among potential early colonists will be enjoyed by those species which achieve 'saturating' densities of propagules across large areas of landscape. The small buoyant propagules

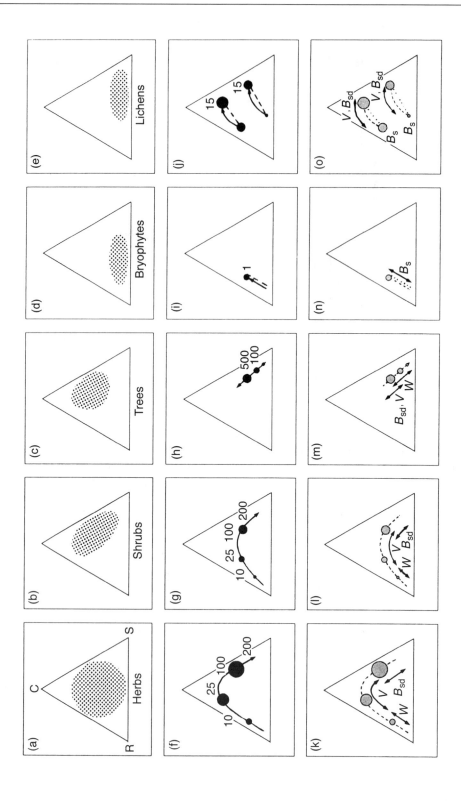

associated with W contain low quantities of mineral nutrients and organic substrates and tend to produce offspring of low competitive ability and persistence in closed vegetation. As successions such as those described in Fig. 12.2k–m proceed, therefore, it is expected that regeneration involving W will not provide an effective mechanism of persistence. Ephemerals dependent upon W will die out and be replaced by perennial herbs and shrubs, which rely upon vegetative expansion (V). Although the single most important contribution of seedling regeneration is likely to be that of establishing founder populations during the initial 'open' stages of succession (Egler, 1954), we forecast that B_{sd} (regeneration involving a bank of persistent seedlings) will play an inconspicuous but important role throughout primary succession (Fig. 15.2m) and in the advanced stages of secondary succession (Fig. 15.2k

and l). This conclusion is based upon the field observation (e.g. Inghe and Tamm, 1985) that many of the herbaceous and woody species that occupy mature ecosystems of low annual productivity (whether of low- or high-standing crop) produce seeds intermittently but maintain populations of seedlings; some of these may be capable of slowly graduating to reproductive maturity should disturbance events permit a sequence of temporary releases from the dominant effects of immediate neighbours. A marked contrast with the uninterrupted successional sequences of Fig. 15.2k–m is evident when attention is turned to the mechanisms of regeneration in the cyclical vegetation changes imposed by management interventions (Fig. 15.2n and o). In the simplest case, that of the arable field (Fig. 15.2n), the dominant regenerative strategy predicted is that associated with a bank of persistent seeds.

Fig. 15.2. (*Opposite*) Models using the theoretical framework described by Grime (1974, 1979) to depict the strategic range of five life-forms (a–e), to represent various successional phenomena (f–j) and to indicate the role of four regenerative strategies in these same phenomena (k–o). (Based on Grime, 1987.) The triangular model focuses on variation in the quality and durational stability of the habitat or niche to which the organism has been attuned by natural selection. Thus three extremes of evolutionary specialization (primary functional types) are recognized, together with a range of intermediate, secondary strategies. The full spectrum of conditions occurring in nature and their associated strategies can be described as an equilateral triangle in which the relative importance of competition, stress and disturbance is represented by three sets of contours. Functional types in the established phase: C, competitor; S, stress-tolerator; R, ruderal. (a–e) The strategic range of various life-forms; (f–j) successional phenomena represented in terms of the triangular model. The strategies of the dominant plants at particular points in time are indicated by the position of arrowed lines within the triangular model. The passage of time in years during succession is represented by numbers on each line and shoot biomass at particular points is reflected in the size of the circles. (f) Secondary succession in a forest clearing on a moderately fertile soil in a temperate climate. Biomass development is initially rapid and there is a fairly swift replacement of species as the community experiences successive phases of competitive dominance by rapidly growing herbs, shrubs and trees. At a later stage, however, the course of succession begins to deflect downwards towards the stress-tolerant corner of the triangular model. This process begins even during stages where the plant biomass is expanding appreciably and reflects a change from vegetation exhibiting high rates of resource capture and loss to one in which resources, particularly mineral nutrients, are efficiently retained or recaptured in the biomass. (g) Secondary succession for a site of lower soil fertility. Here the course of events is essentially the same as that described in (f) except that the successional parabola is shallower and the plant biomass smaller, as a consequence of the earlier onset of mineral nutrient limitation of the initial phase of competitive dominance. (h) Primary succession in a skeletal habitat, such as a rock outcrop. In this case the initial colonists, probably lichens and bryophytes, are stress-tolerators of low biomass and they occupy the site for a considerable period, giving way eventually to small, slow-growing herbs and shrubs. This sequence coincides with the process of soil formation and provides an example of the facilitation model of vegetation succession (Connell and Slatyer, 1977). (i and j) Vegetation responses to major perturbations. The range of models is extended to include loops representing the simple truncation of succession by annual harvesting and fertilizer input in an arable field (i) and the cycles of vegetation change associated with coppicing *Fraxinus excelsior* woodland (j, top) or rotational burning of *Calluna vulgaris* moorland (j, below). (k–o) The role of four regenerative strategies in the successional pathways represented in (f–j). See text for discussion. Functional types in the regenerative phase: W, numerous, small, widely dispersed propagules; V, vegetative spread; B_s, persistent seed bank; B_{sd}, persistent seedlings.

Here it is not difficult to recognize the selective advantage that populations of ephemeral weeds derive from the presence of high densities of persistent seeds in soils subjected to frequent and spatially predictable disturbance. In the coppice and heathland cycles, persistent seed banks appear to play an essentially similar role, but this is limited to phases of seedling recruitment initiated by occasional disturbance and followed by longer intervals in which biomass accumulates and other regenerative strategies (V, B_{sd}) may be expected to come into play.

In order to recognize principles of wide generality (in particular, the contrasting roles of W, B_s V and B_{sd} in vegetation dynamics), the examples selected have been relatively simple in character. It should be emphasized, however, that, where vegetation is the result of a complex and continuing history of fluctuating conditions of environment and management (e.g. ancient hay pastures and meadows of the British Isles), it is not unusual to encounter plant communities in which there appears to be a stable coexistence between populations exhibiting each of the regenerative strategies listed in Table 15.1.

Regenerative strategies in plant communities and larger units of landscape

The search for pattern

Regenerative processes in plant communities may be studied directly by examining the fate of seeds and seedlings, but this can be a difficult and time-consuming task, particularly in species-rich perennial vegetation. As such, it is undertaken with insufficient regularity to yield definitive information on regenerative processes in widely different communities. In the absence of detailed recording, an alternative approach is required if regenerative phenomena are to be built into predictive models of plant community responses. The development and testing of theoretical frameworks, such as the one described above, offers one alternative. Classification of plant species into regenerative functional types (e.g. W, S, B_s, V and B_{sd}) creates the possibility that plant communities may be characterized with regard to the regenerative biology of constituent populations. Recently, information synthesized from field and laboratory observations on the regenerative biology of a large number of species has been made available (Grime et al., 1988). Here we test the propositions that such information may be used as follows:

1. To predict the regenerative potential of two neighbouring grassland communities.
2. To characterize the regenerative potential of a range of plant communities in contrasted habitats within the same local area.
3. To provide insights into the functioning of vegetation in larger units of landscape.

Evidence from calcareous grassland on two adjacent slopes in Millers Dale

Regeneration from seed in artificially created gaps has been documented by Hillier (1984) for two contrasting, unmanaged grasslands (a north- and a south-facing slope) in Millers Dale, Derbyshire. These data provide an excellent opportunity to examine the degree to which the regenerative potential of each community can be predicted by reference to existing published information on the regenerative strategies of the species comprising the established turf. Prior to the experimental manipulations, shoot biomass of species in the established turf was estimated for each slope (Staal, 1979). Using Grime et al. (1988), supplemented where necessary by Brenchley (1918), Chippindale and Milton (1934), Douglas (1965), van Breemen and van Leeuwen (1983), Ryser and Gigon (1985), Darby (1987) and Bakker (1989) for additional information on B_s, these data have been classified with respect to the three seed regenerative strategies (S, B_s and W) for which substantial information is available. Species relative importance (%), and thus the relative importance of each regenerative strategy in the established vegetation, was calculated. Table 15.2 contains the results of this analysis, together with comparable data (based on seedling numbers) for the relative importance of S, B_s and W in the seedling flora colonizing gaps 400 mm in diameter. These were created artificially by removing all vegetation, including bryophytes, down to ground level at two different seasons; the census of seedlings was conducted in these gaps in order to avoid possible canopy-induced seed dormancy in the closed turf (Gorski, 1975; King, 1975; Fenner, 1980) and thereby to encourage full expression of the regenerative potential of the community.

Clearly, for the south-facing slope, the relative proportions of S, B_s and W in the established vegetation closely reflect those in large gaps. In this type of vegetation, therefore, established vegetation may indeed provide a useful indicator of regeneration.

On the north-facing slope, the importance of the regenerative strategy B_s appears to have been severely underestimated by predictions based on the established vegetation. However, additional information on the history and structure of the vegetation may help to explain this discrepancy. Despite a previous existence as grassland maintained by frequent fires (Lloyd, 1968), this site is now derelict and

at present supports a tall-herb community with a large perennial element and dense bryophyte layer. Herb litter and bryophyte material together account for more than 75% of the total standing crop + litter at this site and, although small gaps arise from time to time as the result of the activity of short-tailed field voles (*Microtus agrestis*), vegetation cover is more or less continuous throughout the year. In such circumstances, it seems likely that the buried seed bank will have been augmented annually, with only small losses due to germination, with the result that the seedling flora of the experimental gaps was dominated by species dependent on the regenerative strategy B_s and representing seed banks accumulated over several years. In fact, many subordinate species in the turf may be entirely dependent on the reinstatement of a form of management that creates gaps for their continued survival at the site. The south-facing slope, in contrast, supports a much more open sward, in which natural gaps occur regularly as a result of plant deaths due to frost-heaving and summer droughts.

A number of points arise from this analysis. First, data banks on the regenerative biology of plant species are now sufficiently well developed to permit the beginning of attempts to recognize general principles with regard to regeneration in

Table 15.2. The relative importance (%) of the regenerative strategies S, B_s and W (*sensu* Grime, 1979) in the established vegetation of a north- and south-facing slope in Millers Dale, Derbyshire, and in the seedling flora of 40 cm diameter autumn-cut (September 1978) and spring-cut (March 1979) gaps at each site.

	S	B_s	W
South-facing slope			
Established vegetation	61.0	25.2	13.4
Autumn-cut gaps	62.9	28.6	7.0
Spring-cut gaps	59.9	31.1	11.1
North-facing slope			
Established vegetation	61.9	29.5	0.1
Autumn-cut gaps	27.4	79.5	0.7
Spring-cut gaps	14.9	87.5	0.8

Values for the established vegetation are based on shoot biomass data. Values for the gaps are based on seedling totals recorded in June 1979 in seven replicate gaps cut at each season. Values do not sum to 100% because: (i) some species have more than one regenerative strategy; and (ii) a small number of species could not be classified with respect to regenerative strategy.

selected plant communities. However, there is considerable scope to add to the predictive power of the database with regard both to confirmation of regenerative strategies (particularly B_s) and to an increase in the number of species for which information is available.

Secondly, in skeletal habitats such as that on the south-facing slope in Millers Dale, where environment rather than management is the primary influence on vegetation structure, regeneration phenomena may indeed be predictable from the established vegetation. In some grasslands, however, as on the north-facing slope in Millers Dale, regeneration from seed has an intermittent role in cyclical vegetation processes under human control (cf. Fig. 15.2n and o). Here communities may contain species capable of exploiting different phases of the recolonization process following occasional major disturbances. In these circumstances, knowledge of the history and dynamics of the vegetation is indispensable.

Thirdly, the analyses presented here are based on the seedlings recorded on one occasion only. It is possible that a particular set of environmental or physical conditions influenced the composition of the seedling flora. At this stage, great care is required in making generalizations about regenerative potential by reference to the established vegetation of the community. Nevertheless, the results, particularly those relating to the south-facing slope, are extremely encouraging and we submit that this approach merits further investigation in other types of vegetation.

Regenerative processes in major habitat types in Lathkill Dale

In the previous section, the relative importance of the regenerative strategies S and B_s was shown to be in remarkable agreement with the established vegetation of the two slopes in Millers Dale (Table 15.2). The fact that both support unmanaged grassland raises the possibility that these values characterize a particular habitat type. This idea

is explored further here by means of an analysis of survey data from eight different but neighbouring habitat types in Lathkill Dale, Derbyshire.

Data are taken from J.P. Grime, P.S. Lloyd and S.R. Band (unpublished) and J.G. Hodgson and S.R. Band (unpublished). The relative importance of S, B_s and W has been calculated for the established vegetation in each habitat (Table 15.3), using the method already described for the Millers Dale data. Such analyses are inevitably incomplete at present, due to the large number of species for which information on regenerative strategies is lacking. However, the results in Table 15.3 encourage us to believe that different habitat types may have a characteristic 'regenerative profile', which can be predicted by reference to the established vegetation.

Three striking features are apparent in Table 15.3. First, as might be expected, the regenerative strategy W clearly plays a more significant role in the early successional, skeletal habitats such as cliffs, outcrops and quarry heaps, where much bare ground is available for colonization. Secondly, seasonal regeneration (S) appears to assume much more importance in deciduous woodland, where the window of opportunity for germination and seedling establishment is severely restricted by both light quality and the physical barrier presented by leaf fall. Thirdly, the regenerative strategy B_s (persistent seed bank) is remarkably well represented in the majority of the habitats considered, although some variation between habitats is evident, particularly in the tendency for the importance of B_s to fall in woody vegetation (cf. Grime, 1979) and for higher values to be recorded from disturbed habitats, e.g. outcrops and the partially burned south-facing grassland (cf. Peart, 1984).

The generally high representation of B_s is of particular interest. In many of the habitat types considered, the role of this regenerative strategy is not immediately obvious and it is questionable whether the high values are a feature of the contemporary situation or a legacy of past dispersal. Persistent seed banks may play an unobtru-

Table 15.3. The relative importance (%) of the regenerative strategies *S*, *B*$_s$ and *W* in the established vegetation of eight different but neighbouring habitat types in Lathkill Dale, Derbyshire.

	S	*B*$_s$	*W*	Number of species out of total currently unclassified
Cliffs	38	30	18	9/14
Quarry heaps	44	40	18	17/63
Outcrops	50	45	5	27/102
South-facing calcareous grassland (including scree and recolonizing burned areas)	49	51	4	11/56
Acidic plateau grassland	49	46	0	7/30
Unmanaged, north-facing calcareous grassland[a]	63	31	0	11/50
Scrub	52	21	1	17/61
Deciduous woodland	73	30	1	7/42

[a]Remarkably consistent with the regenerative profile obtained from established vegetation on the north-facing slope in Millers Dale (Table 15.2).
Values are based on frequency data and do not sum to 100% because: (i) the three categories are not mutually exclusive; and (ii) some species could not be classified with respect to regenerative strategy (see final column).

sive role in current vegetation dynamics, e.g. the exploitation of scattered small gaps, in addition to the more obvious part that they play in recovery after major episodes of damage. However, the dispersal history of many plants with persistent seed banks may also account for the high value recorded for B$_s$ in some habitats, e.g. cliffs, outcrops and quarry heaps. It may be significant that both Salisbury (1953) and Grime (1986) have noted that many of the most effectively dispersed species in the British flora have persistent seeds and apparently owe their colonizing success to passive dispersal in soil.

Scaling up

The need to incorporate regenerative processes into regional and global models of land-use and climate change impacts has led us to investigate the possibility of examining the distribution of regenerative strategies at scales compatible with those employed by geographers, climatologists and landscape ecologists. We have been fortunate in having access to the results of a detailed grassland survey (J.G. Hodgson, A.M. Neal and R. Hunt, unpublished) of

part of Lathkill Dale, Derbyshire (Fig. 15.3), in which 166 1 m^2 quadrats were recorded on a 50 m grid within a 1 km^2 study area. Using the technique described earlier, the relative importance of *S*, *B*$_s$ and *W* has been calculated for each surveyed grid point. In Fig. 15.4, the distribution map for *B*$_s$ is shown.

Since the underlying data set from which the map in Fig. 15.4 was produced comprises mainly ungrazed grassland (both acidic and calcareous), outcrops and scree intermixed at a scale only slightly larger than the grid size of the original survey, the lack of any evidence of major divisions between habitat types was not unexpected. However, two areas stand out, one with unusually high values of *B*$_s$ (area A in Fig. 15.4) and one with unusually low values (area B). The former corresponds to an expanse of acidic grassland lightly grazed by cattle. Table 15.3 suggests that acidic grasslands may have generally high values for *B*$_s$. In circumstances where gaps are continually formed in the turf, colonization by germination of buried seeds is to be expected and may be responsible for a noticeable increase in the importance of species with this regenerative strategy in the turf. Area B corresponds to an

Fig. 15.3. An aerial photograph of part of Lathkill Dale, Derbyshire, taken on 12 August 1972 from a height of 2200 m. The dale was sampled on a 50 m grid; records were not made from sample points falling in dense scrub, in woodland or on steep rock-faces. (Crown copyright.)

extensive and well-established bracken patch. Here regeneration from seed is likely to be inhibited by a deep litter layer.

Analyses of regeneration at the landscape scale can only be tentative at present, due to the lack of information for many important species. The time is fast approaching, however, when this objective must be addressed as an essential element in the effort to incorporate regenerative processes into regional and global models.

Regenerative strategies, declining plant diversity and ecosystem function

The main thesis advanced in this chapter is that, at least in skeletal form, we now have a theoretical framework with which to incorporate the regenerative mechanisms of plants into models of vegetation change through space and time. When this framework is used in conjunction with models, such as CSR theory, which reflect traits of the established phase (life history, growth rate, resource dynamics, anti-herbivore defence, decomposition rates) a basis is available upon which to conduct a reductionist analysis of both large-scale and long-term vegetation processes. However, there appears to be no sound reason why the approach should be arrested at this point; it can be argued that knowledge relating to the traits of dominant plants also provides many insights into the functioning of ecosystems. This assertion is supported by a range of field investigations

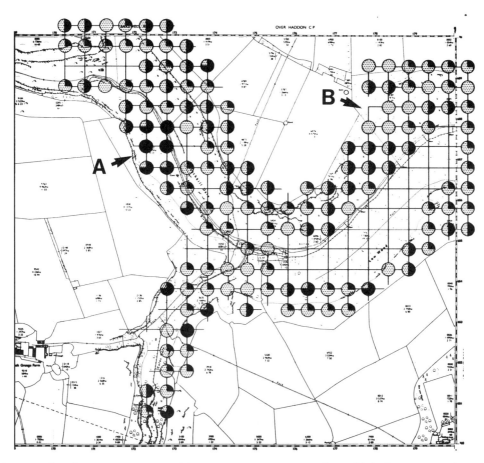

Fig. 15.4. The relative importance (%) of the regenerative strategy B_s in the established vegetation at the 166 sample points recorded in Lathkill Dale in 1980 (see Fig. 15.3 also). Key to pie charts: ◔, 0–20%; ◑, 21–40%; ◕, 41–60%; ◕, 61–80%; ●, 81–100%. (Base map © Crown copyright.)

(e.g. Leps *et al.*, 1982; MacGillivray *et al.*, 1995; Hooper and Vitousesk, 1997; Wardle *et al.*, 1997) in which ecosystem properties and responses to perturbation have been successfully predicted from calculations based upon the functional characteristics and contribution to the plant biomass of the major component species.

If the traits of dominant plants are the main controllers of ecosystem characteristics such as productivity, water relations, mineral nutrient cycling, organic-matter accumulation and resistance and resilience to extreme events, it follows that we should place a strong focus upon the fate of dominants in our efforts to examine the

consequences of declining plant diversity. In recent investigations (Naeem *et al.*, 1994; Tilman and Downing, 1994), emphasis has been restricted to the possible immediate consequences for ecosystem functions of losses in species richness. However, doubts have been cast upon the validity of these studies (Grime, 1997; Huston, 1997) and emphasis is currently shifting (Grime, 1998) to a consideration of the long-term consequences of declining species richness. In earlier sections of this chapter (see Fig. 15.2k–o), the crucial role of seedling regeneration has been identified in the successional and cyclical processes that sustain vegetation types and their attendant ecosystems in natural and semi-natural landscapes. In particular, it is evident that continuing recruitment of 'traditional' dominants in the face of intermittent landscape disturbance often depends upon saturating dispersal of wind-dispersed propagules or maintenance of banks of persistent seeds and seedlings. This alerts us to the strong possibility, reviewed in more detail elsewhere (Grime, 1998), that the most serious repercussion of declining plant diversity may be losses in seed rain and in seed and seedling banks, leading to a progressive decline in the precision with which appropriate dominant plants are recruited into ecosystems that are reassembling after disturbance events. There is now an urgent need for observational and experimental studies that measure the extent to which losses in species richness at community and landscape scale are an indicator of future losses of ecosystem properties arising from failure of 'traditional' dominants to regenerate in landscapes subject to disruptive exploitation and management.

Conclusions

The regenerative process is a critical feature of plant community dynamics, which must be included in descriptive models at the landscape scale. With this aim, analysis of the regenerative process can no longer depend upon bespoke studies of seed germination and seedling survival. Faster progress is required. A well-developed theoretical framework of regenerative functional types is available and the time seems right to attempt to use this framework more actively in the prediction and analysis of regeneration in plant communities. Such a task has become possible due to the increasing availability of data banks on the regenerative strategies of individual plant species. The three analyses presented here show some of the potential of this more adventurous approach. It must be acknowl-

edged, however, that information on the regenerative strategies of less common species is often lacking and there is still much work to be done in collating and adding to the body of information on the regenerative strategy B_{sd}.

Ultimately, descriptive models need a mechanistic input. Our understanding of the operation of regenerative processes in plant communities is well developed. Attempts to predict mechanism in a rigorous manner are the logical next step. While we are still some way from achieving this goal, the analyses presented here convince us that it can be attempted. We hope others will take up the challenge.

Recently, the subject-matter of this chapter has been promoted to the forefront of the debate concerning the consequences for ecosystem function of the worldwide losses observed both in floras and in the species richness of individual plant communities. We conclude that urgent action is required to test the hypothesis that the most serious dangers arising from these processes are impoverishments of seed rain, seed banks and seedling banks, leading to a decline in the precision and effectiveness of seedling recruitment and failure and replacement of 'traditional' vegetation dominants, with coincident losses of ecosystem properties.

Acknowledgements

We are grateful to Dr John Hodgson for generously allowing us access to part of his extensive, unpublished data bank, without which the analysis of habitat types in Lathkill Dale (Table 15.3) could not have been attempted. We are also grateful to Dr Roderick Hunt, Mr Andrew Neal, Dr John Hodgson and many other colleagues at the Unit of Comparative Plant Ecology (UCPE), who were involved in the resurvey of part of Lathkill Dale in 1980. The unpublished data from this survey form the basis of our analysis in Fig. 15.4.

References

Bakker, J.P. (1989) *Nature Management by Grazing and Cutting.* Kluwer Academic Publishers, Dordrecht.

Brenchley, W.E. (1918) Buried weed seeds. *Journal of Agricultural Science,* 9, 1–31.

Chippindale, H.G. and Milton, W.E.J. (1934) On the viable seeds present in the soil beneath pastures. *Journal of Ecology* 22, 508–531.

Connell, J.H. and Slatyer, R.O. (1977) Mechanisms of succession in natural communities and their role in community stability and organization. *American Naturalist* 111, 1119–1145.

Darby, C.D. (1987) The dynamics of buried seed banks beneath woodlands, with particular reference to *Hypericum pulchrum.* Unpublished PhD thesis, Plymouth Polytechnic.

Douglas, G. (1965) The weed flora of chemically-renewed lowland swards. *Journal of the British Grassland Society* 20, 91–100.

Egler, F.E. (1954) Vegetation science concepts I. Initial floristic composition, a factor in old-field vegetation development. *Vegetatio* 4, 412–417.

Fenner, M. (1980) The inhibition of germination of *Bidens pilosa* seeds by leaf canopy shade in some natural vegetation types. *New Phytologist* 84, 95–102.

Gill, D.E. (1978) On selection at high population density. *Ecology* 59, 1289–1291.

Gorski, T. (1975) Germination of seeds in the shadow of plants. *Physiologia Plantarum* 34, 342–346.

Grime, J.P. (1974) Vegetation classification by reference to strategies. *Nature* 250, 26–31.

Grime, J.P. (1979) *Plant Strategies and Vegetation Processes.* Wiley, Chichester.

Grime, J.P. (1985) Towards a functional description of vegetation. In: White, J. (ed.) *The Population Structure of Vegetation.* Junk, Dordrecht, pp. 503–514.

Grime, J.P. (1986) The circumstances and characteristics of spoil colonization within a local flora. *Philosophical Transactions of the Royal Society of London, Series B* 314, 637–654.

Grime, J.P. (1987) Dominant and subordinate components of plant communities: implications for succession, stability and diversity. In: Gray, A., Edwards, P.J. and Crawley, M.J. (eds) *Colonization, Succession and Stability.* Blackwell Scientific Publications, Oxford, pp. 413–428.

Grime, J.P. (1989) The stress debate: symptom of impending synthesis? In: Calow, P. (ed.) *Evolution, Ecology and Environmental Stress. Proceedings of the Linnean Society Bicentenary Symposium, June 1988.* Special issue of the *Biological Journal of the Linnean Society* 37, 3–17.

Grime, J.P. (1997) Biodiversity and ecosystem function: the debate deepens. *Science* 277, 1260–1261.

Grime, J.P. (1998) Benefits of plant diversity to ecosystems: immediate, filter and founder effects. *Journal of Ecology* 86, 902–910.

Grime, J.P., Hodgson, J.G. and Hunt, R. (1988) *Comparative Plant Ecology: a Functional Approach to Common British Species.* Unwin Hyman, London.

Grubb, P.J. (1977) The maintenance of species-richness in plant communities: the importance of the regeneration niche. *Biological Reviews* 52, 107–145.

Grubb, P.J. (1985) Plant populations and vegetation in relation to habitat disturbance and competition: problems of generalization. In: White, J. (ed.) *The Population Structure of Vegetation.* Junk, Dordrecht, pp. 595–621.

Hillier, S.H. (1984) A quantitative study of gap recolonization in two contrasted limestone grasslands. Unpublished PhD thesis, University of Sheffield.

Hooper, D. and Vitousek, P.M. (1997) The effects of plant composition and diversity on ecosystem processes. *Science* 277, 1302–1305.

Huston, M.A. (1997) Hidden treatments in ecological experiments: evaluating the ecosystem function of biodiversity. *Oecologia* 110, 449–460.

Inghe, O. and Tamm, C.O. (1985) Survival and flowering of perennial herbs IV. The behaviour of *Hepatica nobilis* and *Sanicula europaea* on permanent plots during 1943–1981. *Oikos* 45, 400–420.

Keddy, P.A. (1989) *Competition.* Chapman and Hall, London.

King, T.J. (1975) Inhibition of seed germination under leaf canopies in *Arenaria serpyllifolia, Veronica arvensis* and *Cerastium holosteoides. New Phytologist* 75, 87–90.

Leps, J., Osbornovakosinov, J. and Rejmanek, M. (1982) Community stability, complexity and species life-history strategies. *Vegetatio* 50, 53–63.

Lloyd, P.S. (1968) The ecological significance of fire in limestone grassland communities of the Derbyshire Dales. *Journal of Ecology* 56, 811–826.

MacGillivray, C.W., Grime, J.P. and the ISP Team (1995) Testing predictions of resistance and resilience of vegetation subjected to extreme events. *Functional Ecology* 9, 640–649.

Naeem, S., Thompson, L.J., Lawler, S.P., Lawton, J.H. and Woodfin, R.M. (1994) Declining biodiversity can alter the performance of ecosystems. *Nature* 368, 734–737

Noble, I.R. and Slatyer, R.O. (1979) The use of vital attributes to predict successional changes in plant communities subject to recurrent disturbances. *Vegetatio* 43, 5–21.

Peart, M.H. (1984) The effects of morphology, orientation and position of grass diaspores on seedling survival. *Journal of Ecology* 72, 437–453.

Ridley, H.N. (1930) *The Dispersal of Plants Throughout the World*. Reeve, Ashford.

Ryser, P. and Gigon, A. (1985) Influence of seed bank and small mammals on the floristic composition of limestone grassland (*Mesobrometum*) in northern Switzerland. *Bericht Geobotanisches Institut Rubel Zurich* 52, 41–52.

Sagar, G.R. and Harper, J.L. (1961) Controlled interference with natural populations of *Plantago lanceolata, P. major* and *P. media. Weed Research* 1, 163–176.

Salisbury, E.J. (1953) A changing flora as shown in the study of weeds of arable land and waste places. In: Lousley, J.E. (ed.) *The Changing Flora of Britain*. Buncle, Arbroath, pp. 130–139.

Staal, L. (1979) An investigation into the pattern of regeneration within gaps and the type of seed bank on a north-facing slope and a south-facing slope at Millers Dale near Litton. Unpublished doctoral thesis, University of Utrecht, The Netherlands.

Stebbins, G.L. (1951) Natural selection and the differentiation of angiosperm families. *Evolution* 5, 299–324.

Stebbins, G.L. (1974) *Flowering Plants: Evolution Above the Species Level*. Edward Arnold, London.

Thompson, K. and Grime, J.P. (1979) Seasonal variation in the seed banks of herbaceous species in ten contrasting habitats. *Journal of Ecology* 67, 893–921.

Tilman, D. and Downing, J.A. (1994) Biodiversity and stability in grasslands. *Nature* 367, 363–365.

van Breemen, A.M.M. and van Leeuwen, B.H. (1983) The seed bank of three short-lived monocarpic species, *Cirsium vulgare* (Compositae), *Echium vulgare* and *Cynoglossum officinale* (Boraginaceae). *Acta Botanica Neerlandica* 32, 245–246.

van der Pijl, L. (1972) *Principles of Dispersal in Higher Plants*. Springer-Verlag, Berlin.

Wardle, D.A., Bonner, K.I. and Nicholson, K.S. (1997) Biodiversity and plant litter: experimental evidence which does not support the view that enhanced species richness improves ecosystem function. *Oikos* 79, 247–258.

Wilbur, H.M., Tinkle, D.W. and Collins, J.P. (1974) Environmental certainty, trophic level and resource availability in life-history evolution. *American Naturalist* 108, 805–817.

Chapter 16
Gaps and Seedling Colonization

James M. Bullock

NERC Centre for Ecology and Hydrology Dorset, Winfrith Technology Centre,
Dorchester, Dorset, UK

Introduction

Gaps and disturbances

The gap concept is central to plant ecology, particularly in theories seeking to explain species coexistence and community structure (Tilman, 1994; Lavorel and Chesson, 1995; Pacala and Levin, 1997). Fundamental to these theories is the idea that species in a community respond differently to gaps, particularly in terms of their ability to colonize gaps and, from another perspective, their requirement for gaps in the regeneration process. This chapter will review the evidence for such species differences. However, to start with, it is necessary to define gaps. The gap concept is popular partly because it is a vague term, so that ecologists can define as a gap whatever is convenient to their study area or model. They echo Humpty-Dumpty, who said, 'When I use a word, it means just what I choose it to mean.' Forest ecologists have arrived at a number of working definitions of gaps (van der Meer *et al.*, 1994), but these are specific to forests and are more concerned with practical problems of detecting and measuring gaps than with defining a gap in ecological terms. The definition of 'gap' that I shall use is 'a plant-free space' or, more accurately, 'a competitor-free space'. This summarizes the concept as used generally by empirical and theoretical ecologists in that a gap is an area free from competition for a period of time.

'Gap' is often used interchangeably with 'disturbance'. Disturbance is an extremely vague concept, but is used generally to describe the removal or loss of living plant biomass (e.g. Grime *et al.*, 1988; Pickett and White, 1985, p. 7, use too broad a definition, which seems to include any type of perturbation). This mirrors my definition of gaps because it describes the process that leads to the consequence of a competitor-free space. Gaps, by common usage in the ecological literature, are a spatially and temporally constrained subset of disturbances (Fig. 16.1). They are relatively small in scale, ranging in size from a single plant (or part of the plant) to tens of plants. Larger disturbances, such as those caused by fire or storm (see Keeley and Fotheringham, Chapter 13, this volume), encompass many hundreds or thousands of plants. Gaps are also relatively short-lived. They are usually created by a single event and last from less than one to a few (probably two to three) generations of the dominant/canopy species, by which time the gap has become indistinguishable from the undisturbed vegetation. This time frame may exclude larger disturbances, which take a long time to regenerate, but also

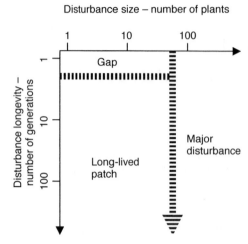

Fig. 16.1. A two-dimensional classification of gaps as a subset of disturbances. Note that the temporal and spatial axes are in terms of plant generations and number, allowing application of the same concepts to all vegetation types. The lines are dashed to indicate that these are not absolute separations, but indicate a zone of transition between the categories.

'patch' disturbances, such as anthills, gopher mounds or badger-sett entrances. These are long-lived and maintained by continuous redisturbance and they develop communities that are compositionally, spatially and dynamically distinct from the undisturbed vegetation (Platt, 1975; King, 1977; Peart, 1989).

These definitions impose boundaries along continua and the positions of the boundaries should be seen as approximate. However, the key point is that gaps are temporary openings (i.e. plant-free spaces) in vegetation, which are part of the everyday dynamics of a community, in contrast to patches and major disturbances, which lead to large temporal and/or spatial disruption of community structure and dynamics.

A functional classification of gaps

In order to understand how gaps affect regeneration in plant communities, it is useful to derive a functional classification of gap types. If a gap is a competitor-free space, gaps can be classified in terms of the type of competition from which the gap forms a (partial) refuge (Fig. 16.2):

1. Canopy gaps are formed by the loss of photosynthetic biomass, which reduces competition for light.
2. Root gaps are the result of the loss of roots from a zone of the substrate, which reduces competition for soil resources.
3. Stem gaps are defined by the absence of a stem or trunk rooted into the substrate.

Canopy gaps are the most commonly studied gap type and are the almost exclusive focus of the enormous literature on gap regeneration in forests. This can be seen in the current definitions of forest gaps, such as 'a hole in the forest extending through all levels down to ... 2 m above the ground' (Brokaw, 1982). Root gaps are poorly studied (Casper and Jackson, 1997), although, if gap creation involves loss of whole plants, the resulting gap will be a root and canopy gap. Stem gaps are a rather obvious, but ignored, type of gap. It is safe to say that all plants require stem gaps for establishment, as it is a physical impossibility for any plant (other than epiphytes) to establish where another plant is already rooted into the substrate. The presence of rooted stems means that a major part of the substrate of most plant communities is physically unavailable as an establishment microsite. Given that a requirement for stem gaps is probably ubiquitous, I shall only consider canopy and root gaps in the remainder of this chapter.

Gap colonization

Gap-colonization ability

If a gap is created in a community, the colonizing species assemblage usually differs from that of the surrounding vegetation in terms of the relative abundances of species and, sometimes, the actual species composition. For example, Bullock *et al.* (1995) created gaps in a temperate grassland by removing all plant parts (stem, root and canopy) in small areas (3–9 cm diameter).

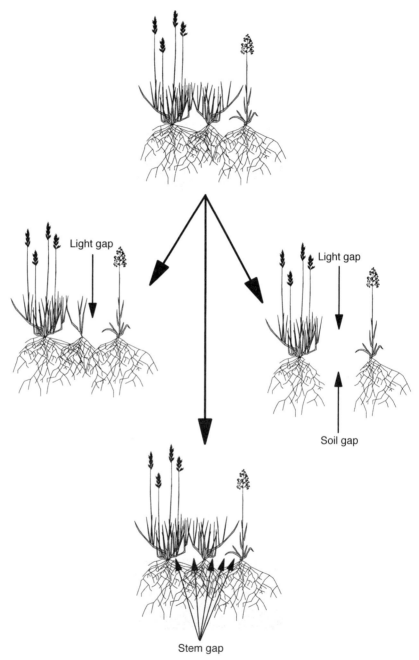

Fig. 16.2. Types of gap.

The seedlings colonizing the gaps over 1 year were of the same species as found in the closed sward, but the relative frequencies were changed, with some being over-represented (by up to a factor of 3: Table 16.1) and others being strongly under-represented (by up to a factor of 70) in the gaps. This is a common finding in all vegetation types, such as temperate grassland (Hillier, 1990; Kotanen, 1997), savannah (Brown and Archer, 1988), desert (Boeken *et al.*, 1995), subalpine grassland

Table 16.1. The gap-colonization abilities of ten grass species and two other taxa colonizing gaps in a temperate grassland (from Bullock *et al.*, 1995). This is calculated as the ratio of the frequency of seedlings in the gap to the frequency of ramets in the surrounding vegetation. A value > 1 means the species is more common inside the gap, whereas a value < 1 indicates a poor colonizer. No species have the same frequency in the gap as outside it. These colonization abilities are different from those in Bullock *et al.* (1995) because here only seedling colonists are considered.

Species	Gap colonization ability
Cynosurus cristatus	2.919
Bromus hordeaceus	2.877
Hordeum secalinum	1.591
Lolium perenne	0.708
Poa spp.	0.495
Herb species	0.388
Alopecurus pratensis	0.331
Other grass species	0.317
Phleum pratense	0.115
Dactylis glomerata	0.043
Agrostis stolonifera	0.026
Festuca rubra	0.014

(Williams, 1992), alpine tundra (Chambers, 1995), temperate forest (Canham, 1989; Peterson *et al.*, 1990) and tropical forest (Brokaw, 1985; Hubbell and Foster, 1986; Whitmore, 1989). There are a few cases where no difference has been found (e.g. Rapp and Rabinowitz, 1985; Hubbell *et al.*, 1999).

The difference between the relative frequency of a species in a recently colonized gap and that in the closed vegetation, which I shall call the species' gap-colonization ability, is the outcome of two processes: (i) the ability of species to disperse seeds spatially or temporally (i.e. the seed bank) into the gap; and (ii) the ability of these seeds to establish new plants in the gap compared with that in the intact vegetation. This concept is clearly linked to Grubb's (1977) 'regeneration niche', and the gap-colonization ability is a description of whether and the extent to which gaps are a better regeneration niche for a species than the closed vegetation. Grubb (1977) defined the regeneration niche as 'the requirements for a high chance of success in the replacement of one mature individual by a new mature individual of the next

generation'. The regeneration niche is determined by a species': (i) amount of viable seed production; (ii) dispersal patterns in space and time; (iii) germination requirements and seasonality; (iv) seedling establishment requirements; and (v) requirements for further development of the plant. Factors (i) and (ii) relate to the first process determining gap-colonization ability described above, while factors (iii)–(v) relate to the second process.

Although this chapter is considering only seedling establishment, it is important to remember that gap colonization occurs by other methods as well. In forests, lateral branch extension by edge trees may be important (Brokaw, 1985; Rebertus and Burns, 1997). Clonal ingrowth is the major mode of gap colonization in grasslands, accounting for between 59 and 99% of colonizing ramets (Rusch and van der Maarel, 1992; Arnthórsdóttir, 1994; Bullock *et al.*, 1995). Species show clear differences in their abilities to colonize gaps by these other modes (Bullock *et al.*, 1995).

Species functional groups

Observed differences in species' gap-colonization ability (by seed) have caused forest ecologists to propose a simple dichotomous classification of forest species. Thus, 'pioneer' ('shade-intolerant', 'light-demanding') species are found as regenerating saplings only in canopy gaps (root gaps are not considered in this classification), whereas 'climax' ('shade-tolerant', 'primary', 'non-pioneer') species are found in both the understorey and gaps (Denslow, 1980; Brokaw, 1985; Swaine and Whitmore, 1988; Reich *et al.*, 1998). Pioneers are said to be restricted by the fact that they can germinate and establish seedlings only in gaps in which the forest floor receives full sunlight, whereas climax species can germinate and establish in canopy shade. However, Whitmore (1989) also stated that climax species show a wide range in the amount of light required for seedling establishment, from species establishing only in light shade to those tolerating deep shade. This suggests that there is a continuum of types, ranging from species needing full sunlight for establishment to those able to tolerate complete shade, and the pioneer/climax dichotomy is artificial and of little use. It seems that only very few forest tree species are true pioneers (Canham, 1989; Schupp *et al.*, 1989; Brown and Whitmore, 1992). Indeed, Brown and Whitmore (1992) stated that pioneers respond only to major disturbances, not to the small canopy gaps typical of forest gap dynamics.

Having said this, it seems to be possible to arrive at working classifications of trees into two groups, based on whether they are mostly found regenerating in gaps ('gap' species) or are equally likely to be regenerating in the understorey as in gaps ('non-gap' species). Hubbell and Foster (1986) looked at the distribution of saplings of 81 tree species in the Barro Colorado Island forest, Panama, and found that 18 were positively associated with gaps, six were negatively associated and 57 showed no apparent specialization. There has been a little work to find differences between 'gap' and 'non-gap' species. Analyses of tropical forest communities have shown that gap species tend to have smaller seeds than non-gap species (Canham and Marks, 1985; Metcalfe and Grubb, 1995). However, the existence of very small-seeded non-gap species (Metcalfe and Grubb, 1995; Grubb and Metcalfe, 1996; Metcalfe *et al.*, 1998) indicates that this is not a perfect relationship, and Grubb and Metcalfe (1996) suggest that alternative life-history solutions to shade tolerance mean that seed size is not a good prescriptive character. Studies in different forest types have shown that gap species have higher growth rates in full light (Denslow, 1980; Kobe *et al.*, 1995; Reich *et al.*, 1998) and a less branched growth form (Coomes and Grubb, 1998) than non-gap species.

Some authors have suggested a more subtle separation of forest tree species in terms of the gap sizes with which they are associated (Denslow, 1980; Augspurger, 1984). This is based on the fact that the composition and species' relative abundances of the colonizing species assemblage have been shown to be affected by gap size (Brokaw, 1987; Brokaw and Scheiner, 1989), although this finding has been contradicted by Dalling *et al.* (1998). However, such gap-size effects have been found in other vegetation types (Williams, 1992; Bullock *et al.*, 1995; Kotanen, 1997).

Although some species in other vegetation types have been called gap dependent, no simple classification system has been adopted outside forest ecology. Functional classifications of grassland plants have tended to be more complicated and not to centre on gap versus non-gap divisions (Grubb, 1976; Gay *et al.*, 1982; Lavorel *et al.*, 1994, 1999; Thompson *et al.*, 1996). In fact, categorization of species as 'gap' or 'non-gap' may be too simplistic. Given that species within a community differ in gap-colonization ability and, to some degree, in their ability to colonize differently sized gaps, it may be more useful to investigate the colonization processes (i.e. dispersal and establishment, as described above) that lead to variation in these abilities before seeking life-history traits linked to these abilities. Enhanced gap-colonization ability through a greater potential to disperse into gaps represents a different regeneration niche to differential seedling establishment in gaps versus closed vegetation. The subsequent sections will investigate how each of these processes contribute to species' gap-colonization abilities.

The gap environment

I have defined gaps as a type of release from resource competition of some form. How do gaps, as located by field ecologists, differ environmentally from the closed vegetation? Obviously, canopy gaps are always found to have higher light intensities (e.g. measured as photosynthetically active radiation) at the soil surface than closed vegetation, in grasslands (Bradshaw and Goldberg, 1989; Aguilera and Lauenroth, 1995; Morgan, 1997) and forests (Vitousek and Denslow, 1986; Coomes and Grubb,

1998; Denslow *et al.*, 1998). The higher light intensities may also deplete the water resource by decreasing surface soil moisture (Denslow, 1980; Collins *et al.*, 1985; Vitousek and Denslow, 1986; Bradshaw and Goldberg, 1989; Aguilera and Lauenroth, 1995; but see Ehrenfeld *et al.*, 1995) and may increase daytime soil temperature, while the lack of a protective canopy may lower night-time temperatures (Thompson *et al.*, 1977; Denslow, 1980; Vitousek and Denslow, 1986). Gap creation by plant death often results in an increase in dead biomass within the gap. Decomposition and mineralization of this biomass sometimes increase soil nutrient concentrations (e.g. phosphorus (Denslow, 1980), nitrogen and phosphorus (Bradshaw and Goldberg, 1989)), but often there are no clear differ-ences between gap and non-gap soils (Denslow, 1980; Collins *et al.*, 1985; Ehrenfeld *et al.*, 1995).

The above distinguishes gap from non-gap. However, at least for canopy gaps, it may be more realistic to consider a continuum from no (or very small) gap to very large gap. Competition is not all or nothing; rather its effects on a plant increase in intensity as the density of competitors increases (e.g. Bullock *et al.*, 1994b). This gap continuum is apparent in the great range of light intensities measured at or near ground level in many vegetation types (Silvertown and Smith, 1989; Ryel *et al.*, 1994; Tang and Washitani, 1995; Valverde and Silvertown, 1997). Increased gap size results in higher light intensities at the gap centre, as shown in Fig. 16.3, which also

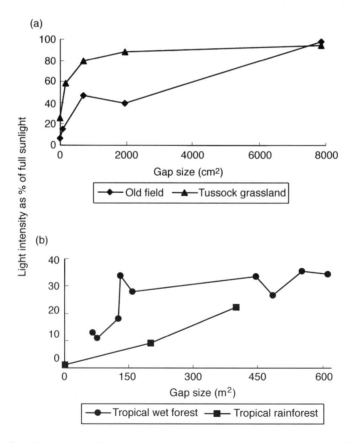

Fig. 16.3. The effect of gap size on light intensity at the gap centre in: (a) an old field (Goldberg and Werner, 1983) and a tussock grassland (Morgan, 1997); and (b) a tropical wet forest (Denslow *et al.*, 1998) and a tropical rainforest (Chazdon and Fetcher, 1984). A gap size of zero is under the canopy (non-gap).

illustrates the continuum from non-gap to large gap, in that the smallest gaps have a very similar environment to that under the canopy.

Environmental changes in root gaps are very poorly studied. In many cases, the creation of a canopy gap will also result in a root gap (e.g. in tree-fall gaps). Comparing tree-fall gaps with the intact forest, Sanford (1989) and Wilczynski and Pickett (1993) found reductions by 48 and 42%, respectively, in the biomass density of fine roots (< 10 mm diameter and < 3 mm diameter, respectively). One would expect root gaps to have increased soil nutrients, but there is no information on this; the changes in soil nutrients in 'total' gaps described above are probably caused by a greater input of nutrients from decaying biomass, rather than a decrease in removal rates by roots. Aguilera and Lauenroth (1995) found that total gaps in an arid short-grass steppe had higher soil moisture than intact vegetation, probably as a result of root removal.

The environmental changes described above are those generally associated with gaps. Gaps are formed by a wide variety of process, such as senescence of individual plants, windthrow, drought-related death, frost-heaving, erosion, herbivory, digging by animals, and animal dung, depending on the vegetation type and site conditions (Brown and Archer, 1988; Goldberg and Gross, 1988; Busing, 1995; Lertzman *et al.*, 1996; Rebertus and Burns, 1997). Both the vegetation type and the way a gap is formed will also affect the gap environment in more specific ways. For example, cattle dung creates gaps with lower moisture content, higher phosphorus and nitrogen concentrations and high amounts of toxins, compared with the intact vegetation (Malo and Suarez, 1995). Removal of above-ground biomass in a saltmarsh increased soil salinity and decreased soil redox potential (Bertness and Hacker, 1994).

Conditions may also vary within a gap. Forest ecologists often divide tree-fall gaps into zones: root, the area where the tree roots are exposed; bole, the area occupied by the fallen trunk; crown, the area occu-

pied by the fallen crown; and undisturbed, the area exposed by the tree fall, but not occupied by any parts of the tree. These differ in terms of soil disturbance (highest in root zones), amount of decomposing biomass (highest in bole zones, but more rapidly available in crown zones) and presence of physical obstructions to seedling growth (highest in bole and crown zones) (Vitousek and Denslow, 1986; Uhl *et al.*, 1988). Peterson *et al.* (1990) divided the root zone further into the pit, wall and mound caused by the eruption of roots from below ground. These zones differed in soil temperature, light intensity (mound > wall > pit) and soil moisture (pit > wall).

Gap effects on seedling establishment

I have shown above that gaps present a reduction in competition for one or more resources (light, nutrients, water) and changes in other environmental factors (e.g. soil temperature), compared with the intact vegetation. It is therefore likely that the ability of a species to establish plants from seed will be higher in a gap than in intact vegetation. However, there are a number of complicating caveats to this simple proposal. The amount of reduction in competition must be sufficient to affect seedling establishment (e.g. gap size may be important); the resource released must be that which limits seedling establishment (e.g. release from root competition is irrelevant if light is limiting); and changes in other factors (e.g. increasing soil temperature) or resources (e.g. decreasing soil moisture) may counteract (or enhance) a reduction in competition. It is these caveats which may cause variation between species in how they respond to gaps in terms of establishment.

It is also important to consider the different stages of seedling establishment that may be affected by gap conditions. 'Germination', 'emergence' and 'establishment' are distinct terms, but they are often used interchangeably by ecologists. Germination is the protrusion of some part of the embryo from the seed-coat. This is easily seen in the laboratory, but it is difficult to detect in the field. Emergence of leaves or cotyledons above the substrate surface – a product of germination and early survival of the seedling – is usually the first sign of a plant that can be seen by the ecologist. Many ecologists (myself included: see Bullock *et al.*, 1994a) are guilty of equating this with germination, but little is known as to how many plants die between germination and emergence. The processes of germination and emergence and the subsequent survival of the seedling are all combined to describe establishment, which is often the only measure of the seed-to-plant transition in field studies. The end of the establishment phase, in the working definitions used by field ecologists, can be anything from the appearance of the first adult leaf to senescence of old plants. The most common end-point is the end of the first growing season after emergence. I shall use this definition in this chapter, although sometimes a longer period may be more appropriate, e.g. for forest trees.

Germination responses to gaps

Laboratory studies have produced much information on species differences in germination cues. Higher light intensity, red : far-red (R : FR) ratio or temperature or greater temperature fluctuations may all stimulate germination, but to a greater or lesser degree (or have no effect), depending on the species (Thompson and Grime, 1983; Pons, 1991; see also Pons, Chapter 10, and Probert, Chapter 11, this volume). Light may, in some cases, inhibit germination (see Pons, Chapter 10, this volume). Various aspects of the chemical environment may also have species-specific effects on germination (see Hilhorst and Karssen, Chapter 12, this volume). Some responses are linked to gap conditions. For example, positive responses to light, higher R : FR ratio and increased temperature fluctuation are often interpreted as cues by which seeds detect and respond to canopy gaps (Thompson and Grime, 1983; Fenner, 1985; see also Pons, Chapter 10, and Probert, Chapter 11, this volume).

Most studies of germination responses to canopy gaps involve laboratory or glasshouse comparisons of full light with some form of artificial shading, using plant leaves, foil or cloth, which either cut out all light or greatly reduce the photosynthetically active radiation. Table 16.2 lists results from six studies of species differences in whether or not full light is required for germination, but, more interestingly, it shows that relatively few species are completely inhibited by shading. Only ten of 71 species studied showed an absolute requirement for full light. Thompson and Grime (1983) studied germination responses to fluctuating temperatures, also in the laboratory, and found that 46 of 112 species tested were stimulated by fluctuating temperature. However, an

Table 16.2. Germination responses of species to full light (gap) versus shade in five studies. Shade was either the complete absence of light ('dark') or light passed through leaves ('leaf') or green cloth ('cloth') to remove most of the photosynthetically active radiation. An absolute light requirement means no seeds germinated in shade. Enhanced by light means germination was increased in full light, but there was some germination in shade. Unaffected by light indicates no difference in germination between the environments. The columns show the number of species showing each response.

Study	Type of shade	Absolute light requirement	Enhanced by light	Unaffected by light
King (1975)	Leaf	0	3	0
Fenner (1980)	Leaf	6	10	2
Silvertown (1980)	Leaf	0	17	10
Augspurger (1984)	Cloth	1	0	17
Pons (1991)	Dark	3	8	2
Milberg (1994)	Dark	0	1	1

absolute requirement for temperature fluctuations was found in only a small proportion of these 46 species (the specific
number is not stated).

Unsurprisingly, given the problems of
detecting germination in the field, there is
much less information on germination
responses to real gaps. Fenner (1978) measured germination of 12 species in three
types of artificial vegetation: bare soil (total
gap), short grass (partial canopy gap) and
tall grass. There were no germination differences between the bare and short-grass
treatments, but all species showed lower
germination in the tall grass compared
with the other treatments. Again, only one
species showed an absolute canopy-gap
requirement; the 11 other species showed
some germination in the tall-grass vegetation.

Fenner's (1978) study also provides
some evidence that root gaps do not affect
germination. The canopy gaps (short grass)
had the same effect as the combined root
and canopy gaps (bare soil). The fact that
nitrate commonly stimulates seed germination (Thanos and Rundel, 1995; Bell *et al.*,
1999) suggests that, if root gaps result in
increased soil nitrate, they could increase
seed germination (this may also be a
consequence of canopy gaps, where
decomposing biomass increases nitrate
concentrations). However, this possibility
has not been explored.

Emergence response to canopy gaps

It is very likely that many canopy gaps are
also root gaps (although the two may not
necessarily cover the same area), as most
gaps studied by ecologists are created by
total plant removal. However, most studies
tend to consider these gaps as canopy gaps
rather than total gaps, that is, species
responses are interpreted in terms of environmental changes due to canopy removal,
rather than those due to root removal. In
this and subsequent sections, I shall retain
this convention because it seems likely that
the major response is to the canopy gap.
However, some examples and the section

on root gaps illustrate the problems created
by this assumption.

Canopy gaps have been shown to
increase seedling emergence compared
with intact vegetation in semi-arid grassland (Aguilera and Lauenroth, 1993), temperate grassland (Johnson and Thomas,
1978; Keizer *et al.*, 1985; Bullock *et al.*,
1994a; Morgan, 1997), tropical forest
(Horvitz and Schemske, 1994) and temperate forest (Valverde and Silvertown, 1995).
Emergence also tends to be greater in larger
than in smaller gaps (Pons and van der
Toorn, 1988; Aguilera and Lauenroth,
1993; Valverde and Silvertown, 1995;
Morgan, 1997).

Although there is a general trend for
increased emergence in canopy gaps, there
are species differences in the degree to
which gaps enhance emergence and in the
effect of gap size on emergence (Silvertown
and Wilkin, 1983; Ryser, 1993). In a comparison of six old-field biennial herbs,
Gross (1984) showed that emergence of the
smaller-seeded species was increased by
canopy gaps, but the larger-seeded species
showed no gap response. *Plantago lanceolata* has larger seeds than *P. major* and the
latter showed a greater emergence response
to gaps (Pons and van der Toorn, 1988). In
a study of two *Solidago* species in an old
field, Goldberg and Werner (1983) found
that the emergence of *S. canadensis* was
greater in large gaps, whereas the larger-
seeded *S. juncea* emerged better in small
gaps and intact vegetation. Interestingly, in
none of the studies listed in this or the previous paragraph was there no seedling
emergence in the intact vegetation; thus, no
species had an absolute gap requirement
for emergence.

A few studies have shown seedling
emergence to be unaffected by canopy gaps
(compared with intact vegetation) or gap
size: the 'gap' tree species *Cecropia obtusifolia* in tropical forest (Alvarez Buylla and
Garay, 1994); the herb *Verbascum thapsus*
in old fields (Gross, 1980); and several
herbs in a chalk grassland (Keizer *et al.*,
1985). Gill and Marks (1991) even found a
negative effect of canopy gaps on emergence of shrubs in an old field. The effect

of canopy gaps on emergence can also depend upon the types of canopy gap and vegetation. Winn (1985) showed that the emergence of a herb, *Prunella vulgaris*, was increased by removal of the herb canopy in deciduous wood, conifer wood and an old field, while removal of the moss layer increased emergence in the woodland vegetation but decreased emergence in the old field. However, Gross (1980) found that there were no differences in the emergence of *V. thapsus* between treatments involving the removal of either biennials/annuals, perennial herbs or perennial grasses, or any combinations of these.

If light is not the main factor limiting emergence, canopy gaps may not increase emergence. O'Connor (1996) found that seedling emergence of the tussock grass *Themeda triandra* in semi-arid savannah was mostly limited by water availability. Seedling emergence was decreased by canopy gaps, because the exposure had reduced the soil water content. When water was added, emergence was increased in canopy gaps compared with the intact vegetation.

The difference between germination and emergence is rarely examined in relation to gaps. Horvitz and Schemske (1994) found that only 0.4% of seeds of the herb *Calathea ovandensis* emerged in shaded forest. Eighty-eight per cent of the non-emerged seeds were dormant and the remainder were dead. This suggests that the main cause of low emergence in shade was a lack of germination cues, rather than post-germination mortality.

Emergence response to root gaps

Aguilera and Lauenroth (1993) studied seedling emergence of the tussock grass *Bouteloua gracilis* in 10 and 20 cm diameter gaps, half of which were canopy gaps only and the remainder canopy and root gaps (using steel tubes to exclude roots). Emergence was increased by root exclusion. Root gaps have a greater effect on emergence than canopy gaps. This was because water was probably the main limiting resource: Aguilera and Lauenroth (1995) found that gaps in this semi-arid steppe community had increased soil moisture levels. Moretto and Distel (1998) also found that root exclosures increased seedling emergence of two *Stipa* species, again in a water-limited grassland.

Canopy gaps and seedling survival

Survival of emerged seedlings is also generally increased in canopy gaps, compared with intact vegetation (Gill and Marks, 1991; Aguilera and Lauenroth, 1993; Bullock et al., 1994a), and is often higher in larger gaps (Goldberg and Werner, 1983; Alvarez Buylla and Garay, 1994; Morgan, 1997). Gap quality may also be important: Gross (1980) found that survival over 1 year of *V. thapsus* was higher, compared with intact vegetation, in gaps where one component of the vegetation (biennials, perennial herbs or perennial grasses) was removed, but was highest when all vegetation was removed. Poor survival in intact vegetation compared with canopy gaps is presumably a direct consequence of low light levels. However, shading has also been shown to increase seedling mortality indirectly, by exacerbating fungal attack through increased humidity (Augspurger, 1984; Fowler, 1988) or by providing cover for rodents or molluscs that predate seedlings (Gill and Marks, 1991; Hanley et al., 1996).

However, there are a number of cases where survival was not different in gaps compared with intact vegetation (Gross, 1984; Ryser, 1993; Garcia Fayos and Verdu, 1998) or in larger compared with smaller gaps (Gray and Spies, 1996). In fact, there are some examples where seedling survival was lower in gaps than in the intact vegetation. This seems to have been caused by water stress in larger gaps and most commonly occurs in arid environments (Johnson and Thomas, 1978; Keizer et al., 1985; Ryser, 1993). Sipe and Bazzaz (1995) found that three maple species showed lower survival in large gaps than in small gaps or intact forest, and survival was lowest in the centres of the large gaps. As with emergence of *T. triandra*, O'Connor (1996) found that seedling survival was decreased by canopy gaps unless water was added, in which case gaps increased survival. Hillier (1990) found that seedling survival of most species was higher in smaller gaps and intact vegetation in a British south-facing dry chalk grassland, but that survival was greater in larger gaps in a nearby north-facing, wetter grassland.

As with emergence, most studies show some seedling survival in intact vegetation, even if this is much lower than in gaps. However, in contrast to the emergence data, there are a few cases where species show an absolute gap requirement for survival (Goldberg and Werner, 1983; van der Toorn and Pons, 1988). A composite, *Rutidosis leptorrhynchoides*, survived only in the largest gaps (100 cm diameter) in a grassland (Morgan, 1997). *C. obtusifolia* seedlings can only survive the first year in gaps over 100 m^2 (Alvarez Buylla and Garay, 1994).

Comparisons of species within communities illustrates this range of survival responses to gaps. Several studies have found between-species variation in the differential between survival in shade compared with that in light (Grime and Jeffrey, 1965; Ryser, 1993). Kobe *et al.* (1995) showed that sapling mortality of ten temperate forest tree species in deep shade ranged between near-zero and > 50%, whereas all showed nearly 100% survival in full sun. In a laboratory experiment, Augspurger (1984) found that 15 forest tree species showed enhanced survival in full light, while three species showed no difference between light and shade treatments. The 15 species showed wide variation in the degree to which survival was enhanced by sunlight. Goldberg (1987) found that some grassland species survived better in gaps than in intact vegetation, while others showed no response to gaps. There was no relationship between seed size and the response to gaps. However, other studies have shown no species differences in their responses to gaps (Gross, 1984; Sipe and Bazzaz, 1995).

Species may also show differential increases in seedling survival with increasing gap size (Goldberg and Werner, 1983; McConnaughay and Bazzaz, 1990). Small-seeded *Plantago major* survived only in the largest gaps in a grassland, while the larger-seeded *P. lanceolata* survived in all gap sizes and intact vegetation, although survival was higher in larger gaps (van der Toorn and Pons, 1988). However, in two separate studies, neither Brown and Whitmore (1992) nor Gray and Spies (1996) found any difference in the effect of gaps on seedling survival of different tree species: survival was higher in gaps than in the understorey, but there was no extra effect of gap size.

It has been suggested, especially for forest trees, that plant species may partition gap zones by showing different establishment in different parts of a gap. Núñez Farfán and Dirzo (1988) compared survival of two trees in the crown and root zones of tree-fall gaps in a tropical forest. One species showed greater survival in the crown zone, while the other responded better to the root zone. Gray and Spies (1996) found a small degree of partitioning between three forest tree species: *Tsuga heterophylla* and *Abies amabilis* showed lower survival in the more exposed north and central parts of the gaps, while *Pseudotsuga menziesii* showed equal survival in all zones.

Root gaps and seedling survival

Aguilera and Lauenroth's (1993, 1995) studies (see above) showed similar effects of root gaps on seedling survival of *Bouteloua gracilis*. Root exclosures increased survival (Aguilera and Lauenroth, 1993), and root gaps in the 1995 study had a greater effect on survival than canopy gaps. In another water-limited system, Moretto and Distel (1998) found that all seedlings of two *Stipa* species died in intact vegetation and canopy gaps, but there was some survival in combined root and canopy gaps. Thus, root gaps were necessary for seedling survival.

McConnaughay and Bazzaz (1991) suggested that the presence of roots in the soil
might be important not only in depleting
water and nutrients, but also in presenting
physical barriers to root growth. A root gap
may therefore provide a physical space for
establishment, as well as an area of
reduced resource competition. However,
they found no effect on survival of the
amount of physical soil space available to
plants, although there were more subtle
effects on morphology and phenology.

Establishment in gaps

Seedling establishment is a result of successful germination/emergence and survival. Differential germination (and,
partially, emergence) between gaps and
closed vegetation is a response to variation
in environmental cues, whereas differential
seedling survival is an effect of changes in
limiting factors – the former is in the control of the plant, but the latter is not. So, if
the establishment process is broken down
into its components, we can better understand the mechanism of a species' response
to gaps. In many cases, seedling emergence
and survival both show positive responses
to gaps (Aguilera and Lauenroth, 1993;
Bullock et al., 1994a). However, there are
other more subtle paths. R. leptorrhynchoides in a tussock grassland (Morgan,
1997), Cecropia obtusifolia in a tropical forest (Alvarez Buylla and Garay, 1994),
Plantago major in a grassland (van der
Toorn and Pons, 1988) and Solidago
canadensis in an old field (Goldberg and
Werner, 1983) emerged in all gap sizes and
in intact vegetation, but survived only in
the largest gaps, thus restricting establishment in these species to large gaps.
Conversely, several species studied by
Gross (1984) and by Silvertown and Wilkin
(1983) showed enhanced emergence in
gaps, but no gap effect on survival – leading again to gap effects on establishment,
but through a different mechanism.
Johnson and Thomas (1978) found
increased emergence of Hieracium species
in gaps, but decreased survival, which

resulted in an overall negative effect of
gaps on establishment. Ryser (1993) found
the same pattern for six chalk grassland
herbs. Gill and Marks (1991) found the
opposite pattern (gaps decrease emergence,
but increase survival) among shrubs in an
old field, which resulted in better establishment in gaps. The same argument applies
to root gaps. Moretto and Distel (1998)
found some emergence of two Stipa species
in intact vegetation, canopy gaps and combined canopy and root gaps, but seedlings
survived only in the presence of a root gap.

In the preceding sections, I have given
many examples of species differences in the
germination/emergence and/or survival
responses to gaps or increased gap size.
However, within a community, these
responses fall within a limited range. There
are hardly any cases where seedling emergence is not higher in canopy or root gaps
than in intact vegetation, or in larger than
smaller gaps. Conversely, there are no cases
where a species showed no emergence in
intact vegetation, and few where a species
had an absolute light requirement for germination. Survival responses to gaps or larger
gaps present a greater range, but the evidence reviewed above suggests that, within
a community, these responses probably
always range either from neutral to positive
or (rarely) from neutral to negative (in
water-limited communities). Furthermore, a
neutral response seems to be rare.

This provides support for the suggestion that it is misleading to divide species
in a community into those requiring gaps
for regeneration and those indifferent to or
inhibited by gaps (see also Zagt and
Werger, 1998; Hubbell et al., 1999): most
species in most communities will establish
better in gaps (and, where gaps have a negative effect, this is common to most species
in the community) and, conversely, most
species in most communities can establish
some seedlings in intact vegetation. The
'gap-size partitioning' hypothesis – that
species show niche specialization for different gap sizes – is also poorly supported
by establishment data. This would require
that establishment peaks at specific gap
sizes for different species, whereas the evi-

dence is for generally qualitatively similar (usually positive) responses to increasing gap in all species within a community, and very few cases of complete exclusion of some species from particular gap sizes. The hypothesis of gap-size niche differentiation has been popular in forest ecology, but some are now rejecting it (e.g. Brown and Jennings, 1998).

However, given that there is continuous quantitative variation in the gap response among species in a community, it is useful to ask whether species differ in any systematic way that might suggest different life-history strategies. It is an attractive proposition that smaller-seeded species should show a greater positive survival response to gaps, as they would not have the reserves available to larger-seeded species to support seedling growth in the highly competitive environment of the intact vegetation. A more biased germination response to gaps in smaller-seeded species would indicate that selection had favoured germination in microsites more conducive to seedling survival. In grasslands, several studies have found that a smaller seed size is correlated with a greater response to gaps in terms of germination (Silvertown, 1980), seedling emergence (Goldberg and Werner, 1983; Gross, 1984; Winn, 1985), survival (Goldberg and Werner, 1983) or total establishment (Burke and Grime, 1996). Other studies have found no relationship between seed size and the emergence (Goldberg, 1987) or survival (Gross, 1984; Goldberg, 1987) response to gaps. It is more rare for studies to present data that allow analysis of whether both emergence and survival are correlated with seed size in the same community. Both correlations were found by Goldberg and Werner (1983), but they used only two species. Studies using more species have shown that neither (Goldberg, 1987) or only one (Gross, 1984) of these correlations occurred. Despite many statements about seed size and responses to gaps in forest communities, only Augspurger (1984) has carried out an analysis of forest trees. She found no relationship between seed size and the gap survival response. Therefore, while there is some evidence for a link between seed size and gap establishment response, it is not as conclusive or abundant as some authors would seem to believe.

Dispersal into gaps

Sources of seeds colonizing gaps

The seeds present in a gap, with the potential to develop into colonizing seedlings, may come from a variety of sources. Seeds colonizing large 'total' gaps (3.75 m^2) in an acid grassland came 45% from the seed bank, 15% in rabbit droppings and 40% from other sources (Pakeman *et al.*, 1998). All species arrived by multiple methods, but differed in the most common method. In another grassland, seedlings establishing in 100 cm^2 'total' gaps came 50% from the seed bank, 30% from short-distance seed rain (from sources up to a maximum of 50 cm away) and the remainder from long-distance seed rain (from > 50 cm away) (Kalameesl, 1999). Bullock *et al.* (1994c) found that most seedlings colonizing grassland gaps were derived from recent seed rain rather than the seed bank. Malo *et al.* (1995) found that seeds dispersed into gaps in a grassland via both rabbit droppings and wind dispersal, while Malo and Suarez (1995) showed that the primary source of seedlings in gaps formed by cowpats was seed transported in the dung.

In forests, an alternative to colonization of gaps through the seed bank is the formation of 'seedling or sapling banks'. Seedlings establish under the canopy and 'sit and wait' until a canopy gap forms in which they are able to grow. Such advance regeneration is a major source of colonists in forest gaps, often accounting for over 90% of saplings (Denslow, 1980; Brokaw and Scheiner, 1989).

Dispersal ability

As well as differential establishment, the second cause of differences in relative abundance of a species between gaps and the intact vegetation is the species' ability to disperse seeds into gaps. In fact, it is becoming common for authors to state that species differences in the ability to get seeds into gaps are more important in determining the composition of the gap community than species differences in establishment in gaps (Dalling *et al.*, 1998; Hubbell *et al.*, 1999). The first aspect of this is whether a species shows a consistent difference in the density of seeds in gaps compared with the intact vegetation. Soil disturbance during gap creation may expose the buried seed bank, thus increasing the availability of seeds. This has not been well studied, although Putz (1983) suggested that large forest gaps have a greater density of pioneer species than smaller gaps, because greater soil disturbance in the former exposes these species in the seed bank. Targeted dispersal into gaps may occur through a number of processes. It has been suggested that wind-dispersed seeds may settle preferentially in gaps through the effects of the broken canopy on turbulence or heating of the gap on convectional currents (Schupp *et al.*, 1989). Schupp *et al.* (1989) showed that large-seeded forest species dispersed by monkeys, toucans and guans were dropped in the canopy, but small-seeded species were dispersed preferentially in and near gaps, because the smaller birds and bats dispersing these seeds were more active in these areas. Conversely, the restricted seed dispersal discussed below may mean that fewer seeds fall into gaps than under parents in the intact vegetation (Garcia Fayos and Verdu, 1998).

Even if dispersal into gaps does not differ from that under the canopy, species may differ in the ratio of the relative density of seeds in a gap to the relative abundance of plants in the vegetation – what I shall call the 'gap-attainment ability'. The number of viable seeds of a species in a gap will be determined by the fecundity of plants of the species in the intact vegetation, the dispersal pattern of seeds from these plants and the presence of seeds in the seed bank. The null model for analysis of species differences in gap-attainment ability is that all species have equal per capita seed production; these seeds disperse evenly over the community and the species' seed banks are evenly distributed over the community and show equal survival. Deviations from each of the assumptions of this null hypothesis occur through two process: (i) spatio-temporal heterogeneity; and (ii) species life-history differences.

1. Species' seed rain is usually spatially heterogeneous (Peart, 1989; Debussche and Isenmann, 1994), because plants are probably never evenly distributed (Tilman and Karieva, 1997), and seed dispersal is usually highly restricted, such that most seeds fall near to the parent (see Willson and Traveset, Chapter 4, this volume). In bird-dispersed plants, seed rain may be uneven, because it is concentrated around roosting trees (Debussche and Isenman, 1994). Because of these processes, Dalling *et al.* (1998) found that the species composition of seedlings in gaps was highly dependent on the proximity of adult plants to a particular gap. Seed banks are also highly heterogeneous (Schenkeveld and Verkaar, 1984; Rusch, 1992; see also Thompson, Chapter 9, this volume), probably causing spatial variation in the identity of gap colonizers. Temporal heterogeneity in seed production may lead to effects of the timing of gap creation on the identity of seeds reaching the gap. The facts that species produce seeds at different times of the year and plants show year-to-year variation in seed production (see Grubb, 1977) mean that the season and year of gap creation may be important.

2. Fecundity differs widely among species within communities (Grubb, 1977), and Peart (1989) found that disturbances in a grassland were colonized mostly by the species with the highest seed production. Fecundity differences will also affect the seed-bank densities, as will the well-described species differences in seed bank longevity (see Thompson, Chapter 9, this volume). Differences in seed dispersal curves, and especially in the median dispersal distance (see Willson and Traveset, Chapter 4, this volume) are likely to affect gap-attainment ability strongly, as distance from a gap is less important for species with longer dispersal curves. For example, while some species in a grassland studied by Rusch and Fernandez Palacios (1995) showed a high association of seedlings with adult plants, others were more evenly distributed, probably because the latter were able to disperse further. Seed-bank and dispersal characteristics may also lead to zonation of seed composition in a gap. Poorer dispersers may be more concentrated at the edge of a gap (e.g. Bullock *et al.*, 1995). The soil disturbance of root zones in tree-fall gaps may cause them to be colonized by species in the seed bank, in contrast to the bole and crown zones, which may be colonized by other means (Putz, 1983).

Despite these possibilities, there has been remarkably little work comparing species gap-attainment abilities and how life-history differences may affect these abilities. It is often suggested that pioneer species in forest have higher gap-attainment abilities than shade-tolerant species by high seed production, long-distance dispersal and the maintenance of a seed bank (Brokaw, 1985; Swaine and Whitmore, 1988). There is actually little evidence for this. However, such characteristics may be a simple consequence of small seed size, which, as discussed above, is associated with a greater response to gaps in terms of seedling establishment. For wind-dispersed seeds, smaller seeds might be expected to have a higher median dispersal distance as a simple consequence of having a lower terminal velocity (Augspurger and Franson, 1987; Greene and Johnson, 1989). However, this is not a simple relationship because, while small seeds are mostly just wind-dispersed, larger-seeded species show a variety of dispersal methods (Leishman *et al.*, 1995), which may result in longer dispersal distances than for wind-dispersed seeds (Willson, 1993). Smaller seed size also tends to be related to greater longevity in the soil (Thompson *et al.*, 1993; Bekker *et al.*, 1998), although Leishman and Westoby (1998) did not find this relationship. It is unclear whether this relationship is a straightforward physical consequence of seed size (smaller seeds are more likely to get buried and survival is increased at greater depths (Bekker *et al.*, 1998)) or of the fact that smaller seeds may require more specific germination cues (see above) or whether this reflects a greater ability for ungerminated seeds to survive in the soil. Finally, within particular growth forms (e.g. herbs or trees), there is a general negative relationship between seed size and the number of seeds produced by a plant (Westoby *et al.*, 1997) – the result of a simple trade-off in resource allocation.

Conclusions

Species differences in gap-colonization ability are a result of different gap-attainment abilities and differential establishment responses to gaps (Fig. 16.4). There is a large amount of work showing that colonized gaps have a changed species community compared with the intact vegetation. However, work on the reasons for these differences has been rather unfocused. In forest ecology, the pioneer/shade-tolerator dichotomy became dogma, despite little evidence for clear species differences in gap-attainment abilities or establishment responses. It is only recently that workers have begun to discard this overly simplistic dichotomy. There is much evidence for species differences in each of the different stages of the establishment response to gaps. However, these differences are often subtle, being quantitative rather than qualitative, and may be complex and inconsistent when different stages of the establishment process are considered together. There is a need for studies to consider the whole establishment process – germination, emergence and survival – and to carry out wide-ranging species comparisons within communities.

Species differences in gap-attainment abilities, through either spatio-temporal heterogeneity or life-history characteristics, may yet prove to be the key to understanding species differences in gap-colonization ability, especially given the evidence that most species within a community will show qualitatively similar establishment

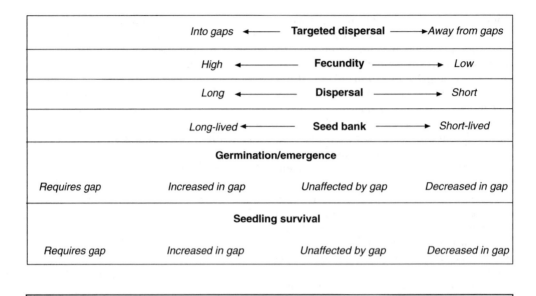

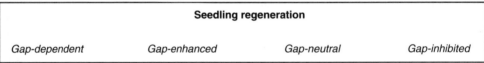

Fig. 16.4. How dispersal and establishment attributes affect the gap colonization ability of a species, and thus whether regeneration will occur only in gaps (gap-dependent), more frequently in gaps (gap-enhanced), equally in gaps and closed vegetation (gap-neutral) or more frequently in the closed vegetation (gap-inhibited).

responses to gaps. There is a need for more work measuring gap-attainment abilities, and to compare these with species' establishment responses to gaps. Such studies must consider different types of gap – whether they are root or canopy gaps, and the gap environment as determined by gap size, gap zone, how a gap was created, etc.

There may yet prove to be a very simple key to categorizing species' responses to gaps. There is evidence that smaller-seeded species will have higher gap-attainment abilities (higher fecundity, greater dispersal distance, longer-lived seed bank), a gap requirement for germination and a greater differential seedling emergence and survival in gaps compared with the intact vegetation than larger-seeded species. However, there is a need for much more

information of the sort described above before this hypothesis can be accepted. Furthermore, it is clear that this categorization will be a continuum rather than a dichotomy and that there will be species that will not conform to it – for example, the very small-seeded species, described by Metcalfe and Grubb (1995), which establish under the forest canopy.

Understanding of these processes and of the causes of species differences in gap-colonization ability is of importance for models that seek to understand community dynamics and species coexistence. Species differences in gap-attainment ability through patchiness in distributions of plants, seed rain and seed bank or temporal heterogeneity in seed production will lead to spatial or temporal variation between gaps within a community in the species colonizing a gap, whereas differences in gap-attainment ability or establishment responses to gaps due to species' life-history characteristics will lead to a repeatable subset of species colonizing gaps. The former would result in non-equilibrium gap-driven community dynamics of the type proposed by Hurtt and Pacala (1995), where history and spatiotemporal patterns are important, whereas the latter would lead to equilibrium dynamics, based on niche partitioning, as modelled by Pacala and Levin (1997).

References

Aguilera, M.O. and Lauenroth, W.K. (1993) Seedling establishment in adult neighbourhoods – intraspecific constraints in the regeneration of the bunchgrass *Bouteloua gracilis*. *Journal of Ecology* 81, 253–261.

Aguilera, M.O. and Lauenroth, W.K. (1995) Influence of gap disturbances and types of microsites on seedling establishment in *Bouteloua gracilis*. *Journal of Ecology* 83, 87–97.

Alvarez Buylla, E.R. and Garay, A.A. (1994) Population genetic-structure of *Cecropia obtusifolia*, a tropical pioneer tree species. *Evolution* 48, 437–453.

Arnthórsdóttir, S. (1994) Colonization of experimental patches in a mown grassland. *Oikos* 70, 73–79.

Augspurger, C.K. (1984) Light requirements of neotropical tree seedlings: a comparative study of growth and survival. *Journal of Ecology* 72, 777–795.

Augspurger, C.K. and Franson, S.E. (1987) Wind dispersal of artificial fruit varying in mass, area and morphology. *Ecology* 68, 27–42.

Bekker, R.M., Bakker, J.P., Grandin, U., Kalamees, R., Milberg, P., Poschlod, P., Thompson, K. and Willems, J.H. (1998) Seed size, shape and vertical distribution in the soil: indicators of seed longevity. *Functional Ecology* 12, 834–842.

Bell, D.T., King, L.A. and Plummer, J.A. (1999) Ecophysiological effects of light quality and nitrate on seed germination in species from Western Australia. *Australian Journal of Ecology* 24, 2–10.

Bertness, M.D. and Hacker, S.D. (1994) Physical stress and positive associations among marsh plants. *American Naturalist* 144, 363–372.

Boeken, B., Shachak, M., Gutterman, Y. and Brand, S. (1995) Patchiness and disturbance: plant community responses to porcupine diggings in the central Negev. *Ecography* 18, 410–422.

Bradshaw, L. and Goldberg, D.E. (1989) Resource levels in undisturbed vegetation and mole mounds in old fields. *American Midland Naturalist* 121, 176–183.

Brokaw, N.V.L. (1982) The definition of treefall gap and its effect on measures of forest dynamics. *Biotropica* 14, 158–160.

Brokaw, N.V.L. (1985) Treefall, regrowth and community structure in tropical forests. In: Pickett, S.T.A. and White, P.S. (eds) *The Ecology of Natural Disturbance and Patch Dynamics*. Academic Press, London, pp. 53–70.

Brokaw, N.V.L. (1987) Gap-phase regeneration of three pioneer tree species in a tropical forest. *Journal of Ecology* 75, 9–20.

Brokaw, N.V.L. and Scheiner, S.L. (1989) Species composition in gaps and structure of a tropical forest. *Ecology* 70, 538–541.

Brown, J.R. and Archer, S. (1988) Woody plant seed dispersal and gap formation in a North-American sub-tropical savanna woodland – the role of domestic herbivores. *Vegetatio* 73, 73–80.

Brown, N.D. and Jennings, S. (1998) Gap-size niche differentiation by tropical rainforest trees: a testable hypothesis or a broken-down bandwagon. In: Newbery, D.M., Prins, H.H.T. and Brown, N. (eds) *Dynamics of Tropical Communities*. Blackwell Science, Oxford, pp. 79–93.

Brown, N.D. and Whitmore, T.C. (1992) Do dipterocarp seedlings really partition tropical rainforest gaps? *Philosophical Transactions of the Royal Society of London Series B* 335, 369–378.

Bullock, J.M., Clear Hill, B. and Silvertown, J. (1994a) Demography of *Cirsium vulgare* in a grazing experiment. *Journal of Ecology* 82, 101–111.

Bullock, J.M., Clear Hill, B. and Silvertown, J. (1994b) Tiller dynamics of two grasses – responses to grazing, density and weather. *Journal of Ecology* 82, 331–340.

Bullock, J.M., Clear Hill, B., Dale, M.P. and Silvertown, J. (1994c) An experimental study of vegetation change due to sheep grazing in a species-poor grassland and the role of seedling recruitment into gaps. *Journal of Applied Ecology* 31, 493–507.

Bullock, J.M., Clear Hill, B., Silvertown, J. and Sutton, M. (1995) Gap colonization as a source of grassland community change: effects of gap size and grazing on the rate and mode of colonization by different species. *Oikos* 72, 273–282.

Burke, M.J.W. and Grime, J.P. (1996) An experimental study of plant community invasibility. *Ecology* 77, 776–790.

Busing, R.T. (1995) Disturbance and the population dynamics of *Liriodendron tulipifera* – simulations with a spatial model of forest succession. *Journal of Ecology* 83, 45–53.

Canham, C.D. (1989) Different responses to gaps among shade-tolerant tree species. *Ecology* 70, 548–550.

Canham, C.D. and Marks, P.L. (1985) The response of woody plants to disturbance; patterns of establishment and growth. In: Pickett, S.T.A. and White, P.S. (eds) *The Ecology of Natural Disturbance and Patch Dynamics*. Academic Press, London, pp. 197–216.

Casper, B.B. and Jackson, R.B. (1997) Plant competition underground. *Annual Review of Ecology and Systematics* 28, 545–570.

Chambers, J.C. (1995) Relationships between seed fates and seedling establishment in an alpine ecosystem. *Ecology* 76, 2124–2133.

Chazdon, R.L. and Fetcher, N. (1984). Photosynthetic light environments in a lowland tropical rain forest in Costa Rica. *Journal of Ecology* 72, 553–564.

Collins, B.S., Dunne, K.P. and Pickett, S.T.A. (1985) Responses of forest herbs to canopy gaps. In: Pickett, S.T.A. and White, P.S. (eds) *The Ecology of Natural Disturbance and Patch Dynamics*. Academic Press, London, pp. 218–234.

Coomes, D.A. and Grubb, P.J. (1998) A comparison of 12 tree species of Amazonian caatinga using growth rates in gaps and understorey, and allometric relationships. *Functional Ecology* 12, 426–435.

Dalling, J.W., Hubbell, S.P. and Silvera, K. (1998) Seed dispersal, seedling establishment and gap partitioning among tropical pioneer trees. *Journal of Ecology* 86, 674–689.

Debussche, M. and Isenmann, P. (1994) Bird-dispersed seed rain and seedling establishment in patchy Mediterranean vegetation. *Oikos* 69, 414–426.

Denslow, J.S. (1980) Gap partitioning among tropical rainforest trees. *Biotropica Supplement* 12, 47–55.

Denslow, J.S., Ellison, A.M. and Sanford, R.E. (1998) Treefall gap size effects on above- and below-ground processes in a tropical wet forest. *Journal of Ecology* 86, 597–609.

Ehrenfeld, J.G., Zhu, W.X. and Parsons, W.F.J. (1995) Above- and below-ground characteristics of persistent forest openings in the New Jersey Pinelands. *Bulletin of the Torrey Botanical Club* 122, 298–305.

Fenner, M. (1978) A comparison of the abilities of colonizers and closed-turf species to establish from seed in artificial swards. *Journal of Ecology* 66, 953–963.

Fenner, M. (1980) Germination tests on thirty-two East African weed species. *Weed Research* 20, 135–138.

Fenner, M. (1985) *Seed Ecology*. Chapman and Hall, London, 151 pp.

Fowler, N.L. (1988) What is a safe site? Neighbor litter, germination date, and patch effects. *Ecology* 69, 947–961.

Garcia Fayos, P. and Verdu, M. (1998) Soil seed bank, factors controlling germination and establishment of a Mediterranean shrub: *Pistacia lentiscus* L. *Acta Oecologica – International Journal of Ecology* 19, 357–366.

Gay, P.E., Grubb, P.J. and Hudson, H.J. (1982) Seasonal changes in the concentrations of nitrogen, phosphorus and potassium, and in the density of mycorrhiza, in biennial and matrix-forming perennial species of closed chalkland turf. *Journal of Ecology* 70, 571–593.

Gill, D.S. and Marks, P.L. (1991) Tree and shrub seedling colonization of old fields in central New York. *Ecological Monographs* 61, 183–205.

Goldberg, D.E. (1987) Seedlings colonization of experimental gaps in two old-field communities. *Bulletin of the Torrey Botanical Club* 114, 139–148.

Goldberg, D.E. and Gross, K.L. (1988) Disturbance regimes of midsuccessional old fields. *Ecology* 69, 1677–1688.

Goldberg, D.E. and Werner, P.A. (1983) The effects of size of opening in vegetation and litter cover on seedling establishment of goldenrods (*Solidago* spp.). *Oecologia* 60, 149–155.

Gray, A.N. and Spies, T.A. (1996) Gap size, within-gap position and canopy structure effects on conifer seedling establishment. *Journal of Ecology* 84, 635–645.

Greene, D.F. and Johnson, E.A. (1989) A model of wind dispersal of winged or plumed seeds. *Ecology* 702, 339–347.

Grime, J.P. and Jeffery, D.W. (1965) Seedling establishment in vertical gradients of sunlight. *Journal of Ecology* 53, 621–642.

Grime, J.P., Hodgson, J.G. and Hunt, R. (1988) *Comparative Plant Ecology*. Unwin Hyman, London.

Gross, K.L. (1980) Colonization by *Verbascum thapsus* (mullein) of an old-field in Michigan: experiments on the effects of vegetation. *Journal of Ecology* 68, 919–927.

Gross, K.L. (1984) Effects of seed size and growth form on seedling establishment of six monocarpic perennials. *Journal of Ecology* 72, 369–387.

Grubb, P.J. (1976) A theoretical background to the conservation of ecologically distinct groups of annuals and biennials in the chalk grassland ecosystem. *Biological Conservation* 10, 53–76.

Grubb, P.J. (1977) The maintenance of species richness in plant communities: the importance of the regeneration niche. *Biological Reviews* 52, 107–145.

Grubb, P.J. and Metcalfe, D.J. (1996). Adaptation and inertia in the Australian tropical lowland rain-forest flora: contradictory trends in intergeneric and intrageneric comparisons of seed size in relation to light demand. *Functional Ecology* 10, 512–520.

Hanley, M.E., Fenner, M. and Edwards, P.J. (1996) Mollusc grazing and seedling survivorship of four common grassland plant species: the role of gap size, species and season. *Acta Oecologica – International Journal of Ecology* 17, 331–341.

Hillier, S.H. (1990) Gaps, seed banks and plant species diversity in calcareous grasslands. In: Hillier, S.H., Walton, D.W.H. and Wells, D.A. (eds) *Calcareous Grasslands: Ecology and Management*. Bluntisham Books, Huntingdon, pp. 57–66.

Horvitz, C.C. and Schemske, D.W. (1994) Effects of dispersers, gaps, and predators on dormancy and seedling emergence in a tropical herb. *Ecology* 75, 1949–1958.

Hubbell, S.P. and Foster, R.B. (1986) Canopy gaps and the dynamics of a neotropical forest. In: Crawley, M.J. (ed.) *Plant Ecology*. Blackwell Scientific, Oxford, pp. 77–96.

Hubbell, S.P., Foster, R.B., O'Brien, S.T., Harms, K.E., Condit, R., Wechsler, B., Wright, S.J. and de Lao, S.L. (1999) Light-cap disturbances, recruitment limitation, and tree diversity in a neotropical forest. *Science* 283, 554–557.

Hurtt, G.C. and Pacala, S.W. (1995) The consequences of recruitment limitation: reconciling chance, history and competitive differences among plants. *Journal of Theoretical Biology* 176, 1–12.

Johnson, C.D. and Thomas, A.G. (1978) Recruitment and survival of seedlings of a perennial *Hieracium* species in a patchy environment. *Canadian Journal of Botany* 56, 572–580.

Kalamees, R. (1999) Seed bank, seed rain and community regeneration in Estonian calcareous grasslands. PhD thesis, University of Tartu.

Keizer, P.J., Van Tooren, B.F. and During, H.J. (1985) Effects of bryophytes on seedling emergence and establishment of short-lived forbs in chalk grassland. *Journal of Ecology* 73, 493–504.

King, J.J. (1975) Inhibition of seed germination under leaf canopies in *Arenaria serpyllifolia*, *Veronica arvensis* and *Cerastium holosteoides*. *New Phytologist* 75, 87–90.

King, T.J. (1977) The plant ecology of ant-hills in calcareous grasslands. III. Factors affecting population sizes of selected species. *Journal of Ecology* 65, 279–315.

Kobe, R.K., Pacala, S.W., Silander, J.A. and Canham, C.D. (1995) Juvenile tree survivorship as a component of shade tolerance. *Ecological Applications* 5, 517–532.

Kotanen, P.M. (1997) Effects of gap area and shape on recolonization by grassland plants with differing reproductive strategies. *Canadian Journal of Botany* 75, 352–361.

Lavorel, S. and Chesson, P. (1995) How species with different regeneration niches coexist in patchy habitats with local disturbances. *Oikos* 74, 103–114.

Lavorel, S., Lepart, J., Debussche, M., Lebreton, J.-D. and Beffy, J.-L. (1994) Small-scale disturbances and the maintenance of species diversity in Mediterranean old fields. *Oikos* 70, 455–473.

Lavorel, S., Rochette, C. and Lebreton, J.D. (1999) Functional groups for response to disturbance in Mediterranean old fields. *Oikos* 84, 480–498.

Leishman, M.R. and Westoby, M. (1998) Seed size and shape are not related to persistence in soil in Australia in the same way as in Britain. *Functional Ecology* 12, 480–485.

Leishman, M.R., Westoby, M. and Jurado, E. (1995) Correlates of seed size variation: a comparison among five temperate floras. *Journal of Ecology* 83, 517–529.

Lertzman, K.P., Sutherland, G.D., Inselberg, A. and Saunders, S.C. (1996) Canopy gaps and the landscape mosaic in a coastal temperate rain forest. *Ecology* 77, 1254–1270.

McConnaughay, K.D.M. and Bazzaz, F.A. (1990) Interactions among colonizing annuals: is there an effect of gap size? *Ecology* 71, 1941–1951.

McConnaughay, K.D.M. and Bazzaz, F.A. (1991) Is physical space a soil resource? *Ecology* 72, 94–103.

Malo, J.E. and Suarez, F. (1995) Cattle dung and the fate of *Biserrula pelecinus* L (Leguminosae) in a Mediterranean pasture – seed dispersal, germination and recruitment. *Botanical Journal of the Linnean Society* 118, 139–148.

Malo, J.E., Jimenez, B. and Suarez, F. (1995) Seed bank build-up in small disturbances in a Mediterranean pasture: the contribution of endozoochorous dispersal by rabbits. *Ecography* 18, 73–82.

Metcalfe, D.J. and Grubb, P.J. (1995) Seed mass and light requirements for regeneration in Southeast Asian rain forest. *Canadian Journal of Botany* 73, 817–826.

Metcalfe, D.J., Grubb, P.J. and Turner, I.M. (1998) The ecology of very small-seeded shade-tolerant trees and shrubs in lowland rain forest in Singapore. *Plant Ecology* 134, 131–149.

Milberg, P. (1994) Germination ecology of the polycarpic grassland perennials *Primula veris* and *Trollius europaeus*. *Ecography* 17, 3–8.

Moretto, A.S. and Distel, R.A. (1998) Requirement of vegetation gaps for seedling establishment of two unpalatable grasses in a native grassland of central Argentina. *Australian Journal of Ecology* 23, 419–423.

Morgan, J.W. (1997) The effect of grassland gap size on establishment, growth and flowering of the endangered *Rutidosis leptorrhynchoides* (Asteraceae). *Journal of Applied Ecology* 34, 566–576.

Núñez-Farfán, J. and Dirzo, R. (1988) Within gap spatial heterogeneity and seedling performance in a Mexican tropical forest. *Oikos* 51, 274–284.

O'Connor, T.G. (1996) Hierarchical control over seedling recruitment of the bunch-grass *Themeda triandra* in a semi-arid savanna. *Journal of Applied Ecology* 33, 1094–1106.

Pacala, S.W. and Levin, S.A. (1997) Biologically generated spatial pattern and the coexistence of competing species. In: Tilman, D. and Karieva, P. (eds) *Spatial Ecology*. Princeton University Press, Princeton, pp. 204–232.

Pakeman, R.J., Attwood, J.P. and Engelen, J. (1998) Sources of plants colonising experimentally disturbed patches in an acidic grassland, eastern England. *Journal of Ecology* 86, 1032–1040.

Peart, D.R. (1989) Species interactions in a successional grassland. I. Seed rain and seedling establishment. *Journal of Ecology* 77, 236–251.

Peterson, C.J., Carson, W.P., McCarthy, B.C. and Pickett, S.T.A. (1990) Microsite variation and soil dynamics within newly created treefall pits and mounds. *Oikos* 58, 39–46.

Pickett, S.T.A. and White, P.S. (1985) *The Ecology of Natural Disturbance and Patch Dynamics*. Academic Press, Orlando.

Platt, W.J. (1975) The colonization and formation of equilibrium plant species associations on badger disturbances in a tall grass prairie. *Ecological Monographs* 45, 285–305.

Pons, T.J. (1991) Dormancy, germination and mortality of seeds in a chalk grassland flora. *Journal of Ecology* 79, 765–780.

Pons, T.L. and van der Toorn, J. (1988) Establishment of *Plantago lanceolata* L. and *Plantago major* L. among grass .1. Significance of light for germination. *Oecologia* 75, 394–399.

Putz, F.E. (1983). Treefall pits and mounds, buried seeds, and the importance of soil disturbance to pioneer tree species on Barro Colorado Island, Panama. *Ecology* 64, 1069–1074.

Rapp, J.K. and Rabinowitz, D. (1985) Colonization and establishment of Missouri prairie plants on artificial soil disturbances. I. Dynamics of forb and graminoid seedlings and shoots. *American Journal of Botany* 72, 1618–1628.

Rebertus, A.J. and Burns, B.R. (1997) The importance of gap processes in the development and maintenance of oak savannas and dry forests. *Journal of Ecology* 85, 635–645.

Reich, P.B., Tjoelker, M.G., Walters, M.B., Vanderklein, D.W. and Buschena, C. (1998) Close association of RGR, leaf and root morphology, seed mass and shade tolerance of nine boreal tree species grown in high and low light. *Functional Ecology* 12, 327–338.

Rusch, G. (1992) Spatial pattern of seedling recruitment at two different scales in a limestone grassland. *Oikos* 65, 433–442.

Rusch, G. and Fernandez Palacios, J.M. (1995) The influence of spatial heterogeneity on regeneration by seed in a limestone grassland. *Journal of Vegetation Science* 6, 417–426.

Rusch, G. and van der Maarel, E. (1992) Species turnover and seedling recruitment in limestone grasslands. *Oikos* 63, 139–146.

Ryel, R.J., Beyschlag, W. and Caldwell, M.M. (1994) Light field heterogeneity among tussock grasses: theoretical considerations of light harvesting and seedling establishment in tussocks and uniform tiller distribution. *Oecologia* 98, 241–246.

Ryser, P. (1993) Influences of neighboring plants on seedling establishment in limestone grassland. *Journal of Vegetation Science* 4, 195–202.

Sanford, L.O. (1989) Fine root biomass under a tropical forest light gap opening in Costa Rica. *Journal of Tropical Ecology* 5, 251–256.

Schenkeveld, A.J. and Verkaar, H.J. (1984) The ecology of short-lived forbs in chalk grasslands: distribution of germinative seeds and its significance for seedling emergence. *Journal of Biogeography* 11, 251–260.

Schupp, E.W., Howe, H.F. and Augspurger, C.K. (1989) Arrival and survival in tropical treefall gaps. *Ecology* 70, 562–564.

Silvertown, J.W. (1980) Leaf canopy-induced seed dormancy in a grassland flora. *New Phytologist* 85, 109–118.

Silvertown, J. and Smith, B. (1989) Mapping the microenvironment for seed germination in the field. *Annals of Botany* 63, 163–167.

Silvertown, J.W. and Wilkin, F.R. (1983) An experimental test of the role of micro-spatial heterogeneity in the coexistence of congeneric plants. *Biological Journal of the Linnean Society* 19, 1–8.

Sipe, T.W. and Bazzaz, F.A. (1995) Gap partitioning among maples (*Acer*) in central New England: survival and growth. *Ecology* 76, 1587–1602.

Swaine, M.D. and Whitmore, T.C. (1988) On the definition of ecological species groups in tropical rain forests. *Ecology* 73, 78–86.

Tang, Y.H. and Washitani, I. (1995) Characteristics of small-scale heterogeneity in light availability within a *Miscanthus sinensis* canopy. *Ecological Research* 10, 189–197.

Thanos, C.A. and Rundel, P.W. (1995) Fire-followers in chaparral: nitrogenous compounds trigger seed germination. *Journal of Ecology* 83, 207–216.

Thompson, K. and Grime, J.P. (1983) A comparative study of germination response to diurnally-fluctuating temperatures. *Journal of Applied Ecology* 20, 141–156.

Thompson, K., Grime, J.P. and Mason, G. (1977) Seed germination in response to diurnal fluctuations of temperature. *Nature* 267, 147–149.

Thompson, K., Band, S.R. and Hodgson, J.G. (1993) Seed size and shape predict persistence in soil. *Functional Ecology* 7, 236–241.

Thompson, K., Hillier, S.H., Grime, J.P., Bossard, C.C. and Band, S.R. (1996) A functional analysis of a limestone grassland community. *Journal of Vegetation Science* 7, 371–380.

Tilman, D. (1994) Competition and biodiversity in spatially structured habitats. *Ecology* 75, 2–16.

Tilman, D. and Karieva, P. (1997) *Spatial Ecology*. Princeton University Press, Princeton.

Uhl, C., Clark, K., Dezzeo, N. and Maquirino, P. (1988) Vegetation dynamics in Amazonian treefall gaps. *Ecology* 69, 751–763.

Valverde, T. and Silvertown, J. (1995) Spatial variation in the seed ecology of a woodland herb (*Primula vulgaris*) in relation to light environment. *Functional Ecology* 9, 942–950.

Valverde, T. and Silvertown, J. (1997) A metapopulation model for *Primula vulgaris*, a temperate forest understorey herb. *Journal of Ecology* 85, 193–210.

van der Meer, P.J., Bongers, F., Chatrou, L. and Riera, B. (1994) Defining canopy gaps in a tropical rain forest: effects on gap size and turnover time. *Acta Oecologica* 15, 701–714.

van der Toorn, J. and Pons, T.I. (1988) Establishment of *Plantago lanceolata* L. and *Plantago major* L. among grass .2. Shade tolerance of seedlings and selection on time of germination. *Oecologia* 76, 341–347.

Vitousek, P.M. and Denslow, J.S. (1986) Nitrogen and phosphorus availability in treefall gaps of a lowland tropical rainforest. *Journal of Ecology* 74, 1167–1178.

Westoby, M., Leishman, M. and Lord, J. (1997) Comparative ecology of seed size and dispersal. In: Silvertown, J., Franco, M. and Harper, J.L. (eds) *Plant Life Histories*. Cambridge University Press, Cambridge, pp. 143–162.

Whitmore, T.C. (1989) Canopy gaps and the two major groups of forest trees. *Ecology* 70, 536–538.

Wilczynski, C.J. and Pickett, S.T.A. (1993) Fine root biomass within experimental canopy gaps: evidence for a below-ground gap. *Journal of Vegetation Science* 4, 571–574.

Williams, R.J. (1992) Gap dynamics in sub-alpine heathland and grassland vegetation in south-eastern Australia. *Journal of Ecology* 80, 343–352.

Willson, M.F. (1993) Dispersal mode, seed shadows, and colonization patterns. *Vegetatio* 108, 261–280.

Winn, A.A. (1985) Effects of seed size and microsite on seedling emergence of *Prunella vulgaris* in four habitats. *Journal of Ecology* 73, 831–840.

Zagt, R.J. and Werger, M.J.A. (1998) Community structure and the demography of primary species in tropical rainforest. In: Newbery, D.M., Prins, H.H.T. and Brown, N. (eds) *Dynamics of Tropical Communities*. Blackwell Science, Oxford, pp. 193–219.

Index